Sensors for Safety and Process Control in Hydrogen Technologies

Series in Sensors

Series Editors: Barry Jones and Haiying Huang

Other recent books in the series:

Resistive, Capacitive, Inductive, and Magnetic Sensor Technologies
Winncy Y. Du

Semiconductor X-Ray Detectors
B. G. Lowe and R. A. Sareen

Portable Biosensing of Food Toxicants and Environmental Pollutants
Edited by Dimitrios P. Nikolelis, Theodoros Varzakas, Arzum Erdem, and Georgia-Paraskevi Nikoleli

Optochemical Nanosensors
Edited by Andrea Cusano, Francisco J. Arregui, Michele Giordano, and Antonello Cutolo

Electrical Impedance: Principles, Measurement, and Applications
Luca Callegaro

Biosensors and Molecular Technologies for Cancer Diagnostics
Keith E. Herold and Avraham Rasooly

Compound Semiconductor Radiation Detectors
Alan Owens

Metal Oxide Nanostructures as Gas Sensing Devices
G. Eranna

Nanosensors: Physical, Chemical, and Biological
Vinod Kumar Khanna

Handbook of Magnetic Measurements
S. Tumanski

Structural Sensing, Health Monitoring, and Performance Evaluation
D. Huston

Chromatic Monitoring of Complex Conditions
Edited by G. R. Jones, A. G. Deakin, and J. W. Spencer

Principles of Electrical Measurement
S. Tumanski

Novel Sensors and Sensing
Roger G. Jackson

Hall Effect Devices, Second Edition
R. S. Popovic

SERIES IN SENSORS

Sensors for Safety and Process Control in Hydrogen Technologies

Thomas Hübert
Lois Boon-Brett
William J. Buttner

CRC Press is an imprint of the
Taylor & Francis Group, an **informa** business

Employees of the Alliance for Sustainable Energy, LLC, under Contract No. DE-AC36-08G028308 with the U.S. Department of Energy have authored this work. The United States Government retains and the publisher, by accepting the article for publication, acknowledges that the United Sates Government retains a non-exclusive, paid-up, irrevocable, worldwide license to publish or reproduce the published form of this work, or allow others to do so, for United States Government purposes.

CRC Press
Taylor & Francis Group
6000 Broken Sound Parkway NW, Suite 300
Boca Raton, FL 33487-2742

First issued in paperback 2018

ISBN-13: 978-1-4665-9654-2 (hbk)
ISBN-13: 978-1-138-89434-1 (pbk)

Visit the Taylor & Francis Web site at
http://www.taylorandfrancis.com

and the CRC Press Web site at
http://www.crcpress.com

Contents

Contributors

Bernd Fellmuth
Physikalisch-Technische Bundesanstalt
Berlin, Germany

Valerio Palmisano
European Commission, Joint Research Centre
Petten, Netherlands

Ulrich Schmidtchen
BAM Bundesanstalt für Materialforschung und prüfung
Berlin, Germany

Eveline Weidner
European Commission, Joint Research Centre
Petten, Netherlands

Foreword: From Phlogiston to the Hydrogen Age in 250 Years

The title, or rather the epigraph, of this foreword is an obvious tribute to the first visionary of the Hydrogen Age, the great Jules Verne, whose iconic prediction in Mysterious Island (1874) that "hydrogen and oxygen, used singly or together, will furnish an inexhaustible source of heat and light" turned out to be prophetic. It is also quite ironic that at the time when Verne envisioned that "water will be the coal of the future," coal was the symbol of progress while today it is a symbol of environmental pollution and climate change.

Jules Verne, of course, made his prediction based on the discoveries of the hydrogen fathers.

One was Henry Cavendish, who in his famous publication "Three Papers, Containing Experiments on Factitious Air" (1766), first described the generation of "inflammable air" as he termed it, from the reaction of metal with acid solutions. Although Cavendish was the first to recognize that "inflammable air" is a discrete substance, he believed it to be identical to the hypothetical substance "phlogiston."

It took Cavendish 15 more years to discover that burning "inflammable air" with "fire air" (discovered by Carl Wilhelm Scheele in 1771/2 and Joseph Priestley in 1774, and later named "oxygen" by Lavoisier in 1777) produces pure water.

This discovery helped Antoine Lavoisier have his Eureka! moment: he reproduced Cavendish's experiment in a closed vessel (a bell jar) for the first time in history demonstrating mass conservation principle and declared that "this phenomenon can be explained without the use of phlogiston." In 1783 he gave "inflammable air" its new name, Hydrogen, and mathematically proved (with the assistance of Pierre Laplace) that water was not an element, but a compound of two gases, hydrogen and oxygen. This discovery completely discredited the phlogiston theory and laid the foundation for modern chemistry, as we know it.

The key point of this preamble is to illustrate that, essentially from its discovery almost 250 years ago, hydrogen has been used and viewed as a tool of progress. This view continues to be true today.

Use of hydrogen as an energy carrier and a fuel circumvents the many environmental, socio-economic and geopolitical concerns inherent to fossil fuels. Due to increasing concerns over the limited nature of fossil fuels, there is a growing desire to harvest and use alternative sources of energy. Hydrogen produced from renewable sources can provide environmentally clean, affordable and secure fuel for electricity generation, transportation and other sectors.

This book is about sensors, particularly sensors used in applications where hydrogen is a chemical feedstock and in applications exploiting hydrogen as an energy carrier and for energy storage. The importance of sensors for monitoring purposes and for ensuring the safety and security of our surroundings is growing every day in ways barely imaginable even a decade ago. Novel sensing applications are constantly being identified, broadening the technology's scope and expanding its impact on modern living. This is particularly true of sensors for hydrogen technologies. As hydrogen technologies emerge onto the consumer market, performance and functionality are important however these are secondary to safety, which is paramount. Mitigating risks associated with hydrogen is essential for instilling confidence in consumers to adopt hydrogen technologies—this can be achieved through the effective use of correct sensors.

In this book the rationale for sensor use and the role sensors play as key enabling devices for both control and safety in hydrogen technologies are outlined. Various facets of sensors are illustrated including practical aspects specific to hydrogen technologies. A comprehensive account of physical and chemical detection principles, sensor design and practical aspects of sensor deployment in hydrogen applications provides the reader with an invaluable insight into modern sensing techniques and their capabilities, limitations and drawbacks. The aim of this book is to disseminate practical information on sensors used in hydrogen technologies and to illustrate how their use will help ensure that the emerging hydrogen global market is safely and efficiently implemented. Reference is also made to regulations, codes and standards, which are essential in ensuring sound equipment design, engineering and protection of public safety. The book is an invaluable compendium of knowledge and expertise gathered by the authors and contributors over many years working in the sensor and hydrogen fields.

Of course discussions exist on the feasibility of and extent to which hydrogen can replace conventional fuels. The release of the world's first serial hydrogen fuel cell passenger car Toyota "Mirai" in November 2014 and a growing number of power-to-hydrogen projects clearly indicate that hydrogen is becoming a central link in the energy chain of a future society where a variety of novel alternative energy options will coexist in harmony. Only hydrogen has the potential to fully connect and integrate renewable energy sources in the economy and enable not only green but also clean technologies. This potential can and will only be reached when it is supported by continued research and development of hydrogen production, storage and conversion technologies. Thus the pioneering work started by Cavendish and Lavoisier 250 year ago, which forms the basis for the frontier research set forth by scientists and engineers today, must continue to pursue a sustainable global economy in which hydrogen will be firmly established as an "inexhaustible source of heat and light" to mankind.

Andrei V. Tchouvelev,
A.V. Tchouvelev & Associates Inc., Canada; President, HySafe; Chair, ISO/TC 197 Hydrogen technologies

Prologue: Current and Future Roles for Hydrogen

Ulrich Schmidtchen, BAM Bundesanstalt für Materialforschung und -prüfung, Berlin

New Roles for a Well-known Substance: Hydrogen

Why write a book specifically about sensors for hydrogen technologies? Is hydrogen significantly different from other gases? Not particularly, at least not as far as its physical and chemical properties are concerned. Hydrogen is a flammable gas, lighter than air just like numerous other gases. Or maybe it is because hydrogen has been discovered only recently and has special applications? This is not true either, hydrogen was one of the first elements to be discovered and this was done by Henry Cavendish in 1766. It was subsequently named by Antoine Lavoisier in 1783 by combining the Greek terms "hydro" (water) and "genes" (to engender). Hydrogen has played an important role in the chemical industry for more than a century (e.g., in the Haber-Bosch process for ammonia production and for the desulphurisation of crude oil) and will continue to do so for the foreseeable future. Of course it goes without saying that sensors for the measurement and monitoring of different process parameters (including gas concentration, pressure and temperature) play an important role in these applications. No, it is not hydrogen in itself which makes it meaningful to write a book like this one. Hydrogen must be seen in a much wider context, specifically its emerging use as an energy carrier in industrial and consumer markets.

Up until the start of the Industrial Revolution over 200 years ago, the demand for mechanical work was partly satisfied by human and animal labour and partly by the power of flowing water or wind, while heat was produced by burning wood. While only the latter process released carbon dioxide, this carbon had been taken recently from the atmosphere and so its release did not contribute to a significant change in the atmosphere's composition.

However as the use of steam engines increased so did the demand for wood. In fact wood usage increased at a rate that threatened to diminish all supplies, and so the mining and combustion of coal started as an alternative to deforestation. As a result of coal use, fossil carbon, which had been deposited and thus sequestered in the ground for millions of years, was suddenly released. The carbon dioxide content of the atmosphere subsequently increased from 0.028 vol% in pre-industrial times to its still increasing current content of around 0.04 vol%. As a consequence, the natural greenhouse effect which already raises the earth's surface temperature from an estimated value of −18 °C to the actual +14 °C, has been significantly enhanced by an anthropogenic component, namely the increased carbon dioxide in the atmosphere. Between 1880 and 2012 the global mean temperature near the surface of the earth has risen by 0.85 °C.

The effects of the emission of greenhouse gases on climate were at first not recognized and even neglected for many decades because they were masked by the general variability of meteorological phenomena. Today it is widely believed that we have started a process which is difficult to stop. The climate effects arising from atmospheric change will extend over the whole 21st century, even if all countries immediately reduce their greenhouse gas emissions drastically which they are rather unlikely to do. Attempts to limit the rise of the global mean temperature to no more than 2 °C in comparison to pre-industrial times will be challenging, to say the least.

There is no single solution for the climate change problem. It goes without saying that we need to fundamentally change how we control and satisfy our energy requirements if we want to preserve our planet. The shift from fossil (oil, gas, coal) to renewable energy sources is probably the single most important measure to counteract global warming. The necessity to do this is enhanced by the depletion of our fossil fuel reserves. Since about 2006 we have not seen an increase in crude oil production, and it is unlikely that the amounts which come into the market will increase significantly soon; rather they will decrease. A similar trend is likely to happen to natural gas production after 2020.

Fortunately we have alternatives. There is no shortage of renewable energy sources. Germany currently generates one quarter of its electrical energy from renewable sources (wind, solar, hydro, and biomass). Globally solar, wind, hydro, biomass, tidal, geothermal, and other energy sources have the potential to satisfy our energy needs. At least in principle. In real life however these energy sources are not as easily harnessed as conventional sources. This is reflected in two aspects in particular:

1. Renewable energy sources are not always available where consumers need the energy. There are excellent wind conditions in the open sea, a lot of sun in the deserts and hydro power in the mountains, but this is typically not where energy demand is concentrated, i.e., in industrial regions and big cities. This means that the energy needs to be transported to the site of use.

2. Renewable sources do not necessarily supply energy at the time when consumers need it. They follow not only regular cycles like the tides or day and night, they are also subject to irregular fluctuations which are difficult to predict, e.g., variations in wind speed and sun intensity. As a result it may become necessary to store energy for use at a later time.

There are, of course, many ways to store and transport energy. Natural gas is often stored in underground reservoirs, and distributed via pipelines to meet consumer demand. An elaborate electrical grid transports electricity from the point of generation to the point of use (and there is little doubt that the grid will become even more important for "smart" electricity distribution in the future). However, there is a lack of large storage capacities for electricity. Electricity storage is required not only for short periods (in order to smooth short-term demand and supply fluctuations, such as the differences associated with daily peak power requirements) but also to match seasonal variations in demand and supply.

Hydrogen is one of a number of options for the storage and transport of energy in the future. As the use of hydrogen as an energy carrier increases, corresponding new processes and technologies will arise. Hydrogen will be used in non-industrial settings by the general public. Sensors for hydrogen and other related parameters will be needed for each of these new applications. Systems for hydrogen detection are required because hydrogen has neither a colour nor a distinctive odour detectable by human senses. The role of detection must be filled by sensors. Since leaked hydrogen can mix with air to produce flammable mixtures, safety is one critical reason for sensor use in hydrogen technologies, but there are many more, such as for process control and environmental monitoring.

Hydrogen as a Fuel

In Europe there is a political target to reduce the emissions of greenhouse gases by 80 % in comparison to 1990, by 2050. Current greenhouse gas emission sources can be attributed to industry, private homes, and transport. Ambitious targets in the transport sector require that CO_2 emissions are to be reduced close to zero before 2050. There is only one feasible possibility to achieve this: replacement of combustion engines with electrical engines running on power from renewable sources. Vehicle manufacturers around the globe are developing and commercialising electrical drive trains. One of the challenges facing the developers of electric vehicles is to ensure a driving range which meets customer expectations. For battery electric vehicles this can be achieved by increasing the size of the battery. There is, however, a physical size and weight constraint for on-board batteries which ultimately limits the achievable electric drive range. Despite on-going developments in battery performance and energy storage capacity, massive improvements in range capabilities are not expected in the forseseeable future. Current commercial battery vehicles and prototypes have a shorter range (typically < 300 km) than conventionally fuelled vehicles. Nevertheless their driving range is very well suited for urban use.

Hydrogen fuel cell vehicles complement battery electric vehicles in that they also have electric drive trains but with significantly longer driving ranges. A fuel cell vehicle is equipped with a hydrogen tank, usually a 70 MPa compressed gas tank, for on-board hydrogen storage. The fuel cell generates electrical power, water, and heat. Fuel cell vehicles currently offer driving ranges of the order of 600 km. Many vehicle manufacturers around the world are developing both battery electric and fuel cell electric vehicles. It is expected that a significant number of fuel cell cars will be commercialized and mass-produced by 2020.

While vehicle performance is certainly an important factor in market acceptance of any new automotive technology, safety and customer perception of safety are also critical factors. Mass commercialisation of hydrogen fuel cell vehicles will only be realised when customers are confident that these vehicles are as safe as conventional vehicles. While hydrogen vehicle manufacturers go to great lengths to ensure their product's safety, they are faced with additional and new challenges. The nature of hydrogen fuel is quite different from what consumers are used to (hydrogen is a light gas which is stored on-board under considerable pressure while conventional fuels are mostly liquids). While a hydrogen storage tank is arguably more resistant to external impacts than a gasoline tank, other safety risks exist which customers are not accustomed to, such as hydrogen leakage and the potential formation of flammable plumes. Alternatively, hydrogen has properties that mitigate potential hazards from an unwanted release. Hydrogen dissipates quickly, especially for outdoor releases. Air or water pollution concerns associated with conventional liquid fuels are not an issue for hydrogen. While hydrogen vehicles may not be more dangerous than conventional ones, probably even less, the risks are different and must be dealt with accordingly.

Sensors will play an important role for the early detection of potentially dangerous situations associated with unwanted hydrogen releases from vehicles. Sensors will also initiate appropriate counter-measures. Many prototype vehicles have hydrogen leak sensors located in the motor compartment and in the passenger compartment. If a hydrogen leak is detected they will initiate a pre-determined sequence of measures to mitigate the consequences, such as opening the windows to ventilate the passenger compartment and to isolate the hydrogen storage system.

Hydrogen as an Energy Storage Medium

In general, electricity is difficult and costly to store, in particular in larger quantities (GWh) over longer periods of time (beyond 48 hours). Various electricity storage options exist, each addressing different capacity and storage time requirements. These include pumped hydro storage, adiabatic compressed air storage, redox flow battery storage systems, battery systems (NaNiCl, NaS), and electrolytic hydrogen production. The latter provides a particularly cost-effective way for long-term storage of intermittent renewable electricity, for which only few other options exist. If it is necessary to store large amounts of electricity for long periods of time (everything exceeding ca. three days) there is almost no alternative to hydrogen.

Salt cavern hydrogen storage is a proven technology and can accommodate about 60 times the electricity equivalent in the same volume as adiabatic compressed air storage. Hydrogen stored in such large storage systems can be used for electricity production using large combined cycle gas turbines, used directly as a feedstock in the chemical industry, or distributed via pipeline to end-users [1]. Hydrogen storage in salt caverns has a number of attractive features including the fact that salt deposits are found in many regions of the world and it is easy to create voids in them. This technique is currently exploited for storage of hydrogen, intended for use as a chemical feedstock for the chemical industry rather than as an energy carrier. In addition to salt deposits widespread porous rock formations ("aquifers") can also be exploited for large-scale hydrogen storage. As a result hydrogen storage capacities can be installed in most regions of the world, which reduces the necessity for transport over extensive distances.

Another option for storing and distributing hydrogen which is gaining more interest recently is to feed the hydrogen into the existing natural gas grid. The volume of this grid is enormous. For example, in Germany alone there are some 500,000 km of pipelines. Together with the 47 existing underground storage facilities they have a combined volume of more than 23 billion m^3. The target for 2025 is to expand the total volume capacity to more than 32 billion m^3. Current German regulations permit addition of 5 % of hydrogen to the natural gas [2]. There are plans to raise this limit to 10 %. Higher concentrations are not feasible because they would change the characteristics of the gas in the pipeline too much, which would lead to problems in terms of materials, sealing, metering, etc. Advantages of this procedure are that the amount of flammable gas in the pipeline is increased and, more importantly, it provides a means to empty the hydrogen storage facilities which would otherwise be full sooner or later. The disadvantage is that the hydrogen is effectively lost, because it cannot be separated from the methane anymore in an economical way. So hydrogen injection into the natural gas grid can only be seen as an intermediate step, not as a final solution for large-scale hydrogen storage.

Of course it would be preferable to use the hydrogen close to the production site. Power stations using combined gas and steam turbines for power generation are one way of achieving this. Filling stations for hydrogen-fuelled vehicles are another, provided there is enough demand from vehicles close to the hydrogen production site.

Hydrogen as a Energy Transport Medium

Large volumes of hydrogen are transported every day on public roadways and in pipelines. Most of this transport is done for the chemical industry which needs hydrogen as a raw material.

When hydrogen is used as an energy carrier in the future some of the existing energy transport media may become superfluous. Fossil fuels, namely oil, natural gas, and coal, are currently transported from the point of extraction, which can be in faraway regions of the world like Siberia, Alaska, or the Persian Gulf, to the point of use. Hydrogen, on the other hand, can be potentially produced anywhere in the world; the production process will of course depend on the raw materials and resources which are locally available. As a result it will not be necessary to transport hydrogen over very large distances, i.e., across continents or oceans.

The fact that both hydrogen and primary renewable energies are more evenly distributed over the globe than oil and gas reserves has not only important technical and economical impacts, but it may also impact geopolitical effects. Many current fossil fuel reserves are located in politically unstable regions of the world. Similarly, transporting of these fuels sometimes leads to bottlenecks in, for example, the Strait of Hormuz. A war or a crisis in these regions immediately jeopardizes the energy security of the oil-dependent nations, and hence their economies. Distributed energy production will eliminate many of the problems associated with localized energy sources and the transport of the energy to the point of use. It is likely that there will be some large-scale central hydrogen production plants in the future; however, they will not be the exclusive source. This will increase the security of energy supply.

With mass market uptake of hydrogen as a fuel for hydrogen-fuelled vehicles, a network of filling stations will be required. Many such stations exist today, some of them producing hydrogen on site by electrolysis or by other processes. However it may be necessary to supplement their supply with hydrogen produced from central facilities. Distribution of the hydrogen to these stations may be by truck or other means.

Conclusion

For more than one century hydrogen has played a key role in the chemical industry. Hydrogen is still used in the Haber-Bosch process, for the desulphurisation of crude oil and for a host of other processes. In the near future we may expect hydrogen to form part of a sustainable energy system. The basis of this system will be the exploitation of renewable energy sources, and hydrogen will play a role as an important energy storage and transport medium.

There is no technology without risk, nevertheless consumers and regulators demand that emerging hydrogen technologies are as safe as conventional energy technologies, and, if possible, safer. There are good reasons to assume that this target can be reached. Hydrogen is similar to natural gas in many respects, and there is extensive experience handling this fuel with a high degree of safety. Both stationary (heating appliances and fuel cells) and mobile (vehicles) applications of hydrogen operate at a high safety level, as can be seen from many recent demonstration projects.

Sensors will form an integral part of the safety and control system of existing and future hydrogen technologies. Since our natural senses are not able to respond to hydrogen, hydrogen sensors are needed to detect unwanted leaks. However other sensors will be also used for monitoring and safety devices in laboratories, workshops, and professional environments but also in private environments like homes or vehicles. As such hydrogen sensors are a key enabling technology that will facilitate the transition towards a hydrogen-inclusive economy.

The Structure of the Book

This book is on sensors which are regularly deployed in technologies and processes related to hydrogen production, storage, distribution, and use. Not all types of sensors are equally suitable for specific hydrogen applications. The information in this book is intended to help the reader understand the basics of sensors, sensing technologies, sensor applications, and to provide guidelines for choosing the right sensors and the use of them correctly. Correct deployment of appropriate sensors demands knowledge of the sensing principle and of the physical or chemical quantities being measured. Because of the properties of hydrogen, the potential for its vastly increased use in a future low-carbon economy and possible hazards associated with its use, special attention is paid to hydrogen sensors. This book will not focus on the details of the hydrogen technologies nor on the many safety-related aspects of these technologies. Many books are already available on these topics [3,4]. Instead the detection principle of hydrogen sensors and other sensor types used, in the dynamic and rapidly developing field of hydrogen technologies, are treated in detail in this book.

In the first chapter a brief overview is presented on basic hydrogen properties and particularly on those properties which are most relevant for safety and for sensing. To illustrate the extensive field of contemporary applications and the exciting possibilities for near future sensor applications, existing and emerging markets using large quantities of hydrogen are mentioned. The role of sensors as devices for monitoring and control of processes and as safety monitoring devices is outlined.

The second chapter gives an introduction to sensing technology and provides the reader with relevant information pertaining to sensor definitions and classification, sensor metrics, and performance parameters, in addition to background information on sensor preparation technologies and techniques. While there are many books available which provide more exhaustive information on each of these topics, e.g., [5,6], the level of detail provided in this chapter is sufficient to appreciate the salient features of sensing and sensing technology which are central to hydrogen safety and monitoring of relevant applications.

Chapter 3 provides a comprehensive overview of emerging and commercially available hydrogen sensors, an explanation of their sensing principle, and important aspects of their performance. A comprehensive and up-to-date account of the theory (physical or chemical principles), design, and practical implementations of hydrogen sensors for use in hydrogen related applications is presented. Similar information on chemical sensors for other gases, such as oxygen and trace components, which are also highly important in hydrogen technologies because of potential hazards to human health, process safety or facility performance, is provided in Chapter 4. In Chapter 5 descriptions of physical sensors for temperature, pressure, gas flow, and fire indication, which are also germane for the safe use of hydrogen, are provided. Standards, codes, and regulatory documents, which provide practical advice and legislative requirements regarding sensor deployment and performance, are described in Chapter 6. This chapter also makes reference to the main procedures for sensor testing in gas standards including precise analytical methods and reference methods. The chapter concludes with a discussion on sensor selection and some installation guidelines are provided.

In Chapter 7 traditional and emerging processes and technologies involving hydrogen are described. The application of sensors in processes for the production of hydrogen,

hydrogen storage, distribution, and the use of hydrogen in stationary and mobile fuel cells is discussed. Furthermore, the use of hydrogen as a coolant and chemical reagent (medium) in various processes is described. The exploitation of sensors for replacing traditional analytical instrumentation is also discussed. Finally supplementary information is provided on hydrogen properties, measuring quantities, and sensor parameters.

Acknowledgements

I would like to express my deep gratitude for the long and most fruitful collaboration with my colleagues Dr. Ulrich Banach, Heidi Lorenz, and Dr. Carlo Tiebe. I thank my institution that it provided me with the resources available to write this book. A thanks goes to all co-authors and contributors for their constructive and enduring collaboration and to the editor Francesca McGowan and to Marcus Fontaine, without which this book would not have been.

T. Hübert

I would like to express my appreciation to the Safety Codes and Standards Program within the DOE Fuel Cell Technology Office for support provided to me for the production of this book. I also want to thank my co-authors and our editor Francesca McGowan for extreme patience, endurance, and professionalism shown throughout the writing this book. Finally I would like to provide a special acknowledgement to Eveline Weidner for her very significant and much appreciated contributions.

B. Buttner

I would like to express my extreme thanks to a number of people for their support and valuable contributions to this book. To my colleagues including Dr. Marc Steen, Dr. Pietro Moretto, Dr. Francesco Dolci, Dr. Andreas Pfrang, Dr. Natalia Lebedeva, Dr. Franco Di Persio, Mr. Robert Houghton and Dr. Johan Blondelle (RTD) and to Mr. Sander van Herwaarden of Xensor B.V. and Prof. Woosuck Shin of AIST for reviewing and contributing to various parts of this book. A special thanks goes to Dr. Eveline Weidner, who was not only a contributor, but was also a major motivating force which brought this book to fruition; on behalf of all the authors our sincerest gratitude and admiration is extended to her. And finally to my family for their patience and particularly to my *Polaris*, for all his unwavering support in good weather and bad!

L. M. M. Boon-Brett

CHAPTER 1

Hydrogen Properties and Technologies of its Production and Use

CONTENTS

1.1 PROPERTIES AND REACTIVITY OF HYDROGEN

1.1.1 Occurence and Basic Properties

Hydrogen is the most commonly occurring element in the known universe accounting for 75 % of its mass. It occurs on earth as a colourless gas, but is more commonly chemically bonded in many inorganic and organic compounds. Hydrogen is an essential constituent of all living organisms.

Hydrogen is the lightest and simplest element. A hydrogen atom (chemical symbol H) has an atomic radius of 23×10^{-12} m, a mass of 1.674×10^{-27} kg, and a standard atomic weight of 1.00794 [7]. Three naturally occuring isotopes of hydrogen are known ^{1}H, ^{2}H (also called deuterium or heavy hydrogen and given the chemical symbol D), and ^{3}H (also known as tritium and given the chemical symbol T). The occurrence ratio of these three isotopes is $1{:}10^{-4}{:}10^{-17}$. Compared with hydrogen and deuterium, tritium is unstable and radioactive with a 12.3-year half-life, and occurs naturally only in trace amounts [7]. Since ^{1}H is by far the most commonly occurring isotope, the properties and reactivity of the heavier isotopes are not described in the following sections.

Pure hydrogen typically exists as a diatomic molecule consisting of two covalently bonded hydrogen atoms and has the chemical symbol H_2 (see Figure 1.1). A hydrogen molecule has two protons and two electrons. Due to the interaction of the rotation of the two protons with the magnetic nuclear spin, two different forms of hydrogen exist — ortho hydrogen and para hydrogen having aligned and anti-aligned nuclear spins, respectively. At ambient conditions hydrogen is a mixture of approximately 75 % ortho hydrogen and 25 % para hydrogen. At temperatures < 80 K, para hydrogen is the more stable form and at 20 K, the thermal equilibrium concentrations are approximately 99.8 % para and 0.2 % ortho hydrogen. Most physical properties differ only slightly between the two spin states. However, there is an energy difference between the two forms, which results in different specific heats and thermal conductivities. The conversion of ortho hydrogen to para hydrogen (e.g., when cooled) is an exothermic process.

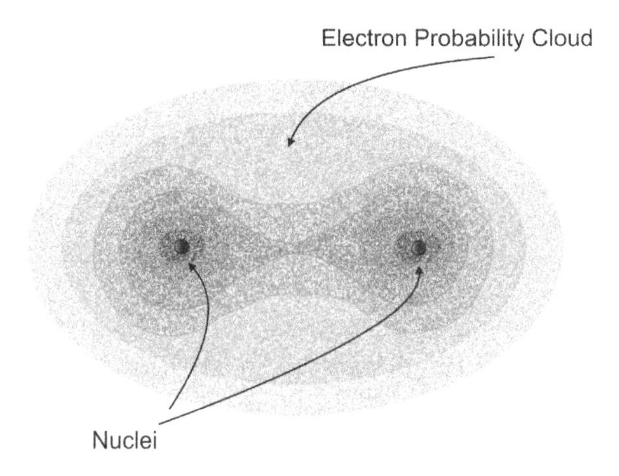

Figure 1.1: Illustration of a hydrogen molecule. (The intensity of the grey area is proportional to the electron density.)

The ionization energy[1] of the hydrogen atom is 13.598 eV or 1312 kJ/mol [7]. This high ionization energy precludes the use of photo ionization sensors[2] for detecting hydrogen, because the energy of common UV light sources is not sufficient for ionization. Hydrogen is known as a strong reducing agent (i.e., it will transfer electrons to other atoms or molecules). On the other hand, the electron affinity[3] is 0.754 eV or -73 kJ/mol, which is consistent with hydrogen also being an oxidizing agent, at least with some materials (e.g., such as in metal hydrides in where hydrogen has acquired a negative or anionic character (see Section 1.1.5.2) [7]. Thus hydrogen can act either as an oxidising or reducing agent in chemical reactions. The electronegativity[4] of hydrogen is 2.2 and is different from the other elements of the first column of the periodic table (e.g., the electronegativities of lithium and sodium are 0.98 and 0.93, respectively), but is more comparable with halogen elements of column 7 (e.g., the electronegativities of fluorine and chlorine are 3.98 and 3.16, respectively) [7]. The binding (or dissociation) energy[5] of the hydrogen molecule is 4.52 eV or 436 kJ [8] and therefore thermal cleavage requires high temperatures. Only 0.081 % of hydrogen will be dissociated at 2,000 K. This increases to 7.85 % at 3,000 K. Further data on properties of atomic and molecular hydrogen are compiled in Table 8.4.2.

With the exception of the noble gases (column 8 elements of the periodic table), hydrogen reacts with all elements.

Hydrogen can bond with oxygen to form water, with metals to form hydrides, and with carbon to form organic hydrocarbon compounds. In chemical compounds, hydrogen can have a positive or negative oxidation state, typically either +1 or -1. The character of the bond (covalent, ionic, or metallic) in hydrides depends on the position of the counter element in the periodic table and on the difference in electronegativity. Additionally, weak so-called "hydrogen bonds" occur as electrostatic attractions or dipole-dipole attractions between molecules in which hydrogen is bound to a highly electronegative atom. The binding energy of a hydrogen bond is relatively low and is in the range of 17 kJ/mol to 170 kJ/mol.

The specific physical and chemical properties of hydrogen determine the conditions for its safe handling and use. Furthermore, many of these properties are exploited for the sensing of hydrogen in gas mixtures; for example, the reactions of hydrogen with oxygen and with metals, described in Sections 1.1.4 and 1.1.5, which are important for industrial processes and for hydrogen sensing platforms.

1.1.2 Thermodynamic States

Diatomic hydrogen is gaseous under normal conditions, however it can exist in a liquid or solid state at low temperatures and elevated pressures. The conditions at which gaseous, liquid, or solid phases can occur are described by three state variables: pressure, temperature, and volume.[6] A phase diagram is commonly used to illustrate different thermodynamic states of a substance defined by these variables. It contains lines of equilibrium (or phase boundaries) between the solid, liquid, and gaseous phase. A two-dimensional phase diagram of hydrogen is shown in Figure 1.2 [9].

---

[1] The ionization energy is the energy required to ionise an atom or molecule, i.e. to remove an electron.

[2] Usually named Photo Ionisation Detector - PID.

[3] The electron affinity is, in contrast to ionization energy, a measure of energy required by a neutral atom or molecule to accept an additional electron.

[4] Electronegativity is the ability of an atom or a functional group to attract electrons (or electron densities) in a chemical bond and is most commonly quantified according to the Pauling scale.

[5] The binding energy is the energy required to split the covalent bond between two atoms of a molecule. It is a measure of the strength of a covalent bond and is typically between 2 eV to 7 eV (200 kJ/mol to 700 kJ/mol).

[6] The SI unit for the pressure, p, is Pascal (Pa), for temperature, T, is Kelvin, and for volume, V, is m^3.

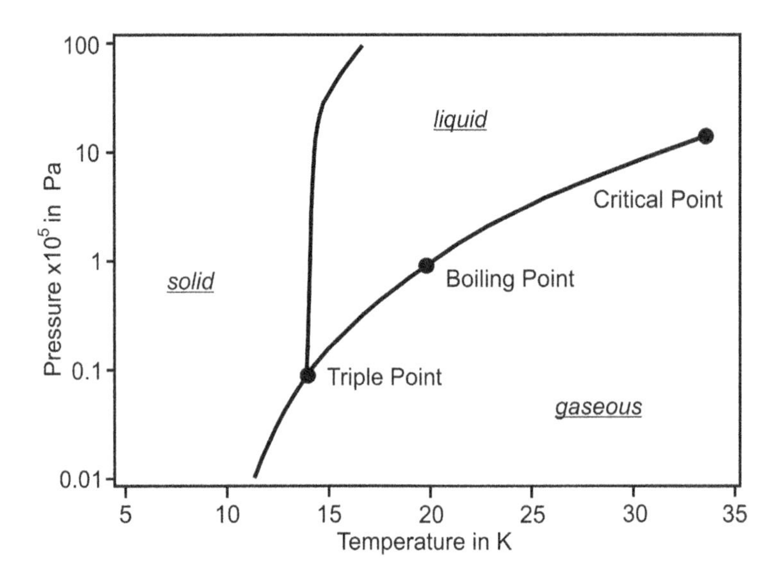

Figure 1.2: Phase diagram of hydrogen [10].

A phase diagram often illustrates the conditions for important parameters such as the normal boiling point (e.g., the boiling point at 101,325 Pa) and the triple point (the temperature and pressure where the liquid, solid, and gas phase are in equilibrium). The critical point is a state where the liquid state and gaseous state of a fluid cannot be distinguished. For hydrogen it is characterized by a critical temperature T_{crit} of 33.145 K (-240.005 °C), a critical pressure p_{crit} of 1.2964 MPa and a critical density ρ_{crit} of 15.508 mol/L. The boiling point of hydrogen is 20.28 K (-252.87 °C) at 101,325 Pa; the triple point is 13.95 K (-259.2 °C) and 7,210 Pa. However, the values from different sources slightly differ.

There are two types of state variables. Intensive variables, such as pressure p, temperature T, specific volume v ($v = V/m$, given in m^3/kg), or the density ρ ($\rho = m/V$=1/v), are thermodynamic properties that do not depend on the amount or size. Extensive variables, such as volume V, mass m (in kg), and amount of substance n (in moles), are physical quantities and as such their values are proportional to the size of the system or the amount of substance in the system. Extensive quantities are the sum of separate non-interacting subsystems which comprise the entire system [11].

The state variables pressure, temperature, and volume are interrelated by the equation of state $F(p,T,V) = 0$. The equation of state for gases including hydrogen can be approximated by the ideal gas law. The ideal gas law equation of state for extensive physical quantities is:

$$p \cdot V = n \cdot \mathrm{R} \cdot T \tag{1.1}$$

And for intensive physical quantities:

$$p \cdot V_m = \mathrm{R} \cdot T \tag{1.2}$$

Where, in addition to the previous mentioned quantities, $n = N/\mathrm{N}_A$ is the amount of substance in moles, N the number of particles, $\mathrm{N}_A = 6.022 \times 10^{23}$ mol^{-1} the Avogadro constant, R = 8314.472 J/kmol K is the universal gas constant, V_m=V/M is the molar volume in m^3/kmol, $M = m/n$ is the molar mass in kg/kmol. From Equation 1.2, it can be deduced that one mole of an ideal gas at a temperature of T = 273.15 K and a pressure

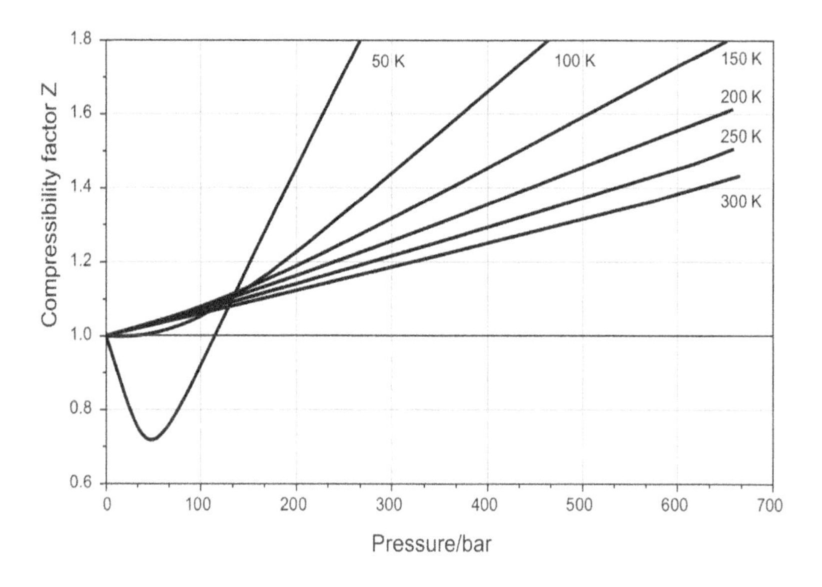

Figure 1.3: Compressibility factor Z for hydrogen [9].

of p = 101,325 Pa will occupy a volume V_m of 22.4 L. Equations 1.1 and 1.2 describe the behaviour of a so-called ideal gas. The ideal gas law assumes that gas molecules are point particles with no cross sectional area, and that they do not exert an attractive or repulsive force on each other. A deviation from this ideal behaviour can be observed for most gases and gas mixtures due to molecular interactions. This is especially evident for conditions deviating from ambient, including higher pressures and temperatures. Several empirically and theoretically derived corrections have been suggested to account for this behaviour. A compressibility factor, $Z = F(p, T)$ gives a sufficient approximation of gas behaviour for non-ideal behaviour in many applications:

$$p \cdot V = n \cdot R \cdot T \cdot Z \tag{1.3}$$

Z becomes more pronounced at elevated temperature and pressure. For example, at 293.15 K the compressibility factor for hydrogen is Z = 1.01 at 1.57 MPa, Z = 1.1 at 15.7 MPa, and Z=1.5 at 78.6 MPa [12]. Figure 1.3 shows that strong deviations from ideal behaviour of hydrogen will occur even at moderately low temperatures and high pressures. Additional discussions on equations to approximate the state of a real gas are given in Chapter 8.

Besides the thermodynamic state variables, the caloric variables including energy E, enthalpy H, entropy S, and the specific heat capacities c_v and c_p are important properties of hydrogen. The isobaric specific heat capacity c_p, which is given as $c_p = \partial H/\partial T = T\partial S/\partial T$ is 14.304 kJ/kg K for hydrogen gas at STP.[7] This value is comparable to other diatomic gases. c_p varies with temperature and can be estimated in the range of 250 K to 1500 K according to an empirical equation, such as Equation 1.4:

$$c_p = C_1 + C_2 \left[\frac{C_3}{T} / \sinh\left(\frac{C_3}{T}\right) \right]^2 + C_4 \left[\frac{C_5}{T} / \cosh\left(\frac{C_5}{T}\right) \right]^2 \tag{1.4}$$

[7]Standard conditions of temperature and pressure for gases (see Section 8.1).

Where c_p is given in J/kmol K and T in K. For hydrogen the values of the constants are $C_1 = 27620$, $C_2 = 9560$, $C_3 = 2466$, $C_4 = 3760$, and $C_5 = 567.6$ [13].

Hydrogen has a temperature dependent Joule-Thomson coefficient, $(\partial T/\partial p)_H$. This coefficient describes the change of temperature when a compressed gas is expanded at constant enthalpy (e.g., adiabatic expansion). Most gases have a positive Joule-Thomson coefficient, and thus tend to cool upon expansion. For hydrogen, the Joule-Thompson coefficient is negative for temperatures above 202 K. Only below this so-called inversion temperature, the expansion of hydrogen leads to a decrease of temperature. One manifestation of the Joule-Thomson inversion is that pre-cooling of the gas is necessary in order to apply expansion de-pressurisation and compression cycles for the liquefaction of hydrogen (similar to helium).

The thermal conductivity of hydrogen gas is 0.187 W/m·K at normal temperature and pressure conditions (NTP)[8] [10]. The thermal conductivity of hydrogen is significantly higher in comparison to other gases, such as air which has a thermal conductivity of 0.0262 W/m K [7]. This property is exploited for the sensing of hydrogen in thermal conductivity sensors, see Section 3.3.

The speed of sound in hydrogen gas is 1294 m/s at NTP, whereas in air it is only 343 m/s [10]. This difference can also be used for hydrogen detection in gases, see Section 3.9.

Because of its low molecular weight, hydrogen gas has a low density of 0.08988 kg/m^3 at STP which is about 14 times less than the density of air, which is 1.205 kg/m^3, at the same conditions. Due to its low density, hydrogen has a tendency to rise when released into air at temperatures above 22 K. Because of its high buoyancy, hydrogen can accumulate at the top of confined spaces before diluting by diffusion or convective (ventilation) processes. Hydrogen has a low viscosity of 89.5 μP at NTP and a high diffusivity. The diffusion coefficient in air is 6.1×10^{-5} m^2/s, whereas the coefficient for methane is 0.196×10^{-5} m^2/s [14]. Therefore it mixes rapidly with other gases upon release. Buoyancy and high diffusity are properties of hydrogen, which have to be considered in the installation of hydrogen sensors. The small size and low viscosity of hydrogen cause also a high diffusivity in solid materials. Permeability through metals, glass, or polymers can result in low-level releases from compressed hydrogen containers.

1.1.3 Hydrogen in Gas Mixtures

Hydrogen can occur as part of a gas mixture, e.g., when it is released into air. It can easily homogeneously mix in gases in all amounts, although upon initial release into an enclosed chamber or room, hydrogen will initially aggregate at the higher points of the enclosure before a uniform distribution occurs via thermal mixing. The amount of hydrogen in a gas mixture can be expressed in units like mass m, volume V, pressure p, or number of moles n (see Section 1.1.2). Since the total amount of a specific component such as hydrogen in a mixture is often not of interest, it is common to express the amount of a specific component in a gas mixture as a fraction of the total amount of the mixture. For gases, this is usually expressed as a volume fraction which is the volume of a constituent of a mixture divided by the sum of volumes of all constituents prior to mixing. For an ideal gas (and for hydrogen at ambient pressures, see Figure 1.3), this is equivalent to the number of moles of the constituent divided by the sum of the number of moles for each constituent in the mixture. The volume fraction varies from 0 to 1 and is a unitless quantity. Volume fractions are commonly converted into volume percent by multiplying the volume fraction by 100 with

[8]"Normal Temperature and Pressure" (see Section 8.1).

the abbreviation "vol%." This unit of measurement is predominantly used in this book. For low concentrations, the volume fraction of a component is commonly expressed in the units ppm or ppm_v (part per million), ppb or ppb_v (part per billion), and ppt or ppt_v (part per trillion). These units are however, not officially part of the International System of Units (SI Units). The quantities ppm, ppb, and ppt are more properly expressed as dimensionless volume ratios of 10^{-6}, 10^{-9}, or 10^{-12}. Alternatively, the International Union of Pure and Applied Chemistry (IUPAC) as well as the International Organisation for Standardization (ISO) also recommended that this notation should not be used but instead be replaced by μmol/mol, nmol/mol, or pmol/mol.

The amount of a gaseous component B in a gas mixture is also commonly expressed as the partial pressure of component (p_B). In an ideal gas, p_B is the volume fraction of the component B multiplied by the total pressure ($p_B = V_B \cdot p$). The amount of hydrogen in a mixture can also be expressed as a concentration, although this is not commonly done for gas phase applications. A concentration is the mass, volume, or number of particles of a substance divided by the volume of the mixture of the components at defined pressure and temperature. An overview of important quantities for the content of a gaseous component is given in Section 8.2.

The presence of hydrogen as a flammable gas in a gas mixture is also often expressed as a percentage of its lower flammable limit (LFL). This is a common display output for many hydrogen detection apparatus, as opposed to the actual hydrogen concentration. In air, the LFL for hydrogen is about 4 vol%; thus 25 % of the LFL would correspond to 1 vol% hydrogen.

1.1.4 Reaction of Hydrogen with Oxygen

The chemical reaction of hydrogen with oxygen is of fundamental importance for many processes including power generation and hydrogen sensing using catalytic combustion. The overall reaction of the hydrogen oxidation can be described as follows [10]:

$$\mathrm{H_2 + \tfrac{1}{2}O_2 \longrightarrow H_2O_{(gaseous)}},\ \Delta_\mathrm{R}H = \text{-241.7 kJ/kmol}$$
$$\mathrm{H_2 + \tfrac{1}{2}O_2 \longrightarrow H_2O_{(liquid)}},\ \Delta_\mathrm{R}H = \text{-286.1 kJ/kmol}$$

The heat of formation for water is added in the second reaction, yielding a higher value for the heat of reaction (enthalpy). The chemical equations illustrate that reaction of hydrogen with oxygen — the oxidation of hydrogen — is unique in that only water is produced and no carbon dioxide is formed. In comparison, the following equation can be established for oxygen reacting with carbon containing fuels under ideal stoichiometric conditions:

$$\mathrm{C_xH_yO_z + (x + y/_2 - z)O_2 \longrightarrow xCO_2 + y/_2H_2O}$$

The mechanism for the hydrogen combustion in air is based on a radical chain reaction and can be described with the following single steps [15]:

$$\mathrm{H_2 + O_2 \longrightarrow 2\,O^{\cdot}H}\ \mathrm{(I)}$$
$$\mathrm{O^{\cdot}H + H_2 \longrightarrow H_2O + H^{\cdot}}\ \mathrm{(II)}$$
$$\mathrm{H^{\cdot} + O_2 \longrightarrow O^{\cdot}H + O^{\cdot}}\ \mathrm{(III)}$$
$$\mathrm{O^{\cdot} + H_2 \longrightarrow O^{\cdot}H + H^{\cdot}}\ \mathrm{(IV)}$$
$$\mathrm{H^{\cdot} \longrightarrow \tfrac{1}{2}\,H_2}\ \mathrm{(V)}$$
$$\mathrm{H^{\cdot} + O_2 + M \longrightarrow HO_2 + M}\ \mathrm{(VI)}$$

According to this mechanism, the reaction starts with the formation of OH radicals[9] (step I). Initiation requires that a minimum ignition energy is supplied. Reaction propagation occurs in steps (II), (III), and (IV), which lead to a chain reaction by the formation of two additional OH radicals. The termination of this chain reaction occurs by radical recombination (step V), radical interaction with neutral molecules in the gas phase, or quenching by reaction of the radical with solid matter (step VI). This mechanism predicts a radical chain reaction with more chain branching than chain terminations, which leads to a rapid increase in the number of OH radicals. The reaction speed results in liberation of energy and an increase of temperature and gas volume. The net reaction of oxygen with hydrogen via combustion is the same as via electrochemical processes. This electrochemical reaction is exploited in fuel cells for power and electrochemical sensors for detection.

1.1.5 Interaction of Hydrogen with Transition Metals

The interaction of hydrogen with metals is of special importance for hydrogen sensing. Many hydrogen sensors use transition metals as an essential hydrogen sensitive component or as a material for electrodes and electrical contacts.

Commonly used transition metals in hydrogen sensors include palladium or platinum and their alloys with silver, copper, or nickel. For example, pellistors contain platinum as a catalyst in an alumina matrix, semiconductor devices use transition metals as dopants, field effect–based and SAW sensors frequently use palladium films as electrodes, and optical sensors often exploit the change of optical properties of thin palladium or platinum films (see Chapter 3). Magnesium-based alloys are also being investigated as the active element in optical hydrogen sensing. Furthermore, the reactivity of hydrogen with transition metals is of interest for many industrial applications, including catalysis, powder metallurgy, and for nuclear and fusion reactors. Surface and bulk interactions with metals are of importance for hydrogen purification, hydrogen storage, and fuel cells.

The interactions between hydrogen and metals can be described in terms of the following phenomena [16,17]:

- Trapping and sticking of hydrogen molecules on metal surfaces
- Physisorption (adsorption) of hydrogen molecules on metal surfaces and interfaces
- Dissociation of the adsorbed hydrogen molecules
- Chemisorption (absorption) of hydrogen including chemical bonding on surfaces
- Surface penetration of hydrogen atoms (or molecules) and dissolution
- Bulk diffusion of hydrogen as a molecule, atom, or ion into the metal lattice
- Hydride formation (MH_x)

The following provides a short overview of hydrogen surface and bulk reactions.

[9] Radicals are atoms or molecules with one or more unpaired electrons, which are here indicated as a dot on top of the symbol for the atoms or molecule.

1.1.5.1 Surface Reactions

Physisorption is the result of van der Waals forces between the surface and an adsorbent and is by definition reversible. Physisorption generally occurs rapidly (nearly instantaneous) and is accompanied by an enthalpy change that is approximately equal to the heat of condensation of the gaseous adsorbent, i.e., about 40 kJ/mol or less. The step following adsorption on the surface is often the dissociation of the hydrogen molecule to form atomic hydrogen. The reaction to produce atomic hydrogen is given by:

$$H_2 \underset{\text{Desorption}}{\overset{\text{Absorption}}{\rightleftharpoons}} 2\,H$$

This reaction is reversible to some extent. The kinetics of adsorption can be described by a Langmuir isotherm, where the fractional coverage Θ becomes:

$$\frac{d\Theta}{dt} = \mathrm{K} \cdot p_{\mathrm{H2}}(1 - \Theta(t))^2 - d\Theta(t)^2 \tag{1.5}$$

Where $\Theta(t)$ is the fractional surface coverage as a function of time and hydrogen partial pressure p_{H2} and K is a constant. This equation can be solved for the steady-state coverage, in which case the rate of change of fractional coverage is zero:

$$\Theta = \frac{\mathrm{K} \cdot p_{\mathrm{H2}}^{1/2}}{1 + \mathrm{K} \cdot p_{\mathrm{H2}}^{1/2}} \tag{1.6}$$

Where K is the ratio of the rate constants of the adsorption and desorption processes.

The heat of adsorption depends on the hydrogen coverage of the surface and can change considerably as the surface becomes completely covered by hydrogen atoms. Also, the work function of the metal changes with increasing coverage by hydrogen. These effects have consequences for hydrogen sensing, e.g., on the sensitivity of the field effect (or work function) based sensing platforms which use palladium or platinum thin films. The heat of adsorption for hydrogen on platinum surfaces is about 45 kJ/mol to 100 kJ/mol and decreases with rising surface coverage by hydrogen [16]. On palladium surfaces, the nearly spontaneous dissociative adsorption of H_2 molecules occurs with little or no activation energy barrier.

The heat of adsorption for hydrogen on palladium surfaces are experimentally determined or calculated to be in the range of 12 kJ/mol to 133 kJ/mol. Heat of adsorption values depend on experimental conditions, the rate of coverage and temperature [18, 19]. The relatively low heat of adsorption can explain the fact that hydrogen sensors based on palladium can work at slightly elevated temperatures or even at room temperatures. The inherent selectivity for hydrogen, fast sorption kinetics, and reversibility of hydride formation explain the prevalence of palladium and palladium-based alloys as sensing materials for hydrogen. The high diffusion coefficient of hydrogen in palladium, at 25 °C (about 10^{-7} cm^2/s), also makes it possible to use this metal as a membrane for hydrogen purification.

Chemisorption is a chemical reaction between the surface atoms of the metal and the adsorbed atom or molecules leading to the formation of chemical bonds. The binding energy is a relevant parameter for the chemisorption of hydrogen on the surface of transition metals. Chemisorption binding energy values of 190 kJ/mol and 345 kJ/mol were calculated in good agreement with experimental results for bonds of hydrogen on the surface of transition metals [20]. The binding energy for the palladium-hydrogen bond is in the range of 271 kJ/mol and 278 kJ/mol while the binding energy for the platinum-hydrogen bond is in the range of 240 kJ/mol to 270 kJ/mol [20].

Reaction of hydrogen and oxygen on a platinum surface

The catalytic oxidation of hydrogen is of particular relevance for combustion processes and hydrogen sensing. The following elementary steps can be distinguished [21], [22]:

Adsorption:
$H_2 \longrightarrow 2\,H_{ads}$ (I)
$O_2 \longrightarrow 2\,O_{ads}$ (II)

Surface reactions:
$H_{ads} + O_{ads} \longrightarrow OH_{ads}$ (IIIa)
$OH_{ads} \longrightarrow H_{ads} + O_{ads}$ (IIIb)
$H_{ads} + OH_{ads} \longrightarrow H_2O_{ads}$ (IV)
$OH_{ads} + OH_{ads} \longrightarrow H_2O_{ads} + O_{ads}$ (V)

Desorption:
$H_2O_{ads} \longrightarrow H_2O\uparrow$ (VI)
$2\,H_{ads} \longrightarrow H_2\uparrow$ (VII)
$2\,O_{ads} \longrightarrow O_2\uparrow$ (VIII)

This reaction initiates at room temperature and was already used by the Doebereiner lighter in the 19th century. This reaction also forms the basis for the hydrogen detection principle of catalytic sensors (see Section 3.2).

1.1.5.2 Hydrogen Solubility in Metals and Hydride Formation

Following the dissociation of the hydrogen molecule on the metal surface, the hydrogen can diffuse into the metal lattice. Diffusion of hydrogen can also take place at grain boundaries and dislocations. Hydrogen can be chemically bonded to form metal hydrides, in which the crystallographic structure of the metal is only slightly changed or a new crystalline phases is formed. Both stoichiometric and non-stoichiometric compounds with variable amounts of hydrogen atoms in the lattice exist. Most metals react with hydrogen to form either metallic, covalent, or ionic hydrides. The type of metal-hydrogen bonding depends on the difference of electronegativities between the metal and hydrogen.[1] The d-block transition metals (Ti, Cr, V,..., Zn) tend to form non-stoichiometric interstitial hydrides. Transition metals with cubic or hexagonal close packing of atoms readily accommodate hydrogen absorption. Hydrogen is incorporated first in the bulk metal in the octahedral or tetrahedral sites followed by the formation of hydride of composition $MH_{\geq 1}$ (e.g., M = Ti, V, Cr, Fe, Co, Ni, Cu,). With increased hydrogen concentration the smaller gaps will be occupied followed by formation of a hydride of composition $MH_{\geq 2}$ (e.g., M = Ti, V, Cr, Zn, Zr, Nb, La, Hf). A higher hydrogen content can be accommodated in $LaH_{2.87}$,$Yb_{2.55}$, and UH_3. As a result of the increased hydrogen incorporation the metal lattice expands in volume up to 30 %. This leads to large stresses at the phase boundaries between the solid-solution and hydride phase, resulting in defects or cracks, which can change the mechanical properties of the material.

Rare earth, alkali, and alkaline earth metals can form stoichiometric hydrides (e.g., YH_3, ScH_3, LiH, NaH). As most transition metals do not react with hydrogen under ambient conditions, intermetallic hydrides, combining rare earth or alkaline earth metals (A) with

[1] The electronegativity of hydrogen is 2.2.

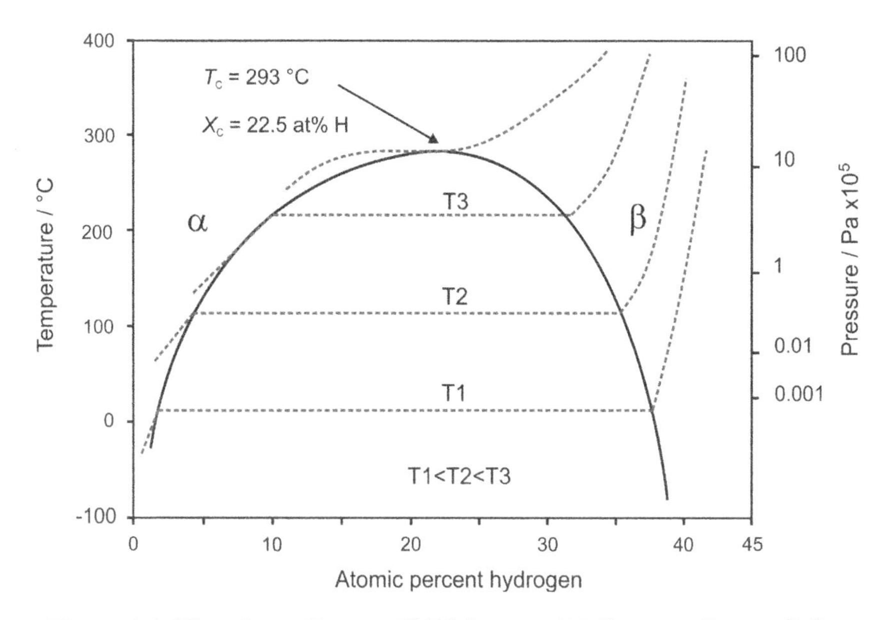

Figure 1.4: The phase diagram H-Pd for $p \geq 100$ Pa according to [24].

transition metals (B) have been extensively investigated. Many types are known, such as A2B and AB5. Examples of A2B ternary metal hydrides include Mg_2FeH_6 and Mg_2NiH_4 [23].

The Palladium-Hydrogen System

The palladium-hydrogen interaction has particular relevance with regards to hydrogen sensors. Palladium forms an interstitial hydride where hydrogen occupies empty spaces in the crystal lattice sites. Hydrogen can be dissolved in palladium to nearly up to an H/Pd ratio of 0.8. The amount of hydrogen absorbed in the metal depends on temperature and the external hydrogen gas pressure. There are two phases in the palladium-hydrogen system. At low concentrations of hydrogen, a solid-solution α-phase forms, where hydrogen randomly occupies octahedral sites of the face centred cubic (fcc) metal lattice. The low-concentration α-phase has a lattice constant close to palladium metal. At room temperature the hydrogen/palladium ratio for this phase is up to 0.017, corresponding to an increase of the lattice parameter of less than 2 %. At higher hydrogen concentrations the β-phase starts to form with a composition of $PdH_{0.67}$, which coexists with the α-phase. The stability of the two phases depends on temperature and hydrogen gas pressure. The two-phase region in the H-Pd system can be observed at temperatures up to 293 °C [24] (see Figure 1.4).

The volume of the newly formed β-phase increases by about 10 % in comparison to the α-phase [25]. This volume change associated with hydrogen absorption and desorption can have an important impact on the ageing behaviour of hydrogen sensors (e.g., causing delamination or crack formation in sensors with thin metal films). However, the changes in volume can be reduced and the first order α to β phase transition can be avoided by alloying Pd with Ag or Ni, which will help reduce the susceptibility of thin metal films to delamination (see Section 3.5).

The incorporation of hydrogen in metals also has an impact on their mechanical properties. The recombination of hydrogen atoms to form H_2 can result in the formation of cavities, strain, and cracks in bulk metals. Micro-perforation can occur. The resulting degradation of mechanical properties, (i.e., reduced tensile ductility, elasticity, strength of metals, and stress cracking) due to the interstitial hydrogen, is called hydrogen embrittlement. Martensitic stainless steel, titanium, nickel, and, under certain conditions also copper, are especially susceptible to hydrogen embrittlement [26,27]. This phenomenon has to be considered when selecting the materials for sensor and manufacturing other components that will be in contact with hydrogen, such as housings or membranes as well as for pipes, fittings valves, and tanks.

1.2 SAFE USE OF HYDROGEN

Ulrich Schmidtchen, BAM Bundesanstalt für Materialforschung und -prüfung, Berlin

1.2.1 Hazards

Under normal ambient conditions hydrogen is a colourless, odourless, and tasteless gas. As with other fuels and chemical feedstock, hydrogen has hazards associated with its storage, handling, and use. These hazards follow from its properties, including hydrogen's reactivity with oxygen, as discussed previously (see Section 1.1). It is important to note that many of these hazards are not specific to hydrogen and are inherent safety concerns associated with all flammable gases and vapours. However due to limited experience and data regarding its use, knowledge of hydrogen hazards and their associated risks is often lacking and the perception of associated risks may be incorrectly exaggerated.

Hazard and risk are terms which will be used regularly in this section. These terms are often used synonymously despite having a different meaning. A *hazard* is a chemical or physical condition that has the potential for causing damage to people, property, or the environment. On the other hand a *risk* is a measure of human injury, environmental damage, or economic loss in terms of both the incident likelihood and the magnitude of the loss or injury [28]. There are three groups of hazards related to hydrogen release or interaction with materials, namely:

Physiological hazards
: These hazards relate to potential health damage associated with direct or indirect contact with hydrogen. Release of hydrogen into air can result in oxygen displacement which can lead to asphyxiation because of inhalation of oxygen-deprived air. Rapid release of high pressure hydrogen can lead to overpressure which can damage the lungs or ears. Direct contact with cold hydrogen gas or liquid can cause hypothermia and freeze burns. While hydrogen flames have a low radiant heat compared with carbon-based fuels, the potential hazards associated with this heat radiation should not be ignored. Furthermore hydrogen flames are almost invisible to the human eye. Close or direct contact with a hydrogen fire can cause burn injuries.

Physical hazards
: Hydrogen can interact with materials including metals or polymers with which it is in contact, inducing embrittlement and mechanical degradation. The low temperatures of liquid hydrogen may increase the brittleness of materials or may induce mechanical stresses when adjoining materials of dissimilar thermal expansion coefficients are used. These concerns are particularly relevant to hydrogen containment and distribution systems, which may fail resulting in hydrogen spills and leaks.

Chemical hazards

Mixtures of hydrogen with air are flammable over a wide range of compositions and are easily ignitable. A hydrogen flame is nearly invisible. The combustion of hydrogen in oxygen or air is very fast and gives rise to a sudden increase in temperature, volume expansion. and a pressure increase in a closed system. This is commonly called "explosion". These consequences can cause health risks and can damage facilities. While the combustion of hydrogen may cause severe accidents it should be noted that such hazards exist for every energy carrier, whether it is oil, gas, coal, wood, or uranium. All of them are dangerous in one way or another simply because they contain energy. The uncontrolled release of this energy may lead to catastrophic results. Due to the specific properties of gaseous or liquefied hydrogen, its use can lead to unwanted events, damages, and injuries. A survey of incidents associated with hydrogen is available from U.S. Pacific Northwest National Laboratory [29], and accidents are registered in the hydrogen incidents and accidents database of the European Commission [30]. Most incidents registered in this database took place in laboratories, fuelling stations, power plants, hydrogen delivery vehicles, and hydrogen storage facilities (data about incidents in facilities of the chemical industry are not readily available). The types of equipment involved in reported incidents are mostly pipes, fittings, valves, hydrogen storage equipment, vehicle and fuelling systems, safety and ventilation systems, and pressure relief devices [4].

1.2.2 Flammability

Hydrogen is a flammable gas and the most important property relevant to safety is its flammability. When hydrogen is mixed with an oxidiser (usually oxygen) a combustible mixture can form which can be ignited by a suitable ignition source. Similar to all flammable gases, the combustion of hydrogen can exhibit different behaviour depending on the ambient conditions, the local environment, the ignition source energy, and on the hydrogen and oxidant concentrations. Propagation of a flame following ignition will only occur when all these conditions are right.

Common terms describing gas combustion processes include flammability, explosion, deflagration, and detonation. Some of these terms are defined and used differently in the literature depending on the target audience. Terminology also depends on whether a text comes rather from the European or the North American area; this is particularly true for the terms "flammability limits / range" and "explosion limits / range."

A gas is flammable when it is capable of being ignited in air, oxygen, or another supporting atmosphere [31]. This can only happen when the fuel concentration is inside certain concentration limits which are usually called explosion limits or flammability limits. The first term is usual in CEN standards and even legally mandatory in Europe, while the second is common especially in North America [32]. These limits are not well-defined natural properties of the mixture but depend on how they are measured which must be done according to exactly prescribed procedures (given, e.g., in EN 1839 [32] or ASTM E 681 [33]). For mixtures of hydrogen and air the limits are approximately 4 vol% and 77 vol% respectively, the differences between the results measured according to the different standards being negligible for practical purposes. The explosion/flammability limits of a flammable gas mixture depend strongly on temperature and pressure. Figure 1.5 illustrates the effect of pressure.

While the lower explosion limit of hydrogen is comparable to that for other gases (e.g., methane is 4.4 vol%), the upper limit of 77 vol% is extraordinarily high (e.g., methane is

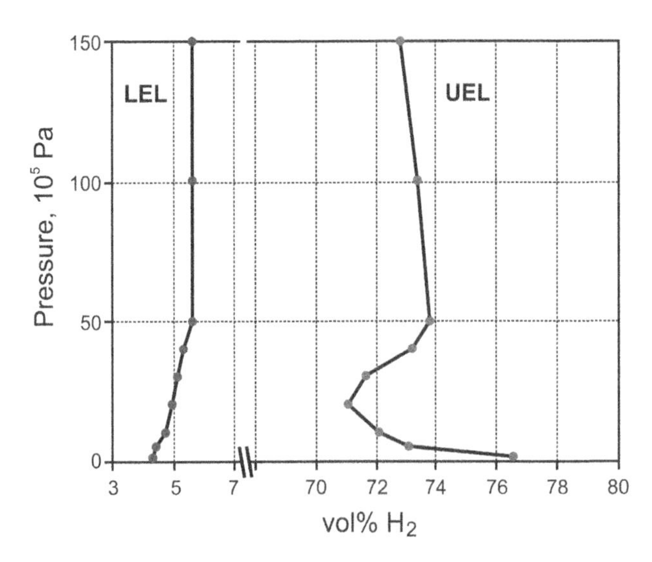

Figure 1.5: Influence of pressure on the lower and upper explosion limit [34].

16.5 vol%). No other gas, with the exception of acetylene, has a higher upper limit. The lower explosion limit typically comes into effect when hydrogen is released from containment into the ambient environment (air). Conversely, the upper limit plays a role when an oxidant is introduced into pure hydrogen. The wide explosion limits of hydrogen in air, combined with the relatively low ignition energy often required, implies that the potential for hydrogen to form flammable mixtures and to be ignited is higher compared to other flammable gases.

It should be noted that the values given in the literature are valid for atmospheric conditions and for mixtures with air containing about 78.0 vol% N_2 and 20.9 vol% O_2. The upper explosion limit increases dramatically when the hydrogen is mixed with pure oxygen or oxygen-enriched air (while the lower limit is almost not affected). A minimum amount of oxygen of 5 %, the so-called "limiting oxygen index," is necessary for hydrogen flame propagation [35].

The combustion of hydrogen in oxygen or air takes place due to a radical chain reaction mechanism (see Section 1.2.2). In a quiescent mixture with no confinements a flammable mixture of hydrogen in air will undergo deflagration whereby the flame relies on heat and mass transfer mechanisms to combust and move into areas of unburned hydrogen. In a deflagration the flame propagates at a subsonic velocity. Flame speed may be accelerated by induced turbulence from obstacles or by confinement. In the extreme case the flame speed may become supersonic resulting in detonation. Detonations are the most destructive mode in which hydrogen can combust. The combustion wave front propagates with a supersonic velocity through the unburned fuel and there is a sudden liberation of thermic energy and a volume expansion which causes a shock wave. The maximum possible flame propagation speed, i.e., the flame front velocity, is 975 m/s for a 29.6 vol% H_2 in air (stoichiometric) mixture. Pressures generated by detonations exceed those of deflagrations with overpressures above 1 MPa (detonation) possible. The laminar burning velocities of hydrogen - air mixtures are characteristically very high with maximum values of about 2.5 m/s to 2.9 m/s for mixtures containing 40 vol% H_2 [12]. This compares with laminar burning velocities of approximately 0.37 m/s and 0.30 m/s for methane and gasoline, respectively.

The mere presence of an ignitable mixture at ambient conditions will not in itself lead to sustained combustion or explosion. The temperature of self-ignition of flammable hydrogen containing gas mixtures is relatively high. Data reported in the literature vary in the range of 500 °C to 585 °C due to different methods of determination [4]. The mixture must be ignited by a source with a sufficient amount of energy. The flammability limits are defined as the concentrations where this so-called ignition energy becomes infinite. Between these limits the ignition energy is a function of the mixture concentration, with a range of values over three orders. Mixtures containing 23 vol% hydrogen have the lowest ignition energy of 0.017 mJ which is very low in comparison to other fuels and about 1/10 of a gasoline-air mixture. Potential ignition sources include sparks from electrical equipment, electrostatic discharges, flames, or catalytically active particles. Although the ignition energy of hydrogen-air mixtures varies with gas composition, most ignition sources have an energy of more than 10 mJ, therefore a flammable hydrogen mixture is very likely to be ignited.

As mentioned previously the determination of explosion limits, ignition energy, auto-ignition temperature, and many other safety-relevant characteristics of hydrogen gas mixtures depends on the measurement procedure. While international standards exist which give exact details of the procedures to make comparable measurement unfortunately these standard methods are not harmonised. For this reason, care should be taken to note which method has been used to get a specific result. In addition, real life conditions of accidents or other events are certainly different from these standard conditions, which mean that the values found in literature may not apply. Accordingly, most safety regulations on allowable hydrogen levels tend to be conservative, that is action levels are set at a fraction of the common hydrogen LEL of 4 vol%.

1.2.3 Specific Safety Issues

While there are many flammable gases, hydrogen is unique because it has the lowest density of all chemical elements. The hydrogen molecule is not only the lightest but also the smallest, which has a number of safety relevant consequences.

1. The gas volume escaping from a leak of a given size is higher for hydrogen compared with other gases (including helium which is frequently used as a leak test gas) as hydrogen's small molecular size leads to a low gas viscosity. Accordingly a containment, which is leak tight for air or natural gas, may not necessarily be leak tight for hydrogen. This may mean that safe operation of a hydrogen facility may require more sensors and at different places than for other gases.

2. Hydrogen has a higher diffusion rate than any other gas. A cloud formed by a inadvertent hydrogen release will disperse quicker than a cloud of another gas, which means that the area where a fire or explosion could occur increases faster than for other gases. At the same time the combustible hydrogen mixture cloud dilutes faster below the critical limit than other gases. Whether this effect makes hydrogen more or less dangerous than other flammable gases cannot be stated unequivocally but depends on the specific conditions and circumstances.

3. As a consequence of its small size, hydrogen has a higher propensity to permeate through solid matter including metals. Consequently a hydrogen pressure vessel will loose a certain amount of gas over time because hydrogen diffuses into the material and permeates to the outside. This must be considered particularly when selecting the materials for hydrogen pressure vessel and pipelines. In polymeric materials this effect will be greater than in metals. While there are ways to reduce hydrogen permeation

(e.g., by creating barrier layers in the material), there is no way to avoid it completely. For high pressure composite material vessels permeation may become more pronounced if the vessel is held at pressure for several months or years. Losses from metallic pressure vessels that are under the same conditions are technically insignificant.

4. The interaction of hydrogen with certain metallic materials (see Section 1.1.5) needs careful consideration. Hydrogen atoms when diffused into certain metallic lattices may accumulate within and at the tip of an existing crack thereby accelerating its growth. Material failure may happen much earlier in the presence of hydrogen than for other gases under the same conditions. This process is a complicated interaction which depends on the type of crystal lattice, micro-structure of the material, gas pressure, condition of the metallic surface, and stress in the material. This makes it difficult to qualify a metallic material as principally suitable or unsuitable for hydrogen use as it depends on the details of the application.

1.2.4 Explosion Protection

Commercial or scientific work with flammable gases must be performed as safely as reasonably possible. Explosion protection is an important measure to achieve this. There is an extensive body of experience on how to prevent explosions or to minimize their effects. There are also national and international regulations and other codes about this topic. For the countries of the European Union the relevant regulations are found in the ATEX directives [36,37] and for North America in the International Fire Code [38] or NFPA 2 (see Sections 6.1.2 and 6.1.3). While the details may be different there is a certain pattern which is followed more or less everywhere with regards to explosion protection. Safety principles consist of prevention, detection, protection, and intervention. Three groups of protection measures are considered.

1. Primary explosion protection to prevent the formation of an explosive mixture.

2. Secondary explosion protection to prevent the ignition of an explosive mixture.

3. Tertiary or "constructive" explosion protection to minimize the effects of an explosion.

Explosion protection is a stepwise process. The three groups of explosion protection measures by no means exclude each other, and the end-user is legally obliged by regulation and legal codes to prevent the formation of an explosive mixture as far as is reasonably possible. If the formation of an explosive atmosphere cannot be fully eliminated the user must take measures to prevent possible ignition of the explosive mixture; the prevailing regulations also become more strict if an explosive mixture cannot be precluded. For any remaining explosion risk the user must take appropriate measures to minimise the effects of a possible explosion.

One strategy of **primary explosion protection** is to prevent the unintentional escape of gas from its vessels or pipes as far as possible. There are several ways to achieve this. One way is to minimise the number of flanges or other connections. Make-and-break connections in particular are a potential release site and should be avoided if alternatives are available. All connections should be preferably welded or soldered if possible. Valves or couplings demand special attention and sealing materials should be chosen appropriately and their ageing effects should also be taken into consideration. Minimization of the amount of combustible gas that can be released is another mitigation strategy.

Another primary explosion protection strategy is to detect a gas leakage sufficiently early and to activate mitigating measures before a flammable or explosive mixture can form.

Since hydrogen has neither a distinctive colour nor smell, sensors play a pivotal role for the detection of unwanted hydrogen releases. However, these sensors must be installed in the right positions for them to be effective (see Section 6.4).

If hydrogen is detected, a common sequence of measures to prevent the formation of an explosion mixture as part of a primary explosion protection strategy is to:

Stop or isolate the gas supply

This should preferably be done by passive elements, for example remotely operated valves which fail into the safe position (in this case the closed state) if external energy is not supplied. Blocking the connection to the gas supply will have the effect that only the gas already in the process can escape. Depending on the details of the application there may also be similar valves separating the main parts from each other to further reduce the amount of gas which can be released.

Provide additional natural ventilation

This usually means that windows in the roof or near the ceiling are opened to let hydrogen escape.

Provide artificial ventilation

The required efficiency of the artificial (or forced) ventilation system is calculated on the basis of the amount of hydrogen which can be expected to be released in case of an accident. Typically, forced ventilation specifies a number of equivalent air exchanges per hour. The ventilation system must also be compatible with (potentially) flammable gas mixtures. Thus, when designing hydrogen facilities, special care should be given to the ventilation system. This is especially true when adapting or upgrading existing facilities to accommodate hydrogen activities. Furthermore, the gas escaping from a natural or forced ventilation system must be vented safely to avoid an external explosion hazard. Care should be taken that the exhaust gas may not be ignited outside by some ignition source.

Another explosion protection technique comprises flooding the work area with an inert gas like nitrogen or carbon dioxide. The appropriateness of this technique depends mainly on the size of the room and whether occupants would be unduly subjected to an asphyxiation hazard. The use of an inert gas to minimize flammable hazards is most conveniently employed in closed unoccupied areas.

Secondary explosion protection aims at preventing the ignition of an explosive mixture. Since leaked flammable gas concentration gradient is usually from 100 vol% at the point of leakage to 0 vol% at a given distance away from the leakage, there will be some region in between where an ignitable mixture is formed. While primary explosion protection may minimize the likelihood of forming an explosive mixture it will not necessarily eliminate them.

Secondary explosion protection focuses on the use of appropriate electrical components and installation which cannot act as an ignition source. There is a wide choice of explosion protected electrical devices, from plugs and switches to lights, depending on the particular nature of the application. Avoidance of electrical components is of course preferable. This can mean the use of pneumatically activated valves instead of electrically activated ones. Similarly, sensors used for primary explosion protection must not act as ignition sources. Accordingly, the relevant regulations applicable on the site of the installation must be taken into account to prevent this (see Section 6.1).

Grounding of the complete installation is one of the foundations of secondary explosion protection. Not only must the hydrogen containing components be grounded, but also other

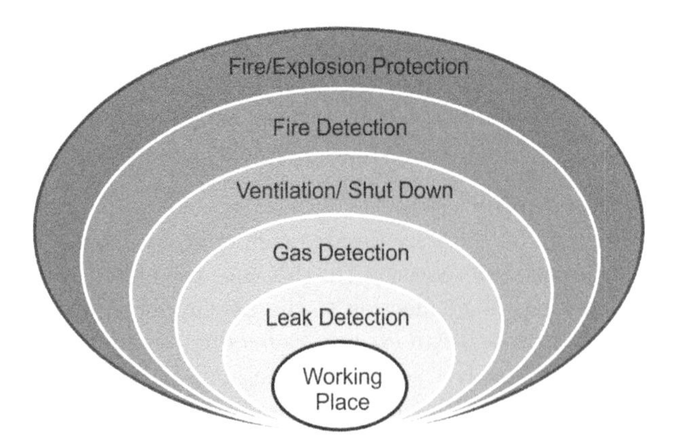

Figure 1.6: Layered protection sequence for mitigating risk associated with hydrogen gas leaks.

items such as heating radiators. Electrical sparks can not only be generated by electrical equipment, but also by persons due to accumulation of static charge. Static discharge can be eliminated by certain combinations of floors and shoes which are especially designed for this purpose. A conductive floor can help to prevent damage from static charge. The use of grounding straps, worn by workers, also minimizes the chance of a worker causing an explosion through an inadvertent static discharge.

While electrical sparks are especially important ignition sources they are by far not the only ones. Mechanical sparks can also cause ignition. Not even the use of spark-free tools is an absolute guarantee against creating them; while the tool material itself will not generate sparks, an object hit with the tool still can.

Another type of ignition source is a hot surface generated, for example, by friction. Compressors and pumps need special considerations because they are not only electrical components, but they are often comprised of rotating machinery which can heat up by friction. This can be especially problematic if they are not working properly. It is recommended to operate such devices outside the space to be protected.

In order to address the hazards posed by hydrogen, manufacturers of fire and gas detection systems work within the construct of layers of protection to reduce the incidence of hazard propagation. Under such a model, each layer acts as a safeguard, preventing the hazard from becoming more severe. Figure 1.6 illustrates such a layered protection sequence for mitigating risk associated with hydrogen gas leaks.

1.3 HYDROGEN AS AN INDUSTRIAL MASS PRODUCT

Hydrogen has been used for more than one hundred years in industrial processes. Global hydrogen production was estimated to be around 53 Mmt (million metric tons) in 2010 and is expected to exceed 60 Mmt by 2016. However, only about 12 % is available for general use on the open market while the rest is delivered to dedicated customers [39]. Nearly 96 % of all hydrogen is currently derived from fossil fuels [40]. Steam reforming of natural gas is by far the most frequently used hydrogen production process, accounting for around 48 % of the total production, followed by partial oxidation of liquid hydrocarbons at 30 %, and gasification of coal at around 18 % (see Figure 1.7). Hydrogen produced by electrolysis of

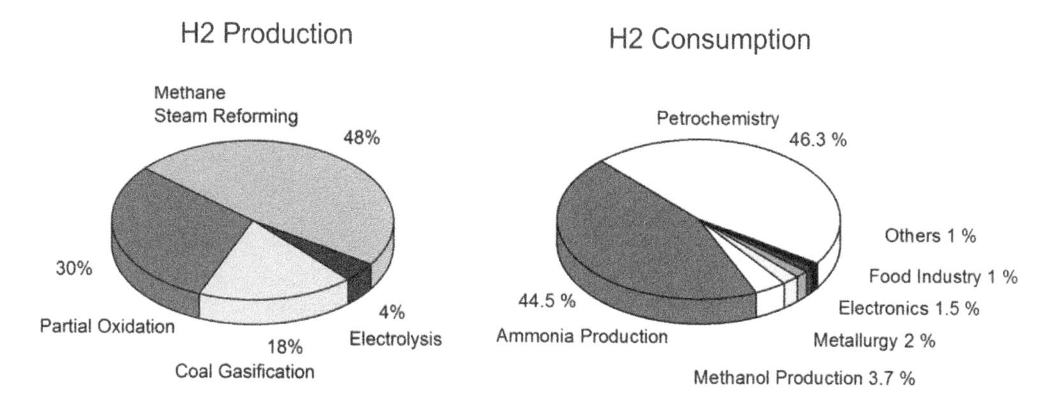

Figure 1.7: Distribution of relative amounts of global hydrogen production and use by various processes, data from [41].

water and as a by-product from other sources accounts for only 4 % of the total production (see Section 1.3). With the exception of electrolysis, these hydrogen production methods also produce carbon dioxide, since the feedstock is carbon based.

Most of the hydrogen produced is used for oil refining in the petrochemistry industry (46.3 %). Hydrogen is also used for ammonia (44.5 %) and methanol production (3.7 %) [41]. It is also applied in metallurgy, glass manufacturing, and the food industry. In this section the main technical processes for producing and using hydrogen are briefly described, and detailed descriptions are given in the literature [42]. The use of sensors in these industries is discussed in Chapter 7.

1.3.1 Contemporary Hydrogen Production Methods

1.3.1.1 Steam reforming

The most important contemporary process for producing hydrogen is the reaction of carbon containing materials with water vapour (steam) in the presence of a catalyst. This process produces "syngas," a mixture of hydrogen and carbon monoxide (CO). The carbon material can be natural gas, methanol, or petrol. The carbon feedstock can also come from renewable sources such as biomaterials. The catalyst is subject to poisoning from sulphur and other compounds that might be present in the feedstock. Accordingly, the source materials need to be processed to ensure sufficient purity for use in the reforming process. Contamination levels can be especially problematic for feedstock generated from biomaterials. The general chemical reaction equation for steam reforming of hydrocarbons (C_nH_m) is:

$$C_nH_m + nH_2O \longrightarrow (n + m/_2)H_2 + nCO$$

The specific reaction for methane (natural gas) steam reforming can be given as

$$CH_4 + H_2O \longrightarrow 3\,H_2 + CO,\ \Delta_R H^o_{298K} = 206.2\ \text{kJ/mol}$$

The methane reforming process is endothermic as indicated by the positive enthalpy (ΔH) for the reaction. As a result, heat must be provided to the process. Accordingly, the process is run at elevated temperatures in a furnace heated by gas combustion. Figure 1.8 shows a photograph of a steam reforming plant.

Figure 1.8: Steam reforming plant combined with PSA hydrogen purification facility (Copyright by Linde AG. Reproduced with permission.).

The process takes place at temperatures of 700 °C to 1100 °C and pressures of 2.5 MPa to 3 MPa; using iron or nickel as catalyst. The amount of hydrogen in the process gas is 86 % to 90 %. In a second step, the so called **water-gas shift reaction**, carbon monoxide is oxidized:

$$CO + H_2O \longrightarrow CO_2 + H_2,\ \Delta_R H^o_{298K} = \text{-41.2 kJ/mol}$$

This reaction is exothermic (energy is released) and occurs at 250 °C to 400 °C on a ferric oxide catalyst. The CO concentration decreases down to around 0.2 vol% to 0.6 vol% and additional hydrogen is produced. About 240×10^9 Nm^3 of hydrogen gas is produced per year from natural gas steam reforming.

The hydrogen gas stream produced by steam reforming needs to be purified to remove by-products and contaminants including carbon dioxide, soot, sulphur components, and traces of CO. **Pressure swing adsorption technology** (PSA) is commonly used for hydrogen purification. PSA is based on adsorption equilibrium at different pressures and on the selective adsorption affinity of the porous adsorber material to specific contaminants in the hydrogen. In the first step, adsorption at pressures of 0.6 MPa to 1 MPa on an adsorber material (molecular sieves or activated carbon) takes place. Carbon dioxide and other contaminants, like methane, nitrogen, or argon are separated from the gas phase as the compressed hydrogen gas passes through the sorbent bed of the PSA. When the adsorber material is saturated, pressure is reduced and the adsorbates are released and the adsorber is purged for regeneration. A simplified process flow diagram for steam reforming and PSA is given in Figure 1.9

Alternatively, purification can be performed by wet scrubbers using amines or methanol as cleansing agents. Cryogenic distillation techniques can be also used to purify hydrogen.

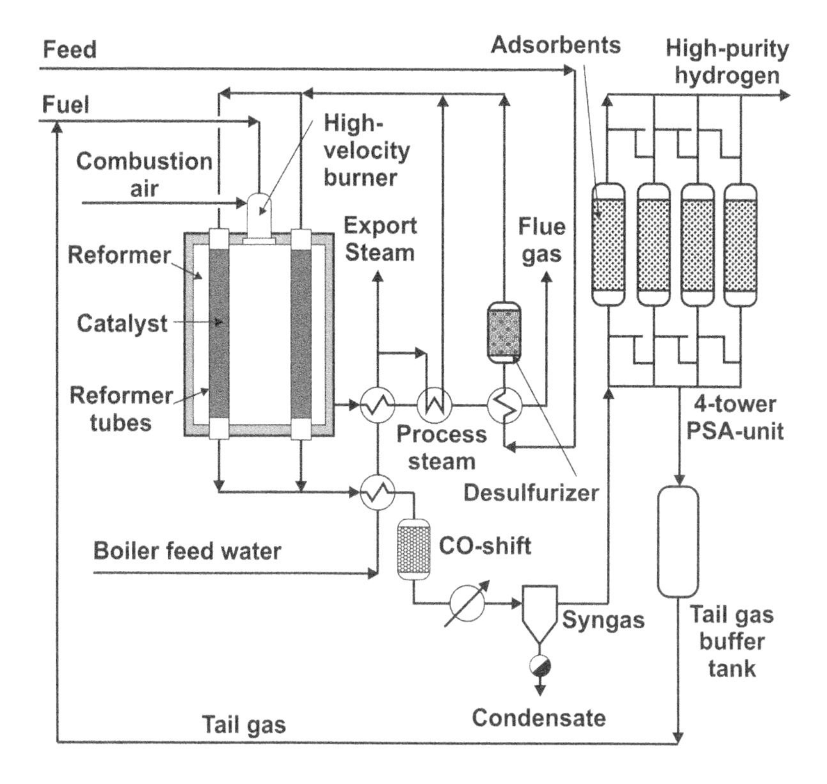

Figure 1.9: Simplified process flow diagram for steam reforming and PSA [44].

Residual amounts of carbon monoxide in hydrogen can be selectively reduced to methane or oxidised to carbon dioxide in a subsequent step, while residual oxygen can be removed by catalytic deoxygenation. The process of converting oxygen into water uses a catalyst which is based on palladium group metals. The formed water vapour is then separated by a dryer [43]. When very high purity hydrogen is required, palladium alloy tubes can be used for purification. The tubes act as a separator for hydrogen at temperatures above 300 °C; the hydrogen permeates through the metal tube, leaving behind all non-hydrogen constituents. Hydrogen of purity up to 99.99999 vol% can be obtained. Typical classes of purified hydrogen gas are given in Table 4.1.

1.3.1.2 Autothermal Reforming

Autothermal reforming is a combination of steam reforming and partial oxidation to increase the hydrogen yield and boost the process efficiency by exploiting energy from the carbonaceous material for the process. Natural gas, petrol, or diesel is typically used as feedstock for this process, however other materials such as cellulose, residential waste, or sewage sludge can also be used. The process is performed at pressures around 10 MPa and at temperatures between 950 °C and 1100 °C. If the sulphur content of the reactants is sufficiently low to avoid poisoning, a catalyst can be used to decrease the operating temperature to between 700 °C and 900 °C. Autothermal reforming can also be used for biomass gasification which is carried out at 400 °C to 700 °C and pressures of 20 MPa to 30 MPa.

1.3.1.3 Coal Gasification

Coal gasification uses the reaction between coal and water at high temperatures to produce hydrogen. The endothermic reaction between powdered coal with steam yields "water gas" at temperatures from 850 °C to 1600°C and at ambient or slightly increased pressure (2 MPa). The reaction can be given by the following equation:

$$C + H_2O \longrightarrow CO + H_2, \ \Delta_R H = +131 \text{ kJ/mol}$$

The process is carried out under autothermal conditions for which the process energy is obtained though the addition of oxygen to drive the simultaneous partial combustion of coal. The gas contains 35 % to 40 % hydrogen, 30 % to 40 % carbon monoxide with the remainder being carbon dioxide and methane. The water-gas shift reaction can then be performed to significantly reduce the carbon monoxide and increase the hydrogen yield. About 90×10^9 Nm^3 of hydrogen are produced per annum from coal globally.

1.3.1.4 Chloralkali Process

The electrolysis of an aqueous sodium chloride solution is performed in order to produce sodium hydroxide and chlorine. Hydrogen is formed as a by-product hydrogen according to the overall reaction equation.

$$2\,NaCl + 2\,H_2O \longrightarrow 2\,NaOH + Cl_2\uparrow + H_2\uparrow$$

The reaction is endothermic and the required energy is provided electrically. The reaction of the chlorine with the hydroxyl groups to form chloride/hypochloride has to be suppressed. Also the mixing of chlorine with hydrogen, which are formed at the cathode and anode, respectively, could result in an exothermic reaction. This reaction has to be suppressed also. Therefore the output of the anode and the cathode are separated. This is typically achieved using three different approaches, including the use of a diaphragm, a membrane, or via the mercury cell process. The membrane and diaphragm provide physical separators. The formation of hydrogen and oxygen in the mercury cell is suppressed due to the over-potential at the electrodes. However, the formed sodium amalgam is subsequently decomposed by reaction with water to form sodium hydroxide and hydrogen. Each year about 1.56 Mmt of hydrogen are produced globally as a by-product of the chloralkali process.

1.3.2 Use of Hydrogen

Hydrogen is a valuable and versatile commodity with a wide range of applications. As stated above, most of produced hydrogen is used in the petrochemistry industry for oil refining. A further part is used for ammonia production while the remainder is used for other diverse processes such as methanol synthesis, production of semiconductors, metals, float glass, and for food processing. The overall global demand for hydrogen is expected to increase by around 5 % to 6 % during the next years [39]. The main hydrogen consuming processes are presented in this section. One growing market for hydrogen use is as a fuel for transportation and stationary power systems (e.g., FCEV and stationary fuel cells). These applications are described in Section 1.4.

Oil Refining

The petrochemical process of hydrocracking crude petroleum for upgrading fossil fuels is one of the largest consumers of hydrogen. Hydrocracking reduces the molecular

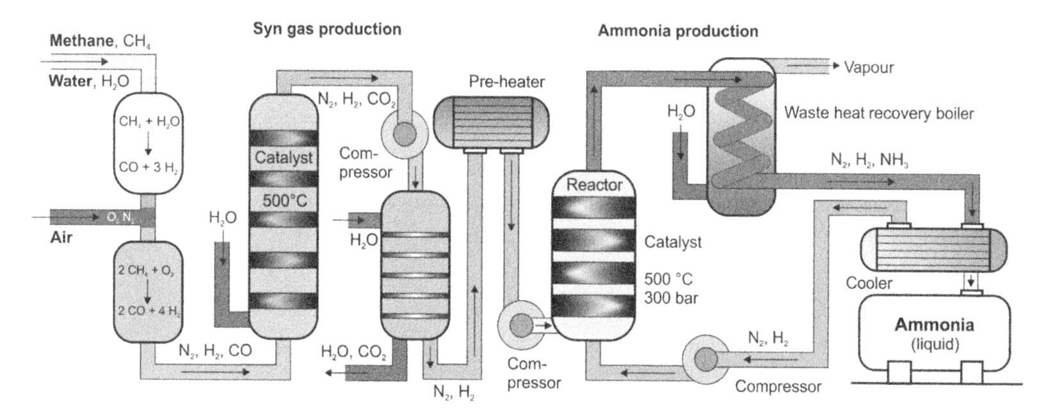

Figure 1.10: Simplified process flow sheet for ammonia production. (according to: http://commons.wikimedia.org/wiki/User:Sven)

weight of heavy gas and oil to substances of a lower molecular weight. In this way crude oil is processed into refined fuels such as gasoline and diesel. The hydrocracking process of heavy oil takes place at temperatures of 250 °C to 480 °C and at pressures of 7.5 MPa to 16 MPa. In the hydrodesulphurization process hydrogen reduces the sulphur content of the fuel, which is removed as H_2S. The process is carried out at temperatures in the range of 300 °C to 400 °C and pressures from 3 MPa to 13 MPa. Cobalt and molybdenum impregnated alumina catalyst is used. Oxygen and nitrogen are simultaneously reduced to H_2O and NH_3. Hydrogen consumption in the oil refining industry is expected to continue grow at an annual growth rate between 4 % to 10 %.

Coal Liquefaction

The Bergius process uses coal and hydrogen for the formation of liquid hydrocarbons, petrol, and diesel. The reaction takes place at pressure of 20 MPa to 70 MPa and temperatures of 400 °C to 600 °C. Aluminium chloride, zinc oxalate, and ferric oxide are used as catalysts.

The Fischer-Tropsch process uses carbon monoxide and hydrogen (produced from coal gasification) for the production of liquid hydrocarbons (e.g., petrol, diesel, oil) according to the following reaction:

$$\mathrm{nCO} + (2\,\mathrm{n} + 1)\,\mathrm{H_2} \longrightarrow \mathrm{C_nH_{2n}} + \mathrm{H_2O}$$

Where n is 10 to 20. The reaction is performed at temperatures from 250 °C or 350 °C and uses catalysts based on Co, Fe, Ni, or Ru on an oxide support.

Production of Ammonia

The Haber-Bosch process uses the reaction between hydrogen and nitrogen to produce ammonia (NH_3).

$$\tfrac{1}{2}\mathrm{N_2} + \tfrac{3}{2}\mathrm{H_2} \longrightarrow 2\,\mathrm{NH_3},\ \Delta_{\mathrm{R}}H = \text{-45.7 kJ/mol}$$

Process temperatures of 450 °C to 550 °C and pressures of 25 MPa to 35 MPa are typical. The catalyst used is based on ferric oxide. Figure 1.10 shows a schematic for the complex ammonia production process. The required hydrogen is typically produced from a reforming process described above in Section 1.3.1. Ammonia is primarily used to produce fertilizers for agriculture. A second major market for ammonia is in the chemical industry.

Methanol Production

Methanol is produced from syngas, which is obtained from fossil or renewable sources such as biogas or sewage sludge according to the following reaction:

$$CO + 2\,H_2 \longrightarrow CH_3OH,\ \Delta_R H = -90.8\ \mathrm{kJ/mol}$$

A copper-zinc oxide-alumina catalyst is used to promote the reaction which is performed at pressures between 5 MPa and 25 MPa at temperatures between 200 °C and 300 °C. Methanol is used in the chemical industry as a solvent or for synthesis, e.g., for formaldehyde or acetic acid production. It can also be used as a fuel for automotive combustion engines or as a hydrogen source in fuel cells [45].

Chemical Products and Plastics

Hydrogen is a fundamental reagent in many synthetic processes for the manufacture of important industrial chemicals, such as alcohols (ethanol, glycol), formic and acetic acid, and hydrogen peroxide, among others. It is also used for the production of polymers, such as polyester, nylon, and polyurethane.

Food Industry

The hydrogenation of vegetable oils under pressure and at temperatures of around 170 °C produces solidified products with a higher melting point (e.g., margarine). The unsaturated vegetable oils which contain carbon-carbon double bonds are partially or completely reduced into C-C-single bonds by reaction with hydrogen using nickel powder as the catalyst.

Metallurgy and Metal Processing

Hydrogen is used in metallurgy as a reducing agent for the production of many metals from ores. Hydrogen is used in the production of aluminium. The direct reduction of iron oxide using hydrogen or hydrogen containing gas mixtures, instead of coal, is an alternative for steel production. Hydrogen is also used in the production of magnesium, copper, nickel, molybdenum, and tungsten. In subsequent processes hydrogen is employed for heat treatment of various metals (hardening). Hydrogen is also used for cutting and welding (atomic hydrogen welding - AHW, e.g., for tungsten).

Glass Manufacturing

Sheets of glass can be made by floating molten glass on top of a bath of molten tin and allowing it to cool slowly. Hydrogen is used in this float glass manufacturing process together with nitrogen to create a reducing atmosphere over the tin bath to prevent oxidation of the metal. Hydrogen is also used in the melting process of fused silica (quartz glass) as a fuel. Its combustion in an oxygen-hydrogen flame generates the required high temperatures of more than 1800 °C. Hydrogen is used for production of optical fibres, for heat treatment of hollow glass products, and for glass welding.

Rocket Fuel

Liquid hydrogen and liquid oxygen were used as the fuel and oxidizer for rocket propulsion in the Space Shuttle Main Engine and for the cryogenic stages of the Ariane 5 rocket.

Cooling Medium

Hydrogen has the highest known thermal conductivity of any gas. As a result hydrogen is used as a cooling medium in gas or steam turbines used in electricity generation.

Semiconductor Industry

Hydrogen is used in the semiconductor industry for manufacturing silicon, such as for saturating dangling bonds of amorphous silicon, as a carrier gas, and for purifying the gas atmosphere (elimination of oxygen). Hydrogen can be used as an electron donor for oxide materials, such as ZnO, SnO_2, or TiO_2. Most of the LEDs produced are manufactured using epitaxy which uses hydrogen as a carrier gas because of its ultra-high purity to avoid contaminations by oxygen and water vapour.

Laboratory Equipment and Chemical Analysis

Hydrogen is used as a carrier gas in gas chromatography and in various analytical instruments, most commonly as a fuel for the detection of combustible gases in flame ionisation detectors (FID) and flame photometric detectors as well as total hydrocarbon analysers.

The industrial production and use of hydrogen in a large scale is mainly based on limited carbon sources and still contributes to unwanted environmental impacts. Also the sodium chloride electrolysis product will probably decrease because of the adverse properties of chlorinated organic materials in the environment. New approaches for carbon free and environmental-friendly technologies are reviewed in the following section.

1.4 EMERGING HYDROGEN MARKETS AND RELATED TECHNOLOGIES

Eveline Weidner, European Commission, Joint Research Centre, Petten

In addition to the well-established market for hydrogen as a feedstock for the chemical industry, the use of hydrogen as an energy carrier is gaining importance in multiple applications. Hydrogen as a fuel or energy storage medium is attracting interest due to its potential to increase energy security and sustainability of the energy supply chain and to reduce greenhouse gas emissions. Power generation from fuel cells in transport and stationary applications is one of the most relevant and investigated aspects of the hydrogen energy chain. These applications are currently still supported through public funding in North America, Asia, and Europe, since a mass market for these technologies does not yet exist. Although originally supported by national governments, some niche applications such as fuel cell forklift trucks have found viable business cases. Hydrogen consumption for non-traditional applications (i.e., outside the petroleum and chemical sectors) is estimated to grow from 0.17 Mmt in 2013 to nearly 3.5 Mmt in 2030 [39].

1.4.1 Fuel Cells

Fuel cells are energy conversion devices which transform chemical energy into electrical energy. Hydrogen based fuel cells are based on the redox reaction of hydrogen with oxygen and consist of an anode and a cathode separated by an electrolyte (see Figure 1.11). Their overall reaction can be stated as:

$$H_2 + \tfrac{1}{2}O_2 \longrightarrow H_2O,$$

and the half-cell reactions for the PEM fuel cell are:

Anode: $H_2 \longrightarrow 2\,H^+ + 2\,e^-$
Cathode: $\frac{1}{2}O_2 + 2\,H^+ + 2\,e^- \longrightarrow H_2O$

Depending on the fuel cell type, different ions are transferred through the electrolyte, which in turn will determine the operating temperature. The electrolyte is impermeable to gases and is not electrically conductive; however, it does conduct ions. Either positively charged or negatively charged ions are transported through the electrolyte from the electrochemical reactions occurring at the anode and cathode. The electrodes are connected by an external circuit and the fuel cell will produce direct current when supplied with fuel. As each cell generates only about 1 V, cells are connected in series into stacks, analogous to the multiple electrochemical cells used in conventional battery configurations. Fuel cell systems are available over a large range of power outputs. Several types of fuel cells have been developed, the most common ones consuming oxygen and either hydrogen or hydrocarbon fuel. An overview of different fuel cell types is given in Table 1.1. Comprehensive descriptions of fuel cells and their applications can be found in the literature, e.g., [46].

Fuel Cell Technologies

Polymer electrolyte membrane fuel cells (PEMFC) are characterized by the proton conducting membrane (e.g., Nafion®, fumapem®) separating the porous electrodes. The electrodes consist of a gas diffusion layer coated with a catalyst material. The so-called membrane electrode assembly (MEA) comprises two flow field plates or bipolar plates, a membrane, catalyst layers, and gas diffusion layers (see Figure 1.11). Both half-cell reactions are catalysed by platinum, which is a key cost driver for this technology. Platinum loadings have already been reduced significantly, and in future recycling of the noble metal will be part of the production chain. As the catalyst is easily poisoned by common chemicals, high purity hydrogen needs to be used. PEM fuel cells are easily scalable in capacity and are therefore deployed for a large range of applications. Their low operating temperature (< 100 °C), high dynamic response, and compact structure make them especially suitable for transport applications.

The alkaline fuel cell (AFC) uses a solution of highly conductive potassium hydroxide (KOH) as the electrolyte. AFCs were the first fuel cells deployed, primarily for military and space applications. Recent advances have renewed interest in this technology, which has some advantages over PEM fuel cells such as higher cell voltages, cheaper catalyst material, and improved robustness against chemical poisons. Typically smaller units up to 100 kW have been installed, however a project for deployment of a 1 MW system in South Korea has been proposed [48].

Phosphoric acid fuel cells (PAFC) operate at higher temperatures (up to 200 °C) than PEM or alkaline fuel cells. Porous electrodes are placed in a concentrated phosphoric acid solution. Noble metal catalysts are used to promote the electrode reactions. Although sulphur species will still poison the catalysts, PAFCs are more tolerant to fuel impurities than PEMFCs. PAFC power plants in the 200 kW to 400 kW range have been deployed mainly in the U.S., Japan, and Germany.

Molten carbonate fuel cell (MCFC) and solid oxide fuel cell (SOFC) operate at high temperatures. The high operating temperature means that non-precious metals can be used as catalysts, potentially lowering costs and also enabling reforming to be carried out internally. Therefore both these fuel cell types can be fuelled by natural gas or biogas.

Table 1.1: General properties and applications of fuel cell types, based on [47]

Fuel Cell Type	**Typical fuel**	**Electro-lyte**	**Charge carrier**	**Electrical efficiency (%)**	**Operating temperature (°C)**	**Power (kW)**	**Applications**
Proton exchange membrane fuel cell (PEMFC)	H_2	Polymer	H^+	40 - 60	50 - 100	< 0.1 - 500	Fuel cell vehicles, back-up power, UPS, portable applications, stationary power generation
Alkaline fuel cell (AFC)	H_2	KOH	OH^-	60 - 70	90 - 100	10 - 100	Space and military applications, chlorine industry
Phosphoric acid fuel cell (PAFC)	H_2	Phosphoric acid	H^+	40	150 - 200	100 - 400	Distributed generation, combined heat and power systems
Molten carbonate fuel cell (MCFC)	Natural gas, biogas	Carbonates	${CO_3}^{2-}$	45 - 50	600 - 700	300 - 3000	Stationary power generation, combined heat, hydrogen, and power generation (CHHP)
Solid oxide fuel cell (SOFC)	Natural gas, biogas	Y-zirconia	${O_2}^-$	60	700 - 1000	1 - 2000	Auxiliary power, electric utility, distributed generation, combined heat, and power systems (CHP)

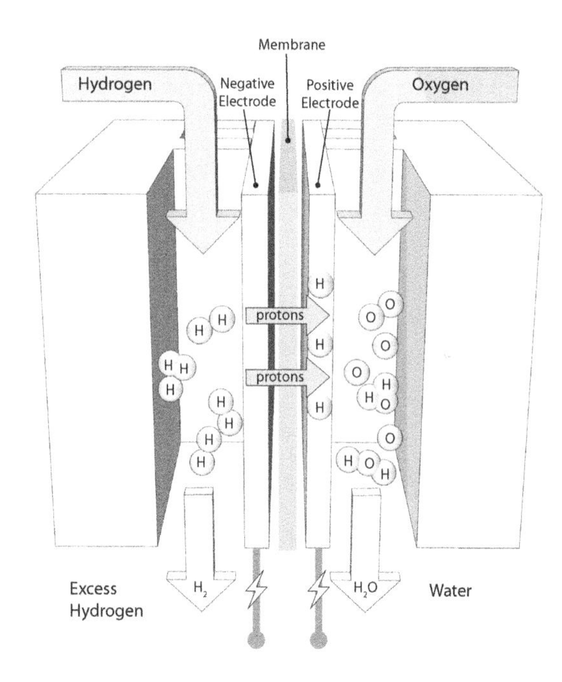

Figure 1.11: Polymer electrolyte membrane electrode assembly.

Fuel Cell Applications and Markets

Fuel cell systems are commercially offered in a wide power range from 10 mW to more than 1 MW. Fuel cells can replace batteries, combustion engines, and thermal power generation systems. Applications therefore range from providing mW level portable power to electricity and heat production in large co-generation power plants. Depending on the requirements, different types of fuel cells are being used (see Table 1.1).

At present, the largest share of the market for fuel cells on a MW basis is for stationary fuel cells. A total of 215 MW of fuel cells were shipped in 2013 (more than 50 % of which for the Asian market, followed by North America and Europe) [49]. The stationary fuel cell market can be roughly divided into residential combined heat and power (CHP) units below 5 kW, commercial/industrial primary power generation CHP in the 200 kW to MW range, and back-up power applications (1 kW to 5 kW). Stationary fuel cells operate at about 50 % fuel-to-electricity efficiency, but the overall efficiency can be > 80 % for co-generation units when their thermal energy is exploited. Fuel cell CHP systems have good potential to be a commercially viable option for distributed generation. The waste heat of fuel cells can be utilized to provide both residential heating and hot water or can be used for distributed heating for larger units. In Japan more than 100,000 micro-CHP units (around 1 kW) have been installed in homes. Due to the success of this initiative, the initial national funding is being phased out, and sales in the millions of units are expected by 2020. In Europe several demonstration projects are also deploying micro-CHP units. Fuel cell systems providing more than 400 kW power output are commercially available and are being installed in facilities, for example in waste water treatment, which have a ready supply of biogas.

Fuel cells are also providing back-up power in telecommunication towers due to their high reliability, which is also of primary importance for provision of emergency power. They are therefore being deployed in data centres, hospitals, and government agencies. Fuel cells are also suitable for providing power in remote locations; however, either hydrogen or natural gas have to be supplied. All fuel cell types have been deployed for stationary applications; however, high temperature fuel cells, operating between 600 °C and 1,000 °C need long warm-up times and are therefore less appropriate for providing back-up power.

The market for portable fuel cells has been explored but they are not readily commercially available, limited by the current high costs for fuel cells for consumer electronics, and safety concerns of bringing hydrogen into enclosed environments such as aircrafts.

A reduction of greenhouse gases emissions and urban pollution from the transport sector is possible through replacement of combustion engines by electric motors running on power from renewable sources. Most vehicle manufacturers are developing electrical drive trains, and battery electric vehicles are gaining increasing market shares. Fuel cell electric vehicles (FCEV) can be seen as complementary to the battery electric vehicle (BEV). Due to the typically long recharging times of batteries BEVs are more suited to urban transport, whereas FCEVs can be implemented for larger vehicle sizes and driving ranges above 500 km. PEM fuel cells are chosen for transport applications due to their technical maturity, low operating temperatures, and relatively small size and weight. The power range of fuel cells for the transport sector is 20 kW to 250 kW, ranging from auxiliary power units to bus drive trains. Vehicle manufacturers, often called original equipment manufacturers (OEMs) in the industry have formed alliances in order to prepare common technological platforms for their FCEVs. Although in general FCEVs are in a pre-commercial and prototype stage, some OEMs now offer FCEVs to the market in small batches on a commercial basis. Initial markets include Japan, Korea, California, Germany, and Scandinavia, with other regions being targeted at later stages, depending on infrastructure roll-out. Different initiatives, such as the German H2 Mobility [50] and H2USA in the U.S. [51], have been established to help support the build-up of the necessary hydrogen refuelling infrastructure. In Europe, Germany is a leader in hydrogen refuelling station deployment, with plans to implement 400 stations within the next decade. In California, the California Fuel Cell Partnership [52] is supporting the construction of up to 60 fuelling stations by 2017. Around 15 stations are set to be operational in Scandinavia by 2015, to be complemented by 30 smaller satellite stations [53]. Fuel cell bus demonstration projects are being operated in several countries (e.g., [54]). Fork lift trucks are seen as a near commercially viable market, with more than 4800 units having been deployed in the U.S. [55]. Other fuel cell applications for mobility include two- and three-wheeled vehicles such as scooters, heavy duty vehicles, trains, ferries, and smaller boats, manned and unmanned light aircraft and submarines.

1.4.2 Hydrogen Production from Renewable Energy Sources

Hydrogen can be produced from a large variety of feedstocks by many different processes, nevertheless currently > 95 % of hydrogen is derived from fossil fuels (see Figure 1.12). An overview of the production of hydrogen for industrial use is given in Section 1.3. Renewable energy sources (RES) such as biomass/biogas, solar energy, and wind energy can be used to produce hydrogen with low associated greenhouse gas emissions (for an overview see [56]).

Water Electrolysis

The electrochemical decomposition of water to hydrogen and oxygen is referred to as water electrolysis. It can be considered as the reverse of the process taking place in a fuel cell, and there are corresponding electrolyzer technologies for PEM, alkaline, and solid oxide fuel cells. The reaction takes place within the cells of an electrolyzer. Similar to fuel cells, electrolysers consist of an anode, a cathode, an electrolyte, and a catalyst material. Electrolyzers range

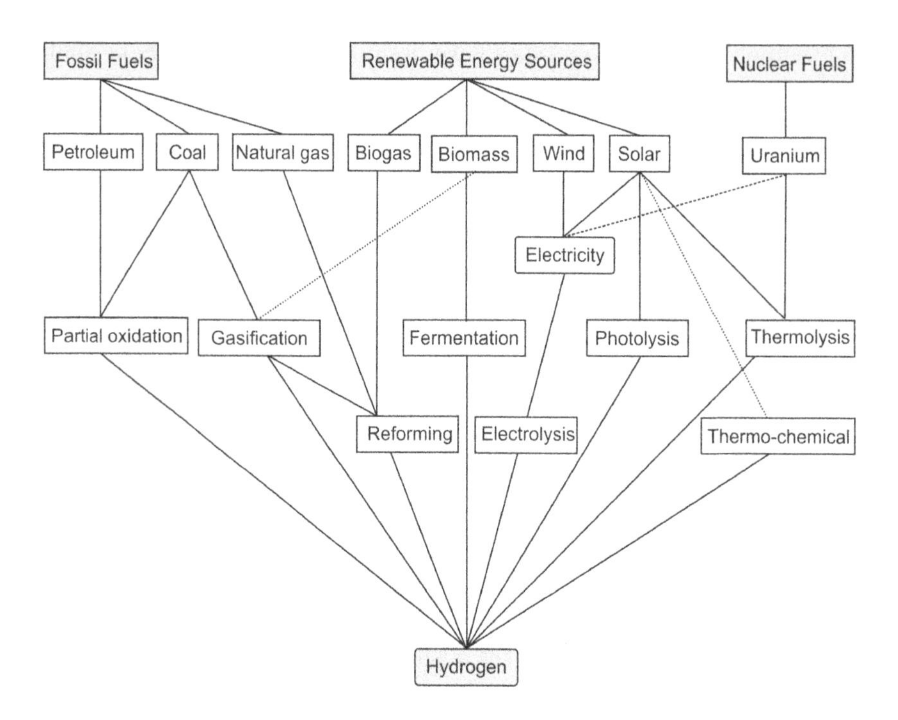

Figure 1.12: Overview of main hydrogen production routes.

from small units for provision of high purity hydrogen for laboratories to installations in the MW range. Hydrogen has been produced at an industrial scale through water electrolysis for over one hundred years. In the past hydrogen was used mainly for fertilizer (via ammonia) production, for example in Norway and Egypt. At present electrolysis is often used for the on-site production of hydrogen for fuel cells or hydrogen vehicle fuelling stations. Renewable or green hydrogen can be produced if the electric energy is provided by a renewable energy source. Currently the possible conversion of excess electricity to hydrogen (power to gas) has received much attention, with several demonstration plants running or under development (for an overview see [57]). A possible future scenario would be hydrogen production coupled to wind parks (e.g., NREL wind-hydrogen project [58]), supplying hydrogen refuelling stations, the chemical industry, or injecting hydrogen into the natural gas grid. Research is under way to examine the durability of electrolysers operating under variable loads when they are coupled to fluctuating renewable energy sources RES.

There are three main types of electrolysers, alkaline, polymer electrolyte membrane and solid oxide electrolysers. The overall water splitting reaction is the reverse of the fuel cell reaction and can be given as:

$$H_2O_{liq.} \longrightarrow H_2\uparrow + \tfrac{1}{2}\,O_2\uparrow\ \Delta H^\circ{=}285.8\text{kJ/mol.}$$

For PEM electrolysis (see Figure 1.13), the half cell reactions are:

Anode: $H_2O \longrightarrow 2\,H^+ + \frac{1}{2}\,O_2 + 2\,e^-$
Cathode: $2\,H^+ + 2\,e^- \longrightarrow H_2$

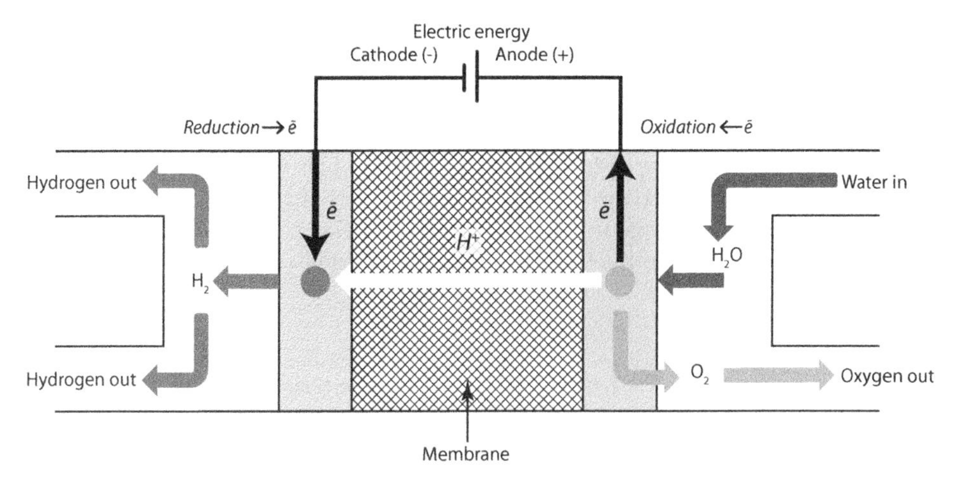

Figure 1.13: Polymer electrolyte membrane electrode assembly.

while for an alkaline electrolyser, the half-cell reactions are:

Anode: $4\,OH^-(aq) \longrightarrow O_2(g) + 2\,H_2O(l) + 4\,e^-$
Cathode: $2\,H_2O(l) + 2\,e^- \longrightarrow H_2(g) + 2\,OH^-(aq)$

The energy requirement of the water splitting reaction is the sum of the electricity and heat needed to operate the electrolyzer. A high electrical efficiency of the water splitting reaction is possible at elevated temperatures, as part of the energy can be supplied thermally. Electricity requirements range from 4 kWh/Nm^3 to 5 kWh/Nm^3 H_2 depending on the technology. This results in an efficiency of around 65 % to 75 % (based on lower heating value of hydrogen). High pressure electrolysers, working at pressures of up to 35 MPa [59], reduce the energy needed downstream for compression when high pressure hydrogen is needed.

Alkaline electrolysis

Alkaline electrolysis is a proven technology for the industrial scale production of hydrogen. The electrolyte is a liquid alkaline solution, commonly a solution of 20 % to 30 % KOH. Most of the conventional alkaline water electrolysers are designed to run at a temperature around 80 °C to 90 °C. A membrane between the two electrodes separates the product gases but is permeable to the hydroxide ions and water molecules. Alkaline electrolyzers operate at low current density, leading to a larger spatial requirement for the system. The operating pressure is typically below 3 MPa. When coupling electrolyzers to fluctuating renewable energy sources, the partial load range is of particular importance. The lower partial load range of alkaline electrolyzers is limited to 20 % to 40 % of the nominal load due to issues with gas contamination. Hydrogen back-diffusion to the oxygen evolution chamber, which can occur at low loads, must be prevented for safety and efficiency reasons. Alkaline electrolysers are, however, the most advanced commercial electrolyser technology.

Polymer electrolyte membrane electrolysis

In a PEM electrolyzer, the electrolyte is a solid proton conducting material. Water reacts at the anode to form oxygen and positively charged hydrogen ions, which move across the membrane to the cathode. At the cathode the hydrogen ions combine with electrons from the external circuit, forming hydrogen gas. The polymer electrolyte membrane provides a high proton conductivity and a low crossover of gas. The mechanical strength of the membrane also enables high pressure operation. PEM electrolyzers are more compact compared to alkaline electrolysers, operating at high current densities. This enables a reduction of costs. This technology is well suited to dynamic operation, as PEM cells can be designed to withstand widely varying voltage inputs. Dynamic operation capabilities are beneficial for coupling a PEM electrolyser to fluctuating RES such as wind parks.

Solid oxide electrolysis

Solid oxide electrolysers (SOEC) use a solid ceramic material as the electrolyte, which selectively transports negatively charged oxygen ions at elevated temperatures. Solid oxide electrolyzers must operate at temperatures high enough for ionic conduction to occur in the solid oxide membranes (about 600 °C to 800 °C). SOEC can be run in various modes, with or without an external heat supply. Heat from various sources can be used to decrease the amount of electrical energy needed to produce hydrogen from water and can therefore reduce the cost of producing hydrogen. Solid oxide electrolysers can be used for the co-electrolysis of steam and CO_2 to syngas. Currently this technology is not commercially available, but the concept has been proven at short stack level. The main issue barring the deployment of this technology is the limited durability of the ceramic materials at high operating temperatures.

Solar production of hydrogen

Solar energy can be used to provide the energy for water splitting. The single-step thermal dissociation of water, also known as water thermolysis, requires however, high temperatures (> 2000 °C) for hydrogen production. Finding suitable reactor materials capable of withstanding the high temperatures as well as thermal cycling has so far remained challenging. Water can also be split into hydrogen and oxygen through thermochemical cycles, which can significantly lower the heat requirement. Heat sources such as concentrating solar power systems are combined with chemical compounds in cyclic reactions that reuse the reactants. As an example, the sulphur-iodine cycle generates hydrogen from water with an efficiency of approximately 50 %. This production method is still at the development level, and the high temperatures requirement (around 1200 °C) still poses severe challenges on the design of a system, in particular on the durability of the materials used. Photoelectrochemical hydrogen production, where sunlight produces current in a photoelectrode driving the hydrogen and oxygen evolution reactions at opposite surfaces of the electrode, has been demonstrated, but still remains at basic research level. Low efficiency and poor stability of the photocatalyst remain limiting factors for its implementation. For an overview of solar hydrogen production routes see, e.g., [60].

Nuclear production of hydrogen

The thermal energy produced in a nuclear reactor can be used to provide heat for several hydrogen production routes, ranging from SMR to thermochemical cycles or electrolysis. Depending on the reactor type, different production processes can be considered. Advanced reactor concepts operating at very high temperatures could deliver the process heat needed for water thermolysis. Concepts integrating high temperature electrolysis in currently operating reactors have been proposed [61,62].

Biomass and biogas based hydrogen production

Hydrogen can be produced from biomass by a variety of methods, the most important being combustion and gasification and biological conversion processes such as fermentation. Biomass gasification yielding syngas is the most advanced technique, but large production plants have not yet been implemented. A water-gas shift reaction (see section 1.3.1) can increase the hydrogen concentration in the syngas, which must still be purified further by PSA or other means. As the hydrogen content of biomass is low, the overall efficiency of the process is limited.

Biogas is produced by anaerobic breakdown of organic matter. It consists of a mixture of methane (CH_4) in the range of 50 % to 75 %, carbon dioxide (CO_2) in the range of 25 % to 45 % and other minor components. The presence of these "minor" components may often necessitate extensive purification of the biogas before it can be fed into a reformer. The biogas can be used directly as a renewable fuel source for thermal power generation or it can be converted into hydrogen. Conventional reforming technologies, such as steam methane reforming (SMR), partial oxidation of methane (POX), or autothermal reforming (ATR) can be used for producing hydrogen from biogas, once it has been processed to the appropriate purity level [63].

1.4.3 Hydrogen Distribution, Storage and Dispensing

Hydrogen Distribution

The distribution and storage of hydrogen is challenging due to its low volumetric energy density. The energy per unit volume of hydrogen at ambient conditions is 2.70 kWh/m^3 compared to methane which is 10.2 kWh/m^3. In addition, its ability to permeate most (metal-based) materials severely limits the available options for containment. To increase its energy density hydrogen is normally compressed, both for transport and storage purposes. The compression of hydrogen by the various available options (mechanical compression, membrane compression, ionic liquid compressors, electrochemical) consume abouts 10 % to 15 % of the hydrogen energy content, depending on the target pressure. The liquefaction of hydrogen is also used to increase energy density (by around 800 times compared to gaseous hydrogen at atmospheric pressure). Liquid hydrogen can be stored at ambient pressures. The process of liquefaction currently consumes more than 30 % of the energy content of the liquefied hydrogen, but further optimisation is possible, with a theoretical limit of $\approx$ 10 %. Liquid hydrogen (LH2) is formed below 20 K, therefore special cryogenic tanks are needed for storage.

Tube trailers and pipelines can be used for the transport of gaseous hydrogen. Liquid hydrogen can be distributed by insulated road tankers. Ships have been proposed for long-distance delivery of liquid hydrogen. Tanker trucks have a capacity of up to 4000 kg of liquid hydrogen. Boil-off losses on the order of 0.2 % to 0.4 % per day may add to the costs of this distribution method. For use in hydrogen refuelling stations the hydrogen gas has to be re-compressed. Pipelines have been used to distribute hydrogen for decades. Several thousands of km of pipelines have been installed globally, mainly for the chemical and petro-chemical industries. Suitable materials have to be used, since plastic natural gas pipes could be permeable to hydrogen. In the long term, the transport of large amounts of hydrogen over long distances is only feasible by pipeline; however, the high installation costs as well as the energy required for compressing and pumping hydrogen need to be considered. Compressed gas hydrogen trailers are a cost-efficient distribution method for shorter distances and lower amounts. Currently 20 MPa trailers are mainly used, which typically contain 300 kg of hydrogen, but efforts are under way to increase the transported quantities by increasing the pressure level. Transport of compressed hydrogen is strictly regulated by international and regional regulations (e.g., European Agreement concerning the International Carriage of Dangerous Goods by Road).

Hydrogen storage

The most mature methods to store hydrogen are as a pressurized gas in high-pressure tanks or as a cryogenic liquid in dewars or insulated tanks (stored at an internal temperature of −253 °C). Compression of hydrogen to about 20 MPa for storage in steel cylinders is the most commonly used storage method, but for some applications, such as fuel cell electric vehicles, higher storage densities are needed. The highest gravimetric energy storage densities can be achieved through liquefaction or cryo-compression of hydrogen; however, the complexity of the storage system is higher and boil-off losses may be incurred. The various storage methods all differ in terms of their storage density (gravimetric and volumetric), pressure level, and energy requirement associated with the storage (e.g., condensing and cooling to a liquid state, pressurization of a gas, or adsorption) process and subsequent hydrogen release when in demand.

Compressed gas vessels for automotive applications
To enable a reasonable vehicle driving range of about 600 km, around 5 kg to 10 kg of hydrogen is required on-board of a light duty FCEV. As there are stringent weight and volume constraints associated with the on-board storage system, high pressure and cryogenic hydrogen storage methods have been considered by vehicle manufacturers. However, high-pressure hydrogen storage has become the industry standard. These systems are designed for a pressure level of 70 MPa for passenger vehicles and 35 MPa for buses. On-board storage consists of storage tanks made from carbon fibre reinforced composite cylinders with an internal liner made of metal (e.g., aluminium) or plastic (e.g., high density polyethylene).

Liquid hydrogen
Hydrogen in liquid form has a much higher energy density than in its gaseous form even at high pressures. In a liquefaction plant hydrogen is compressed and cooled in a multi-step heat exchange process. Liquid hydrogen can be stored in stationary tanks, e.g., at liquefaction or refuelling stations, or for distribution of hydrogen in tankers. Liquid hydrogen storage is also implemented for industrial use, and maritime large scale transport of liquid hydrogen is being considered [64].

Cryo-compressed hydrogen storage
Higher hydrogen density storage at lower pressures can be achieved with cryogenic gaseous hydrogen. Thermally insulated pressure vessels, with an operating pressure of up to 32 MPa, have been developed for hydrogen. The venting losses are much lower than those encountered for liquid hydrogen storage since venting occurs only when pressure build up in the tank rises above the maximum allowable working pressure (MAWP). This storage technology has a higher system complexity and additional monitoring of vacuum stability of the thermal insulation is needed to assure higher performance. Tests are continuing to establish the technical feasibility and safety of this storage technology.

Hydrogen storage in chemical compounds
Hydrogen can be stored at low pressures in solid or liquid chemical compounds, based

on, e.g., magnesium hydride (MgH_2), sodium boron hydride ($NaAlH_4$), or organic compounds such as N-ethylcarbozol. Storage in liquid organic compounds has been proposed for the distribution of hydrogen over long distances [65], and metal hydride storage cartridges are commercially available for portable applications. For stationary storage, a MgH_2 based system has been deployed, with high efficiency due to the use of a phase-change material which utilizes the heat released during hydrogenation [66]. For transport applications, none of the solid-state storage materials managed to fulfil all of the stringent requirements of the automotive industry. Few systems have been proven to work reliably beyond the laboratory scale, as long term stability and thermal management remain as technical hurdles. Research is currently focusing more on other early markets, such as providing hydrogen for portable fuel cells, unmanned flight vehicles, or fork lifts.

Hydrogen Dispensing

The deployment of fuel cell electric vehicles (FCEVs) will require an extensive fuelling infrastructure. Due to the many options of hydrogen production and delivery, the most appropriate solution to supply hydrogen to a refuelling station will depend on local constraints and deployment strategies. Different station concepts have been realized, either based on delivery of gaseous or liquid hydrogen or on-site production via electrolysis or reforming of natural gas. The storage requirement and delivery by truck can be reduced by producing hydrogen on-site. The various national deployment strategies aim to provide a convenient and low-cost network of hydrogen stations to users. To meet this goal, rollout plans have been created to coordinate the deployment of FCEVs and hydrogen infrastructure build-up, geographically and over time. Such plans are being developed by public-private partnerships around the world (e.g., [50]). Some 140 hydrogen refuelling stations are in operation worldwide (90 under construction) to fuel some 400 FCEVs and 100 buses used in demonstration projects.

A hydrogen refuelling station basically consists of a number of high pressure storage tanks, compressors, transfer lines, and the dispenser which includes a flexible hose and the fuelling nozzle. Vehicles can be refuelled within a few minutes, limited by heat generated by the build-up of pressure in the tank. According to the standards ISO 15869 [67] and SAE J2601 [68], the temperature within the vehicle storage tank shall not exceed 85 °C during refuelling. Pre-cooling of the hydrogen (down to -40 °C) enables shorter refuelling times, in the order of a few minutes. The temperature requirement becomes more problematic with back-to-back fills at the dispenser owing to the heat produced from compressing the hydrogen prior to dispensing.

Hydrogen is also delivered to materials handling vehicles such as forklifts at indoor refuelling points. The refuelling of forklifts requires on-site storage, piping, and hydrogen dispensers. While hydrogen storage is placed outdoors, the hydrogen dispensers are often deployed inside warehouses for this application.

Hydrogen for Energy Storage

The integration of fluctuating energy sources into the existing energy grid is challenging due to the mismatch between supply and demand and high power sources which can strain the transmission and distribution systems. The conversion of excess electricity to hydrogen (and vice versa) can help facilitate the integration of large shares of intermittent renewable sources into the electrical grid. Water electrolyzers can produce hydrogen from excess or low-cost electricity, either connected to the grid or in off-grid installations (such as islands, wind parks, or solar parks). Hydrogen can then be either used directly as a chemical feedstock, as a fuel for transport, be fed into the natural gas grid, or be converted back to electricity during periods of large demands. Although power-to-power storage may not prove economically viable, the hydrogen can be provided to different markets, such as FCEV and the chemical industry, rather than being reconverted into electricity. Large scale and longer term storage of energy is also possible by storing hydrogen in salt caverns or other suitable geological formations.

CHAPTER 2

Introduction to Sensor Technology

CONTENTS

2.1 TERMS AND DEFINITIONS

The word "sensor" is derived from the Latin term "sensus," which means sense, perception, or observation. Numerous modern definitions for the term sensor have been proposed which accentuate different aspects of sensing. These include the following definitions, which have been selected for the specific purpose of this book:

- A sensor is a device which receives and responds to a stimulus [69].
- A sensor is a device that measures a physical or chemical quantity and converts it into a signal that can be read by an observer or by modern electronic instruments.
- A sensor contains the sensor element and possibly associated circuit elements [70, 71].

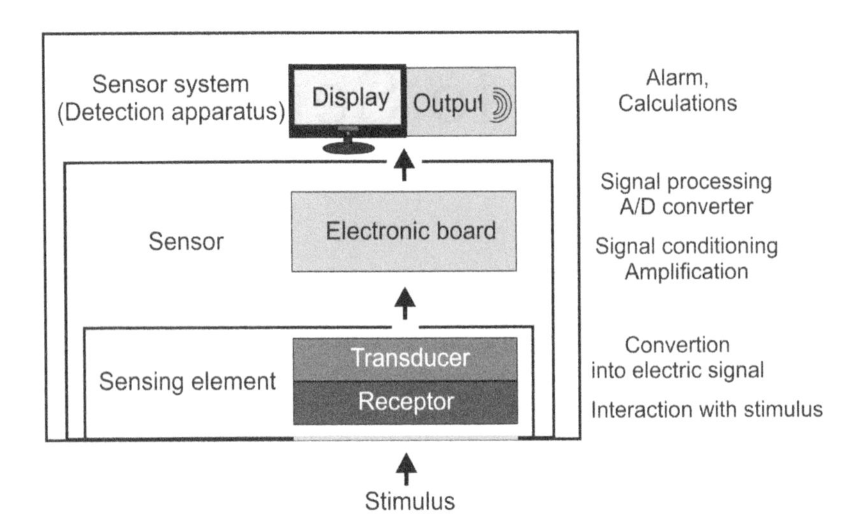

Figure 2.1: Illustration of the functional distinction between a sensing element and sensor. A sensor is often incorporated into a higher level instrument (e.g., a detection apparatus).

- A sensor is a small device that, as the result of a chemical interaction or process between the analyte gas and the sensor device, transforms chemical or biochemical information of a quantitative or qualitative type into an analytically useful signal [72].
- A sensor is an element of a measuring system that is directly affected by the phenomenon, body, or substance carrying the quantity to be measured [73].

Unfortunately, there is no universally accepted definition for the term "sensor." Most definitions are imperfect; they are either too general or non-inclusive, and frequently contradictory examples can be identified for a specific definition. For example, many of the definitions listed above can be applied to laboratory scale analytical instruments, which are not generally viewed as sensors due to size, cost, and complexity of operation. There is however, no clear distinction between analytical instrumentation and sensors. It is further noted that numerous terms tend to be used interchangeably to describe a sensor. This is because the sensor community is quite broad and ranges from the suppliers of sensors, which includes fundamental technology developers, manufacturers of sensors, and instrument developers, to the end-users, which include facility engineers, facility managers, and code officials, and accordingly there can be different perspectives as to what a "sensor" is. For example the last definition listed above comes from the Joint Committee for Guides in Metrology (JCGM) Guide [73], but may be considered more specific to a sensing element, as defined below and illustrated in Figure 2.1.

Some of the more common terms used as synonymous for "sensor" are sensing element, detector, analyser, detection apparatus, and sensor systems. In addition, terms such as indicator, smart sensors, and sensor arrays are used in the sensor literature. Although no universally accepted definition has been established for these terms, some distinctions from among these various terms can be made, which are discussed below. In this book we will endeavour to distinguish between sensing element, sensor, detection apparatus, and sensor system with respect to functional complexity as described below and illustrated in Figure 2.2 for a hydrogen detection apparatus.

Sensing Element

The sensing element is the component that performs the basic sensing operation and provides a measurable, continuously changing physical quantity (e.g., electrical signal) in correlation to the presence or magnitude of the stimulus, e.g., hydrogen concentration [74]. The sensing element includes a receptor and one or more transducer elements. The receptor is the site of interaction with the stimulus to be detected. For chemical sensors the stimulus is a chemical constituent, also known as the analyte. The receptor is the site of the chemical interaction, which is then converted by the transducer into an analytical quantity that is usually electrical in nature. Thus, two distinct functions are associated with the sensing element - it is the site for the interaction of the stimuli at the receptor and the generation of electrical signal by the (electrical) transducer.[1]

Sensor

The sensor is the combination of one or more sensing elements together with electronic circuits for basic control of the sensing element, measurement of the transducer output, and possibly some electronic pre-processing. Electrical pre-processing (see Section 2.6.1) is often necessary because the electrical signals outputted by the sensing elements may not always be conveniently measured with common electrical measurement devices (e.g., standard multimeters). This can arise when the sensing element output is a small change in voltage or current, has an excessively high resistance, or has a low output impedance. For these reasons, an electronic circuit is incorporated to convert the output of the sensing element into a usable signal which can then be accessed by the end-user or interfaced into a data acquisition or control system. The electronic circuit can also buffer the output of the electronic transducer in addition to providing signal amplification and electronic transformations, e.g., transimpedance amplifiers. Electronic circuitry can also control functional parameters of the sensor and its sensing element(s), e.g., heater control. The relationship between a sensing element and a sensor is illustrated in Figure 2.1. Essentially, the sensing element generates an electrical signal in response to the stimulus, while the sensor provides an output accessible to the end-user. In contrast to sensor, the term *indicator* is sometimes used to describe a tool that verifies the presence of a chemical constituent but is commonly not reversible and generally provides semi-quantitative information.

Detection Apparatus

The sensor output will typically be incorporated into a higher-level instrument system—the detection apparatus, which can be either a handheld portable device or a fixed site device. Typically, a modern detection apparatus is a stand-alone system that does not require any additional electronics for operation, although it will often be interfaced to other systems. A detection apparatus for hydrogen (or other chemicals) can contain additional sensing elements or sensors to compensate for, or monitor changes in ambient temperature, pressure, humidity, or other physical parameters. The output of a detection apparatus signal can either be a digital display or an electrical signal or both. The units of the outputted signal are typically either the quantity to be detected (e.g., vol% for gas concentration), or a standard electrical signal (e.g., 4 mA to 20 mA or 0 V to 5 V). If there is a digital display it will usually show the signal as the quantity to be detected. Alternatively, the interface to control functions are typically electrical signals. A detection apparatus may include additional features,

[1]The terms measuring element, measuring head, detection element, or converter are used for sensing element.

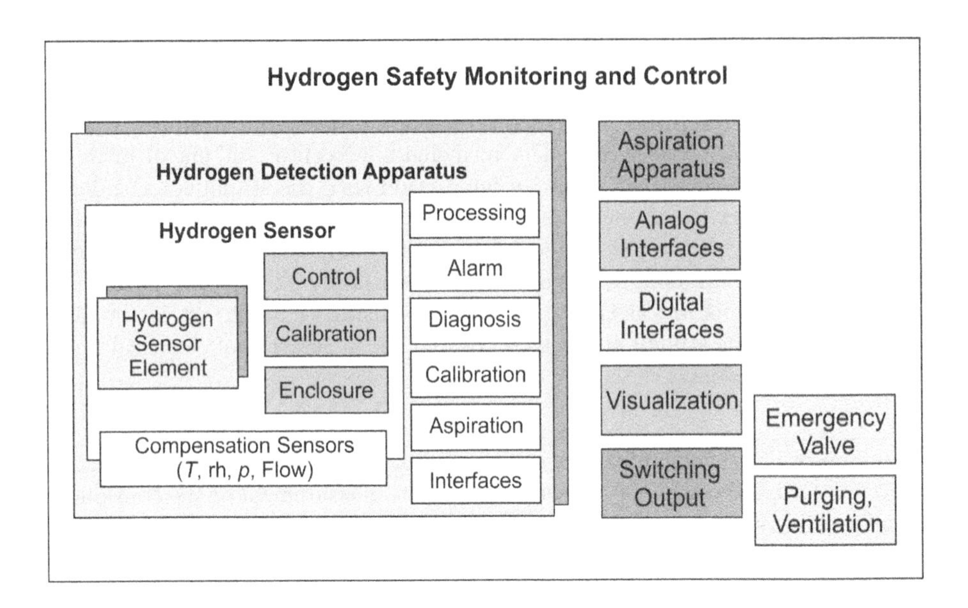

Figure 2.2: Schematic of a detection apparatus.

such as an audible alarm, that is activated when a specific stimulus level is reached. There can also be relays (electronically activated switches) that can control or activate other systems. A relay functions as an actuator to transduce the sensor signal into an action (usually mechanical) [75]. The activation of a relay is triggered when the target stimulus is detected to be above a pre-set threshold level. A detection apparatus can have multiple relays that are activated at different stimulus thresholds. In a hydrogen detection apparatus, functions commonly controlled by the relays include automated communication to first responders, activation of ventilation systems, or to initiate an emergency shutdown of surrounding hydrogen systems. Detection apparatus may have the capability for adjustment of alarm or relay activation set points and routine maintenance functions such as calibration. A detection apparatus is often referred to as a gas detector or as an *analyser*, i.e., an instrument which conducts quantitative analyses.

Sensor System

A sensor system can be viewed as an enhancement of a detection apparatus which can perform advanced monitoring and control tasks, although this is a subtle distinction and the terms sensor system and detection apparatus are often used interchangeably. One distinction is that sensor systems can encompass more than one detection apparatus to achieve appropriate safety or process control requirements through, for example, sensor redundancy. A sensor system with multiple detector apparatus may also have several measuring points, e.g., for area monitoring, in which multiple detection apparatus or sensors are deployed in different locations throughout a facility or process but are interfaced to a single data processing unit. As with a detection apparatus, a sensor system can include relays, and thus perform from a central location all functions of a detection apparatus described above.

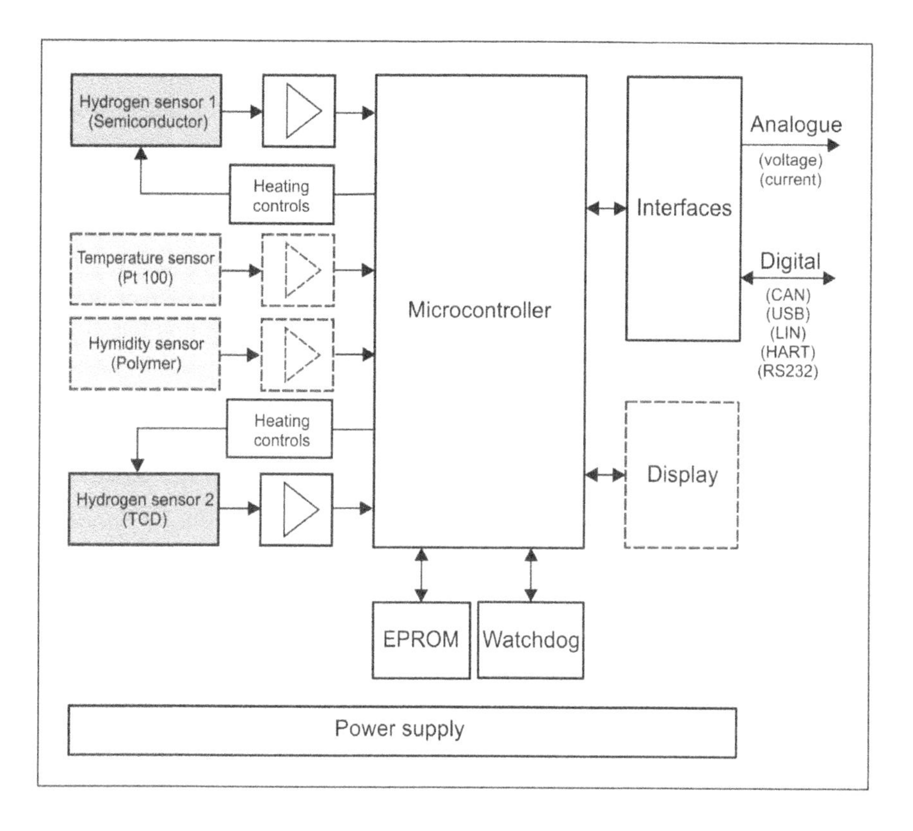

Figure 2.3: Schematic of a hydrogen detection apparatus with temperature (T) and relative humidity (rh) compensation.

Smart Sensor

When a sensor or detection apparatus includes additional features such as recognition and signalling of failure states, autocalibration, error handling, and other self-testing features, it is often referred to as an *intelligent sensor* or *smart sensor* [76]. These functionalities are often performed with a microprocessor. Smart sensors can integrate signals from other sensor types; often these are environmental parameter sensors (e.g., temperature, pressure, and humidity). The signals from environmental parameter sensors can be used for the compensation of variation in the primary sensing element response induced by changes in these environmental parameters. Smart functionalities increase the performance and reliability of the sensor system. Although a smart sensor can contain complex internal signal processor algorithms, these background functions are automatic and usually will be blind to the end-user. An illustration of the functionalities of a smart sensor for hydrogen detection is shown in Figure 2.3.

Sensor Array

Another adaptation of chemical sensors is the configuration of multiple sensors into a sensor array [77]. A sensor array is often used for the investigation of gas mixtures that may simultaneously contain one or more components of interest. A sensor array is a collection of gas sensing elements forming a single structural unit while providing signals that correlate to the individual constituents of a gas mixture. Typically each sensing element in the sensor array responds with a different sensitivity to each constituent in the gas mixture, as opposed to a selective response of an individual sensing element to one specific constituent in the gas mixture. Pattern recognition algorithms are used to correlate the combined output of sensors in the array to the gas composition. The identification algorithms are typically performed by a microprocessor. The incorporation of a sensor array and pattern recognition algorithms into a detection apparatus produces an instrument that is commonly known as an electronic nose. An electronic nose provides information about the composition of a gas mixture through the use of multiple sensing elements analogous to the olfactory mechanism of the human (and other mammals,) nose. In each case, identification is based upon prior exposures of known gases or odours.

2.2 SENSOR CLASSIFICATION

The classification of sensors provides a convenient means of grouping sensors based upon a common property into a small number of categories or types. The classification system may seem arbitrary, and a specific sensor type can sometimes be grouped into multiple categories within a classification system. Nevertheless, classification can quickly provide a better understanding of a sensor's function or detection principle, and hence provide an understanding of its expected performance and limitations. This in turn assists end-users in the selection of the best sensor model for a particular application and provides guidance on how to optimise the use a sensor. Sensor classification can range from very simple to complex, depending on the purpose or criteria. A general classification system has been proposed to cover various sensor types, notably the distinction among chemical, biological, and physical sensors:

- A chemical sensor is a device that transforms chemical information ranging from the concentration of a specific sample component to total composition analysis into an analytically useful signal.
- A physical sensor is a device that provides information about physical properties of the system.
- A biological sensor is a device that provides information about biological matter, e.g., microorganism, cells or organelle, metabolites, enzymes, antibodies, its concentration, states, and activity.

A formalized classification system for chemical sensors was proposed by the Commission on General Aspects of Analytical Chemistry of the International Union of Pure and Applied Chemistry (IUPAC) [74]. The IUPAC chemical sensor classification system categorizes chemical sensors based on the transduction mechanism associated with the chemical interaction on the sensing element to produce a measureable response. In this book, the term "sensor platform" refers to the transduction principle of the indicated sensing element or sensor. The IUPAC sensor classification system is summarized in Table 2.1 [74].

Other sensor classifications based on the transduction mechanism have been proposed. A classification of physical sensors was given by Fraden [69] and is presented in Table 2.2.

However these classifications systems are not unequivocal in that there are specific sensor types that do not conveniently fall into an identified category or may fit into multiple categories. Nor are these systems necessarily the best classification system from a practical

Table 2.1: IUPAC classification of chemical (gas) sensors

Class	Description and subclasses
Optical	Sensors which transform changes of the following optical phenomena as a result of interaction of the target gas with the receptor.
	Absorbance Reflectance Luminescence Refraction index Light scattering Optothermal
Electrochemical	Sensors which transform the effect of electrochemical interaction between the target gas and the electrode.
	Voltammetric / amperometric Potentiometric Chemically sensitive field effect transistor (CHEMFET) Potentiometric solid electrolyte
Mass sensitive	Sensors where accumulation of the target gas changes the mass and properties of the receptor.
	Piezoelectric Surface acoustic wave
Magnetic	Sensors based on the change of paramagnetic properties of the target gas.
Thermometric	Sensors which measure the heat effects of a specific reaction or adsorption of the target gas.

Table 2.2: Classification of physical sensors

Sensor division	**Object of measurement / sensing principle**
Acoustic	Wave amplitude, velocity, phase, polarization, spectrum
Electric	Charge, current voltage, potential, electric field (amplitude, phase, polarization, spectrum), impedance, permittivity
Magnetic	Magnetic field (amplitude, phase, polarization, spectrum), magnetic flux, permeability
Mechanical	Position (linear, angular), orientation, acceleration, force, moment, torque, stiffness, stress, pressure, mass, density, viscosity Speed of flow, rate of mass transport, shape, roughness
Magnetic	Sensors based on the change of paramagnetic properties of the target gas
Optical	Sensors which measure the heat of effects of refraction index, emissivity, reflectivity, absorption
Radiation	Type of radiation, energy, intensity
Thermal	Temperature, heat flux, specific heat, thermal conductivity

point of view. In addition to the transduction mechanism, sensor classifications have been based on the other descriptors. These can include the target parameter or property (e.g., a hydrogen sensor, a pressure sensor), sensor fabrication method (e.g., micro-fabricated sensors), means of application (e.g., point sensors, open path detectors, or wide area monitors), material of sensing element (metal oxide sensor), or the type of the electrical response (e.g., amperometric sensor, capacitative sensor). Sensors have also been classified according to the target market or application, such as safety, automotive, agricultural, or environmental sensors. Sensors can also be classified according to the nature of its signal, such as, for example, an *absolute* or gauged (*relative*) measurement. Typically this is in respect to an absolute physical scale or in comparison to a reference. For example a resistive thermometer is based on the direct correlation between electrical resistance and temperature, and would be an absolute sensor. In contrast, a thermocouple, which is based on a reference temperature (see Section 5.1), is considered as a relative sensor. A more common example would be pressure sensors which are commonly configured to provide either an absolute (pressure above a perfect vacuum) or a gauge (pressure relative to ambient pressure) output.

Another sensor classification based on functionality pertains to their transportability. Categories include *portable*, i.e., hand carried and battery operated, *transportable*, i.e., mobile with the assistance of carts or vehicles, and *stationary* sensors, i.e., fixed site operation for permanent surveillance. This classification system is not typically applied to sensing elements but is more germane to sensors or detection apparatus.

Clearly multiple classifications can be used to describe a specific sensor. Thus, a hydrogen sensor (classified based on its target parameter) may also be described or classified as an electrochemical sensor (based on its transduction mechanism) and as a safety sensor (based on its application/market).

2.3 SENSOR RESPONSE

2.3.1 Sensor Dynamics

The response of a sensor is usually an electrical signal resulting from an external stimulus, i.e., the magnitude or change in magnitude of a chemical or physical quantity in the environment surrounding the sensor. The sensor signal depends on the strength and profile of the stimulus. Typically, in gas measurements, the sensor responds to a change in gas concentration. One common laboratory assessment method is to measure the sensor signal in response to an idealised step change of stimulus. Other stimuli profiles are the ramp function in which the magnitude of the stimulus changes at a constant rate, a Dirac impulse in which the stimulus is applied as a short-duration pulse, or an oscillating impulse. For gas measurements, the step change corresponds to an instantaneous change in the concentration of the target analyte that interacts with the sensor. Such step changes are of course unrealistic, but a common method for laboratory assessment of a sensor response. In this model, the input signal (stimulus) $x(t)$ is given by:

$$x(t) = \begin{cases} x = 0 & \text{for } t < 0 \\ x = 1 & \text{for } t \geq 0 \end{cases} \tag{2.1}$$

A typical sensor response due to a step change in the magnitude of a stimulus is illustrated in Figure 2.4. This behaviour can be observed for many different sensor types, including hydrogen sensors. The sensor response profile can be empirically considered as composed of several parts. There is usually a lag between the application of the stimuli and the observation of the corresponding response of the sensor. Part of this lag (i.e., the dead

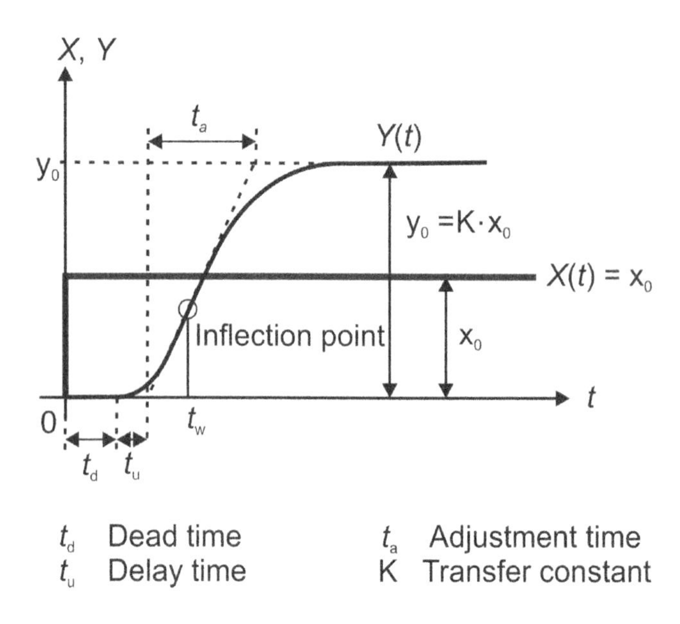

Figure 2.4: Sensor output response Y to a step change x_0 in the stimuli $X(t)$.

time, t_d) is controlled primarily by the time required for the stimulus to actually reach the sensing element. For a gas sensor, this would be the time needed for the physical transport of the analyte from the sensor input surface to the surface of the sensing element. The delay time (t_u) is related to the interaction between the stimulus and the receptor. In a gas sensor, this interaction can be associated with the partitioning of the analyte from the gas phase on to the receptor site of the sensing element. This partitioning can be either a surface or a bulk interaction. The delay time in a gas sensor is, in part, controlled by mass transport properties associated with gas diffusion. The gas will essentially not reach the sensor surface as an idealized step change, but rather with a concentration profile that has a leading edge that increases to the step value over a finite time. During the delay time the signal rises in response to the stimulus (the adjustment time, t_a) and then levels off to a nearly time-invariant response. The monotonic change of the sensor response from one time-invariant value to another during the adjustment time will have an inflection point at which the rate of change in the slope of the tangent line to the graph of the function changes its sign. If well defined, this point can be used for an advanced assessment of the sensor response by real-time analysis performed with a microprocessor.

Naturally, it is important to keep the sensor's total response time ($t_d + t_u + t_a$) as short as possible. In some sensor types, such as temperature and pressure sensors, the response time can be quite short (< 1 s). Conversely, gas sensors typically have response times ranging from 5 s to 60 s, although sensor types with faster response times are being commercialized [78]. More details about the determination of gas sensor response times are discussed below in Section 2.4.2. In Figure 2.4, the time-invariant sensor output for a fixed stimulus value is a steady-state response, the magnitude of which has been termed the "final indication" of the sensor [79]. Since there is some noise and drift associated with every response of a real sensor, the criteria for verifying that a sensor signal has reached an appropriate time invariant state has to be defined. The criteria is based upon defining an acceptable change in the sensor response over a prescribed period of time [79].

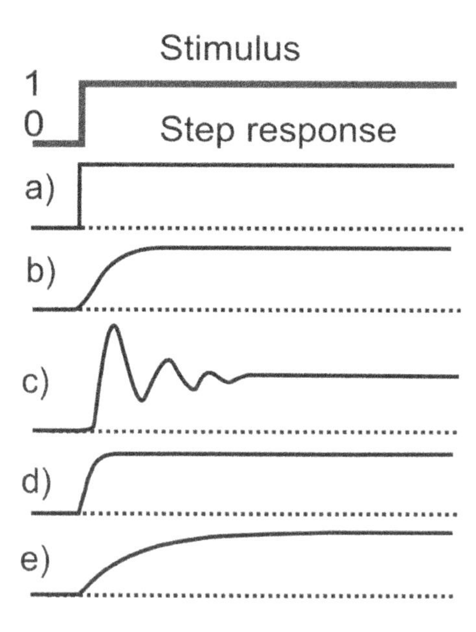

Figure 2.5: Step response of a sensor.

The output of a sensor can exhibit different types of time-dependent profiles in response to a hypothetical step change of the stimulus, some of which are illustrated in Figure 2.5. Curve a) shows the ideal response of the sensor to the ideal step change, in which the sensor response tracks closely the gas concentration profile (e.g., a zero $t_d + t_u$). However, sensor responses usually observed in practice are characterised by either the curve b) which has already been discussed above, or the curves c) to e).

The dynamic behaviour of a sensor in response to a stimulus can be described in terms of control theory [80, 81]. This discipline of engineering deals with the behaviour of dynamical systems, including the mathematical modelling of the relationship between the input and output signals of black box systems.[2] The control theory is applied to numerous systems including electrical elements, such as a first order low-pass filter based on a RC circuit (resistor-capacitor circuit) or a second order low-pass or band-pass filters based on an RLC circuit (resistor-inductor-capacitor circuit). Sensors can be considered as single-input single-output (SISO) systems and as proportional elements (P elements), where the output signals are directly proportional to the input [82]. The mathematical expression for the time dependent relationship between a stimulus or input signal $X(t)$ and the corresponding sensor output signal $Y(t)$ is a "conversion function." Since conversion functions are higher order differential equations whose analytical solution is sometimes difficult or even impossible to solve, Laplace transforms of the time dependent functions $f(t)$ into frequency dependent functions $\mathcal{F}(s)$ are often performed for easier algebraic treatment and visualisation. The transformation from the time domain to complex frequency domain is useful not only for theoretical considerations of the sensor, but also for the analysis of the sensor output in frequency domain, and is often advantageous for process control applications. For a linear and time-invariant system, this approach provides a transfer function $\mathcal{G}(s)$ as [83]:

[2]Black box refers to a system whose inner workings and mechanism are not known; the "input stimuli" and the "output reactions" are the only characteristics observed.

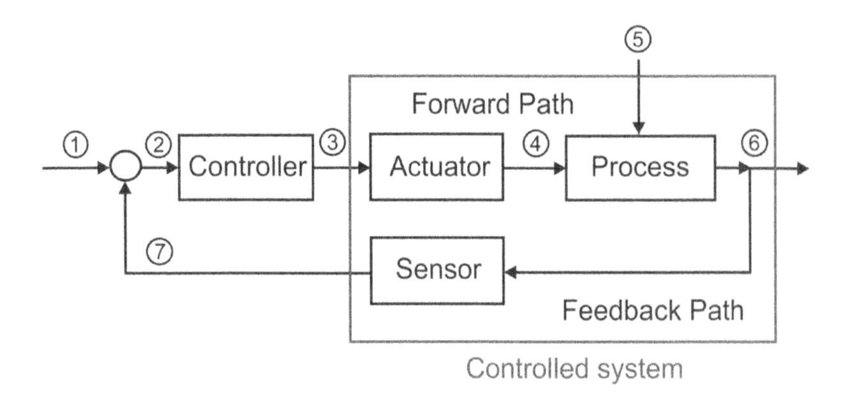

Figure 2.6: Simplified schematic of a feedback control system (control loop). (1) is the reference input (set point), (2) is the error signal, (3) is the manipulated variable, (4) is the control signal, (5) is the disturbance quantity, (6) is the controlled variable, and (7) is the feedback variable.

$$\mathcal{G}(s) = \frac{\mathcal{L}\{y(t)\}}{\mathcal{L}\{x(t)\}} \tag{2.2}$$

Where $\mathcal{L}\{y(t)\}$ and $\mathcal{L}\{x(t)\}$ are the Laplace transforms of the input $x(t)$ and output $y(t)$ signals, and $\mathcal{G}(s)$ is a function of the complex frequency s, $s = \sigma + j\omega$. The inverse Laplace transform gives a response function $a(t)$.

The differential equations, the transfer functions, and the step response functions used for the formal description of transient behaviour of various sensor responses as a function of time are compiled in Table 2.3. The time response shown in Figure 2.5, curve b), is analogous to a first order low-pass filter, which corresponds to a proportional element with a first order time delay (PT1 in Table 2.3). Curves c) to e) are comparable to second order low-pass filters with a damping ratio of D = 0.2 for curve c), D = 1 for curve d), and D = 5 for curve e) (PT2 in Table 2.3).

Modern industrial processes are highly complex, rendering manual control problematic or even impossible. An automated process control, which is dependent upon inputs from sensors, is essential for efficient operation. The use of sensors enables the automatic monitoring, control, and regulation of facilities. A simplified schematic of a feedback control system (control loop) with input from a sensor is given in Figure 2.6. The use of advanced signal processing, such as that offered by signal transfer theory, can improve technological operation. A detailed description of such methodology is beyond the scope of this book.

2.3.2 Static Sensor Characteristics

The empirical correlation between a change in magnitude of the stimulus and the time invariant sensor signal (e.g., the final indication) provides a means for the quantitative determination of the stimulus. The exact functional relationship between the stimulus X and the sensor signal Y typically has to be determined by a calibration procedure. In a calibration chamber, the sensor is exposed to various known values of the target stimulus; for a gas sensor this would be different concentrations of the target analyte. For other sensor types, the stimulus could be a defined temperature, pressure, or other appropriate quantity. The functional relationship between the reference or calibration samples and

Table 2.3: Transfer behaviour of three model elements describing a step response

	Differential equation	Transfer function	Step response	Curve in Figure 2.5
P	$y(t)$=K$\cdot x(t)$	$G(s)$=K	$a(t)$=K$\cdot x_0$	a)
PT1	$T\frac{dy(t)}{dt} + y(t) = \mathrm{K}\cdot x(t)$	$G(s) = \frac{K}{T\cdot s+1}$	$a(t)$=K$\cdot x_0$[1-exp(-t/T)]	b)
PT2	$T_1^2\frac{d^2y(t)}{dt^2} + T_2\frac{dy(t)}{dt} + y(t) = \mathrm{K}\ \cdot x(t)$	$G(s) = \frac{K}{T_1^2\cdot s^2\cdot 2DT_1+1}$	$a(t)=\mathrm{K}x_0 - \frac{Kx_0}{\sqrt{1-D^2}}\exp(-\frac{D}{T_1}t)\ \times$ $\sin(\frac{\sqrt{1-D^2}}{T_1}t + \arctan\frac{\sqrt{1-D^2}}{D})$ $0 < D < 1$	c)
PT2			$a(t)=\mathrm{K}x_0 - Kx_0(1+\frac{t}{T_1})\ \exp(\frac{t}{T_1})$ D=1,	d)
PT2			$a(t)=\mathrm{K}x_0 - \frac{Kx_0}{\sqrt{D^2-1}}\exp(-\frac{D}{T_1}t)\ \times$ $\sinh(\frac{\sqrt{D^2-1}}{T_1}t + \mathrm{arctanh}\frac{\sqrt{D^2-1}}{D})$ $D > 1$	e)

D=$\frac{T_2}{2T_1}$

the sensor signal is called the calibration function $Y = F(X)$. In practice, users typically want to derive the amount of the stimuli (e.g., a gas concentration) from the sensor signal. Accordingly the inverse function, or so-called analysis function $X = F^{-1}(Y)$, is used. The transformation of the calibration function to the analysis function is easily performed for a sensor whose response exhibits a linear relation between the stimulus and response. More complex relationships (e.g., polynomic, exponential) may make it more difficult to derive analytically derive the analysis function from the calibration function. The establishment of a reliable correlation between the stimulus and the sensor signal is of essential importance and a prerequisite for the use of a sensor for the quantitative determination of the magnitude of the stimulus. A description of the procedure to establish the relation between the sensor response and the stimuli value, that is the sensor calibration protocol, is given in Section 6.2.

The mathematical functions used to describe the relationships between the amount of the stimulus X and sensor output Y are good approximations of the real response behaviour. The simplest function is a linear relationship:

$$Y = Y_0 + \mathrm{K}(X - X_0) \tag{2.3}$$

Where X is the value of the stimulus, X_o is a reference value, Y is the sensor indication, Y_0 is the linear y-intercept corresponding to the sensor signal observed at reference stimulus (X_0) level, and K is the slope (as discussed below, the slope of the calibration curve is equal to the sensor sensitivity).

For gas sensors, X_0 is either zero concentration of the test gas or some reference concentration (e.g., some oxygen sensors use the concentration of oxygen in air as the reference concentration). The sensor response is not necessarily zero for a stimulus of X_0. A linear dependency of quantities makes signal handling easier but such a relationship is almost always only valid over a limited measuring range. The measuring range for a sensor is also empirically determined, and is usually specified by the manufacturer. Deviations from linearity will often become more pronounced at the high or low extremes of a sensor's measuring range. Fundamentally, however, a true linear dependency of quantities in nature is very unlikely, although the use of a linear function is often an excellent approximation to correlate a sensor response to the magnitude of the stimuli, at least over a finite range.

In many cases a simple linear approximation of a sensor response is not applicable. A linear fit to non-linear data will result in a non-random distribution of residuals (the distance between measured and calculated values) or different scattering of data (inhomogeneity of sample variances). The sensor response can sometimes be better approximated by a higher order polynomial, such as the third order polynomial expression shown in Equation 2.4:

$$Y = \mathrm{A}'X^3 + \mathrm{B}'X^2 + \mathrm{C}'X + \mathrm{D}' \tag{2.4}$$

Y is the sensor response, while A', B', C', and D' are empirical constants. Of course a linear relationship is simply a first order polynomial. Typically, the use of a higher order polynomial relationship to describe the sensor response is empirical. As with a linear relationship, the polynomial relationship is valid only over a specified measuring range. One approach that can sometimes be used to extend the sensor measuring range is to demarcate the sensor response into regions with unique calibration functions for each region. Such *spline approximations* can be used piecewise in intervals to approximate the relationship between the value of the stimulus and the sensor response. Linear functions as well as polynomial functions of second, third, or higher order are used in splines. The end-points of each range are connected via knots to give a continuous function. Splines are very flexible and can even give good approximations for non-monotonic changes of data values over a broad range of stimuli values. Polynomic spline approximations are used for several commons sensor types. For example, empirical polynomial relationships are used to estimate the temperature dependence in a resistive platinum thermometer (see Section 5.1), as illustrated in Figure 2.7. With such a relationship, the measured temperature has an uncertainty of 0.15 K. Although an inspection of Figure 2.7 may indicate that, there is a linear relationship between resistance and temperature, a simple linear fit of the resistance data with temperature results in significantly greater measurement error relative to the higher order polynomic equation.

Although convenient, linear or polynomic modelling of a sensor's response is often inappropriate. Linear scaling is obviously difficult when considering extremely wide ranges of magnitude for the stimulus. For example, relevant hydrogen or oxygen concentrations can range from ppb levels (10^{-7} vol%) up to the pure gas (100 vol%); a range that covers 9 orders of magnitude. Such a broad range for the stimuli magnitude of concentrations is nearly impossible to meaningfully display on a linear scale. Furthermore, no known gas sensor has been shown to have a linear response over this full range. In many circumstances, a logarithmic representation is used to equate the sensor response (Y) to the value of the stimulus (X), e.g.:

$$Y = \mathrm{A}' + \mathrm{B}' \ln X = \mathrm{A}' + 2.303\mathrm{B}' \log(X) \tag{2.5}$$

A' and B' can be empirical constants, although for sensor platforms such as potentiometric sensors, B' can be predicted from the Nernst equation, while A' can be ascertained from standard potentials and reference conditions - see Section 4.2, for an example.

Several sensor types fundamentally exhibit a response that is logarithmically related to the magnitude of the stimulus; the most common is the pH electrode. The pH electrode is a potentiometric device that shows a logarithmic relationship to the hydrogen ion concentration in solution (pH is actually defined as the negative logarithm of the hydrogen ion activity). Similar to the pH electrode, potentiometric gas sensors (e.g., some oxygen sensors, Section 4.2 and some electrochemical hydrogen sensors, Section 3.4.3) exhibit a sensor response that is proportional to the logarithm of the concentration of the target analyte. The transformation from a linear scale to a logarithm scale over several orders of magnitude is shown in Figure 2.8. Thus, one major advantage of a logarithmic representation is that data over several orders of magnitude can be easily shown in a single plot. Alternatively, the ability to resolve changes in the magnitude of the stimulus is smaller for a logarithmic relationship than a linear plot. It is noted that since the logarithm of zero (0) is undefined, log plots cannot use a datum of zero concentration.

Other mathematical relationships between the sensor response and the stimuli exist. A power-law dependence is characterized by a sensor response that is proportional to the analyte concentration raised to a power:

$$Y = \mathrm{A}' \cdot [X]^{\beta} \tag{2.6}$$

Where Y is the sensor signal, $[X]$ is the concentration of the analyte (e.g., the magnitude of the stimulus (X), which for a gas is typically the partial pressure, while A' and β are empirical constants. Typically the exponent β ranges between 0.5 and 1.0, but is usually less than 1.0. This behaviour is observed on numerous gas sensor types, including metal oxide sensors (Section 3.5) [84] and conductometric polymer sensors. In these sensors, the sensor signal y is often the conductance (G) or the resistance (R) referenced to a standard value (G/G_0 or R/R_0). Power-law behaviour is such that a plot of the log of the sensor final indication versus the log of the concentration ($[X]$) of the analyte is linear having a slope equal to negative β and a y-intercept equivalent to the logarithm of A. This behaviour is illustrated in Figure 2.9. Unlike potentiometric sensors whose output follows a logarithmic relationship over several orders of magnitude of stimuli levels, the range for sensors with a

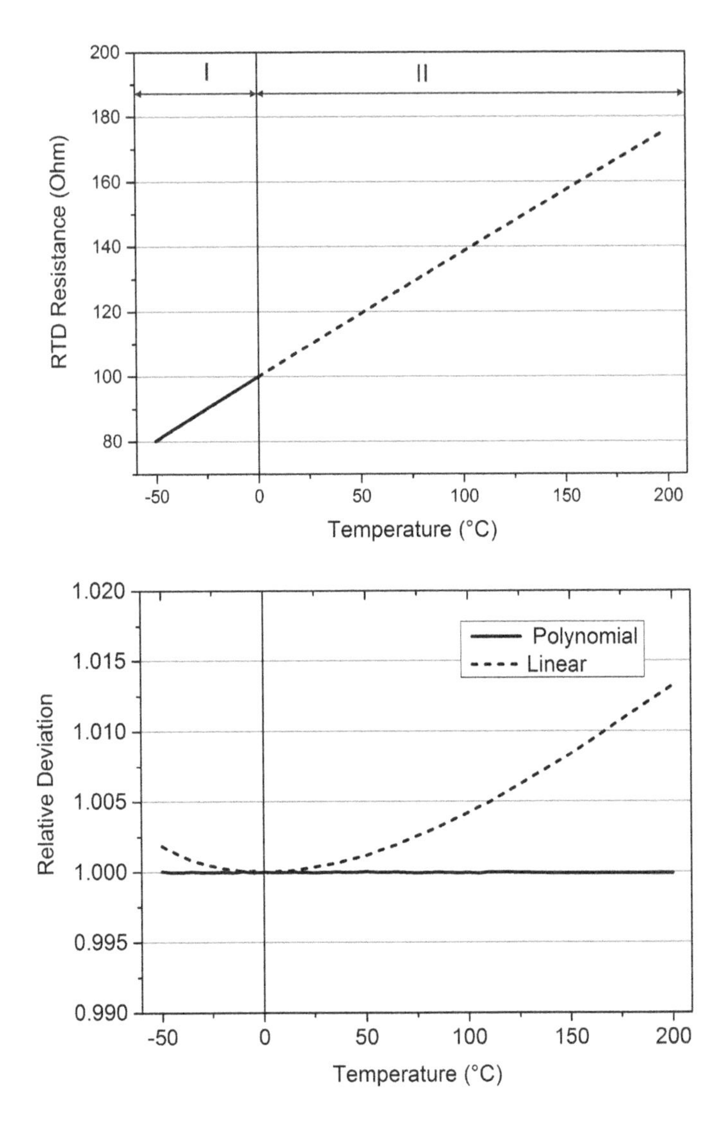

Figure 2.7: Polynomial fit for a Pt-RTD device with two regions (splines) connected by a knot at 0 °C. Top: Resistance vs. temperature of the Pt-RTD as calculated from the empirical equation. I: $R_T = R_0[1 + \mathrm{A}'T + \mathrm{B}'T^2 + \mathrm{C}'T^3(T - 100)]$, II: $R_T = R_0[1 + \mathrm{A}'T + \mathrm{B}'T^2]$. The parameters A', B', and C' are empirical constants (A' = 3.9083×10^{-3} °C^{-1}, B' = -5.775 $\times 10^{-7}$ °C^{-2}, C' = -4.183 $\times 10^{-12}$ °C^{-3}). Bottom: Using the polynomial and an assumed linear fit, the relative deviation ($R_{calculated}/R_{measured}$) is significantly lower for a higher order polynomial.

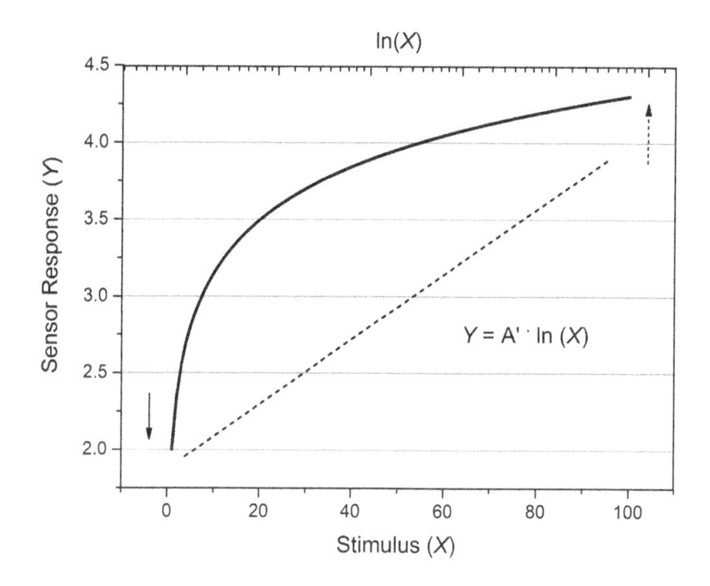

Figure 2.8: Hypothetical response of a potentiometric oxygen sensor operating at 600 °C with an internal air reference (see Section 4.2) as an example of Equation 2.5. The sensor response is shown for oxygen fractional concentration ranging from 10^{-6} to 1 (10^{-4} vol% to 100 vol%). Solid line: Sensor response plotted on a linear scale. Dotted line: Sensor response plotted on a logarithmic scale.

power law dependence is significantly narrower, typically covering less than 2 or 3 orders of magnitude.

Another common mathematical relationship to describe experimental data is the exponential function:

$$Y = \mathrm{A}' \cdot \exp\left(\mathrm{K} \cdot X\right) \tag{2.7}$$

Where Y is the sensor response to the stimulus X, while A' and K are empirical parameters. A response that is an exponential function of the gas concentration is not commonly observed for gas sensors, including the hydrogen sensors described in Chapter 3 and the other gas sensor types described in Chapter 4. However, in some sensor types, the final indication for a fixed stimuli level can show a temperature dependence that can be approximated by an exponential function. Such exponential behaviour is characterized by the logarithm of the sensor response for a fixed stimuli level being linearly dependent with temperature (e.g., temperature is the independent variable in Equation 2.7).

2.4 SENSOR PERFORMANCE PARAMETERS

2.4.1 Overview

The selection of a sensor is based upon its ability to meet various performance parameters (or metrics) as they pertain to an intended application. The most important and obvious performance parameters pertain to the metrological characteristics of the device, which are those parameters that are directly related to the ability of the chosen sensor to perform the

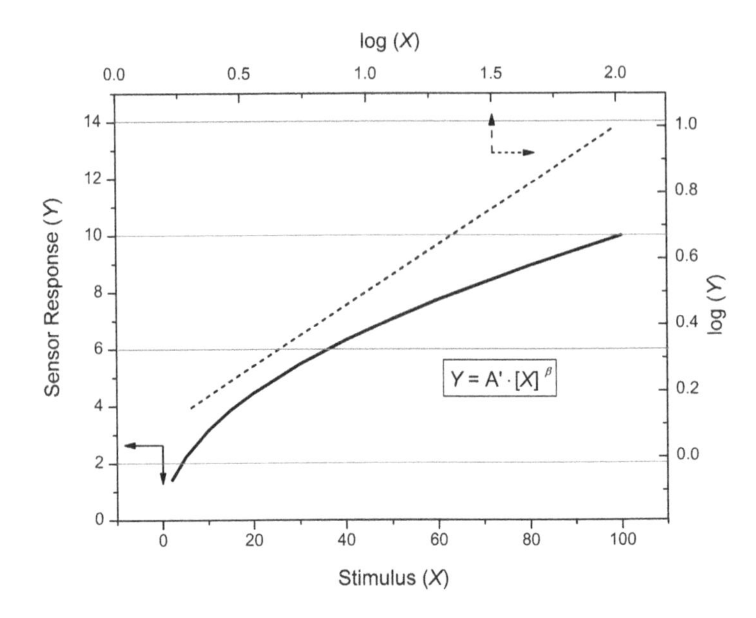

Figure 2.9: Hypothetical sensor power law dependence for A' = 1 and $\beta = 0.5$ as per Equation 2.6. Values for A and B are empirical parameters that will vary with specific sensor type and target analyte. Solid line: Plot of the "sensor response" vs. stimulus. Dotted line: Plot of the log of the sensor response vs. the log of the stimulus.

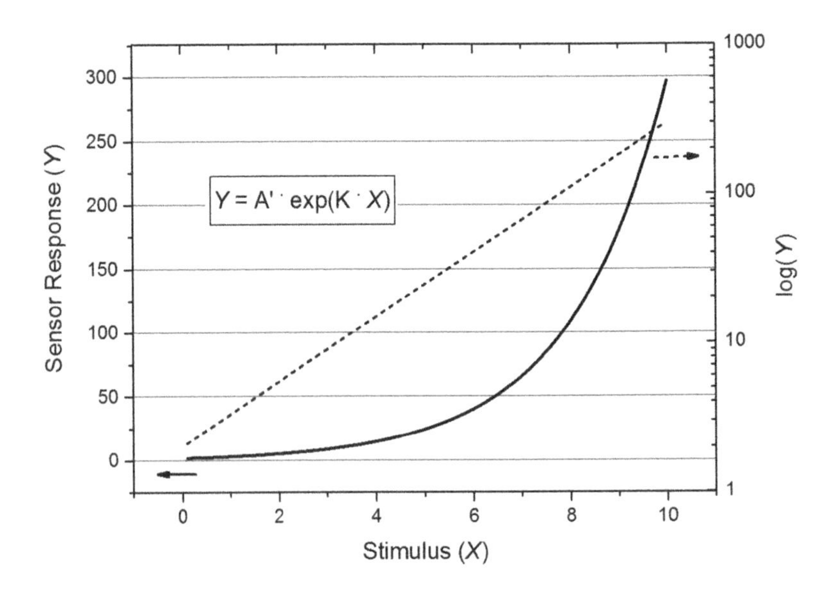

Figure 2.10: Hypothetical sensor response that is exponentially dependent upon the magnitude of the stimulus as per Equation 2.7 (A' = 0.1 and K = 0.5). A semi-logarithm plot (e.g., log (Y) vs. X) produces a straight line with a slope of k/2.303 and an intercept of log (A').

required analytical functions. There are, however, other considerations in the sensor selection process, which can be identified as "logistic" considerations [85]. Logistic considerations are independent of the metrological requirements and can be demarcated into deployment and operational parameters. Logistic parameters are not related to a sensor's ability to reliably detect or accurately quantify the target analyte but rather to other factors which ultimately determine the suitability of a technology for a specific application. Deployment parameters deal primarily with installation of the sensor into a specific system and are typically one-time, up front considerations. Capital cost of the sensor is a main deployment consideration. Alternatively, operational parameters are those metrics that correspond to operation of the sensors once they have been installed and typically are relevant to ongoing or recurring considerations; maintenance and calibration requirements are important operational considerations. The following sections briefly discuss main examples and concepts for these three categories of parameters. A more comprehensive list of various metrics along with definitions and supplemental information on the abovementioned categories are provided in Chapter 8. It is noted that the metrological parameters pertain only to analytical specifications of the sensor (or the specifications required by the application) and are totally decoupled from cost considerations. Costs are deployment and operational considerations.

2.4.2 Metrological Parameters

Measuring Range

One of most important performance metrics for a gas sensor is measuring range. The measuring range (also called the dynamic range) of a sensor is the range of values between two extremes in which the stimulus can be measured within a specified uncertainty. For a chemical sensor, one extreme value is the limit of quantification (LOQ). The LOQ is the lowest amount of analyte which can be quantitatively determined with suitable precision and uncertainty. Sensor signals are usually electrical in nature, and every electrical signal has an element of noise associated with it (e.g., a random fluctuation of response around a mean value). The LOQ is often defined as the stimuli magnitude which induces a sensor response that is ten times the noise in the sensor background signal, e.g., the sensor response at a reference stimuli level, although criteria other than the response that is ten times the noise level may be invoked [86]. Sensor noise is often measured as the standard deviation of the sensor response when exposed to the reference stimuli level. The upper extreme is often limited by a saturation of the sensor response where a further increase of the quantity to be sensed does not produce a proportionate change in the sensor signal. The measuring range is usually specified by the manufacturer. It is usually necessary (and proper) to evaluate other sensor metrological parameters only within the sensor's specified measuring range; one exception for a gas sensor would be to verify functionality or survivability of the sensor following exposure to a gas concentration beyond its specified upper limit, e.g., verification that a hydrogen safety sensor will still accurately quantify hydrogen in its specified measurement range following a brief exposure to pure hydrogen. In ISO 26142 a procedure to evaluate this property has been reported [79].

Accuracy and Uncertainty

The accuracy of a sensor indication is a topic of interest. This term characterizes the closeness of agreement between a measurement result and the accepted reference value of the quantity. Instead of reference value the term true value is often used, although a true value is hardly ever absolutely known since there is always some uncertainty in any physical quantity and in most measurements. Accuracy is more a qualitative concept and usually the term "uncertainty" is more properly used to describe measurement errors. Uncertainty is a parameter that characterizes the dispersion of values attributed to a measured quantity, often this is based on multiple determinations of the quantity. Uncertainty is sometimes incorrectly equated to the deviation of the sensor indication from the reference value (e.g., the difference between a sensor reading and the concentration of a test gas). A relative deviation can be related either to the actual reference value or to the measuring range (full span). However, the proper determination of uncertainty requires the multiple measurements under identical conditions so as to statistically ascertain the numeric uncertainty range. More details on uncertainty and statistical treatment of the sensor response is presented in Section 6.2.

Additional important metrological parameters are *repeatability* and *reproducibility*, which characterize the ability of a sensor to perform a required function under stated conditions for a stated period. Although having apparent similar meanings, repeatability and reproducibility have distinct definitions. Repeatability refers to the closeness of a measured parameter when the measurement is repeated on one system often with a single operator whereas reproducibility refers to the closeness of multiple identical measurements with results obtained from different laboratories. Reproducibility can be demonstrated by inter-laboratory round-robin tests.

Sensitivity

The sensor analytical performance is also often described by its sensitivity. Sensitivity is defined as the ratio of the sensor signal to the stimulus (e.g., the slope of a calibration curve, or more properly, the first derivative of the sensor response with respect to the stimuli: $(\partial Y/\partial X)$. It is perhaps worthwhile to discuss the concept of sensitivity as the term properly pertains to sensors, and gas sensors in particular. For a sensor with a linear response $(Y = \mathrm{K} \cdot X)$, the sensitivity is independent of concentration $(\partial Y/\partial X = \mathrm{K})$ but is dependent upon concentration for non-linear responses. This can be illustrated by considering a sensor response that is described by a polynomial relationship such as Equation 2.4 (Y = A'X^3 + B'X^2+C'X + D'). In this case, the sensitivity would be 3A'X^2 +2B'X +C', and thus will vary with the value for the stimuli. It should also be noted that in the literature, a phrase such as "increased sensitivity" is often used synonymously with "increased signal" or "better lower limit of detection." This is not formally a proper use of the term sensitivity. An improved sensitivity does not necessarily imply a better sensor response since a larger signal may also be accompanied by increased noise (e.g., random fluctuation of the signal) or drift (systematic change, either positive or negative, of the signal).

Selectivity and Cross-sensitivity

A sensor response can often be generated from stimuli other than the intended target. This is called cross sensitivity and is of specific importance for many gas sensors.

An ideally selective sensor will only respond to the target stimuli. Although this is never absolutely true, some sensor types have very selective response functions. Many temperature sensors and pressure sensors, for example, have outputs that are almost totally dependent upon the target stimuli. The response of a chemical sensor, however, can often be significantly affected by stimuli other than the target analyte, which are typically chemical in nature. This is certainly true for hydrogen sensors.

Most hydrogen sensor platforms will exhibit some cross-sensitivity to gases other than hydrogen. The effect is often small and chemical interferences can often be eliminated by design of the application; however, this is not always true and chemical cross-sensitivity can be problematic if ignored. For hydrogen sensors, the chemical cross-sensitivity coefficient is often given as an apparent hydrogen concentration induced by a specific concentration of an interfering chemical. Chemical interferences on hydrogen sensors are not always a linear process, in that the presence of an interfering chemical will not always induce a response that mimics the response induced by hydrogen (e.g., a response that might be interpreted as indicating the presence of hydrogen). Some chemicals can induce a negative sensor response. Others may lower the sensitivity to hydrogen, in which case they can be called inhibitors. Some chemicals are also classified as poisons, in that exposure to the chemical will permanently alter the response of the sensor. The nature of the interfence will be dependent upon both the sensor platform type and the identity of the chemical interferant. Refer to Chapter 3 for details on the various hydrogen sensor platforms [87].

Environmental Conditions

Variation in environmental parameters (e.g., T, p, and rh) can also induce a sensor response that could potentially be interpreted as arising from the target stimuli. In chemical sensors, and hydrogen sensors in particular, temperature-induced variations are the most significant. The transduction mechanism of the various sensing element types is often temperature dependent. The temperature dependence can affect the calibration curve in that the response of the sensor to a specific stimuli level may change with temperature. Temperature variations may also affect the magnitude of the sensor background signal, which can lead to a false indication of presence of the stimuli. Many commercial sensors compensate for the sensing element temperature dependence, often with the incorporation of a miniature temperature sensing element. Sensor manufacturers often quantify the sensor temperature dependence as an apparent stimuli magnitude per Kelvin (e.g., vol%/K). The impact of the temperature dependence of the sensor response should be negligible in the measuring range of the sensor.

The chemical sensor response can also be affected by variation in ambient pressure; however, this cannot always be properly classified as cross-sensitivity in that variation in pressure will induce variation in partial pressure of the target analyte and thus may actually be a proper indication of the amount of analyte in a mixture as a partial pressure but not its fractional or vol% value in a mixture (see Section 1.1). However, many regulations and alarm levels are based upon vol%, and fluctuations in pressure may affect the accuracy of sensor vol% reading. Fortunately, for safety sensor applications, ambient pressure fluctuations tend to be small. Moisture or humidity can be viewed as a specific example of a chemical interferent. As with chemical interferences, the impact of variations in environmental parameters will be dependent upon the sensor platform type. Finally, it is noted that when viewed as factors that affect the sensor output, variations in environmental parameters are metrological parameters since they affect the ability of the sensor to accurately perform the desired measurement. However, many sensors also have a working temperature, pressure, or relative humidity range, usually specified by the manufacturer over which they can operate; this specification is more properly classified as an operational parameter.

Time of Response and Recovery

Another critical metrological consideration is the sensor response time. Sensor kinetics is an important consideration for many safety and process control applications. The response time of a sensor can be defined as the interval between the time when a step change in the

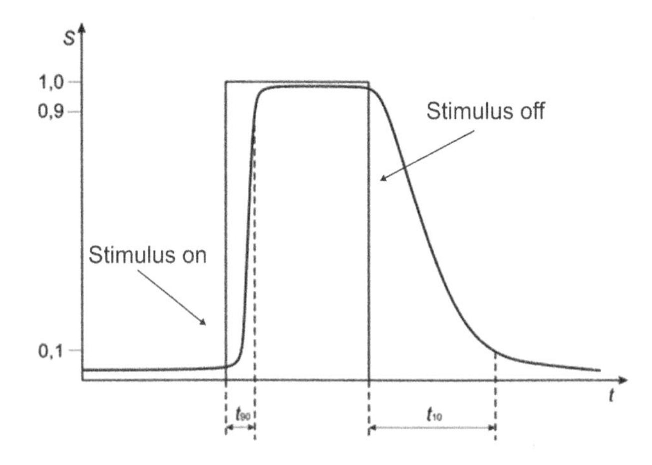

Figure 2.11: Sensor response time t_{90} and recovery time t_{10}.

magnitude of the stimuli (e.g., a change in hydrogen concentration) occurs and the time when the sensor response reaches a specified percentage of the maximum response induced by the stimuli. This is often set at 90 % of the final indication and is accordingly called t_{90}. Conversely, the time of recovery is the interval between the time when the stimuli is switched off and the sensor achieves a response that is within a specified percentage of the original sensor signal. This is often taken as to be within 10 % of the sensor signal before exposure and is called t_{10} [82]. The terms t_{90} and t_{10} are illustrated in Figure 2.11. Other response time protocols have been defined (e.g., t_{63} or t_{10-90}).

A quantitative or numerical specification should be assigned to each performance parameter. For example, a 1 s response time refers to the performance metric "response time" having the specification of 1 s. However, numeric specifications often need to be clearly defined; there are, for example, numerous definitions of response times including t_{90} or t_{50}, and various means to measure the parameter [78]. A reported sensor response time can vary significantly depending on both the definition of the parameter and the test protocol used to determine it.

2.4.3 Logistical Metrics - Deployment Parameter

The use of sensors is often guided by health and safety requirements as established by facility managers or other stakeholders. Codes and other government regulations (see Section 6.1) may also mandate the use of sensors. Thus, it is often not a question as to if a sensor should be used, but which one. Clearly the metrological requirements of a sensor should be the primary consideration for selecting a specific sensor, but there are other practical considerations. It is useful that the selected sensor should be user-friendly and easy to maintain by the end-user. More importantly, the capital cost of the sensor and installation costs must also be considered in relation to the expected increase in safety protection and improvements in process stability. Cost of ownership for a sensor is often the most important consideration by many facility managers.

The sensor should fit the local conditions in relation to size of the device, its placement, environmental conditions, and available electrical power. Sensors and detection apparatus often form part of a safety or process control system. The alarm set points of safety sensors, including hydrogen. are a critical parameter. Common alarm levels used for hydrogen safety sensors include a low level warning (typically set between 10 % and 25 % of the LEL) and a medium level alarm (typically set between 25 % and 50 % of the LEL), although the actual alarm set points can vary depending upon the application. The sensor or detection apparatus must often control other systems within the facility (e.g., ventilation system, system shutdown, remote monitoring interfaces, fire alarms, and other notification systems to emergency first responders). For this reason, appropriate electronic interfaces are necessary to ensure correct communication to these supplemental systems.

Operational Parameters

Operational parameters are those metrics that pertain to operation and maintenance of the sensor or detection apparatus once it has been installed. These parameters are in general ongoing or recurring considerations. The most obvious operational consideration refers to the maintenance requirements for the selected sensor, with sensor calibration being most common maintenance procedure. Sensor calibrations have to be performed periodically throughout the duration of the sensor deployment. Calibration requirements include the protocols and frequency for calibrating the sensor to assure accuracy and/or compliance to regulations. Calibration may be simply a verification of sensor operation or an actual validation of sensor reading with a certified gas standard coupled with adjustment of the sensor readout such that it is within the manufacture's specification for accuracy. The sensor operational lifetime is also a consideration, since this will dictate how often the sensor will need to be replaced. The total cost of ownership for a sensor will consist of the maintenance cost plus the original capital cost discussed above as a deployment metric. Initial savings for inexpensive technology may result in a greater overall cost because of increased maintenance requirements.

There are other operational considerations. These include the working temperature, pressure, and humidity range for the sensor. Robustness against mechanical shock and vibration can be an important factor, especially in heavy industrial applications and for mobile applications. Robustness against fluctuations in the electrical supply can be critical if the location does not have a reliable or stable power source. Power fluctuations should not damage the sensor. Furthermore, warm-up time (e.g., the time required for the sensor to produce reliable measurements once it is powered up) can be important since the sensor would not be fully functional until after the required warm-up time, and this may compromise the overall safety system. Uninterruptable power supplies (UPS) alleviate the impact of power outages and fluctuations. Other operational metrics are discussed in Section 8.5.

Prioritizing Performance Parameters

The manufacturer usually assigns specifications for their products, while the end-user defines the requirement for his application. An end-user must determine which metric is most critical for his application, and then define what the specification should be. The importance for each parameter and its corresponding numeric specification will be application specific. In some cases it may be that no specification for a particular parameter has been defined by the manufacturer. Thus end-users may have to perform qualification testing of a proposed sensor for a specific critical parameter. A critical parameter for one application may have negligible importance for another. As an example, a process control sensor may be expected to precisely and accurately control hydrogen concentration to within a specified range and thus would have stringent analytical resolution requirements (e.g., the ability to quantify small changes in concentration) with minimal importance assigned to the lower detection limit. In contrast, safety sensors deployed as area monitors must have a lower detection limit at a fraction of the alarm threshold but would not have severe requirements to quantitatively resolve small changes in hydrogen concentrations. Hence the application defines the sensor requirements — accordingly it is necessary to explicitly assess sensor requirements for each application. Not every parameter will necessarily have strong relevance for a specific application, but this often times must be determined by the project engineer and it is important that the relative importance of all metrics is considered for each application (see Section 6.4). Thus, a universal (short) list of critical sensor metrics with assigned specifications relevant for all applications is of limited value, and may indeed be counterproductive e.g., a 1 s response time may add cost and complexity to a sensor yet may be of limited value for a given application. The incorporation of parameters not discussed above or listed in Section 8.5 may also be necessary for unique applications with specialized requirements.

2.5 CONTEMPORARY TECHNOLOGIES FOR PREPARATION OF SENSORS

2.5.1 Overview

For commercial viability it is not only important that a gas sensor meets the performance requirements of the end-user. Fabrication must also be easy, reproducible, affordable and ideally scalable to facilitate mass production. Despite their structural and operational diversity many gas sensing elements have the following basic elements: a gas sensing layer or element, electrical connections (electrodes) both of which are supported on a substrate. The fabrication of sensing elements is often similar to the fabrication of electronic components such as resistors, diodes, or transistors. Many (micro) fabrication techniques may be adapted for the fabrication of different sensor types and may be classified as thin film, thick film, and micro-mechanical engineering techniques. Figure 2.12 gives an overview of the most widely used fabrication techniques and the sensor types that they are used for. In the following sections these three techniques are briefly explained however more detailed description are available in relevant books [88–91].

2.5.2 Thin Film Sensor Fabrication

Advanced micro-electromechanical manufacturing technology is capable of fabricating sensing elements that are integrated in monolithic circuits on a single crystal sheet (or wafer) of a semiconductor material, e.g., of Si, SiC, GaN or GaAs. Using this technology it is possible to produce several hundred identical units (chips) on a single wafer. This technology is widely exploited for sensing element fabrication by preparing thin films of gas sensitive material on inert substrates, such as alumina (sapphire, corundum), silica, or oxide glass. The sensing elements can have dimensions of less than one millimetre with a film thickness from about 5 nm to some microns. The fabrication steps typically occur under clean-room conditions with some steps under vacuum or in an inert atmosphere to ensure that structures of high purity are produced.

Sensing elements on semiconductor surfaces are often prepared by doping or epitaxy growth. Doping is the process of introducing foreign atoms into the basic semiconductor material, e.g., silicon, which results in a semiconductor with an n-type or p-type conductivity. Doping can be performed by diffusion, ion implantation, or electrophoresis. Epitaxy is a method for the formation of crystallographically orientated layers on a crystalline substrate. The epitaxial layer can be prepared using either liquid or gaseous precursors.

Sensing films as well as auxiliary conductor paths or passivation layers, may be precipitated on the surface of a semiconductor or other substrate by physical vapour deposition

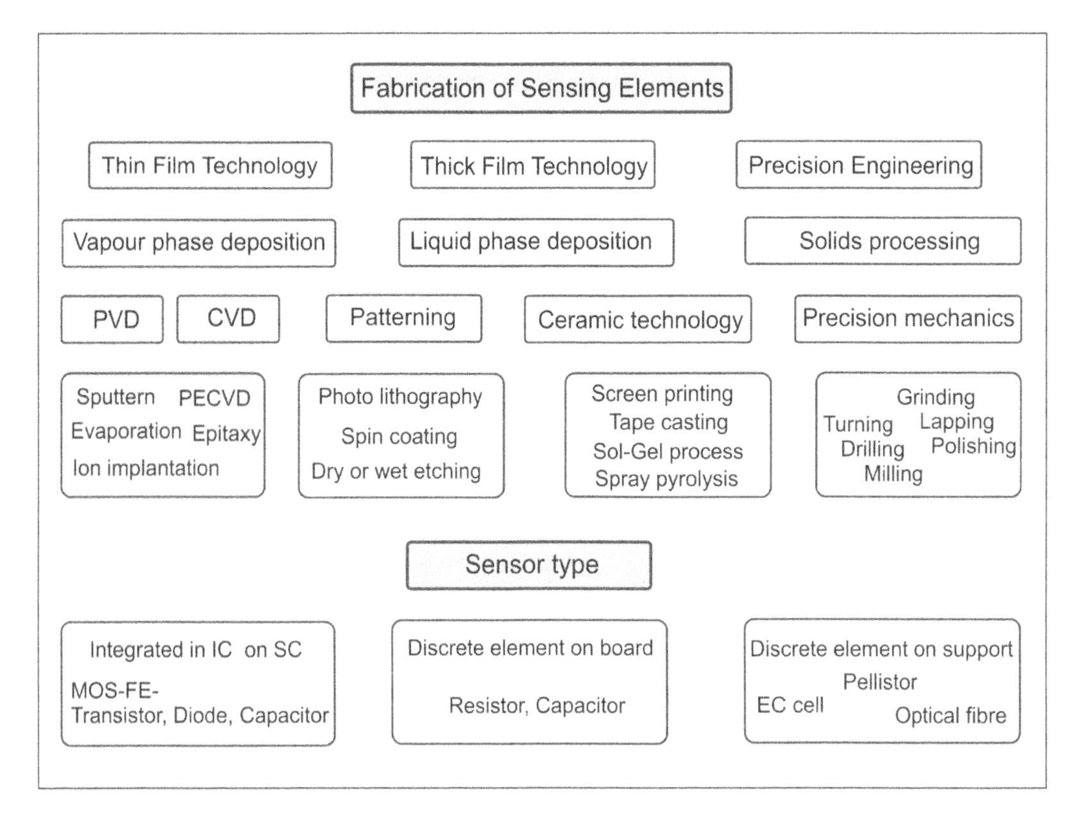

Figure 2.12: Schematic of preparation techniques for sensing elements and sensors.

(PVD) or chemical vapour deposition (CVD). Alternatively liquid-based methods, such as spin coating or spray coating, may also be used.

Physical vapour deposition (PVD) includes processes in which the sensing material to be deposited, e.g., metals, alloys, oxides, polymers, is present as a solid target which is evaporated and transported, via the gas phase, to the substrate where it precipitates. An external energy source is used to evaporate the target sensing material which is subsequently deposited on the cool substrate. Thermal evaporation, electron beam evaporation, arc evaporation, or pulsed laser ablation are evaporation techniques used in PVD. Sputter deposition is another PVD processes in which the target material is bombarded by ions or a plasma resulting in ejection of sputtered atoms which are deposited as a thin film on the substrate. Thin film resistive platinum temperature sensors (see Section 5.1), thin film metal oxide hydrogen sensors (see Section 3.5), and optical platinum film glass fibre hydrogen sensors (see Section 3.8) are often manufactured by PVD.

Chemical vapour deposition (CVD) includes processes where volatile components are evaporated and react in the gaseous phase to form the product that is precipitated on the surface of the substrate, which is often heated. Several variants of this process are in use in industry. The gas phase reaction is stimulated by heat or plasma and is carried out at temperatures ranging from 200 °C to 1300 °C in vacuum or at pressures up to 1 bar. Gas sensitive thin films of ZnO or SnO_2 have been made from appropriate chlorides or organic compounds. This process is also of great importance for the manufacturing of glass fibres for optical sensors. Tubes of pure glasses (silica) are coated by thin layers of material with specific optical properties and then pulled out as a string. Passivation films of SiO_2 are formed from dry or wet oxidation of the silicon surface or by gas phase reactions of silanes or tetraethoxysilane and oxygen. Si_3N_4 films are fabricated from silanes and ammonia.

Liquid-phase epitaxy includes chemical deposition methods using liquid components whereby organic or inorganic films are precipitated by spin, dip, or spray coating techniques [92]. Either the applied liquid already contains the components of the film in solution or the film is formed from precursor in solution such as monomers which polymerise during film processing. Inorganic films or hybrids, i.e., organic-inorganic films, can be prepared via the sol-gel-process. Starting from a precursor solution of the metal ion containing compounds, such as tetraethoxysilane, tetrabutyltitanate, zinc nitrate, partial or complete hydrolysis of these compounds is undertaken to form a sol which contains nano-scaled, colloidal particles. During film formation the small particles combine by poly-condensation and polymerisation to form a gel. Drying and heat treatment drives out the solvents and the organic reaction products. Gas and humidity sensitive layers of doped tin oxide, zinc oxide, silica, or titania, have been fabricated by this process [93]. Electrochemical deposition methods can also be applied for sensor fabrication and include cathodic deposition of metallic conductive paths, anodic oxidation of silicon, aluminium or titanium, or electrophoresis.

Following deposition, patterning of thin films is often required for which photolithographic processes are applied. First an organic film of a light-sensitive photoresist material is applied over the whole thin film area as a sacrificial layer, e.g., by spin coating. The photoresist film is then exposed to UV light, X-rays, or an electron beam which changes its chemical properties in exposed areas. These areas may become soluble (positive resist) or insoluble (negative resist) to the subsequent treatment with a photoresist developer. The exposure to radiation is performed through a metallic mask which transfers a geometric pattern onto the resist layer. During subsequent treatment with a photoresist developer the desired parts of the film are dissolved (stripped). The organic mask, which protects the areas of the substrate which it covers from the etching process, is removed in a later step. Following this treatment the stripped areas are subjected to further treatment steps such as metallization or etching.

Etching is a process to remove material from the surface. Wet etching uses acids, such as hydrofluoric acid for isotropic etching and bases, such as potassium hydroxide for anisotropic (oriented) etching. Electrochemical processes can also be used for etching. Dry etching processes use a plasma under reduced pressure in the presence of chemicals such as carbon tetrachloride as a gaseous sources. Alternatively reactive ion etching methods use sulphur hexafluoride for etching of silicon. Etching processes have been used for the preparation of thin silicon membranes e.g. for pressure sensors. Lasers or electron beams are also used for local abrasion, e.g., for a so-called trimming of metallic thin film resistors. Finally the wafer or the substrate is divided by scratching, breaking, or sawing to obtain the miniaturised sensing elements.

2.5.3 Thick Film Sensor Fabrication

Thick film fabrication is used for manufacturing many sensing elements. It is used also in the fabrication of electrodes, terminals, heater, resistors, capacitors, and insulators as well as for the formation of passivation layers. The films have a thickness typically of about 5 μm to 50 μm and are printed on ceramic substrates. Alumina (Al_2O_3) is widely used due to its stability, processability, and high thermal conductivity. However for special applications, aluminium nitride (AlN) or beryllium oxide (BeO) have also been used as substrates. Substrates are fabricated via ceramic process engineering procedure in which fine grained inorganic raw materials are shaped by different technologies, such as pressing, slip casting or film casting. The subsequent processing steps include drying and a high temperature heat treatment for sintering the grains to form a dense or porous hard solid. The ceramic process is also important for manufacturing sensing layers, such as of zirconia for oxygen detection, and support structures such as dust filters, mechanical supports, and housings.

For preparation of thick films, inks are used for different printing techniques. An ink consists usually of small particles of the desired material (metals or oxides), an organic binder (polymer), and further compounds for conditioning. It may contain additional components such as an inorganic binder (low melting glass powder) and solvents which act as surface active substances in order to generate the specific viscous properties. The ink is usually deposited via a screen-printing process; however, other printing processes such as pad printing or offset printing can be used. The screen is made by a silk or metallic mesh stretched over a frame. Open sections in the mesh permit the passage of the ink to the substrate and hence form the shape of the intended print. In the printing process the screen is placed above the substrate and the ink is placed on top of the screen. A moving squeegee (flood bar) is used to spread and push the ink through the holes of the mesh. After the printing process the print is heated up to temperatures up to 200 °C to evaporate the solvents. The substrate is further heated up to around 1100 °C to decompose the organic binder and to sinter the active gas sensing material. No impurities are permitted during the screen-printing and the heat treatment processes. For this reason manufacturing is usually performed in a clean room to avoid cross-contamination and to maintain a high process stability. This technology has been used to fabricate gas sensitive semi-conductive layers, e.g., thick film doped tin oxide or zinc oxide sensors.

2.5.4 Micro-mechanical Engineering

Micro-mechanical engineering encompasses branches of mechanical engineering, optical engineering, and electrical and electronics engineering for the manufacturing of machines, tools, and components that have exceptionally low tolerances in shape and functional parameters. It has application in manufacturing of micro-electromechanical systems (MEMS), as well as in sensors. Micro-mechanical engineering deals with precise materials processing and manufacturing of small components, e.g., packages. It includes metal processing such as turning, drilling, milling, and processing of brittle ceramics via grinding, polishing, and lapping. Micro-mechanical engineering and MEMS techniques in particular have been used to fabricate gas sensing element platforms such as those based on mechanical functions (surface acoustical wave sensors, cantilevers, etc., see Section 3.7)

2.5.5 Sensor Housing and Protection

Whereas thin and thick film technologies are the main techniques for preparing sensing layers, several additional technologies are deployed for sensor processing and the manufacturing of auxiliary sensor components.

Advanced sensors, e.g., for hydrogen, ammonia, temperature, and humidity (see Sections 3.5 and 4.3) have already been manufactured as integrated circuits containing both the sensing elements as well as the associated microelectronic components for signal processing on a single small plate, called a "chip." In a process step called "chip bonding" a single chip has to be fixed onto a substrate. This can be done by gluing or soldering. In a gas sensor the chip supporting the sensing element is either mounted on a ceramic substrate with the required conductive tracks (prepared by thick film technology) or it is positioned on a printed circuit board (PCB). A fibre reinforced plastic board provides the mechanical support for the discrete electronic components of the PCB including the conductive tracks.

Electrical connections are often formed by wire bonding, in which thin wires are used to form electrical contacts between the integrated circuit (chip) and other electronic components. Typically gold, coper or aluminium alloy wires are bonded onto the surface of an electrical contact pad by the application of heat, pressure, and ultrasound.

Sensors in integrated circuits are packaged in housings to protect against damage and to provide necessary electrical contacts. A large number of package types exist. Some package types are standardized in design and dimensions, e.g., the transistor outline (TO) package, while other housing types are proprietary designs of the manufacturer. The package depends on the sensor physical design requirements. Usually sensing elements or sensors are contained inside a housing that is often made from an organic polymer materials such as polyurethane or poly-acrylate. Other more robust materials are used for sensor housings, such as stainless steel, aluminium alloys, or ceramics (alumina, mullite, or cordierite).

Electronic integrated circuits are usually hermetically sealed, while gas sensors require an opening to ensure that there is a chemical interaction with the ambient atmosphere. Meshes, membranes, and filters are used to protect against the ingress of dust and moisture, while still allowing gas to access the sensing elements. The filter may consist of a fine mesh metallic grid, porous metals, or plastics. Filters can also contribute to selective gas sensing due to their hydrophobic (e.g., fluoropolymers, PTFE) or hydrophilic (polyamide, PA 6) properties. Filtering materials, such as a molecular sieve, can absorb some interfering components in the gas phase, e.g., humidity. Membranes such as Nafion, a sulfonated tetrafluoroethylene polymer, are used also in electrochemical sensors as an electrolyte or conductor. Due to Nafion's acidic groups the polymer is very permeable to water and alcohols and is therefore used as a proton conductor and proton exchange membrane. However Nafion does not conduct anions or electrons.

2.6 SENSOR DATA ACQUISITION AND EVALUATION

2.6.1 Overview

Sensing elements are, for the most part, devices that produce an electrical signal in response to an external stimulus. For a gas sensing element, exposure to the target gas induces a change in the outputted voltage, current, resistance (or conductance), electrical charge, or frequency of the transducer. In the following, an overview on the use of modern electronic components will be presented for control and processing of the sensing element signal. It will be also described how the resulting signal can be interfaced to detection apparatus or sensor systems to provide a useful end-user information. A comprehensive overview of signal processing cannot be presented in this book; however, details on sensor electronics and control can be obtained from literature [69, 81, 94]. Signal processing, beginning from the primary electrical signal to usable output (e.g., a digital display, data storage, smart actions such as ventilation activation), can be divided into several separate steps and will be discussed below. Various steps of signal processing are illustrated in Figure 2.13 for a gas sensor.

2.6.2 Signal Processing

The first and perhaps the most critical step in sensor signal processing is the proper control of the electrical output generated by the sensing element. Any degradation of the signal integrity, such as generation of noise by electromagnetic interferences, signal loss due to line impedance, or caused by poorly designed electronic interface will be manifested and amplified in each subsequent step in sensor signal processing. **Signal conditioning** comprises procedures such

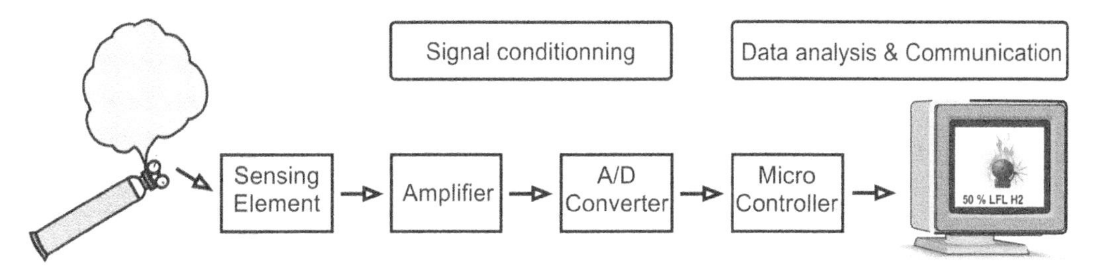

Figure 2.13: Schematic of sensor signal processing.

as amplification and filtering to make the sensor signal suitable for analogue-to-digital (A/D) conversion. Signal conditioning also includes offset (e.g., baseline) correction, averaging (for noise reduction), and linearization (for simpler operation). For a sensing element that responds with a non-linear function of the stimulus, linearization can sometimes be performed using a correction element that reproduces an inverse sensing characteristic. The combination of both results in a signal which better approximates a linear dependence to the stimulus.

Sensing elements are incorporated as components of a readout circuit which in many cases transfer these changes into a primary DC or AC voltage or current signal. A voltage signal is preferred when the resistance is low and a current signal is given preference when the resistance is high.

Conductometric, catalytic, and thermal conductivity sensors each experience a change in resistances when exposed to the stimulus and are often connected to a Wheatstone bridge for precise measurements of small resistance changes (see Figure 2.14). The bridge circuit has two pairs of arms. Each pair is composed of two resistors in series acting as a voltage divider. One arm includes the resistive element associated with the sensing element. An output signal is the difference between the centre points of the two voltage dividers. The sensing element's response to a stimulus is the change of the output signal voltage. The Wheatstone bridge is easy to implement and very sensitive to small changes in resistance. Furthermore, the background signal (e.g., the signal in the absence of the target analyte) can be nulled (zeroed). This bridge circuit is also used in resistive thermometers, strain gauges, as well as in pressure and flow sensors. There are many adaptations of the fundamental circuit of a Wheatstone bridge: Thomson bridge for small resistance, Wien bridge for capacitance, Maxwell-Wien bridge for inductance, Schering bridge for capacitance and loss factor of capacitors, and Wien-Robinson bridge for frequency. These bridges can be used for AC zero measurements, but these are not commonly used for controlling gas sensors [81].

The primary sensor signal may often be of low magnitude, which has to be amplified before sending it to an evaluation unit. Therefore the Wheatstone bridge can be combined with an amplifier, as shown in Figure 2.14. Instead of directly measuring the output signal with a voltmeter or data acquisition (DAQ) system, an **amplifier** is often incorporated into the measurement circuitry to amplify the differential voltage signals of the Wheatstone bridge. There are, however, limitations of the Wheatstone bridge. Thus, precision amplifiers, based on integrated circuit technology, are often used directly to control sensing element signals. In sensor technology, an operational amplifier (op-amp)[3] is commonly used because it can be configured to provide a high DC and AC amplification.

[3]An op-amp is a very low-cost device that can be defined as an integrated circuit (IC) that operates as a voltage amplifier. An op-amp has a differential input. That is, it has two inputs of opposite polarity. An op-amp has a single output and a very high gain, which means that the output signal can be much higher than input signal.

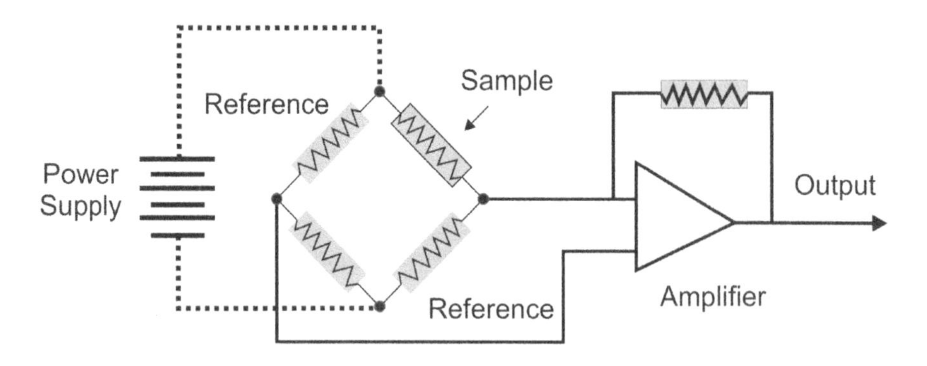

Figure 2.14: Wheatstone bridge for measuring resistance changes modified with a differential amplifier.

In addition to voltage amplification, op-amps can perform other versatile signal processing functions, such as adding an offset voltage to a signal, integration, and differentiation of signals. They can discriminate against noise or interferences. Op-amps can "buffer" the output of the sensing element, which is often a low-impendance electrical signal that can be easily corrupted by line loss and even by signal measurement devices with low input impedance. The input impedance of an op-amp is much greater and thus less prone to signal corruption. Therefore it is recommended that active control electronics, such as the op-amp buffer circuit, be physically placed as close to the sensor element as possible. This can be achieved by integrating the sensing element and control circuitry into a single platform as described in Section 2.5. Op-amps can also convert an electrical signal of one type into another. For example, a current to voltage transducer is regularly incorporated into the control circuitry of amperometric gas sensors to convert the current produced by the electrochemical reaction into a voltage. This is often done because voltage is more convenient to measure than a current.

Filtering of sensor signals is performed because not all signal frequencies contain valid information. **Filters** can be used to change the characteristics of an alternating current (AC) signal in terms of frequency, amplitude, or phase. Thus unwanted parts of the signal can be suppressed. Special filter designs and considerations are provided in the literature [81].

Although analogue circuitry can be used to perform many functions, most detection apparatus will incorporate an **Analogue-to-Digital Converter** (ADC), a device that converts continuous analogue input signals into discrete signals, discrete in time and magnitude. Main parameters of ADCs are resolution (8 bits or 24 bits), the voltage range, and the conversion rate (1 MHz to 2 GHz). The number of bits (typically given by 2 raised to an integral power) together with the voltage both define the minimum resolution of the ADC. For example, a 0 V to 10 V 16 bit (216 = 65536) will resolve to 0.000153 Volts (10/65536). Digital conversion with modern A/D converters can maintain excellent resolution of the original signal, but it is not a perfect replication. Nevertheless, A/D conversion has many advantages. A/D conversion can, e.g., be configured to reduce signal noise by rapid averaging of the analogue signal.

Digitized data can be converted into frequency signals and there are many techniques to perform it. The transformation into frequency gives modulated signals where the frequency presents the information contained in the original data. This transformation is useful for wireless data communication. It can be often presented either as a function of frequency or as variable pulse length signals where the length of the pulse is a transform of the original primary signal. Even though not yet very common, frequency modulated signals offer several advantages for sensor signal processing and communication.

Advanced sensor systems have multi-sensing capabilities that can include gas sensors as well as sensing elements for temperature, humidity, gas flow, or pressure. A **multiplexer** can be used to accommodate the output of multiple sensing elements. Multiplexers sequentially select one input signal at a time from a choice of analogue or digital input signals for further processing, such as A/D conversion. The data logging is nearly simultaneous, although in fact there is a small lag between logging the input from different sensors. The lag is typically on the order of milliseconds, and thus can be ignored for most gas sensing applications.

The digital signal can be interfaced to a **microcontroller**. This is a device which contains a central processing unit (CPU) and performs data analysis and transfer, as well as sensor control. The micro controller can perform compensation of signal non-linearity, noise, drift, and cross-sensitivity as well as compensation for changes in environmental parameters. Microcontrollers can perform other important functions such as self-testing, self-identification, self-validation, self-calibration, and self-diagnosis. Further functionalities within the CPU include an internal clock as a source for timing operations to ensure the correct sequence and to put a time stamp on recorded files. The digitized signals can also be saved as an electronic file for record keeping and post-measurement workup on a computer. Within the CPU the random access memory (RAM) is used for temporary storage of data while an EPROM (erasable programmable read-only memory) contains the program for sensor control and data processing.

The **communication processing**, i.e., transferring outwards of the information which is received and processed by the detection apparatus or sensor system, takes place via different transfer protocols. Common data transfer protocols are based on analogue or digital voltage or current signals or as other data, e.g., numbers or commands [95]. For this a digital-to-analogue converter is employed. The transformation of a digital signal into a **standard analogue signal** gives many advantages for subsequent signal and data processing. Commonly used analogue output signal platforms include standard current signals in the range of 0 mA to 20 mA or 4 mA to 20 mA and the voltage signals from 0 V to 10 V or 2 V to 10 V. Relative to voltage outputs, the advantages of the current outputs are much less sensitive to electromagnetic interference, and an automatic compensation of conduction losses by the current loop. The elevated zero point of the 4 mA to 20 mA current output and likewise with the 2 V to 10 V voltage signal enables instrument fault detection, e.g., a cable break or other fault is indicated by a 0 mA or 0 V reading.

The transfer of data as numbers or commands have the advantages that the signal-to-noise ratio, the influence of temperature, component tolerance scattering, and aging effects are lower. The signal processing is easier to perform, the data are easier to store and to transfer, e.g., by microcontroller. Standard communication is commonly performed via serial protocols such as RS 232 and RS 485. One distinct advantage of serial communication is that a single string of data may include multiple data components coming from, for example, other sensors, measurement parameters, and identification information. The resulting data strings are often easier to store and to transfer.

The transmission of digital output signals is compatible with communication via **bus** systems. Different bus systems for external data exchange are available, e.g., universal systems such as USB 1.1 or USB 2.0. Other specialized transfer protocols are often used in specific industries, such as Controller Area Network (CAN) bus in automotive applications and

Highway Addressable Remote Transducer Protocol (HART) in measuring instrumentation. Wireless data transfer is becoming more common.

Sometimes the sensor's or detection apparatus' analogue signals may need to be logged by an external data acquisition system (DAQ). Commercial DAQ systems for personal computers (software and hardware such as data acquisition cards) are available. This is often for research or independent evaluation purposes, but can also be a part of a larger control system of which the detection apparatus is only one small element. DAQ and control systems are commercially available from different vendors in numerous formats, including, for example, LabVIEW® (National Instruments Inc.) and programmable logic controllers (PLCs) supported by numerous PLC manufacturers; PLC is standardized per IEC 61131-3 [96]. However, many control systems for buildings are based on custom software. Results of sensing can be visualized in displays. Additionally to visual indication, alarms can be indicated by acoustic signals, and countermeasures can be initiated by the sensor system.

A component of sensors is the power supply for sensing element circuit, heating and sensor electronics, either by batteries for portable systems or from the grid in case of stationary deployment.

2.6.3 Data Processing

The data elaboration can be performed in the sensor, if sufficient electronic capability (microcontroller) is inserted and the data elaboration procedure is fixed. Otherwise a data analysis in several steps can be executed beginning with data pre-processing, which refers to any transformation performed on the raw sensor output data prior to building the main analysis model. The goal of data pre-processing is typically two-fold [97,98]:

- Reduction of noise or removal of information that is known to be irrelevant for the task of measurement (e.g., interferences, drift) and
- Numerical preconditioning of the data, such as scaling or normalization. The selection of a suitable data pre-processing approach can have a significant impact on the sensor performance.

Pre-processing techniques can be grouped into three categories: (1) baseline correction, (2) scaling, and (3) dynamic feature extraction. Somewhat related to baseline correction is the issue of drift compensation.

The objective of baseline correction techniques is to remove background noise from the raw sensor responses. Three types of baseline correction techniques are widely used: differential, relative, and fractional techniques, always in relation to a baseline or reference value, e.g., the sensor response from clean air or from a defined analyte concentration.

Differential techniques subtract from the sensor response (y_s) a baseline value (y_0) and can be used to remove additive noise or interferences (y_s-y_0). Differential techniques are typically used for piezoelectric sensors, where the response is a frequency or phase shift with respect to a reference analyte (and/or an uncoated reference sensor), and for MOSFET sensors, where the response is a voltage shift in the $I(V)$ curve. Relative techniques compute the ratio between the sensor response and the sensor baseline value (y_s/y_0). The relative technique is commonly used with metal-oxide sensors. Fractional techniques subtract the baseline value and then divide by the baseline value, which yields a per-unit response: (y_s-y_0)/y_0.

When signals from more than one sensor are to be processed, and if additional signals are used in order to compensate the impact of other parameters (e.g., from temperature and pressure), a multivariate method for data evaluation can be applied. In this case a scaling of sensor data by normalization and auto-scaling may be appropriate. Normalization scales each sensor output value to the range [0, 1] by subtracting the minimum value and then dividing by the overall measurement range of the sensor response. In contrast, auto-scaling normalizes each sensor output value by subtracting the sample mean value and then dividing by the standard deviation, both computed across the entire database. Auto-scaling is more robust to outliers than normalization. Moreover, robust statistics may be used to reduce the sensitivity to outliers.

One of the most serious limitations of current sensor application is the inherent drift of individual sensors, which results in a slow change of the sensor response when exposed to the same analyte under identical conditions (see Section 2.4). Drift is caused by unwanted effects such as ageing of the sensor and poisoning from the environment. Drift compensation may be performed in such a way as described for differential, relative, and fractional techniques of baseline correction. It can be considered as a re-calibration of sensor response either from time to time or event driven (see Section 6.2). If the drift behaviour of the sensors response is known from preceding investigations, it can be fitted and incorporated in the data correction procedure.

In most cases only the equilibrium signals or response maxima were evaluated. The sensor output signal after a certain time or constancy of values is used. However, the analysis of the transient response of the sensors to sudden changes in the sample concentration (or temperature) may have advantages by a faster result of measuring and possibly in higher information content. Thus for example the sensor is exposed to a change of analyte composition or to a short analyte pulse and the gas sensor transient response is recorded. Also a modulation of operation temperature of the sensing element can be used to capture response data. Either all data were used or a parameter such as the initial or a steady state slope can be derived.

Contemporary gas sensors contain not only a sensing element for the analyte (e.g., hydrogen), they may include additional sensing elements for ambient parameters such as temperature, gas flow or humidity for compensation of the impact of these parameters on the accuracy of the gas sensor. In order to evaluate all these data (and also further aspect, e.g., the results of drift correction) a linear regression may not be sufficient and a multivariate calibration of the sensor output signal is necessary. For this procedure several mathematical procedures, such as multi linear regression methods (partial least-squares (PLS) or principal-components regression) or principal component analysis (PCA), artificial neuronal network (ANN), or Fuzzy logic, are available in software programs which can be integrated in the sensor or used in a PC based data evaluation. A visualization of results can be given in numbers and values or in graphs and bar charts, profiles, or polar plots on displays.

Even though the sensor platform establishes basic capabilities, advanced sensor electronic and data processing software have a significant impact on the sensor performance and flexibility to adapt sensors to specific applications.

CHAPTER 3

Hydrogen Sensors

CONTENTS

Gas sensors are widely used in hydrogen technologies. Fuel cell systems, for example, use hydrogen, oxygen, and relative humidity sensors for monitoring the gas streams and providing data for optimising operation of the system and preventing damage to the fuel cell. Hydrogen production systems, e.g., electrolysers, use hydrogen and oxygen sensors for monitoring gas quality online. Systems generating, storing, or using hydrogen also require sensors which are capable of detecting low concentrations of leaked hydrogen. Such hydrogen leak sensors are critical components for safety as they help avoid or mitigate hazardous situations arising from hydrogen combustion or explosion. In this chapter different technologies for detecting hydrogen are described with a short review of their working principle, the history of their development, their deployment, and state of the art. As new markets for hydrogen sensors are rapidly evolving, additional information is provided on novel and emerging hydrogen sensing technologies. Finally a comparison of the performance of available technologies is made and tips on choosing the optimal hydrogen sensor for a particular application are provided. In Chapter 4 other gas sensors, including oxygen and relative humidity sensors, are similarly described.

3.1 CLASSIFICATION OF HYDROGEN SENSORS

Section 2.2 discussed various classification schemes for sensors, in particular chemical sensors. The IUPAC system classifies chemical sensors according to the operating principle of the transducer and the resulting sensor classes are summarised in Table 2.1. Adapting the IUPAC definition for chemical sensors we can define a hydrogen gas sensor as a device which transforms the presence and/or amount of hydrogen in a gas mixture into an electrical signal such as voltage, current, or resistance. Changes in the hydrogen concentration result in a change in the magnitude of the electrical signal. As discussed in Section 1.1 the unique properties of hydrogen (e.g., its high thermal conductivity, its solubility in metals, and its flammability) mean that a wide range of principles are suitable for its detection giving rise to various classes of sensor which are described in detail in this chapter.

The IUPAC system of chemical sensor classification is used extensively by others [99–101]. Nevertheless this system is not strictly adhered to in this book for classifying hydrogen sensors. Instead a more pragmatic classification is proposed, reflecting how hydrogen sensors are marketed and using terminology commonly used by sensor manufacturers in their sensor descriptions and documentation. Table 3.1 outlines the classification system adopted for hydrogen sensors in this book. At times the classification system used here results in particular sensor types being classified differently from its classification according to the IUPAC system. For example, catalytic hydrogen sensors are defined in this book as those sensors which, as a result of the catalytic combustion of hydrogen and the associated heat of reaction, give rise to an increase in temperature which is transduced into an electrical signal. On the other hand thermal conductivity hydrogen sensors give rise to an electrical signal, not as a result of a chemical reaction but by the cooling effect hydrogen can have on a heated element. While the operating principle of both these sensors is clearly very different, the IUPAC system classifies both as thermometric sensors which can give rise to some confusion, particularly when making comparisons between different sensor types. The classification system proposed in this book better reflects how hydrogen sensors are typically marketed by manufacturers making it easier for hydrogen engineers and scientists to understand, differentiate, and choose between the different sensor types.

Table 3.1: Hydrogen sensor classification system.

Sensor types	Subtypes (where relevant)
Catalytic	Catalytic pellistor sensors, Thermoelectric sensors, Pyroelectric sensors
Thermal conductivity	Thermal conductivity sensors
Electrochemical	Amperometric sensors, Potentiometric sensors, Mixed potential sensors
Conductometric	Semiconducting metal oxides sensors, Semistor sensors, Thin metal film sensors
Field effect based	Metal-insulator-semiconductor capacitor sensors, Metal-semiconductor diode sensors, Metal-insulator-semiconductor transistor sensors
Electromechanical	Bulk acoustic wave sensors, Surface acoustic wave sensors, Microcantilever sensors
Optical	Chemochromic sensors, Fibre optic sensors, Open path sensors
Sonic	Leak detection sensors, Concentration sensors

3.2 CATALYTIC HYDROGEN SENSORS

The propensity of hydrogen and other combustible gases to react exothermically with oxygen on a catalyst and to release heat is the basis of the detection principle of a number of hydrogen sensor types. These are generally classified as catalytic sensors. Catalytic pellistors, thermoelectric and pyroelectric sensors are three types of catalytic sensors that have been developed for hydrogen detection. Catalytic sensors detect hydrogen and other combustible gases calorimetrically, that is through the heat released from the catalytic combustion reaction of the flammable gas.

In the absence of an ignition source or catalyst, combustible gases will spontaneously oxidise in air only at or above the autoignition temperature. The autoignition temperature of hydrogen in air is 585 °C [7]. This is quite high relative to other combustible gases and makes hydrogen difficult to ignite in air using heat alone (of course ignition by other sources, such as open flames or even electrostatic sparks, may still initiate hydrogen oxidation). In the presence of a suitable catalyst the ignition temperature can be lowered in a phenomenon termed catalytic combustion. A catalytic hydrogen sensor consists of an appropriate catalytic metal, e.g., platinum, coated on to the surface of a sensing element which is typically operated at around 300 °C to 500 °C. This is significantly below the hydrogen autoignition temperature, but because of the surface catalyst coating, hydrogen will nevertheless react exothermically with oxygen to form water in a complex reaction which can be simplified as:

$$H_2 + 2\,Pt \longrightarrow 2\,Pt{-}H$$
$$Pt{-}H + Pt{-}O \longrightarrow 2\,Pt{-}OH$$
$$Pt{-}OH + Pt{-}H \longrightarrow 2\,Pt + H_2O$$

The release of heat associated with the exothermic combustion reaction of the target gas being detected raises the temperature of the sensing element. The amount of heat generated from the reaction is dependent upon the standard heat of combustion (also known as the enthalpy of combustion) of the target gas and the amount of that gas reacting on the

Table 3.2: Standard heat of combustion of gases [7]

Gas	$\Delta_c H^\circ$ / kJ/mol
Hydrogen	285.8
Carbon monoxide	283.0
Methane	890.8
Ethane	1560.7
Propane	2219.2
Butane	2877.6

sensor surface. Table 3.2 gives the standard heat of combustion of several gases that are often detected using catalytic sensors. Different methods can be employed to transduce the temperature rise into an electrical signal, and accordingly there are a number of subclasses of catalytic sensors as mentioned above.

3.2.1 Catalytic Pellistor

Pellistors are by far the most common, commerciall -available type of catalytic hydrogen sensor. Also known as catalytic bead sensors, combustible gas sensors (CGS) or catalytic solid state sensors, catalytic pellistors are a well-established technology that have been used for decades for the detection of combustible gases. The name "pellistor" is a portmanteau of the words "pellet" and "resistor." A pellistor gas sensor typically consists of two ceramic "pellets" or beads, each with an embedded platinum coil. The coils serve as internal heaters to maintain the pellistors at the appropriate operating temperature. The resistance of the embedded platinum coil also functions as an internal resistance temperature detector (RTD) (see Chapter 5). Most commonly, the beads are made of porous γ-alumina. One bead is activated by impregnating it with a suitable catalyst that preferentially catalyses the oxidation of hydrogen. Typically the catalyst is a noble metal such as platinum or palladium. The second bead has no catalyst on its surface and thus it is inactive such that a combustible gas will not oxidise on its surface (i.e., both beads are below the autoignition temperature, but only the activated bead will induce catalytic oxidation). The inactive bead acts as a compensating or reference element. Changes in the ambient temperature, relative humidity, or the thermal conductivity of the gas being monitored will have similar influence on both beads. Thus any impact that changes in these parameters have on the sensor output can be compensated. Since exothermic combustion will only occur on the active pellistor, its temperature will increase relative to the inactive reference pellistor upon exposure to hydrogen. Typically, the resistance of metals increases monotonically with temperature. Thus, the resistance of the embedded coil will increase when hydrogen is oxidised (combusted) on the surface of the active pellistor. Conversely, the temperature of the reference bead and hence the resistance of the embedded coil remains unchanged. The change in the relative resistance between the two platinum coils is typically measured using a balanced Wheatstone bridge. The two platinum coils form two arms of a Wheatstone bridge circuit. The Wheatstone bridge becomes imbalanced when the active element is exposed to hydrogen. To a good approximation, the net response is linearly proportional to the hydrogen concentration, at least up to the lower flammability limit (LFL). Figure 3.1 illustrates a schematic diagram of the typical configuration of such a catalytic pellistor Wheatstone circuit.

Traditionally alumina-based pellistor sensors are produced by immersing a platinum coil into an aqueous solution of aluminium nitrate, $Al(NO_3)_3$, or aluminium hydroxide, $Al(OH)_3$.

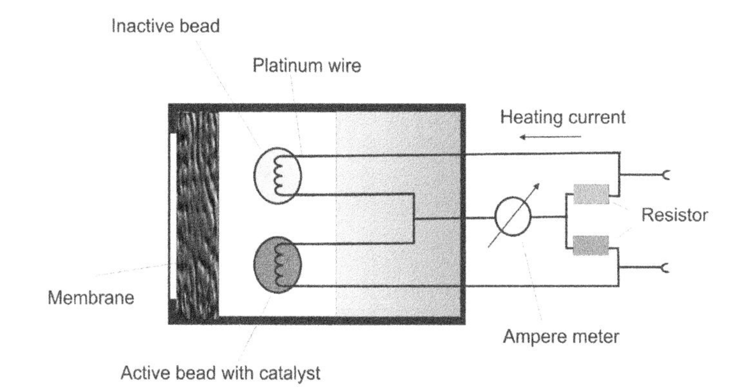

Figure 3.1: Schematic of catalytic pellistor sensor adapted from [102].

A current is subsequently passed through the coil, heating the aluminium compound and converting it to alumina (Al_2O_3) in a process called calcination. The immersing and heating steps are repeated until a bead with the desired dimensions is formed. In addition to its high thermal stability and exceptional chemical resistance, alumina is selected because it has a high thermal conductivity, which ensures uniform temperature across the pellet and good heat transfer between the heating coil and the catalyst. The catalyst is typically applied to the active pellistor as a solution of the catalytic material after which it is heated in air to dry and activate it [103, 104].

Figure 3.2 shows images of the active and inactive beads of a commercial hydrogen pellistor sensor taken with an optical microscope at two different magnifications. The beads are approximately 0.5 mm in diameter and the images show a distinct difference between the inactive pellistor (on the left), which appears as a heterogeneously coloured conglomerate of particles, and the active pellistor (on the right), which has a more homogenous silver/grey coloured surface attributable to the deposited catalyst layer (a mixture of platinum and palladium in this case).

The discovery and initial development of catalytic sensors is credited to Jonson as early as 1923 [106]. This type of sensor has been widely used for methane detection in mines [107]. Pellistors evolved from bare platinum coils that functioned as the heater, the catalyst and the temperature measuring device. However, as a bulk metal, the catalytic activity of platinum is not very high for oxidation reactions and for this reason it was necessary to heat the coil to temperatures exceeding 800 °C to 1000 °C to catalyse the desired reaction [108]. Heating to this high temperature was detrimental to the sensor lifetime due to evaporation of the platinum, which resulted in the gradual degeneration of the coil. Coil degeneration led to a gradual drift in the baseline response of the sensor. At the end of the 1950s Baker developed and patented the first real pellistor sensor by embedding a platinum coil in an alumina or silica pellet [103]. In addition to providing mechanical support to the delicate coil, the pellet also acted as a substrate upon which fine catalyst particles, with a very high surface area, were deposited thereby allowing operation at lower temperatures. This change in sensor design also reduced drift in the baseline resistance that plagued the simple coil design.

The use of pellistors specifically for detection of hydrogen was first reported by Jones et al. [109] in 1989. Beads of alumina were used to support palladium-based catalysts and a measuring range up to 2 vol% hydrogen in air and an operating temperature of circa 200 °C was achieved. Pellistor response, equivalent to the out-of-balance potential across the Wheatstone bridge, has been shown to be dependent not only on the concentration c_B,

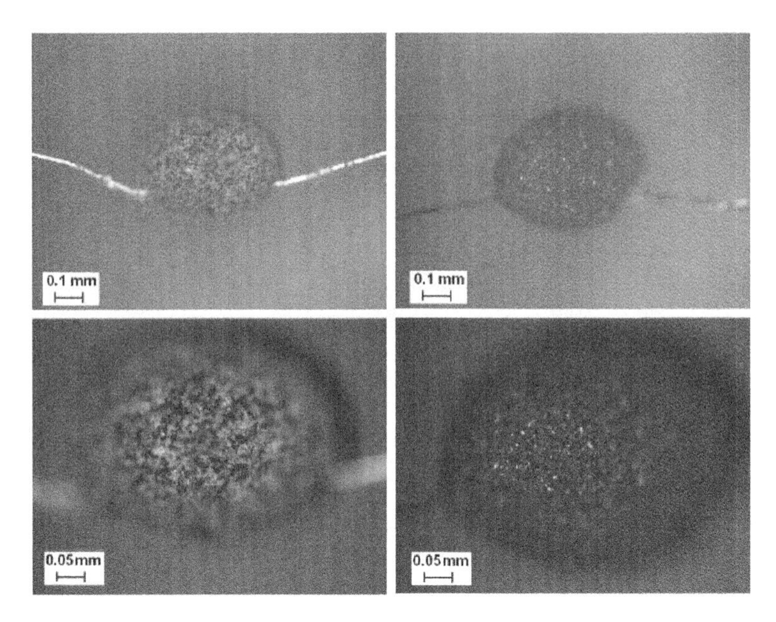

Figure 3.2: Optical microscope images of the inactive bead (left) and catalyst coated active bead (right) of a commercial hydrogen pellistor sensor [105].

and the heat of combustion of the gas, $\Delta_c H^\circ$, but also on its diffusion coefficient, D, [110] according to:

$$V = \mathrm{K} \cdot \Delta_c H^\circ \cdot D \cdot c_\mathrm{B} \tag{3.1}$$

Where K is a constant.

One issue associated with using pellistors for hydrogen detection is their inherent cross-sensitivity to other combustible gas species such as hydrocarbons and carbon monoxide. Strategies to improve sensor selectivity to hydrogen involve the use of catalysts noted for their high activity to hydrogen, such as palladium or platinum. In addition, chemical filters or molecular sieve coatings have been used to hinder the diffusion of other combustible gases to the sensor surface. Deposition of a polytetrafluoroethylene (PTFE) film by radio frequency (RF) sputtering on to the sensor surface has been performed with success for this purpose and also to reduce diffusion of species which could poison or inhibit the activity of the hydrogen sensor's catalyst [111]. The PTFE provided resistance to hexamethyldisiloxane (HMDS) and iodine which are typical poisons for catalytic-type sensors.

The PTFE film was shown to have a negligible effect on the sensor response indicating that it did not hinder the diffusion of hydrogen or oxygen to the catalyst. Increasing the number of catalytic sites is another technique often deployed by manufacturers to decrease the impact of poisons and interferents on sensor performance. This can be done by simply increasing the size of the pellet or by the use of a porous or finely divided catalyst [112], which increases the surface area of the catalyst and favours the oxidation of gases, such as hydrogen, with a higher diffusivity.

Pellistor sensors are typically deployed for the detection of combustible gases. Since the sensor operates at high temperatures the potential for pellistors to ignite the gases at flammable concentrations is a concern. For this reason the heated beads of catalytic pellistor sensors are mounted behind a flame arrestor which not only prevents hot surface ignition of the gas mixture but also provides protection to the delicate sensor structure. Figure 3.3

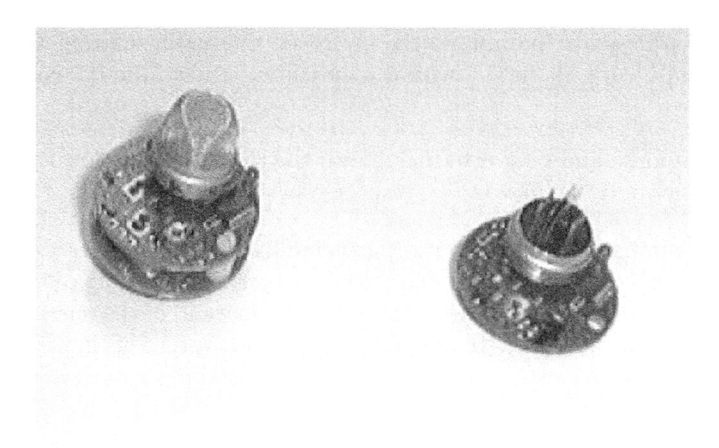

Figure 3.3: Photograph of a catalytic pellistor sensor with the metallic mesh enclosure in place on the left and with the mesh removed on the right [105].

shows a photograph of a catalytic pellistor hydrogen sensor with a stainless steel mesh flame arrestor intact and an identical sensor with the flame arrestor removed. Detectors with catalytic sensors certified for operation in flammable environments are commercially available.

In recent years the application of microfabrication techniques for the development of so-called planar pellistors have dominated the literature on catalytic pellistor hydrogen sensors [113]. Microfabrication techniques facilitate miniaturisation of the sensing element in addition to offering several other improvements that address some of the inherent shortcomings of this gas sensing technology [114]. Gas microsensors, which were first reported in the early 1990s [115], typically have micrometer-scale dimensions and are manufactured using techniques developed and widely applied in the silicon microfabrication industry. Zanini et al. [116] describe the sensor design, fabrication process, and performance characteristics of a combustible gas micropellistor with a platinum catalyst. Micropellistors are usually fabricated using a silicon substrate on which a silicon nitride passivation layer is formed. Two platinum resistors are patterned out of a thin platinum layer deposited on the substrate by an electron beam or sputter deposition. The silicon nitride layer acts as a thermal insulation between the resistor and the silicon substrate. A layer of chromium or titanium is usually deposited before the platinum resistor to improve adhesion of the platinum to the substrate. A membrane layer of silicon nitride is formed over both resistors - the thickness of the membrane is minimised to reduce heat loss from the heater and to facilitate very fast response times and low power consumption. A catalyst material (platinum or palladium) is deposited over one of the resistors [117, 118]. Highly sensitive micropellistor hydrogen sensors have also been prepared using an alumina substrate with platinum and palladium-doped metal oxide catalysts [119, 120]. Reported improvements in performance metrics of micropellistor hydrogen sensors [113] include:

- Lower power consumption due to the small size (200 μm $\times$ 200 μm sized heater) and low thermal mass of the microsensing element ($<$ 60 mW compared to 300 mW - 1000 mW for conventional catalytic pellistors). Heat loss by thermal conduction is minimised by etching away the substrate underneath the heater creating a suspended, free-standing structure.

- Higher sensitivity attributable to the improved thermal insulation and reduced heat losses (40 mV/vol% H_2 in air) within a measuring range from 0.2 vol% to 2 vol% H_2 in air.
- A rapid response and recovery time due to the small size and low thermal mass (t_{90} and t_{10} times of 0.36 s and 1.29 s, respectively, or faster, are achievable).

Micromachined planar pellistor sensors have additional advantages. Manufacturing of conventional catalytic pellistors is a resource-intensive process involving coiling and electrical connection of the delicate resistive wire, deposition (calcination) of the refractory bead on the wire, and deposition of the catalyst material onto the bead. Variations in pellistor-to-pellistor dimensions, electrical properties, and thermal properties are unavoidable and this negatively impacts sensor-to-sensor repeatability and reliability. High precision microfabrication techniques offer improved device repeatability while batch fabrication has the potential of providing large volumes of devices at lower unit cost. Cost is an important consideration in a future hydrogen inclusive economy as large numbers of these sensors will be required in markets which are very price sensitive.

However catalytic micropellistors are not without their drawbacks. Independent testing of one model type of hydrogen catalytic micropellistors has highlighted the occurrence of microstructural changes in the surface layer of the active pellistor following repeated exposures to hydrogen in air mixtures [105]. Observed blistering, cracking, and detachment of the catalytic layer from the resistive filament was attributed to the repeated expansion of the catalyst layer when exposed to hydrogen giving rise to mechanical stresses which resulted in the surface changes. There was a corresponding degradation in performance, particularly a loss of sensitivity. Shortcomings were also observed with respect to mechanical robustness, as several micropellistor units failed to give any response because the suspended heater filament structure was damaged, as verified by optical microscopy. As an active area of research and development, manufacturers are continually improving the performance of micromachined catalytic sensors whereby instability issues, previously observed, may be resolved in newer sensor designs.

Although there are now numerous models of micropellistors currently commercially available [121], the market for hydrogen detection is still dominated by conventional catalytic pellistors. The broad selection of commercial catalytic pellistor sensors and their widespread deployment in existing applications using hydrogen are a testament to their technology maturity level. A comparison of key performance specifications of commercial products [122] shows that catalytic pellistors are used for measurement of hydrogen concentrations up to 4 vol% H_2 in air, see Figure 3.4. Above this concentration the sensor's response is no longer linear to changes in the hydrogen concentration and the response eventually becomes saturated. Catalytic pellistor sensor response times range between 8 s to 30 s; however, micropellistor products claim significantly faster response times (ca. 1 s) [122]. The inactive or reference bead minimizes the impact of changes in ambient temperature and relative humidity and some products have specified operating ranges of -40 °C to +80 °C, 0 % RH to 100 % RH, and 800 hPa to 1200 hPa. The high temperature of the sensing element suggests that operation under condensing but otherwise normal ambient conditions should not pose problems for pellistor use. However the sensor circuitry should be protected against such conditions.

Generally speaking, operation of pellistor sensors requires power in the range 0.3 W to 1 W, which is considerably greater compared with other hydrogen sensing technologies—this is of course related to the need to heat the catalyst to its operating temperature. However, commercial micropellistor products provide superior performance in this respect with claims

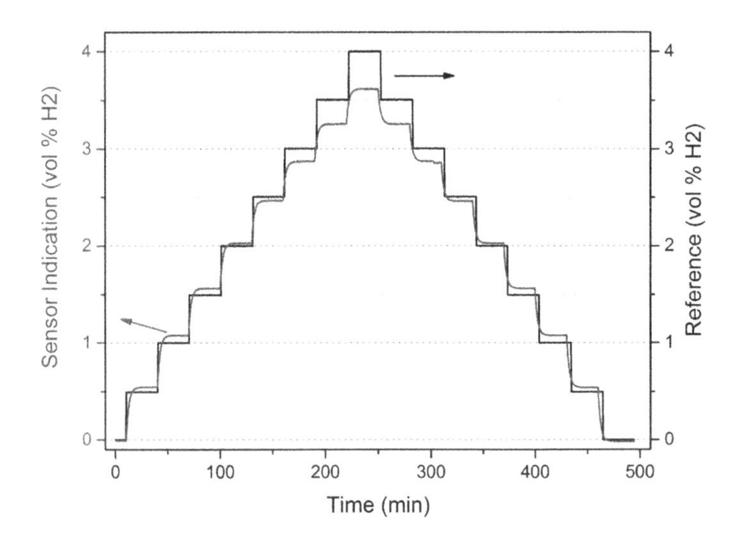

Figure 3.4: Response of a commercial catalytic pellistor to stepwise increase and decrease of hydrogen concentration in air between 0 vol% and 4 vol%.

of power consumption rates of 30 mW or lower. There are other considerations associated with the use of pellistor sensors. One major concern is the potential deleterious effect other gaseous species can have on the functionality of the catalyst which can reduce the operating lifetime of the sensor or permanently affect its operation. For this reason regular calibration is required to correct for signal drift and change in sensitivity. The recommended calibration intervals are often provided by the sensor manufacturer. Furthermore, catalytic sensors require a stoichiometric amount of oxygen for the reaction with hydrogen to operate correctly. Operation at lower oxygen concentrations or under anaerobic conditions can permanently alter the response of many commercial pellistors [123]. This precludes the use of pellistor sensors in applications where operation under oxygen-deprived conditions may be required, even if these conditions are temporary.

3.2.2 Thermoelectric

Thermoelectric sensing is another platform which exploits the heat released from a catalytically driven exothermic reaction between hydrogen and oxygen for the detection of hydrogen. Similar to pellistor catalytic sensors, the oxidation reaction occurs on the surface of a catalytic material and results in a temperature rise of the catalyst and the supporting material. The temperature rise is converted into an electrical signal by a temperature sensing element, which for pellistor catalytic sensors is a platinum resistance wire. Thermoelectric sensors, on the other hand, use the thermoelectric or Seebeck effect for this purpose. The thermoelectric or Seebeck effect converts a temperature difference across a thermoelectric material directly into a voltage. Tin oxide and other similar metal oxide semiconductor materials demonstrate a Seebeck effect and are used in thermoelectric gas sensors. A temperature gradient across the thermoelectric material is achieved by covering part of its surface with a catalyst. When the sensor is exposed to a combustible gas, the gas is oxidised on the surface of the catalyst, raising its temperature. A thermoelectric voltage is created between the "cold" and the "hot" region of the material and the voltage is, to a very good approximation, proportional to the temperature difference between the two regions/junctions according to the relationship:

$$V = \alpha(T_{hot} - T_{cold}) \tag{3.2}$$

Where V is the thermoelectric voltage, α is the Seebeck coefficient and T_{hot} and T_{cold} are the temperatures of the hot and cold regions, respectively.

The thermoelectric effect is widely used for temperature measurement as it is the physical basis for the operating principle of thermocouple devices (see Section 5.1.5). The thermoelectric principle was first reported for gas sensing by McAleer and co-workers in 1985 [124]. Their work described a novel gas sensor in which a thermovoltage was generated across a porous semiconductor when one region was heated by the catalytic combustion of a flammable gas. The flammable gas was hydrogen, which was oxidised on the surface of a platinum–palladium mixed catalyst deposited on a thermoelectric porous tin-oxide pellet. The authors cite that an advantage of their sensor is its ability to operate without requiring an external heat source—an important consideration for applications requiring battery powered devices (e.g. for mobile and/or remote gas detection). Experimental data was used to derive the Seebeck coefficient for tin oxide and was estimated to be 180 $\mu V/K$. This value substantially deviated from the expected value (around 930 $\mu V/K$) which McAleer attributed to the fact that the sensor design and microstructure were not optimised. Despite this, the sensor developed under this pioneering work demonstrated the feasibility of using the thermoelectric effect for gas detection and specifically its suitability for hydrogen detection.

A number of research groups continue to develop and optimise this platform for hydrogen detection. Optimisation has taken several forms including the use of improved thermoelectric materials, thin film designs, doping of the thermoelectric material, and miniaturisation. Researchers at the National Institute of Advanced Industrial Science and Technology (AIST) in Japan have led these efforts. Their initial research highlighted performance improvements of thermoelectric sensors using a film design compared with McAleer's pellet design. Sensor film design strategies included deposition of thick thermoelectric material films on a substrate [125] by screen printing and deposition of a thin thermoelectric film [126] by radio frequency (RF) sputtering. The thin film design is preferred due to better adherence of the thermoelectric material to the substrate, a faster response time, and a high signal to noise ratio even at ambient temperatures. A sensor based on a lithium-doped thin nickel oxide (NiO) film had a response time of 36 s compared to 55 s for the thick film counterpart. Platinum has been used exclusively as the catalyst in thermoelectric sensors due to its preferential catalysis of hydrogen oxidation at low temperatures (< 100 °C). The main thermoelectric materials of interest are nickel oxide (NiO) and silicon-germanium (SiGe). Alkali dopants (lithium and sodium) have been used to control the conductivity of the thermoelectric material enhancing the response of the sensor [127].

In recent years, this technology has been shown to be conducive to miniaturisation [128]. A micromachined thermoelectric sensor utilising a microhotplate enables rapid changing of the operating temperature with reduced power consumption (0.34 W), fast response times (45 s), a linear measuring range between 0.01 vol % to 3.0 vol % hydrogen and an operating temperature of 100°C. Further improvements to the thermoelectric sensor have been realised with an extremely wide linear measuring range reported from 0.5 ppm to 5.0 vol % [129]. Other groups followed this thermoelectric sensor design using bismuth-telluride as the thermoelectric material and γ-Al_2O_3 supported platinum as the catalyst. Four bismuth-telluride "legs" were arranged in a novel design and connected in series to enhance the total Seebeck coefficient; this new sensor design could detect 0.5 vol % to 3.0 vol % hydrogen in air at room temperature with a response and recovery time of 30 s and 55 s, respectively.

Figure 3.5: Photograph of a commercial thermoelectric hydrogen sensor, courtesy of NAST Co., Japan [132].

Investigations into the application of microfabrication techniques suggest that this is feasible for mass production of microthermoelectric hydrogen sensors [130]. Furthermore a prototype thermoelectric hydrogen sensor module has been developed and field tested for one year at a hydrogen fueling station in Japan [131]. This sensor showed a linear response to hydrogen in the concentration range of 0.001 vol% to 4 vol % with a response time of < 2.5 s. Despite being a promising technology, compatible with microfabrication, commercial thermoelectric hydrogen sensors are not commonplace. However a thermoelectric hydrogen sensor, shown in Figure 3.5 and fabricated by NAST Co., has been commercially available in Japan since 2011.

3.2.3 Pyroelectric

The pyroelectric effect is the ability of certain materials to generate an electric voltage when they are heated or cooled. Pyroelectricity is derived from the Greek term for "heat" and arises from electric polarisation in some ferroelectric materials when they are subjected to a change in temperature. Ferroelectric materials exhibit the pyroelectric effect such that when the temperature of such materials is changed, polarization will either increase or decrease, giving rise to a charge displacement. The pyroelectric effect can be exploited in gas sensors to convert the temperature changes associated with the combustion of the analyte into an electrical signal. The determining value for the extent of polarization is called the pyroelectric coefficient. Polarisation decreases at temperatures approaching the Curie temperature of the pyroelectric material. The Curie temperature is the temperature above which the polarisation drops to zero, making this property an important consideration in the choice of pyroelectric material for sensors since the Curie temperature of the piezoelectric material limits the maximum operating temperature of the device. Pyroelectric sensors for hydrogen detection have been reported in the literature albeit it to a much lower extent compared with catalytic pellistor or thermoelectric sensors.

In 1984 Hall published results on the detection of low concentrations (0.03 vol%) of hydrogen in air [133]. The device used a lead zirconate ceramic doped with Fe, Nb, Ti, and U as the pyroelectric material and black platinum as the catalyst. One year later D'Amico and co-workers reported the detection of 0.01 vol% hydrogen using lithium tantalate ($LiTaO_3$) as the pyroelectric material and palladium as the catalyst. More recently a pyroelectric sensor fabricated by silicon micromachining techniques was described, which reduced the thermal inertia of the sensing element and increased the sensitivity to hydrogen [134]. A thin film of ferroelectric lead zirconate titantate (PZT) was used as the dielectric material. Platinum clusters were deposited onto the active surface of the PZT using a micropipette. When operating at room temperature the sensor demonstrated a near linear response to hydrogen concentration in the range of 0.5 vol% to 3.5 vol% with no signs of saturation. The sensor could be modified to broaden the detection spectrum of combustible gases by integrating a microheater into the structure and operating at higher temperatures. Of course the selectivity to hydrogen shown by this sensor when operating without a heater illustrates a clear advantage of this technology which could have important ramifications for its further development and commercialisation—namely a simple sensor design, ease of fabrication, and low power consumption during operation. Despite these potential advantages pyroelectric hydrogen sensors have not been widely developed within the research community and no known commercial products are available.

3.3 THERMAL CONDUCTIVITY HYDROGEN SENSORS

Thermal conductivity is a measure of a material's ability to transfer heat. When a hot and a cold body are separated by a material, the thermal conductivity of that material determines the heat energy transfer rate between the bodies. For a given temperature difference, materials with higher thermal conductivity will transfer the heat at a faster rate compared with materials of lower thermal conductivity. Fourier's Law relates the rate of heat transfer by conduction to the thermal conductivity of the material through which the heat is transferred. Fourier's Law is given by:

$$q = \lambda \cdot A \cdot \frac{dT}{dx} \tag{3.3}$$

Where q is the conductive heat transfer rate, in W, A is the cross sectional area perpendicular to the direction of heat flow, in m^2 and dT/dx is the temperature gradient in the direction of heat flow, in K/m. The thermal conductivity, λ, is typically expressed as W/mK, however another commonly used unit in the US is Btu/ h·ft·°F.

Thermal conductivity is a temperature dependent transport property which has been widely used for gas detection and measurement for over a century [135]. The technique is based on the comparison of the thermal conductivities of different gases. Table 3.3 lists the thermal conductivities of a number of gases and illustrates the variation of this property with temperature.

Thermal conductivity sensors, also known as katharometers, can sense changes in the thermal conductivity of a (pseudo-)binary gas mixture. When the mixture comprises gases with sufficiently different thermal conductivities, measurements can also elucidate the concentration of the components, provided the system has been calibrated for the specific (pseudo-)binary mixture. Thermal conductivity sensors often work by comparing the thermal conductivity of the gas mixture to be measured with that of a reference gas. Figure 3.6 illustrates the components of such a thermal conductivity sensor. The sensor consists of two cells—a measurement cell and a reference cell. The gas to be measured flows through the measurement cell while the reference cell is sealed and contains the reference gas.

Table 3.3: Thermal conductivities of various gases at different temperatures [7]

	Thermal conductivity, λ /(W/m · K) × 10^{-3}				
Gas	200 K	300 K	400 K	500 K	600 K
Hydrogen	131.7	186.9	230.4	280	–
Air	18.4	26.2	33.3	39.7	45.7
Nitrogen	18.7	26.0	32.3	38.3	44.0
Oxygen	18.4	26.3	33.7	41.0	48.1
Helium	119.3	156.7	190.6	222.3	252.4
Carbon monoxide	–	25.0	32.3	39.2	45.7
Methane	22.5	34.1	49.1	66.5	84.1
Ammonia	–	24.4	37.4	51.6	66.8
Carbon dioxide	9.6	16.8	25.1	33.5	41.6

Each cell contains a heated thermoresistor element which may be a thin coiled wire or an inert pellistor (pellet resistor), similar to the inactive pellistor of a catalytic pellistor sensor (see Section 3.2.1). An electrical current is passed through each thermoresistor element to heat them to the same operating temperature. When the test gas has the same thermal conductivity as the reference gas both heated elements lose heat at the same rate so that their temperature and hence their electrical resistance is equivalent. Under this condition the Wheatstone bridge is balanced and the sensor shows a zero response. When a test gas, whose thermal conductivity is different to that of the reference gas, flows through the measuring cell the measurement element will lose heat at a different rate compared to the reference element. Consequently the temperature of the measuring pellistor will rise or fall resulting in a change in the resistance of the thermoresistor element. This resistance change is detected as an imbalance in the Wheatstone bridge. In the two cell design, effects on sensor response due to changes in the temperature of the sensor housing can be compensated when such temperature changes affect both the measuring and reference elements equally. However sensor response can also be affected by changes in gas flow (convection) over the measuring thermoresistor element with higher flow rates enhancing cooling of the thermoresistor. This principle is comparable to the operation of a thermal mass flow sensor (see Section 5.3). The effect is not compensated in the two cell sensor design where gas only flows over the measuring element and not the reference element. There have, however, been efforts to mitigate the influence of convective effects by adopting designs where gas diffusion dominates as the gas transport mechanism to the sensing element [136]. In many cases such designs can result in slow response times.

Relative to most other gases hydrogen has a very high thermal conductivity, see Table 3.3. When air is used as the reference gas, this technique is very effective for detecting hydrogen in air because of the significant difference in their thermal conductivities. This makes it possible to detect relatively low concentrations of hydrogen in air. Conversely gases with thermal conductivities similar to air cannot be detected easily, e.g., ammonia, oxygen, nitrogen, and carbon monoxide. A thermal conductivity device was patented for continuous determination of hydrogen content in gas mixtures for airship applications already in 1904 [137]. The properties of katharometers and aspects of their practical application to the analyses of binary and multi-component gas mixtures, including hydrogen, have been described in the mid-1960s [138].

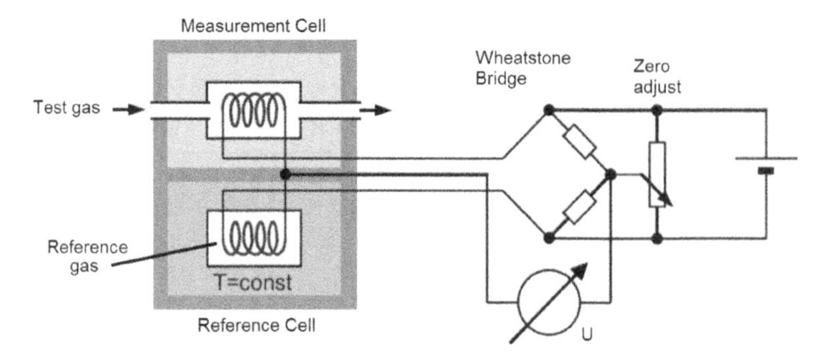

Figure 3.6: Schematic of a thermal conductivity sensor for gas concentration measurement showing the measurement and reference cells connected in a Wheatstone bridge circuit (adapted with permission from [102]).

Thermal conductivity sensors measure the mean thermal conductivity of a mixture consisting of several gaseous components. The averaged thermal conductivity, λ_m, can be given in an approximation as:

$$\lambda_m = \sum_{i=1}^{n} = \chi_i \cdot \lambda_i = \chi_1 \cdot \lambda_1 + \chi_2 \cdot \lambda_2 + \chi_3 \cdot \lambda_3 + \cdots + \chi_n \cdot \lambda_n \tag{3.4}$$

Where χ_i is the molar ratio and λ_i the thermal conductivity of each gaseous component. For gas analysis in many cases only the relative changes in comparison to a reference gas, e.g. air, are considered.

Using χ_{air}=1-χ_{H2}, for a pseudo two component system with hydrogen it gives:

$$\begin{aligned} \frac{\lambda_m}{\lambda_{air}} &= \lambda_m^{rel} = \chi_{air} + \chi_{H2}\frac{\lambda_{H2}}{\lambda_{air}} \\ &= (1-\chi_{H2}) + \chi_{H2} \cdot \lambda_{H2}^{rel} = 1 - \chi_{H2} \cdot (1-\lambda_{H2}^{rel}) \end{aligned} \tag{3.5}$$

and

$$\chi_{H2} = \frac{\lambda_{air}^{rel} - 1}{\lambda_{H2}^{rel} - 1} \tag{3.6}$$

Defining the sensor signal as S=(λ_{air}^{rel}-1), it gives

$$S = \chi_{H2} \cdot (\lambda_{H2}^{rel} - 1) \tag{3.7}$$

Equation 3.7 gives a linear relation between a sensor signal and the hydrogen concentration. However the determination of further components, e.g., CO_2, in the gas mixture becomes challenging:

$$\begin{aligned} \lambda_m^{rel} &= (1-\chi_{H2}-\chi_{CO2}) + \chi_{H2}\frac{\lambda_{H2}}{\lambda_{air}} + \chi_{CO2}\frac{\lambda_{CO2}}{\lambda_{air}} \\ &= 1 - \chi_{H2} \cdot (1-\lambda_{H2}^{rel}) - \chi_{CO2}(1-\lambda_{CO2}^{rel}) \end{aligned} \tag{3.8}$$

The sensor signal is then:

$$S = \chi_{H2}(1-\lambda_{H2}^{rel}) - \chi_{CO2}(1-\lambda_{CO2}^{rel}) \tag{3.9}$$

This consideration shows that complex gas mixtures are difficult to analyse by thermal conductivity sensors. Although the thermal conductivity of carbon dioxide is 10 times lower than the thermal conductivity of hydrogen, it can be demonstrated that a thermal conductivity sensor would indicate about 0.065 vol% less hydrogen for every percent CO_2 that is present in a hydrogen in air mixture.

Designing and manufacturing two cell thermal conductivity sensors, which can make consistent measurements, present a number of challenges. Firstly, it is difficult to produce two cells with sufficiently similar characteristics to form a bridge circuit that can be balanced without the need for static and dynamic correcting circuit elements. Furthermore the two cells should be maintained at the same temperature as far as possible; however, this is difficult to achieve when the heated filament of the measurement cell inherently varies in temperature in order to provide the necessary bridge imbalance and sensor response. An alternative thermal conductivity sensor design uses a single cell which offers solutions to these challenges and which, as a result, has gained prominence in recent years.

This design, which has also been applied to hydrogen detection, comprises a single platinum heater structure (the "hot" element) from which heat is transferred to a "cold" element [139]. The extent of heat transferred depends on the thermal conductivity (and hence composition) of the gas flowing between the two elements. Sensor signal transduction is calculated from the power required to maintain a constant temperature difference between the "hot" and "cold" elements. This sensor design has been successfully integrated into silicon-based micromachined sensors and often integrates separate temperature and relative humidity measurement devices to compensate for changes in these ambient parameters [140]. Figures 3.7 and 3.8 show photographs of a commercial micromachined thermal conductivity hydrogen sensor. It includes a Pt100 temperature sensor and a humidity sensor and features an Application-Specific Integrated Circuit (ASIC) that measures and biases the sensing elements, and communicates digitized values to a microcontroller. The microcontroller handles the measurement protocol, calculates various parameters, including the hydrogen concentration, and communicates this externally via RS-232.

Thermal conductivity sensors are one of the few hydrogen detection technologies which do not depend on a chemical reaction or interaction between hydrogen and the sensing material to operate. Instead detection is based on a physical property of hydrogen gas. This fundamental difference in the detection principle gives rise to important differences in the performance of thermal conductivity sensors relative to other sensor classes and demands specific considerations for their deployment. For example, it is possible to use thermal conductivity sensors for measurement of hydrogen in gas matrices which do not contain oxygen, e.g., in nitrogen. Unlike many other sensing principles the operating principle of thermal conductivity sensors does not require the presence of oxygen. It is however important to calibrate the sensor in the gas matrix in which it will measure hydrogen concentration. The sensor response needs to be corrected to account for differences in the thermal conductivities of the reference gas and the background matrix gas. In a different matrix the zero-reading baseline can shift due to the difference in thermal conductivities. This characteristic of thermal conductivity sensors is illustrated in Figure 3.9 where it shows a decreasing sensor baseline response for background gas matrices with decreasing air to nitrogen ratios. This effect is due to the slightly lower thermal conductivity of nitrogen relative to air which, at decreasing air ratios, lowers the heat loss from the measuring pellistor causing a drop in sensor response. As can be seen from Figure 3.9, however, the net sensor response (defined as the sensor response curve final indication minus the sensor response curve baseline) to the 1 vol% hydrogen target gas remains unaffected. For this reason calibration of the thermal

Figure 3.7: Photograph of the XEN-TCG3880 thermal conductivity gauge with Pt100 temperature sensor, used for the XEN-5310 hydrogen sensor (photograph courtesy of Xensor Integration B.V. [141]).

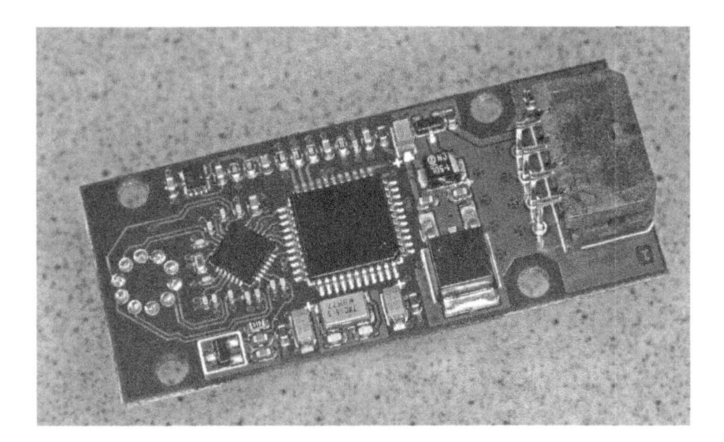

Figure 3.8: Photograph of the XEN-5310 hydrogen sensor, with a XEN-TCG3880 mounted on the reverse side. This side includes a humidity sensor, an analog-to-digital signal processing ASIC, and a microcontroller. XEN-5310 has analog + digital output (photograph courtesy of Xensor Integration B.V. [141]).

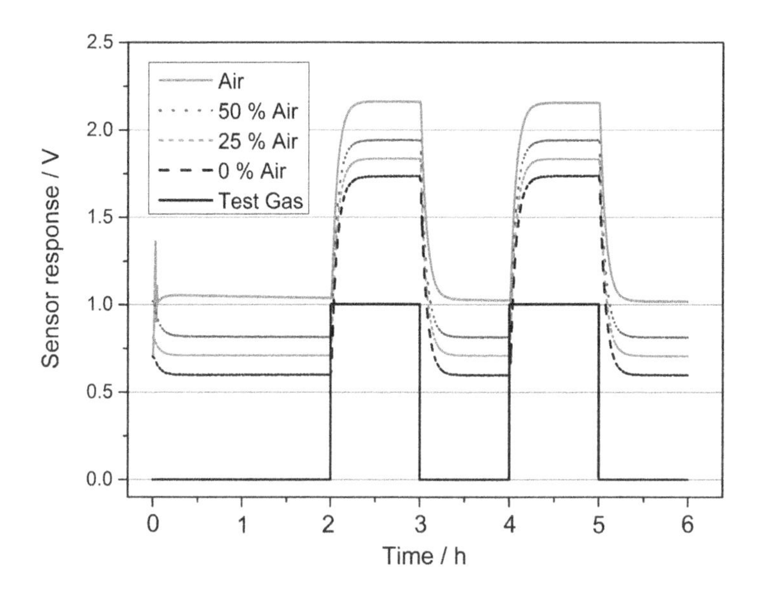

Figure 3.9: Response of a thermal conductivity sensor to 1.0 vol% hydrogen different background gas matrices (21 vol% oxygen, 10.5 vol% oxygen, 5.25 vol% oxygen, and 0 vol% oxygen - in all cases the balance gas was nitrogen). Adapted with permission from [123].

conductivity sensor's response to the target gas, in the background matrix gas in which the target gas is to be measured, is necessary to ensure an accurate concentration measurement.

Furthermore, since there are no catalysed reactions involved in the response mechanism of thermal conductivity sensors, it can be expected that the presence of trace impurities, such as H_2S or HMDS (which are often deleterious to the correct functioning of other sensor types) will not significantly influence the response of the thermal conductivity sensor. For this reason the long-term stability and lifetime of this class of sensor is often superior. Nevertheless the integrated relative humidity sensor may be affected by some interferents and so the overall device output cannot be considered completely immune to such species.

While sensitivity of a sensor to gas species other than the target gas is often undesirable in some specific cases it can be advantageous. As indicated previously the thermal conductivity sensor platform is not specific to hydrogen—it will measure changes in the thermal conductivity of any gas mixture which can then be correlated to its composition. Due to its relatively high thermal conductivity, when compared with most other gases, changes in hydrogen concentrations in a gas mixture will induce a relatively large sensor response. Referring to Table 3.3 it can be seen that the thermal conductivity of helium is somewhat similar to that of hydrogen—accordingly thermal conductivity sensors will have a significant cross-sensitivity to helium. In specific applications helium may be used as a surrogate gas to replace hydrogen, for example in vehicle safety crash tests or system leak-tightness checks. The cross-sensitivity of this sensor platform implies that it may be used to detect leaks, regardless of whether hydrogen or helium is used. The versatility of a thermal conductivity sensor in such a specific application has been successfully demonstrated [142].

During operation the sensing element of thermal conductivity sensors needs to be heated to allow heat flow measurement. Operating temperatures of below 300 °C are typical [140].

This temperature is less than the operating temperature of catalytic combustion sensors making thermal conductivity sensors comparatively less power hungry devices. Their lower power consumption can be an important consideration for applications requiring battery powered sensors (e.g., mobile gas sensing devices or remote sensing). Furthermore since their operating temperature is significantly less than the autoignition temperature of hydrogen in air [143], their use has additional safety benefits with respect to their deployment in potentially combustible environments.

In a recent market survey existing commercial off-the-shelf (COTS) hydrogen sensors have been identified [121]. The survey yielded some 400 sensing products suitable for hydrogen detection of which 16 % used the thermal conductivity principle. This places thermal conductivity sensors third in the ranking with respect to commercial availability, behind catalytic sensors and electrochemical sensors. Thermal conductivity hydrogen sensors are a popular choice in many applications due to their wide commercial availability and robust performance. Many commercial hydrogen sensing products incorporate this detection technology either exclusively or in combination with other sensing element types. When combined with another sensor type the advantages of one can be used to compensate for the disadvantages of the other. Such dual technology sensor products (e.g., combination of a thermal conductivity and metal-oxide sensor) allow tailoring of the hydrogen sensing technologies to meet the specific performance requirements of a particular application. Thermal conductivity sensors often have a relatively high lower detection limit (> 0.03 vol%) for hydrogen. However combining it with a sensing platform with accurate detection capabilities at low concentrations, e.g., metal oxide sensor, can yield commercial hydrogen detection devices with an exceptionally wide dynamic measuring range.

3.4 ELECTROCHEMICAL HYDROGEN SENSORS

Hydrogen is an electrochemically active chemical. More specifically it is generally considered to be a reducing gas which means that it tends to "donate" or lose its electrons to other molecules. Electron transfer can be directly between molecules or mediated by electrodes in an electrochemical cell. This property is exploited for electrochemical sensors. An electrochemical gas sensor is a device with two or more electrodes, which upon establishing an electrochemical interaction with a target gas, generates an electrical signal in response to its presence. The signal is most commonly a voltage or a current, the magnitude of which is dependent upon the gas concentration. Electrochemical gas sensors are one of the most common gas sensing platform types for hydrogen making up about 29 % of the market [121]. In addition to hydrogen, there are commercially available electrochemical sensors for a large number of analytes, including carbon monoxide, nitrogen oxides, oxygen, sulphur compounds, alcohols, aldehydes, and other gases, some of which are discussed in Chapter 4. There have been several recent general reviews on electrochemical gas sensors [144, 145] as well as reviews that focus specifically on hydrogen electrochemical sensors [102, 146].

There are two main electrochemical hydrogen sensor types: amperometric sensors and potentiometric sensors. There are also developmental electrochemical technologies for hydrogen detection, such as the mixed potential sensor, which are not yet commercially available but are actively being researched. Conductometric sensors, which are also called resistance sensors, are sometimes classified as electrochemical sensors [146]. The inclusion of conductometric sensors as an electrochemical platform is a bit arbitrary since the transduction mechanism to induce a change in conductance does not necessarily involve the transfer of charge from the analyte to the active sensing material as implied by an electrochemical classification. Accordingly, conductometric hydrogen sensors are treated separately in Section 3.5 of this book. This chapter will focus on potentiometric and amperometric hydrogen sensors. There will also be some discussion on emerging specialized platforms currently under development.

3.4.1 Amperometric

Amperometric sensors are by far the more common of the two main electrochemical sensor platform types with regards to hydrogen gas detection. Amperometric hydrogen sensors are commercially available from numerous manufacturers and are widely used in commercial gas detection equipment, especially for safety applications.

Structure of the Amperometric Gas Sensor

The structure and integration of the main components of an amperometric sensing element are shown in Figure 3.10, which also illustrates its basic operation. As in any electrochemical cell, an amperometric gas sensor structure comprises two or more electrodes immersed in an electrolyte. Electronic circuitry external to the sensing element is used to control the electrochemical parameters and to measure the sensor response. Unlike a typical electrochemical cell, amperometric gas sensors require a gas-electrolyte-electrode interface, the so-called *triple interface*.

This interphase is usually achieved with a gas diffusion electrode (GDE) structure for the working electrode (WE). The WE is where the electrochemical reaction of interest takes place, which for a hydrogen amperometric sensor is the oxidation of gaseous hydrogen to hydrogen ions dissolved in the electrolyte. For sensor applications, the WE is sometimes called the sensing electrode. The GDE consists of a noble metal catalyst placed on the inside surface of a porous (or permeable) hydrophobic membrane such that the catalyst is in contact with the electrolyte. The outside surface of the GDE is exposed to the environment being monitored for hydrogen, which is typically air. The target analyte must diffuse through the GDE to react on the WE. The hydrophobic nature of the GDE membrane material allows for gas diffusion to the WE, while at the same time it prevents the aqueous electrolyte from leaking to the outside surface. For hydrogen sensing, high surface area platinum black is usually selected for the WE, but the actual choice of catalyst material and its morphology is an active area of research [147].

Operation and Signal Generation

In amperometry, the analytical information is obtained through an electrochemical reaction of the target analyte on the WE. In this reaction the target molecules undergo a change in oxidation state, which involves the transfer of electrons between the molecule and the electrode, and can either be an oxidation, in which the target molecule "loses" one or more electrons, or a reduction in which the target molecule "gains" one or more electrons. The oxidation of one chemical must always be accompanied by the reduction of another within the same electrochemical cell. The "counter" reaction to the analytic reaction at the WE occurs at a second electrode called the counter electrode (CE); if an oxidation reaction occurs at the WE, then the CE reaction must be a reduction, and vice versa. The reactions at either the WE or CE are called the half cell reactions and the net cell reaction is the combination of the WE and CE half cell reactions. Thus, an electrochemical sensor consists of at least two electrodes. In practice, amperometric sensors often include a third electrode, called the reference electrode (RE). Although many amperometric sensors operate in a two-electrode mode, the addition of an independent RE provides significant flexibility and improved stability in the operation of the amperometric gas sensor. The

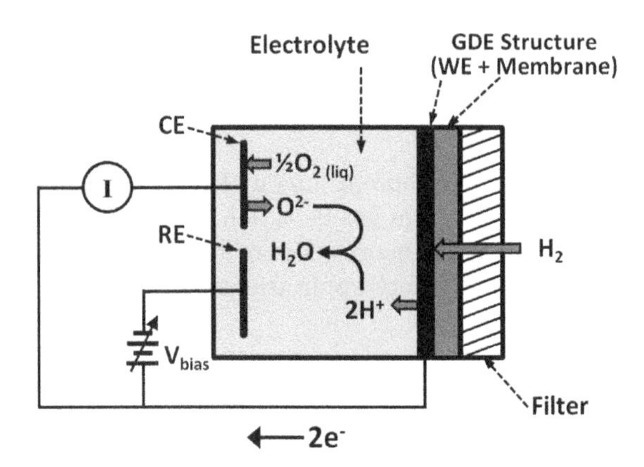

Figure 3.10: Illustration of an amperometric hydrogen sensor. The potential of the working electrode (WE) is set by a combination of the reference electrode (RE) potential and the applied potential (V_{bias}), while an amp meter, I, can be used to measure the sensor signal (i.e., the current between the counter electrode (CE) and the WE); these functions are best performed using a potentiostat circuit, see Figure 3.11.

three-electrode configuration is common for analytical applications. It is a fundamental electrochemical principle that the exchange of electrons between the chemical reactant and an electrode is controlled by their relative potentials [148]. The RE provides an independent and stable internal reference potential, which when coupled with an electrochemical control circuit called a potentiostat (see Figure 3.11) provides precise control of the potential of the WE. The potentiostat allows the potential of the WE to be precisely set by the end-user or by the sensor designers to optimize the amperometric sensor response to the desired electrochemical reaction. Although time dependent potential wave forms can be used to vary the potential of the WE, in most amperometric gas sensor applications the WE is typically maintained at a constant operating potential, which is selected so as to efficiently drive the electrochemical reaction of interest.[1] The establishment of sensor operational parameters is almost always performed by the sensor or instrument developers, and thus is not visible to the amperometric sensor end-user. However, the adjustment of the potential of the WE is one strategy employed by sensor developers to improve sensor selectivity. The potentiostat circuit, shown in Figure 3.11, also includes an amplifier circuit, which transduces the electrochemical current into a readily measured voltage. An overview of electronic control of sensor signals is discussed in Section 2.6.

Amperometric hydrogen sensors generally use a proton-conducting electrolyte. The electrolyte in amperometric gas sensors is often an aqueous system with dissolved salts (e.g., potassium nitrate) or dilute acids (e.g., sulphuric acid). Solid-state proton-conductors such as Nafion(a sulfonated fluoropolymer) have also been used as the electrolyte, although Nafion requires some water to function. Alternatively, solid-state perovskite structures containing cerium or zirconium have been shown to exhibit sufficient proton conductivity above 600 °C for use in amperometric hydrogen sensors [149]. Ionic transport in perovskite electrolytes is

[1]Electroanalytical methods based upon time-dependent potentials (e.g., cyclic voltammetry, linear range voltammetry, step potential) are routinely used in laboratories [148] These methods usually require instrumentation that is significantly more sophisticated than that afforded by conventional gas sensor technology.

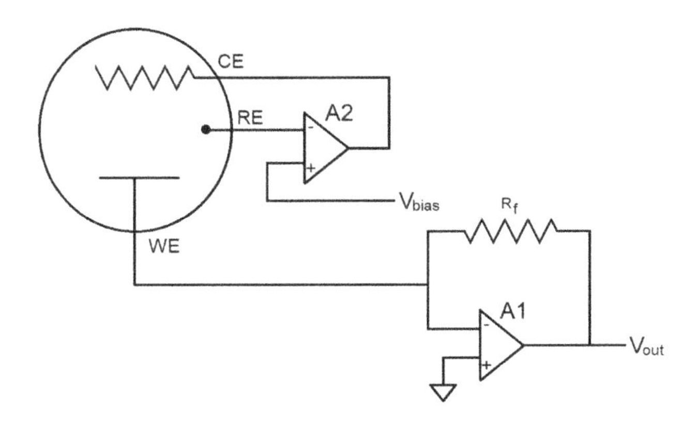

Figure 3.11: Simple potentiostat circuit for control of an amperometric sensor. The components labeled A are operational amplifiers. A1 controls the WE potential while A2 transforms the sensor current into a voltage. V_{bias} is user-controlled, but is most commonly set by the manufacturer. (Adapted with permission from Custom Sensor Solutions, http://www.customsensorsolutions.com.)

not limited to just protons but includes the various isotopes of hydrogen (e.g., hydrogen ^{1}H, deuterium ^{2}H with one proton and one neutron, and tritium ^{3}H with one proton and two neutrons), opening up the possibility of designing sensors for highly specialized applications, such as the analysis of tritiated gas streams, where a quantitative understanding of isotropic composition is important [150].

The use of aqueous electrolytes has a significant impact on the working temperature and humidity range for the sensors. Aqueous electrolytes have a typical working temperature range from approximately -10 °C to +40 °C. They will tend to dry out due to evaporation at higher ambient temperatures and will freeze if the temperature is below the lower limit. To prevent sensors from drying out hygroscopic salts are used, which will equilibrate with ambient humidity to maintain a proper electrolyte level within the sensor. Retention of the water and hence the electrolyte level within the sensor is necessary to ensure ionic contact is maintained between the electrodes. Sensor dehydration can still occur after prolonged operation under dry conditions (e.g., RH < 10 %), which can permanently damage the sensor. Also prolonged operation under high humidity levels, such as high relative humidity at elevated temperature or condensing conditions, can lead to excessive liquid water uptake by the internal electrolyte, resulting in a buildup of internal hydrostatic pressure that can physically damage the sensor housing. This can result in leakage of the often corrosive liquid electrolyte. For these reasons, operation of amperometric sensors with a liquid electrolyte is restricted to an relative humidity range from around 10 % or 20 % up to around 90 %. Although prolonged operation of the amperometric gas sensor at extremely low or high RH can adversely impact the structural integrity of the sensor, the sensor signal is not significantly affected by fluctuations in RH during regular operation. Even though water is still required, the working temperature range of amperometric sensors can be extended to about -40 °C to +80 °C using Nafion as a solid-state proton conductor. Alternatively, high temperature solid-state proton conductors such as the perovskite identified above do not have any water requirements nor are sensor signals significantly impacted by changes in ambient humidity. The high temperature proton conductors can also operate at elevated ambient temperatures. Furthermore, the use of solid-state conductors eliminates the risk associated with leakage of the liquid electrolyte. Although solid electrolytes offer some advantages relative to aqueous electrolytes, most commercial hydrogen amperometric sensors are still based on either an aqueous electrolyte or a solid electrolyte that requires the presence of some water (e.g., Nafion).

Linearity of the Sensor Response

As will be shown below, the electrochemical reaction at the WE generates an electrical current that is proportional to the gas concentration. Since the half cell reaction of hydrogen at the WE is an oxidation, the CE half cell reaction must be a reduction. The reduction reaction is driven by the electrons supplied by the oxidation of hydrogen at the WE and delivered to the CE through the external circuit. In an aqueous cell, one likely CE half cell reaction is the reduction of dissolved oxygen to produce oxide ions (an electrochemical process involving the transfer of electrons from the electrode to the oxygen molecule) that quickly react with hydrogen ions to produce water (a chemical process that does not involve the transfer of electrons):

$$\begin{array}{llll} WE: & 2\,H_2 & \longrightarrow & 4\,H^+ + 4\,e^- \\ CE: & O_2 + 4\,e^- & \longrightarrow & 2\,O_2{}^- \\ & 2\,O_2{}^- + 4\,H^+ & \longrightarrow & 2\,H_2O \\ Net: & 2\,H_2 + O_2 & \longrightarrow & 2\,H_2O \end{array}$$

The net cell reaction is identical to the direct gas phase reaction of hydrogen with oxygen. However, electrical energy is produced in an electrochemical cell while thermal energy is produced in the direct reaction.

In the most common design for an amperometric sensor, the rate of the half cell reaction at the WE is controlled by the diffusion of hydrogen through a planar membrane barrier associated with the GDE. Diffusion control produces a sensor signal that is proportional to the partial pressure of hydrogen, p_{H_2}. This proportionality can be demonstrated by considering electrochemical and mass transport processes. Faraday's Law states that the total charge, Q, that is passed through an electrode will be equal to the product of the number of moles of reacted chemical, N, the number of electrons involved in the reaction, z, and Faraday's constant, F:

$$Q = z\mathrm{F}N \tag{3.10}$$

For a hydrogen sensor, N is the number of moles of hydrogen reacted at the electrode and n is 2. The current, I, generated by the flux of charge, Q, is given by the differential form of Faraday's Law:

$$I = \frac{dQ}{dt} = z\mathrm{F} \cdot \frac{dN}{dt} \tag{3.11}$$

The flux of hydrogen through the gas permeable membrane to the electrode-electrolyte interface is controlled by diffusion processes. Specifically, for one-dimensional diffusion across a membrane, Fick's first law states that the flux of hydrogen is proportional to the concentration gradient of hydrogen across the membrane:

$$\frac{dN}{dt} = m_O[N_O^* - N_O] \tag{3.12}$$

Where N_O^* is the analyte concentration at the outer surface of the gas permeable membrane (e.g., the bulk ambient concentration normally given as p_{H_2}). N_O is the analyte concentration at the WE, and m_O is a constant that incorporates various constants associated with diffusion. The difference between N_O^* (or p_{H_2}) and N_O is an approximation for the concentration gradient across the membrane. It is assumed that all of the analyte that reaches the electrode will undergo electrochemical reaction such that the concentration of N_O is zero. Owing to the fast reaction of hydrogen oxidation on a platinum electrode and the negligible equilibrium concentration of hydrogen at the electrode (as predicted by the Nernst equation and controlled by the applied potential), this assumption is quite reasonable. Equation (3.12) reduces to:

$$\frac{dN}{dt} = m_O[N_O^*] \tag{3.13}$$

Combining Equation (3.11) and Equation (3.13) yields:

$$\frac{dN}{dt} = \frac{I}{z\mathrm{F}} = m_O[N_O^*] \tag{3.14}$$

Rearranging and combining constants yields a direct proportional relationship between the electrochemical current and the concentration of the analyte:

$$I = \mathrm{K} \cdot [N_O^*] \tag{3.15}$$

Thus, in a properly designed amperometric sensor, the rate of hydrogen oxidation is measured as an electrical current that is proportional to the concentration of hydrogen in the test gas. Since diffusion is a temperature dependent property, Equation (3.14) is valid only at a constant temperature and accordingly the sensor signal (current) will typically increase with temperature. The current can be conveniently measured as a voltage (V) drop across a resistor (R) that is externally connected between the WE and CE of the sensor. This feature is incorporated into the potentiostat circuit shown in Figure 3.12. From Ohm's law, Equation (3.14) becomes:

$$V = I \cdot R = \mathrm{K} \cdot R \cdot [N_O^*] = \mathrm{K}' \cdot [N_O^*] \tag{3.16}$$

In addition to hydrogen, numerous other gases and vapours, such as sulphur compounds, nitrogen oxides, alcohols, and aldehydes, are electroactive. Such species could undergo electrochemical reactions at the WE, and thus could interfere with the measurement of hydrogen. One strategy to minimize the cross-sensitivity of the sensor to other chemicals is to block their access to the WE. Many of the potential interferents can be effectively removed from the sample gas with a general purpose filter such as activated carbon. Such a filter is often attached to the outer surface of the GDE. The working life of the filter depends on the cleanliness of the ambient environment; a longer life can be expected for filters used in a clean environment (e.g., an environment with a low concentration of reactive gases) compared to environments that contain high concentrations of background vapours. Since hydrogen is not effectively absorbed by activated carbon, the use of the filter does not affect sensor performance. Activated carbon is, however, ineffective for trapping low molecular weight carbon compounds such as carbon monoxide (CO) or methane (CH_4). Methane, however, will not interfere with the hydrogen measurement using an amperometric hydrogen sensor because the operating potential of the sensor is too low for methane oxidation. Carbon monoxide on the other hand, can be problematic and is often an interferent for amperometric hydrogen sensors. Thus it can be difficult to make accurate hydrogen measurements in the presence of carbon monoxide using this type of sensor. Fortunately for safety applications, carbon monoxide must be present at extremely low levels (e.g., the Occupational Safety and

Health Administration's (OSHA) Permissible Exposure Limit (PEL) for carbon monoxide is 50 ppm), while allowable hydrogen levels can be as high as 4 vol%. However, even at low levels, carbon monoxide can block active catalyst sites and thereby adversely affect the calibration curve of an amperometric hydrogen sensor. Carbon monoxide binding on the WE will lower the sensitivity of the sensor to hydrogen, thus resulting in the need for a slightly higher hydrogen concentration than the actual set point concentration to activate an alarm; that is, the presence of carbon monoxide may cause a false negative alarm (see Section 6.4.2). The impact of carbon monoxide on an amperometric sensor is usually reversible and the sensor will typically recover its normal operational characteristics a short time after exposure has stopped. An alternative strategy to mitigate the effect of carbon monoxide was demonstrated by Chao et al. [151] using a thin, nonporous hydrogen permeable membrane, such as FEP (fluorinated ethylene propylene) Teflon® for the GDE. Hydrogen can readily diffuse through the FEP membrane, while other gases, including carbon monoxide, are effectively impermeable in this material. The use of the membrane essentially eliminated the impact of carbon monoxide on the sensor response. The sensor response time was, however, somewhat lengthened since the hydrogen diffusion was slowed with the incorporation of the FEP membrane.

Metrologic Performance

Amperometric hydrogen sensors are commercially available from numerous manufacturers. These are primarily based on aqueous electrolytes or with Nafion; sensors with solid-state high temperature electrolytes are not yet commercialized. The current commercial sensors typically have a range from 1 ppm to 10 ppm H_2 at the low end and 1 vol% to 4 vol% H_2 at the upper end. The sensor response is linear but often deviates from linearity at higher concentrations. The response also tends to be more stable and repeatable at lower hydrogen concentrations. The response time of the sensor is typically less than 30 s, with most manufacturers specifying around 15 s. Many sensor models, however, tend to drift during deployment and therefore need periodic, sometimes frequent, calibration. Furthermore, a new sensor will need up to 1 hour to stabilize upon powering up for its first time, but warm-up times are much shorter for subsequent power-ups, providing the unpowered state was for a limited duration, e.g., a short-term power outage.

Amperometric gas sensors are physically small, have good sensitivity and typically have a broad linear range. Operation of room temperature devices requires very low power. The sensors are stable with lifetimes of up to 2 years being routine, although this is considerably less than the United States Department of Energy (U.S. DOE) target for sensor lifetimes of 10 years [152]. Shortcomings include limited selectivity (e.g., CO may affect the sensor), a restricted temperature range due to the liquid electrolyte, and a dependence on barometric pressure. Alternatively, humidity fluctuations have a nearly negligible effect on the sensor signals, although extended operation at extreme low or high relative humidity can affect long-term stability. To some extent, the temperature and pressure dependencies can be compensated via electronics or microprocessor controlled corrections. Many amperometric gas sensors require oxygen for long-term stability, and thus should not be used in nitrogen or other inert atmospheres without prior validation for a specific model.

3.4.2 Potentiometric

Potentiometric sensing is a common and well-established analytical method. pH electrodes and other ion-selective electrodes are based on potentiometry, and are typically designed for liquid phase analysis. However, gas phase potentiometric sensors are also common.

For example the oxygen sensor based on zirconia is usually configured to operate as a potentiometric sensor when used as a lambda sensor (see Section 4.1.4); it is also one of the most commercially successful sensors. Potentiometry is based on the concentration-dependent potential controlled by the electrochemical equilibrium between the working electrode (WE) of the sensor and a specific electroactive reactant. The actual potentiometric signal is measured as the potential difference between the WE and a reference electrode (RE) when both electrodes are separated by an ion-conducting electrolyte, as illustrated in Figure 3.12. This relationship is a logarithmic function of the concentration of the analyte as given by:

$$E = E_B^o + \frac{RT}{zF} \ln \frac{a_B}{a_o} \tag{3.17}$$

Where E is the measured potential, E_B^o is the standard potential, R is the gas constant, T is the temperature in Kelvin, and the parameter "a" refers to the activity of the analytes; the subscript "o" refers to a standard condition and the subscript "B" refers to the target analyte. The activity, a_B, is the product of the mole fraction of component, c_B, and a scaling factor called the activity coefficient, γ, that is $a_B = \gamma \cdot c_B$. γ is not a constant, but is a function of concentration, pressure, and temperature. However, at low concentrations γ is essentially unity and activity can be approximated as the concentration of the analyte, which for an ideal gas is given as the partial pressure of the gas (p_B). The reference potential can be a second electrode in equilibrium with a known concentration of the analyte, analogous to the design and operation of the potentiometric oxygen sensor (Section 4.1.4), which typically uses an air-reference ($p_{O_2} = 0.21$ bar at sea level). For a hydrogen sensor, a_o would be equal to a reference concentration (activity) for hydrogen. E_o would be zero since the difference in the reference potential would be identical since each cell reaction would have the reference potential for hydrogen. In this case, Equation (3.17) would reduce to the Nernst equation for a concentration cell:

$$E = \frac{RT}{zF} \ln \frac{(p_{H_2})_B}{(p_{H_2})_o} \tag{3.18}$$

Since it can be difficult to maintain the reference cell at constant p_{H_2}, an internal reference electrode that is independent of hydrogen is sometimes used.

Potentiometric sensors have near-zero current passing through the electrodes. Thus one advantage of the potentiometric platform is that since it draws negligible current, a potentiometric measurement will not measurably affect the concentration of the analyte. Owing to the logarithmic dependence, potentiometric sensors typically have a broad dynamic range. Some sensors can cover up to 8 orders of magnitude (e.g., from 10^{-4} vol% to 100 vol%) or more. The sensor response throughout this range is not linear with concentration, but rather is linear with the logarithm of the concentration. However, since the logarithm of "zero" is undefined, the output of the sensor will be unstable when the concentration of the analyte is zero. Thus a background concentration of the analyte is necessary for stable operation of a potentiometric sensor. It is possible that this background reference can be from the hydrogen normally present in ambient air, which is nominally 0.6 ppm [153]; however, this may fluctuate depending on location, and any fluctuation in the reference gas would adversely affect the signal. It may also be too low to establish a meaningful reference potential. Thus, a reference potential independent of hydrogen activity may also be necessary to maintain stability. For example pH electrodes and other ion-selective electrodes often use a saturated calomel electrode for the reference electrode. Thus, potentiometric hydrogen sensors tend to be applied more commonly to process control applications which

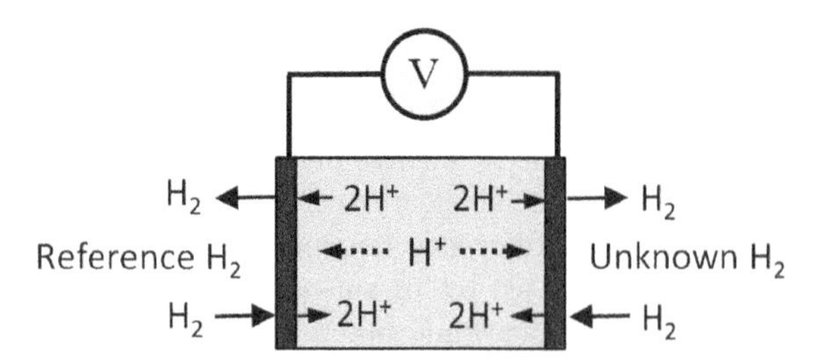

Figure 3.12: Schematic of a potentiometric hydrogen sensor based on a hydrogen ion (proton) conductor in equilibrium with a reference hydrogen concentration and an unknown hydrogen concentration. Since a reference hydrogen concentration is hard to maintain, a self-contained reference electrode independent of hydrogen is often substituted for the reference half cell.

often operate at high p_{H_2}, as opposed to safety applications which would have trace or even zero ambient p_{H_2} under normal situations. Sensor calibration requires at least two concentrations of the analyte; typically one concentration should be a low concentration, while the other should be several orders of magnitude higher. It is also advisable to use a third calibration point if possible, close to the concentration of interest or likely to be encountered in deployment. This third point should have a test gas concentration in between the two required calibration concentrations.

Much of the current research on potentiometric hydrogen sensors is on the choice of electrolyte. Various solid electrolytes have been investigated. A low-temperature potentiometric sensor has been reported using Nafion [154], which is a solid-state proton conductor. As a proton conductor it is well suited for poteniometric hydrogen sensors. Other materials have also been investigated, many of which are based on high-temperature solid-state ionic conductors based on the perovskite family of crystal structures (e.g., ABO_3 where A is a divalent cation such as Ba^{2+} or Sr^{2+}, and B is a tetravalent cation such as Zr^{4+} or Ce^{4+}) [155]. Potentiometric hydrogen sensors based on numerous other solid-state ionic-conductors have also been reported (e.g., Table 4 in reference [146]). Recent developments in the synthesis of intermediate temperature proton conductors based on tin-doped pyrophosphate solid electrolytes open the possibility of constructing hydrogen sensors that will operate above the temperature of polymer solid electrolytes and well below the high temperatures required for the families of perovskite solid electrolytes [156]. Before the discovery of proton conduction in the doped pyrophosphate materials, the lack of proton-conducting solid electrolytes between approximately 150 °C and 500 °C was often referred to as the "Norby Gap" [157].

The use of a reference electrode with a hydrogen independent potential is challenging for high-temperature electrolytes. The saturated calomel electrode, which is often used in ion-selective electrodes, works well at ambient or slightly elevated temperatures but is not amenable for high temperature operation. The identification of a stable reference potential needs development for potentiometric hydrogen sensors based on high temperature proton conductors. Despite their promise potentiometric hydrogen sensors are not nearly as commercially mature as amperometric hydrogen sensors.

3.4.3 Mixed Potential

Research into advanced electrochemical sensor platforms is on going. One of the most promising platforms is the mixed potential gas sensor [158]. The concept is based upon different fundamental electrochemical reaction rates between specific electrode structures and the analyte. One significant advantage of this platform is the ability to tune selectivity of the sensor by choosing electrode materials that have the desired properties with respect to the analyte of interest and the potentially interfering contaminant. In an electrochemical cell, a polarization potential will develop as the reactant concentration is increased; this can be alleviated by increasing the area of the electrode or by using electrode materials with faster reaction rates (typically defined by the exchange current) for the specific reactant; thus the polarization potential for a specific reactant is very much dependent upon the electrode material.

The basic phenomenon was first reported by Fleming [159] and later suggested for the detection of carbon monoxide on an oxygen sensor. The fundamental theory behind the phenomena has been evolving since then. Researchers at Los Alamos National Laboratory have made extensive advances in elucidating the fundamental theory of this class of sensors beginning in 1999 [150]. Mukundan and co-workers demonstrated that the potential which developed between two metal electrodes (e.g., gold and platinum) upon exposure to oxygen and oxygen with carbon monoxide was directly correlated to the difference in polarization curves for platinum and gold that developed in the presence of carbon monoxide. As no electric current is permitted to flow between the electrodes, the oxidation and reduction reactions reach a steady state and a voltage is produced that is logarithmically proportional to the concentration of carbon monoxide. Typically, these structures are fabricated using oxygen ion conductors and thus are based upon the oxygen-oxide electrochemical equilibrium mediated by each electrode. Yttria stabilized zirconia (YSZ) is often used as the electrolyte, which is a pure oxide ion conductor at temperature above about 500 °C. The polarization curves for each electrode are a convolution of the simultaneous oxygen reduction and carbon monoxide oxidation. It was further shown that this difference was dominated by the gold electrode which exhibited a significant increase in overpotential upon exposure to carbon monoxide relative to that observed on platinum. This work led to defining the requirements for a mixed-potential hydrogen sensor, including the use of an electrode which has minimal overpotential associated with hydrogen oxidation (e.g., platinum) and an electrode with a large overpotential associated with hydrogen oxidation. Gold meets the high over potential requirement in that gold has a much lower exchange current to hydrogen and oxygen than platinum. However, it is also necessary that the material is stable at the operating temperature, and gold YSZ interface is not sufficiently stable for practical sensor applications at the required sensor operation temperature. Other electrode materials are being investigated, including indium tin oxide (ITO), which meets both the poor oxygen exchange kinetics (i.e., it behaves like gold) and generates a large mixed potential voltage in the presence of hydrogen. In addition it exhibits a robust thermal stability requirement (i.e., it is a ceramic material). A prototype mixed potential hydrogen sensor based on platinum and ITO is now being developed [160] and it is being considered for commercialization.

3.5 CONDUCTOMETRIC HYDROGEN SENSORS

The electrical properties, including conductivity, of some materials are very sensitive to changes in the chemical composition of their surrounding environment. The dependence of such material's electrical properties with the surrounding chemical environment is used as the detection principle for many gas sensors. Changes in electrical conductance or resistance, which is the reciprocal of conductance, of gas sensitive materials can often provide analytical information regarding changes in the chemical composition of the environment to which the material is exposed. For sensor applications the change of resistance is induced ideally by a

Figure 3.13: Photograph of some commercially available conductometric metal oxide gas sensors, (source: BAM).

single chemical species (e.g., hydrogen) or class of chemicals (e.g., flammable gases). In this section the operating principle and detection mechanism of various conductometric sensors are briefly described, with a description of their application to hydrogen sensing.

3.5.1 Semiconducting Metal Oxide

The resistance of metal oxides varies with changes in the surrounding gas environment. This relationship was first reported in 1950 [161] and since this time it has been successfully exploited to become one of the most widely applied technologies for gas sensing. It is possible to detect a wide range of gas types using this technology including reducing, oxidising and/or combustible gases. Depending on the metal oxide sensing material used and the target gases to be detected the electrical resistance of the metal oxide varies as a function of the target gas concentration. Also known as resistive, chemiresistive, solid state semiconductor or metal oxide (MOX) sensors, many different metal oxide materials have been used in conductometric semiconductor sensors.

The potential use of metal oxides for gas sensing was demonstrated by Seiyama in 1962 who discovered that the adsorption and desorption of gases on a thin film of ZnO at temperatures greater than 400 °C invoked a measurable change in the electrical conductivity of the ZnO. Around the same time Taguchi patented a SnO_2 based detection device for combustible and reducing gases in Japan [162]. Taguchi went on to found the Figaro Engineering Incorporated enterprise that commercialised the so-called Taguchi Gas Sensor (TGS), which still makes up a large proportion of conductometric gas sensors available on today's market. A photograph of some metal oxide conductometric sensor products is shown in Figure 3.13. Since hydrogen is a reducing and combustible gas, this technology lends itself to the detection of hydrogen as will be outlined later in this section.

Sensing Element Components

Semiconducting metal oxide sensors are prevalent commercially because their simple structure makes them easy, and hence inexpensive to fabricate. They are also noted for their ease

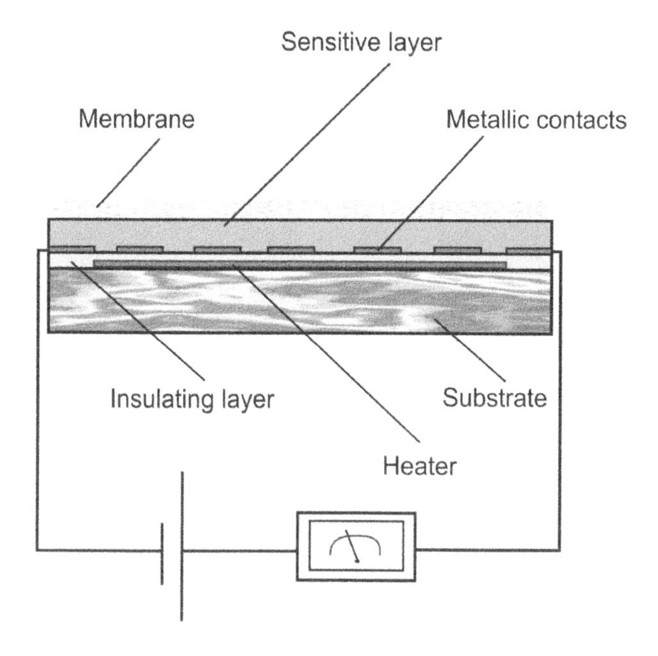

Figure 3.14: Schematic of a metal oxide gas sensor showing the major components. Adapted with permission from [163].

of operation. The sensor's sensing element comprises the following main components as illustrated in Figure 3.14:

- A metal oxide sensing material
- A heating element
- A substrate
- Electrodes

The performance of a semiconducting metal oxide gas sensor can be tailored by varying the material, structural, and operational aspects of these components. This is also true for hydrogen sensors where a large body of research and development work has been performed illustrating how their parameters can influence sensor response and how sensor design and fabrication methods can be optimised to enhance the accurate and selective detection of hydrogen.

Metal oxide sensing material

Metal oxide materials used for gas sensing are typically solid-state polycrystalline semiconductors. The electrical conductivity of many metal oxide materials is controlled primarily by two parameters—the temperature of the material and the surrounding chemical environment. Both of these parameters affect the mobile charge carrier density on the surface of and within the crystalline material. At room temperature, most metal oxide materials are essentially insulators (i.e., materials with a very high resistance), but their resistance decreases dramatically at elevated temperatures. A variety of metal oxide materials have been investigated as materials for conductometric sensors, and include those derived from pre-transition metals (e.g., Al_2O_3, MgO),

Table 3.4: Band gap values of metal oxides including some used for hydrogen sensing [164–166]

Metal Oxide	**Band Gap / eV**
Al_2O_3	6.9
Ga_2O_3	4.8
SnO_2	3.5
NiO	3.5
MoO_3	3.5
TiO_2	3.2
ZnO	3.2
In_2O_3	2.8
WO_3	2.7
Fe_2O_3	2.2

transition metals (e.g., TiO_2, Fe_2O_3, FeO, NiO, Cr_2O_3) or post-transition metals (e.g., SnO_2, ZnO, Ga_2O_3, In_2O_3, Sb_2O_5). In general, post-transition metal oxides, as well as some transition metal oxides are the preferred materials for gas sensing due to their favourable electrical properties, including their band gap, see Figure 3.15.

The electrical properties of a metal oxide material can be described by the band gap model. As with any semiconductor, mobile electrons can be generated in the metal oxide material by thermally exciting electrons from the valance band (a range of lower energy states where electrons are tightly bound and are unavailable to conduct charge) to the conduction band (a range of higher energy states in which electrons are more loosely bound and are thus free to roam and hence conduct charge). The temperature required to do this is controlled by the band gap. The band gap (illustrated in Figure 3.15) is equivalent to the energy required to excite an electron from the valence band into the conduction band where it becomes mobile and available to conduct charge. Band gap values for a number of metal oxides, including some used in hydrogen sensors, are presented in Table 3.4. Metal oxides suitable for gas sensing typically have band gap values between 3 eV and 4 eV. This is because the excitation of electrons from the valance band to the conduction band can be performed at moderately elevated temperatures for materials having a band gap in this range. Metal oxides formed from pre-transition metals typically have a wide band gap (often > 6 eV) and thus these materials often behave as insulators, rendering them unsuitable for gas sensing. If the band gap is too large electrons can only move into the conduction band at very high temperatures. Consequently thermal degradation or material decomposition can occur before a reasonable conductivity is achieved. Some thermally stable metal-oxides with a moderately high band gap, e.g., Ga_2O_3, have been used as gas sensitive materials; however a high operating temperature is required.

Interaction with the surrounding chemical environment also affects the mobile carrier density within a metal oxide material; this interaction forms the basis for gas sensing. In general, many reactive gases can either donate or extract electrons into or out of the conduction band of the metal oxide crystal thereby altering its resistance. The selectivity of metal oxide sensors is inherently limited since a wide variety of gases can exchange electrons with the metal oxide material. However, selectivity depends on numerous factors, including the metal oxide material itself, the temperature, the mode of operation, the incorporation of catalysts into the metal oxide material, and the

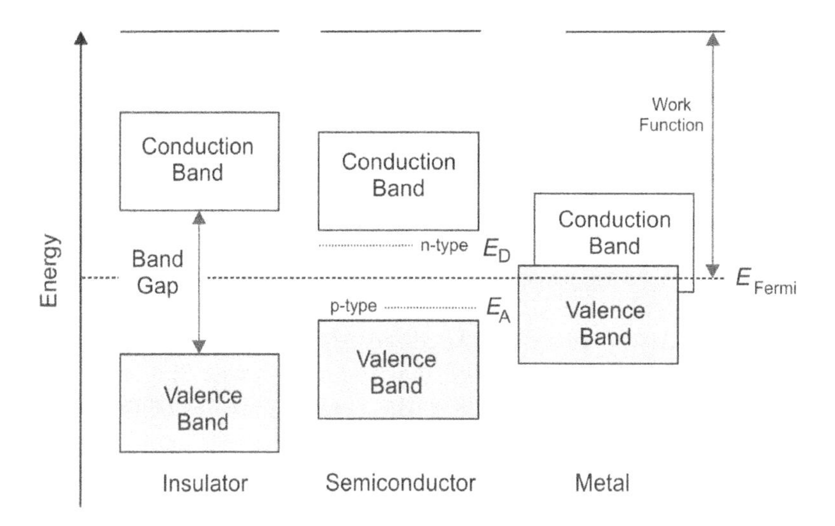

Figure 3.15: Illustration of the band gap model showing the relative energy separation between the valence band and conduction band for solid crystalline materials. Insulators have a large energy gap between the valence and conduction bands so that, even at moderate temperatures, electrons cannot be elevated into the conduction band. In conductors there is no band gap, as the conduction and valence energy bands overlap. Semiconductors have a band gap between that of insulators and conductors (typically between 0.1 eV and 3 eV), such that excitation of electrons into the conduction band can be achieved at moderately elevated temperatures.

use of chemical filters to remove undesirable constituents from the gas sample before it reaches the sensing material. By adjusting and tailoring these factors a sensor's selectivity can often be optimized for a specific analyte.

The sensing elements are typically constructed as a film of the metal oxide deposited onto the insulating substrate material. The metal oxide may be deposited onto the surface of the substrate as either a thick film (generally in the range 5 μm to 20 μm) or a thin film (typically 100 nm to 150 nm).

Heating element

The operating temperatures of metal oxide sensors are typically in the range of 150 °C to 600 °C; the specific temperature depends on the type of metal oxide and to some extent the target analyte, since the nature of the interaction between analyte and the sensing element is affected by temperature. A Joule effect heating element is needed to raise the temperature of the oxide layer to promote surface reactions and to remove water produced from these reactions. In metal oxide sensors the heater element is formed typically onto the underside of the supporting substrate while the active metal oxide thin film material is applied to the upper surface. For microfabricated sensors it may be preferential to fabricate the heater on the topside of the sensing layer for technological reasons. Platinum is commonly selected as the heater material and a voltage is applied across the heater to control the temperature.

Substrate

Metal oxide sensing films are formed on to the surface of a supporting substrate. Despite being an "inert" component of the sensing element, the choice of substrate has been shown to have an important influence on sensor stability. This is a direct consequence of the elevated operating temperature. The use of a substrate material and a metal oxide with similar thermal expansion coefficients is desirable to minimise mechanical stresses induced by thermal expansion at the interface between the two materials [167]. Silicon, oxidised silicon, glass, and alumina (Al_2O_3) are substrate materials commonly used for depositing the sensing layer upon. Al_2O_3 has other beneficial characteristics, including high electrical resistance, excellent thermal stability, and good adherence properties with metal oxides, which make it a popular substrate. Anodisation of aluminium metal yields a porous Al_2O_3 substrate (Anodised Aluminium Oxide - AAO) with a high specific surface area. It is speculated that the metal oxide film formed on the AAO also holds much of the larger specific surface area thereby benefiting sensor transient response [168].

Electrodes

For gas detection, a voltage is applied across the electrodes and the change in the resistance in response to the analyte is measured. The electrodes are integrated onto the oxide material to measure the electrical conductance, G, or the resistance, R, of the metal oxide layer. Electrode materials are preferably noble metals, such as platinum, silver/palladium, gold, or aluminium; however, less expensive metals, such as Cu, Mo, and Ni are also used. The choice of electrode material depends on the metal oxide used and the gas to be detected.

Sensing Mechanism

The aforementioned components are integrated to form the sensing element and to control its operation. The resistance of the metal oxide can increase or decrease in the presence of the target gas depending on the nature of the target gas (i.e., whether it is oxidising or reducing) and on whether the oxide is an n-type or p-type semiconductor. The addition of dopant atoms to a semiconductor modifies the number of charge carriers. The movement of charge carriers can be described by the mobility of both electrons and electron holes. The valency of the dopant, compared with the bulk material, determines whether additional electrons or electron holes are created. Either will increase the conductance of the material. A doped semiconductor with additional electron holes is referred to as "p-type." A doped semiconductor with additional free electrons is known as "n-type." The resistance of p-type semiconducting metal oxides increases in the presence of reducing gases, whereas it decreases in the presence of oxidising gases. Conversely, the resistance of n-type metal oxides increases in oxidising gases but decreases in reducing ones. The change in sensor resistance is due to interactions of the target gas on the surface of the metal oxide. The magnitude of the resistance change can be correlated to the concentration of the target gas. The change of conductance (i.e., the reciprocal of the resistance) of a metal oxide semiconductor can often be approximated by a power law:

$$G = G_0 \cdot c^{\beta} \tag{3.19}$$

Where G is the conductance of the metal oxide in the test gas, G_0 is the conductance in the absence of the analyte, c is the analyte concentration, and β is an empirical coefficient. Conductance depends on the specific conductivity of the material, σ, its cross-sectional area,

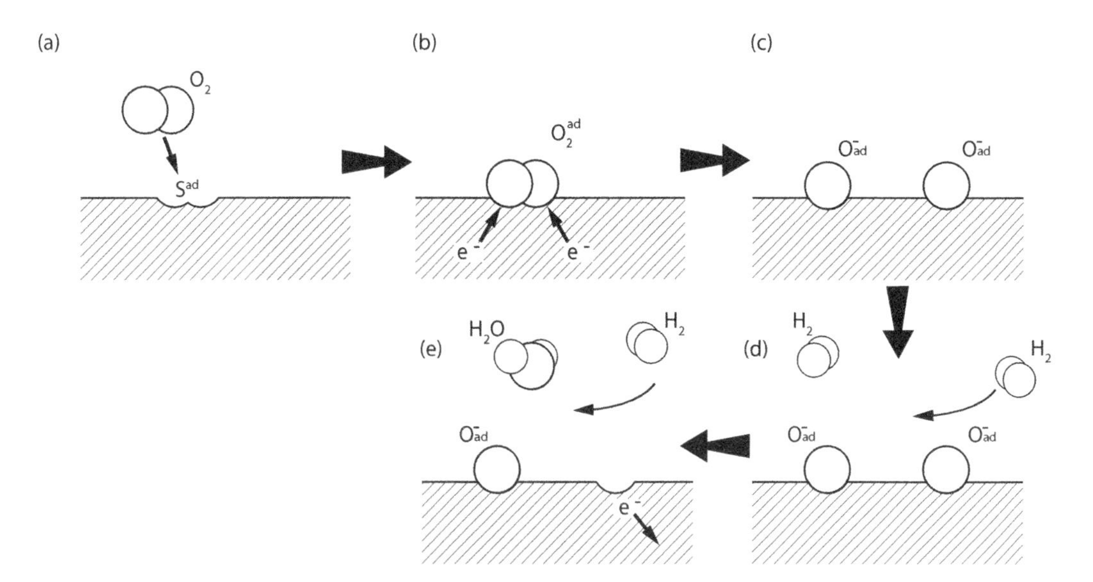

Figure 3.16: Illustration of the hydrogen-oxygen interactions on the surface of a metal oxide (with O^- as the dominant adsorbed species).

A, and its length, l:

$$G = \sigma \frac{A}{l} \tag{3.20}$$

The SI unit for G is in Siemens, S, while σ is S/m. The specific conductivity of a material depends on the number and mobility of the charge carriers and can be described as:

$$\sigma = e(n\mu_n + p\mu_p) \tag{3.21}$$

Where e is the elementary charge, n is the electron density, μ_n is the electron mobility, p is the density of electron holes, and μ_p is the mobility of electron holes.

Sensing reaction mechanism

The interaction of the target gas with the metal oxide follows a complex reaction mechanism involving several intermediate species. The sensing mechanism has been investigated by many groups for different target gases. Several papers postulating the mechanism are available [169–172]. A widely accepted theory explaining the mode of operation of conductometric metal oxide sensors in air is illustrated in Figure 3.16. Oxygen molecules adsorb onto adsorption sites, S_{ad}, at the surface of the metal oxide. Electrons in the conduction band are transferred to the adsorbed oxygen molecule, resulting in a deficit of electrons at the surface of the oxide creating a so-called depletion layer. The depletion layer can be considered as a region in which there is a reduced number of charge carriers (for n-type semiconductors these are electrons). The oxygen molecule is ionised upon transfer of electrons from the metal oxide. Ionised oxygen can exist on the surface as either a molecular species (O_2^-) or an atomic species (O^- or O^{2-}). The predominance of each species is temperature dependent. At temperatures below 150 °C the molecular form dominates on the surface while at higher temperatures the atomic form is prevalent (see Figure 3.17). The extraction of the electrons by adsorbed oxygen from the metal oxide raises its electrical resistance. In the presence of a reducing gas this process is reversed. When a reducing gas, such as hydrogen,

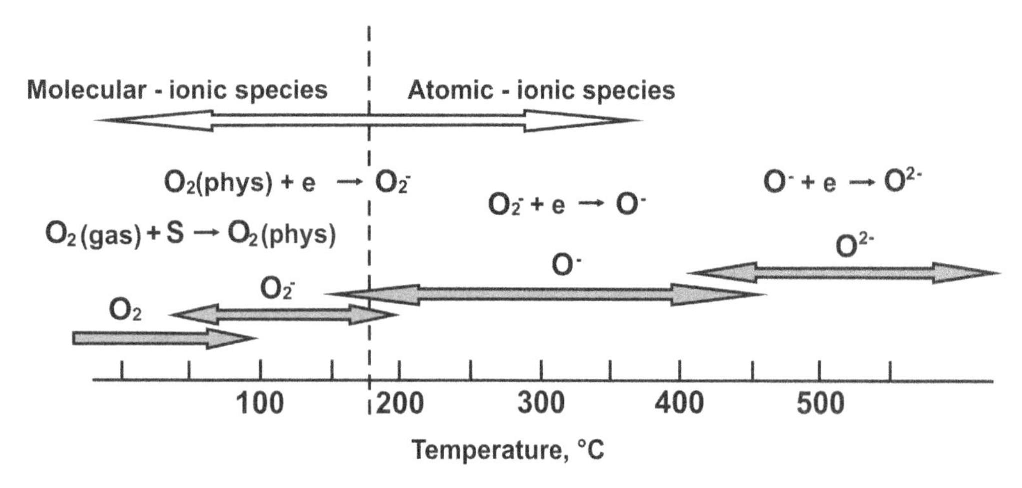

Figure 3.17: Illustration of the dependency of oxygen species present adsorbed on the surface of metal oxide semiconductors on temperature, used with permission from [175].

is introduced it is oxidised by the adsorbed oxygen ions. The oxygen ions are thus removed from the surface and electrons are injected back into the metal oxide thereby increasing its electrical conductivity.

Operating temperature

The response of metal oxide films to gases exhibit a bell-shaped variation with sensor operation temperature. The surface reactions, such as adsorption, desorption, and the oxidation/reduction of the analyte, are thermally activated [99]. Based on its material properties, each metal oxide has an optimum response temperature to a particular analyte, which in part determines the operating temperature of the sensor when optimized for a specific analyte (see Figure 3.18). Typical operating temperatures of these sensors lie in the range of 200 °C to 450 °C. In some respects operation of conductometric metal oxide sensors at elevated temperatures is advantageous as it minimises the negative impact water (vapour) may have on the sensor response and it promotes the analytic chemical reactions which occur on the metal oxide surface. However, a direct consequence of a high operating temperature is an increased power requirement for heating, which in many applications is undesirable. It has been reported that the use of catalytic dopants, e.g., platinum, can significantly lower the operating temperature [173] thereby reducing power consumption. For applications where high temperatures are expected, e.g., gas sensors for fire detection, a metal oxide with a wide band gap and high thermal stability is preferred, e.g., Ga_2O_3.

Optimisation of Sensor Performance

For metal oxide gas sensors several parameters of the sensing layer have been shown to influence sensor response. Such parameters include the choice of substrate material, the choice of metal oxide, the thickness of the metal oxide layer and its microstructure, the grain size, the dopants used, and the operating temperature. Extensive research has been performed to investigate and quantify the impact on sensor performance induced by variations in these parameters; for a review see [175, 176]. Development of modern metal oxide conductometric sensors focuses on design and fabrication strategies to exploit these parameters for optimising sensor performance.

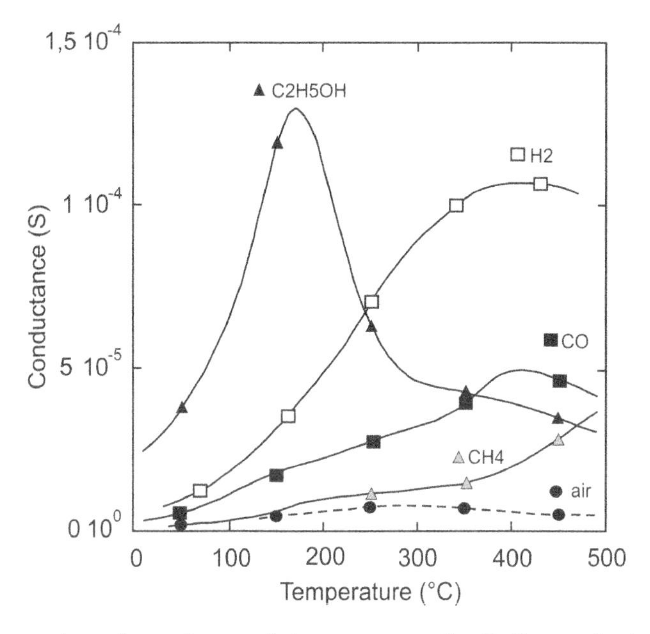

Figure 3.18: Temperature dependence of the response of a SnO_2 conductometric sensor to different gases [174].

Optimisation of sensitivity through the use of catalysts

The sensitivity of MOX sensors can be enhanced through the use of catalysts on the surface of the metal oxide. Noble metals are effective oxidation catalysts and additives, such as palladium or platinum, can promote the desired chemical reactions leading to a higher sensitivity and lower response times. The role of the catalyst can be explained through the concept of spillover, a term describing the dissociation and transport of a species from the catalyst particle into the material, in this case the metal oxide (see Figure 3.19). By lowering the activation energy of the interaction between the analyte and the metal oxide surface, the spillover effect enables operation at lower temperatures. This effect is also referred to as chemical sensitization. Electronic sensitization occurs through the use of impurity dopants in the semiconducting material [177]. The width of the depletion layer is affected by the use of additives and dopants, which can increase the sensitivity to the target gas and lower the operation temperature.

Optimization of selectivity

Selectivity describes the ability of a sensor to respond to the target analyte regardless of the presence of other chemicals. Ideally a gas sensor developed for a specific target analyte should not respond to other species that may be incidentally present. The sensitivity of the sensor to other gases is referred to as cross-sensitivity. The selectivity of metal oxide sensors is potentially poor since metal oxides can show reactivity with many gas species. However, numerous strategies are adopted to increase a sensor's selectivity towards a specific gas. For example the operating temperature of a metal oxide sensor can be adjusted to improve its selectivity as different gases will show a maximum response at unique temperatures, see Figure 3.18. Other measures include modification of the surface morphology and the use of filters. Filters can act in a passive manner, by absorbing species other than the analyte of interest. Filters are

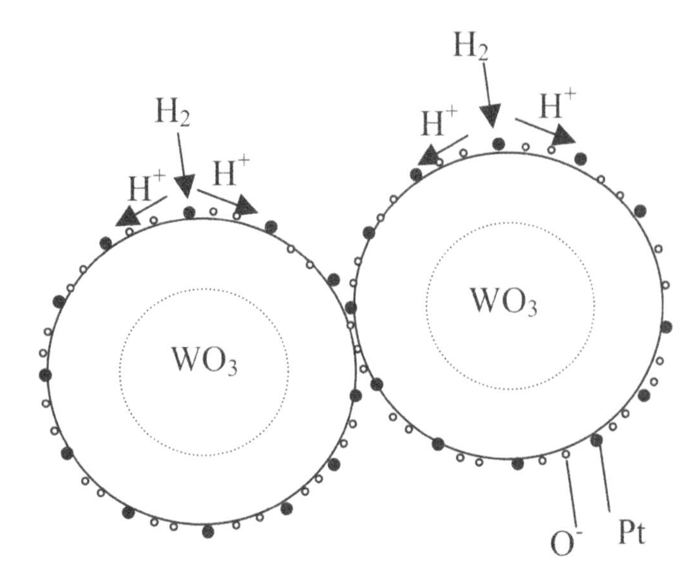

Figure 3.19: (a) Illustration of the spillover effect of Pd on the adsorbed species on the surface of a metal oxide (WO_3) hydrogen sensor [178].

typically comprised of a porous structure, such as charcoal or zeolites. However, this type of filter may become saturated and at some point will need to be regenerated or replaced in order to maintain stable sensor performance. Gas filtering layers deposited directly on the surface of the metal-oxide are used to enhance selectivity. Thin porous layers of silicon dioxide have been used for this purpose.

Another option to increase selectivity is to operate the sensor in dynamic temperature mode [175]. The operating temperature plays a decisive role on the state of oxygen species adsorbed on the metal oxide surface. As mentioned previously, different physisorption/chemisorption reactions are prevalent, depending on the temperature. Modulation of the operating temperature enables a "freezing-in" of a high temperature state, resulting in O^{2-} or O^- species remaining at the surface even at low temperatures. As the reactivity of these ionic species is much higher than that of molecular oxygen ions, the response of the sensor to the target species increases. The characteristic transient response of different gases to the temperature cycles can be determined, enabling improved selectivity. Drawbacks for temperature cycling include a significant increase in the complexity, and hence cost, of the instrument and increased physical stresses on the sensing element due to thermal cycling. Achieving sufficient mechanical stability for this operation mode can be challenging due to the different thermal expansion coefficients of the various materials used, which can lead to stress, cracking, and delamination. However, heating pulses can also diminish the influence of humidity [179] as water molecules are desorbed from the surface, which improves sensor performance. The detrimental effect of humidity on the response and drift of MOX sensors is well known [176]. In particular sensors operating at lower temperatures need to take variations in relative humidity into account, as the formation of water on the surface of the sensors increases the conductivity. Operation temperatures above 400 °C reduce the influence of water molecules considerably, but at a cost of a higher power consumption.

Optimization of microstructure and nanostructuring

The morphology and microstructure of the metal oxide material also influences conductometric type gas sensor performance. The average grain size and distribution of grain sizes have a profound influence on the sensor performance characteristics of polycrystalline metal oxide sensors, especially regarding their durability. An optimization of the microstructure of the metal oxides enhances the transport properties and reduces the drift of the sensor. In gas sensor applications polycrystalline metal oxide materials (solids comprising crystallites or grains of various sizes and orientations) and nanocrystalline metal oxide materials (polycrystalline material with grain sizes in the nanometer scale) exhibit advantageous properties which make them particularly suitable for this purpose. These poly- and nanocrystalline materials have a very high surface area, there are many available techniques for their fabrication, and they have superior mechanical, chemical, and electrical stability.

Increasingly single crystalline metal oxides (where the entire crystal structure is continuous with no grain barriers) including one-dimensional (1D) metal oxide nanostructures such as nanorods, nanowires, nanotubes and nanobelts, are being investigated because of a potential enhancement of specific properties. While polycrystalline metal-oxide structures may suffer from lack of device-to-device reproducibility in sensor response due to variations in the grain size distribution, reproducibility of 1D single crystal structures can be attributed to the homogeneity of these structures. The change in resistance of nanstructures due to interaction with the target gas is determined by the individual contact resistances as well as by the depth of the depletion layer in the individual particle. When the depletion layer approaches a sufficient depth (the Debye length) it envelops the whole particle and conductance changes can be large. In nanosized grains or particles the charge carriers are in surface states and few thermally activated carriers are available for conduction. Small crystallite sizes are therefore advantageous for an enhanced response. For 1D structures, the depletion layer can cover the entire nanowire, and the charge carrier mobility is restricted to 1D channels, which contributes to their promising properties for gas sensing. However, there are still challenges associated with the fabrication and separation processes of 1D single structures for gas sensing. Their integration into a sensor system is challenging. Therefore this type of metal oxide sensor is currently not yet commercialised; however, because of their promise their development is being actively pursued. Accordingly polycrystalline structures still dominate the metal oxide gas sensor market.

Metal Oxide Hydrogen Sensors

Metal oxide conductometric sensors for hydrogen detection are widely reported. Commonly investigated metal oxides for hydrogen sensing include SnO_2, TiO_2, Fe_2O_3, In_2O_3, ZnO, NiO, WO_3, Ga_2O_3, and MoO_3. Excellent reviews of hydrogen metal oxide sensors [180] reveal that SnO_2 is the most widely investigated sensing material. Molecular hydrogen is adsorbed on the surface of the metal oxide where it dissociates into hydrogen atoms. The hydrogen atoms react with chemisorbed oxygen ions to form hydroxyl groups which can further react to form water and oxygen vacancies (V_{O^*}) according to the proposed hydrogen surface reaction [181]:

$$H_2 \longrightarrow 2\,H$$
$$2\,H + 2\,O^{2-} \longrightarrow 2\,OH^- + 2\,e^-$$
$$2\,H + 2\,OH^- \longrightarrow 2\,H_2O(g) + 2\,V_{O^*} + 2\,e^-$$

Electrons produced in the reactions are liberated in the conduction band of the semiconductor and as a result the conductivity of the semiconductor increases. It has been suggested that the conductivity also increases as hydrogen atoms diffuse into the metal oxide lattice, contributing additional electrons into the conduction band. Upon the removal of hydrogen from the sensor's environment, the oxygen ion surface layer is regenerated as molecular oxygen adsorbs and forms O^{2-} ions, filling the oxygen vacancies and removing electrons from the conduction band:

$$O_2 + 2\,V_{O*} + 4\,e^- \longrightarrow 2\,O^{2-}$$

Hydrogen metal oxide semiconducting sensors are available in thick film and thin film variations. Compared with thin film counterparts, thick film metal oxide sensors have higher power consumption due to their higher thermal mass. Typical values for power consumption for thick film sensors are 200 mW to 1 W compared with 30 mW to 150 mW for thin film micromachined devices [182]. Thin film metal oxide hydrogen sensing layers may be integrated into a thermally controlled Micro-Electro-Mechanical Systems (MEMS) device platform, commonly termed a microhotplate [183]. Microhotplate gas sensors comprise a suspended membrane micromachined from the bulk silicon substrate. The microheater, metal oxide hydrogen sensing layer and electrodes are formed on the membrane. In the microhotplate structure the microheater and metal oxide sensing layer are thermally isolated from the bulk substrate making rapid heating/cooling of the sensing layer possible including dynamic temperature control. Due to their amenability to mass fabrication, thin film microhotplate gas sensors are potentially cheaper to produce in large numbers using microfabrication techniques [184]. Furthermore thin film hydrogen structures expedite response kinetics yielding fast responses to changes in gas concentration. A comparative study on the response and recovery times of thin film and thick film metal oxide sensors revealed respective response times to 0.1 vol% hydrogen of 1.4 s ± 0.6 s and 13.6 s ± 0.7 s [185]. Despite these advantages thin film gas sensing structures often become saturated, even at moderate gas concentrations, thereby limiting their dynamic measuring range.

The thickness of the sensing layer has been shown to have an effect on the response of SnO_2 sensors to hydrogen [186]. The sensor response decreased as the film thickness increased between 80 nm and 300 nm, see Figure 3.20. Interestingly the response of the same sensor to CO did not vary significantly with film thickness. This allows for an opportunity to enhance the hydrogen selectivity by adjusting the film thickness. Similarly the substrate material of a metal oxide sensor has also been shown to influence the response to hydrogen. AAO supported 25 nm thin TiO_2 films revealed a hugely increased conductance following exposure to hydrogen [168] when compared with identical films supported on a SiO_2 substrate. Response and recovery kinetics were improved as well. Additionally the AAO based sensor showed improved mechanical and chemical stability. In both cases the operating temperature was 500 °C. Porous silicon substrates have also been used to support TiO_2 and ZnO films for hydrogen detection; in this case, the sensing elements demonstrated good sensor kinetics even at operating temperatures as low as 40 °C [187].

Titanium dioxide nanostructures have been studied by several groups for the detection of hydrogen (e.g. [188]). Metal oxide nanotubes can be easily fabricated and are usually arranged as high surface area arrays that are then coated with a noble metal layer or particles as illustrated in Figure 3.21. A large change in resistance due to exposure to hydrogen is based on a change in Schottky barrier height between the metal oxide and the noble metal [188]. Highly ordered titanium oxide nanowire arrays showed a response to low hydrogen concentration (< 10 ppm) with a response time of 6 s; however, the operating

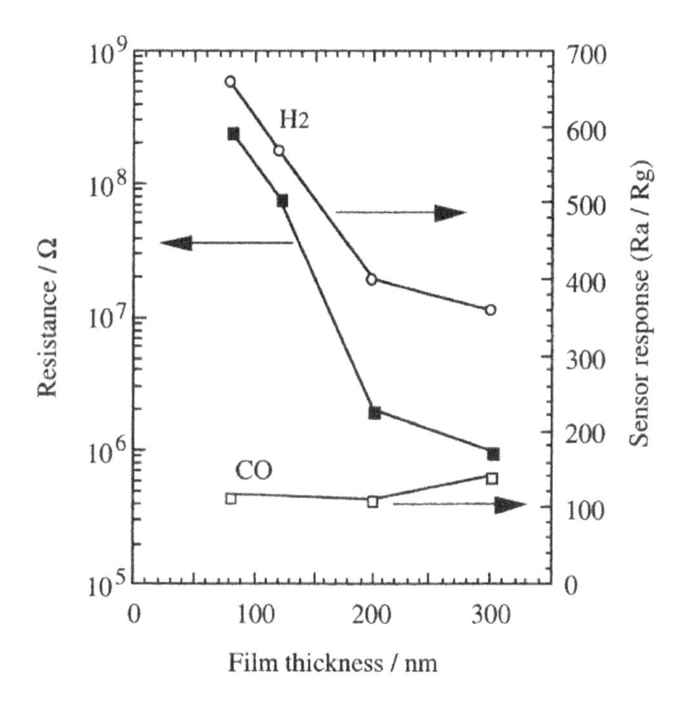

Figure 3.20: Variation in the electrical resistance and sensor response of a SnO_2 thin film sensor to 0.08 vol% H_2 and 0.08 vol% CO depending on film thickness [186].

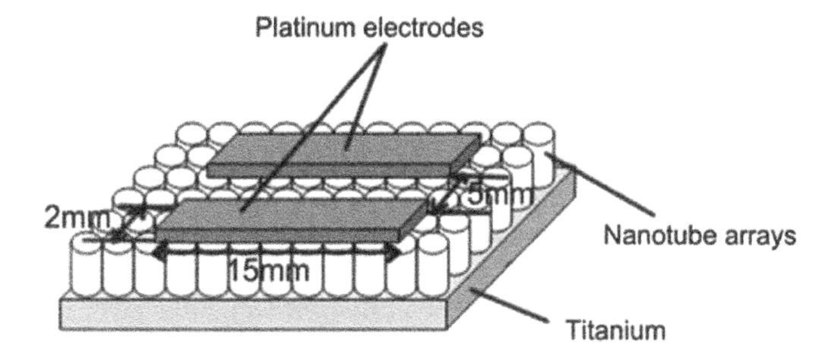

Figure 3.21: Illustration of TiO_2 nanotube arrays grown on a titanium foil base with a pair of platinum electrodes. Used with permission from [188].

temperature of 265 °C is relatively high [189]. A zinc oxide nanorod array with palladium nanoparticle decoration on a flexible substrate of polyimide has also been described, claiming a detection limit of 0.2 ppm. The sensor operated at room temperature, with long response and recovery times of > 100 s [190]. For comparison, the typical operating temperature for conventional tin oxide sensors is typically above 350 °C and often greater than 550 °C, with response times often on the order of 30 s to 60 s. Nevertheless even if a reduction in operating temperature is possible through the use of nanostructured materials, the combination of room temperature operation and short response times has not yet been achieved.

Metal oxide sensors account for around 13 % of commercially available hydrogen sensors (see Section 3.10.2) and most products indicate tin oxide as the hydrogen sensing material. Typically metal oxide based sensors are small and the sensing element is often microfabricated using thin- or thick-film fabrications methods. Products boast a fast response time, t_{90}, typically between 1 s and 15 s. In general this technology is capable of detecting low concentrations of hydrogen in air with some models specifying a lower detection limit of just 0.001 vol% (10 ppm). Laboratory testing of some metal-oxide sensors has, however, indicated their tendency to saturate at modest hydrogen concentrations (< 2 vol% hydrogen in air) even when these concentrations are within the products specified measuring range [191]. Testing also revealed a strong influence of temperature and humidity on the sensor response to hydrogen. Despite their cross-sensitivity towards other hydrogen containing compounds, such as hydrocarbons, metal oxide sensors remain a competitive technology for hydrogen sensing for many applications.

3.5.2 Semistors

So-called "semistors," also known as one-electrode semiconductor sensors, are another conductometric type gas sensor which exploit changes in the resistance of the gas sensitive material for detection. However the structure and transduction mechanism differs somewhat from the previously described conductometric type sensors.

The sensing mechanism of semistors was first explained and the design optimised for methane detection [169]. The structure of the methane semistor (or semiconductor based pellistor) is reminiscent of the previously described catalytic pellistor sensor (see Section 3.2.1), namely a coiled resistive platinum wire embedded in a porous alumina pellet structure. In semistors the pellet is made from a polycrystalline gas-sensitive semiconducting material such as tin oxide. Similar to catalytic pellistors the platinum coil has two functions: it acts as a heater (Joule effect) to heat the pellet up to its working temperature and also acts as the sensing electrode.

In semistors the resistance of the bulk gas sensitive semiconducting material decreases in the presence of a reducing gas. The decrease in resistance of the bulk material creates electric "shorts" between the turns of the platinum coil resulting in an overall reduction in the sensor resistance. Similar to catalytic pellistor sensors, planar designs of this type of sensor are also possible [192]. In principle semistor sensors are non-specific and will respond to all reducing combustible gas species. Specificity to a particular species can be enhanced by covering the surface of the metal-oxide pellet (often based on SnO_2) with a filter layer which preferentially allows the passage of the target gas. As described for pellistor catalytic sensors, the use of a bridge circuit design is also possible to negate potential influences ambient temperature and humidity may have on sensor response [192]. With just a single electrode their structure is simpler and more robust than film metal oxide sensors. Commercial products for hydrogen and other gases are available on the market.

3.5.3 Metal Films

The electrical resistivity of some metals and alloys changes markedly on the absorption of hydrogen. This phenomenon is exploited in thin metal film sensing elements. Thin metallic films, based on, e.g., yttrium, magnesium-nickel, and palladium, have been investigated for the purpose of gas sensing. Palladium containing materials are preferred for hydrogen sensing due to hydrogen's high solubility in palladium at ambient temperatures and due to palladium's catalytic activity [193].

The adsorption of hydrogen in palladium is a two-step process involving the adsorption of hydrogen molecules on the metal surface; the dissociation and the subsequent diffusion of hydrogen atoms into the bulk metal (see Section 1.1.5). In an equilibrium state, the amount of hydrogen atoms in palladium is proportional to the square root of partial pressure of hydrogen in the ambient atmosphere (Sievert's law). The hydrogen atoms cause changes in the crystal structure of the metal and a growth of charge carriers. This results in an increase of electrical resistivity due to a larger number of scattering events between the charge carriers. Ideally, the change of electrical resistivity follows the dependency on the square root of the hydrogen amount in the ambient atmosphere.

Metallic resistor sensors are fabricated by deposition of a gas sensitive film, e.g., by electron beam evaporation of metals, onto a substrate such as silicon, ceramic, or glass, between two electrical contacts. The deposited films have a thickness from about 5 nm to 400 nm and cover the full area or meanderings between the electrodes.

The absorption of hydrogen causes an increase of the crystal lattice of the palladium. A phase transition from a solid solution of hydrogen (α-phase) to palladium hydride (β-phase) takes place at increased hydrogen pressure (see Section 1.1.5). The volume expansion and the phase transition is reversible with a decreasing amount of hydrogen, but as a consequence of repeated absorption and desorption of hydrogen, changes in the microstructure of the metal film can occur potentially leading to stress and crack formation and delamination from the ceramic substrate [105]. Thin buffer layers have been introduced to mitigate structural deformation of palladium thin films arising from hydrogen exposure.

Hydrogen causes a significant increase in the electrical resistivity of palladium capped films of magnesium-nickel or yttrium [194]. The films are capped with a thin covering layer of palladium to prevent oxidation of the metal and to promote dissociation of hydrogen. Heating of the yttrium films from 50 °C to 100 °C is required to reduce the sensor response time. This was achieved using a microelectromechanical system (MEMS) microhotplate which also had additional advantages over conventional designs, including superior signal strength, reduced size, and reduced power consumption. The measuring ranges for hydrogen was 0.01 vol% to 1 vol%. A sensor based on sputtered Mg-Ni films coated with palladium, operating at room temperatures and with an exceptionally large measuring range from 0.01 vol% to 100 vol% H_2 has been reported [195].

Another approach to overcome unwanted structural changes is the use of metal alloys, including those based on Pd-Mg, Pd-Ag, or Pd-Cr [196]. Palladium nickel alloys are of particular interest because the alloys show lower volume changes upon hydrogen absorption; however, they have lower sensitivity to hydrogen. The incorporation of nickel (up to 20 %) into palladium suppresses the α- to β- hydride phase transition which contributes to a non-steady change of electrical properties and crack formation in the film [197]. Palladium-nickel thin film based sensors can detect hydrogen in the range of 0.001 vol% to 100 vol%. The non-ohmic change of the electrical properties of the Pd-Ni thin films is employed for detection of low hydrogen amounts (0.001 vol% to 0.5 vol%) in a MOS structure (Schottky diode) (see Section 3.6.2). On the other hand, for the hydrogen range of 0.1 vol% to 100 vol% H_2, the change of electrical resistance serves as the primary sensor signal. The sensor can operate at pressures from 0.005 bar to 35 bar in an atmosphere of air or nitrogen. Oxygen is not required for hydrogen detection as is typical for other sensing principles. The resistance of the thin metal alloy film changes with temperature. Therefore, the temperature of the sensing film has to be measured and controlled by a heater resistor. The operating temperature of the sensing element is set above the ambient temperature to prevent condensation of water under high humidity conditions. The sensor has an accuracy of 1 % H_2 for hydrogen measurements in the range 10 vol% to 100 vol% H_2. However a drift of sensor signal of 0.4 vol% H_2 per week has been observed. The response time, t_{90}, is lower than 30 s. The sensor is very specific for hydrogen and protected by a membrane hardly influenced by other gaseous components such as carbon monoxide, hydrogen sulphide, or chlorine [198]. The sensor is commercially available as a handheld or stationary device.

Nanostructured materials have been applied to conductometric metal film hydrogen sensors. Two different sensing mechanisms can be distinguished in metal nanostructures. This distinction is based on whether the nanostructured metal coating is continuous or discontinuous. Continuous metal film sensors, described above, show an increased resistance due to the additional scattering of electrons following absorption of hydrogen. Conversely, discontinuous structures show a marked decrease in their electrical resistance in response to interaction with hydrogen. The decrease in resistance can be attributed to the change in volume of the metal particles upon the formation of metal (palladium) hydride. Nanoparticles enlarge when exposed to hydrogen, which brings them in contact with each other, thus creating additional pathways for electrical current. In spite of the higher intrinsic resistivity of the hydride material, this closing of physical gaps results in a large increase in the device conductivity. This process is reversible, as the particles shrink and the electrical conductivity drops when the hydrogen concentration decreases. Discontinuous films offer several advantages such as response times below 1 s, operation at close to room temperature and lower detection limits of 0.001 vol%. A hydrogen sensor based on discontinuous palladium nanoparticles (30 nm) coated on a metal film has been commercialized, although it does not seem to be available any longer. The surface of the sensor is heated to a temperature of 80 °C to reduce its cross-sensitivity to other gases. The sensor can be operated between - 30 °C to 120 °C, has a measuring range of 0.001 vol% to 4 vol% H_2 with a response time of < 5 s. This sensor uses an uncoated reference element to compensate for environmental changes [199].

3.6 FIELD EFFECT BASED HYDROGEN SENSORS

Field effect sensors emerged as new gas sensitive devices in the mid-1970s. Lundström and co-workers, at the Linköping Institute of Technology in Sweden, were the pioneers of this sensing technology [200, 201]. The so-called field effect describes a change in the surface electrical conductance of a conductor or semiconductor material under the influence of an externally applied electric field. When a voltage is applied to the structure it generates an electric field which modulates the charge carrier density in the surface layer of the semiconducting material or changes the polarization at the interfaces between the insulator and the semiconductor or conductor material.

Field effect-based electronic units generally have a double layer (metal and semiconductor - MS) or triple layer (metal, insulating material and semiconductor - MIS) structure. In many cases the insulating layer is an oxide material and the term MOS structure is correspondingly used. The field effect is also influenced by absorption of some gases, particularly hydrogen, into the metal layer. In hydrogen sensitive MIS structures hydrogen molecules are adsorbed on the metal surface, dissociate into atoms, and diffuse through the metal where they

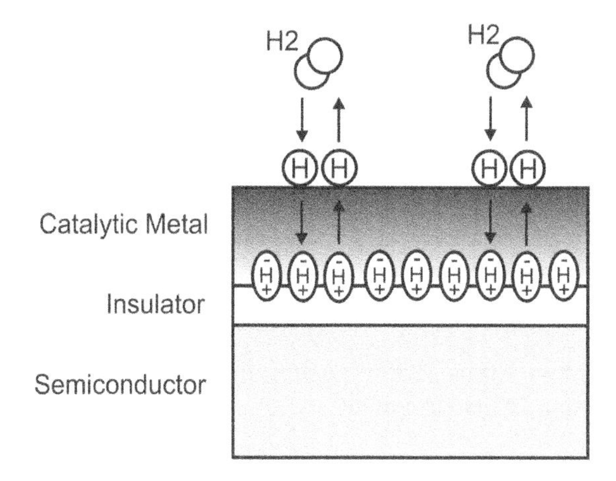

Figure 3.22: Triple-layer structure of field effect-based sensing elements.

accumulate at the metal/oxide interface. The hydrogen atoms at the interface are polarised giving rise to a dipole layer which has an impact on the charge carrier density at the semiconductor/insulator interface (see Figure 3.22). This gives rise to a measurable change either in the capacitance-voltage (C-V) or current-voltage (I-V) characteristics of the structure. The magnitude of the change can be correlated to the hydrogen concentration such that any transducer which can respond to these changes can be used to measure the hydrogen concentration. Such transducers include metal-oxide-semiconductor capacitors, metal-(oxide)-semiconductor (MS/MOS) Schottky diodes, and metal-oxide-semiconductor field effect transistors (MOSFET). These sensing devices, and their development specifically for hydrogen detection, are discussed in more detail in the following sections.

3.6.1 Metal-Oxide-Semiconductor Capacitors

A metal-oxide-semiconductor capacitor uses the field effect-based shift in capacitance to measure hydrogen concentration. A MOS capacitor consists of a simple arrangement of an insulating oxide layer sandwiched between a catalytic metal layer and a semiconductor substrate. The insulating oxide layer is typically around 100 nm thick, which is thick enough to prevent tunnelling between the metal and semiconductor, facilitating build-up of charge on both sides of the oxide. The absorption of hydrogen and the reactions at the metal-oxide interface cause a change in the C-V plot which correlates to the hydrogen gas concentration (see Figure 3.23).

A conventional capacitor consists of a dielectric between two metallic electrodes and shows a linear growth of capacitance with increasing voltage (in its operation range). When a voltage is applied charges accumulate on each electrode. In a MOS capacitor one of the metallic electrodes is replaced by a semiconductor layer which has a limited supply of charge carriers. Depending on the nature and magnitude of the voltage applied, the resulting charge generated in the semiconductor can arise at different depths. Accordingly, a MOS capacitor structure generates a characteristic S-shaped C-V plot which can be divided into three regions as shown in Figure 3.24.

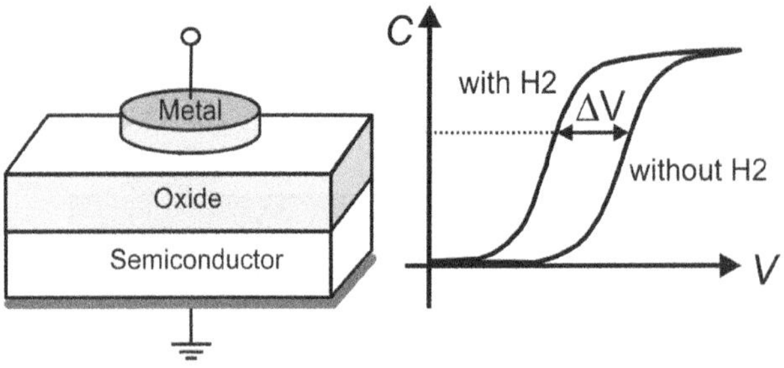

Figure 3.23: Schematic of an n-type MOS capacitor and C-V curves demonstrating the effect of hydrogen absorbed from the environment.

Accumulation region

When a positive voltage is applied to the metal electrode the n-type semiconductor material is more negative relative to the metal electrode. The charge carriers (electrons) accumulate at the semiconductor-oxide interface (see Figure 3.24a). In the accumulation region, the capacitance, C_O, per square unit is dominated by the contribution of the oxide layer which is given by:

$$C_O = \frac{\epsilon_{Ox}}{x_O} \tag{3.22}$$

Where ϵ_{Ox} is the permittivity of the oxide and x_O is the thickness of the oxide layer.

Depletion region

When the applied positive voltage is reduced and becomes negative, the electric field repels the majority charge carriers (electrons) from the interface and a depletion layer, devoid of any charge carriers, is formed. If the applied voltage becomes more negative the thickness of the depletion layer increases. The depletion layer capacitance, C_D, is:

$$C_D = \frac{\epsilon_S}{x_d} \tag{3.23}$$

Where ϵ_S is the permitivity of the semiconductor and x_d is the thickness of the depletion layer (see Figure 3.24b). As a result, the total capacitance of the structure decreases and is given by the series addition of the oxide layer capacitance, C_O, and the depletion layer capacitance, C_D:

$$\frac{1}{C} = \frac{1}{C_O} + \frac{1}{C_D} \tag{3.24}$$

Inversion region

If the applied voltage becomes more negative, the depletion layer continues to increase to a maximum thickness before inversion occurs. Beyond this point the applied negative voltage not only repels the majority charge carrier (electrons) in the depletion layer but increasingly attracts minority charge carriers (holes). The maximum width of the depletion layer, $x_{d,max}$, before inversion is given by:

$$x_{d,max} = \sqrt{\frac{4\epsilon_S \Phi_F}{qN_A}} \tag{3.25}$$

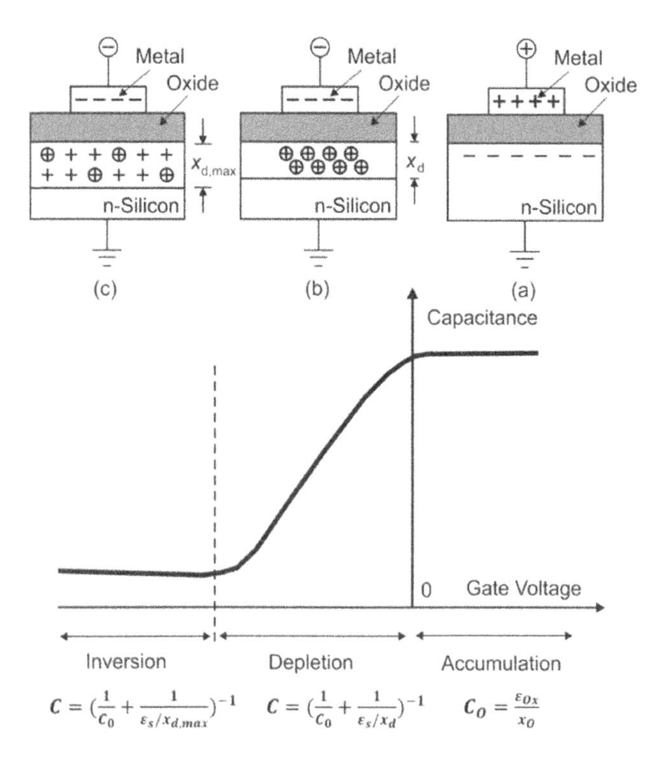

Figure 3.24: Capacitance-voltage (C-V) characteristic of an n-type MOS capacitor.

Where Φ_F is the Fermi potential[2] and qN_A is the charge density in the depletion region. When the negative applied field is sufficiently large there is an equivalent number of holes at the semiconductor interface surface as there were electrons originally and the semiconductor enters the inversion region (see Figure 3.24c). In the inversion region a further lowering of the applied negative voltage is compensated by additional positive charges (holes) in the semiconductor, the capacitance remains nearly constant and is given by:

$$C = (\frac{1}{C_O} + \frac{1}{\epsilon_s/x_d, max})^{-1} \tag{3.26}$$

MOS capacitor devices can be used for hydrogen detection as hydrogen influences the C-V plot and shifts the flat band voltage[3] by a magnitude which correlates to the hydrogen concentration.

Hydrogen sensors based on miniaturised MOS capacitor structures were first reported by Steele in 1976 [202]. Various metal/oxide/semiconductor combinations were fabricated and their performance and properties assessed. A direct comparison of palladium and platinum as the metal electrode in a MOS capacitor sensor showed that palladium is more suitable for hydrogen detection at low concentrations in air while platinum is more effective at

[2]The Fermi potential is a term from quantum physics and can be considered as the energy level with 50 % chance of being occupied at finite temperatures. The Fermi level is situated in the band gap of insulators and semiconductors.

[3]A voltage at which there is no electrical charge in the semiconductor and, therefore, no voltage drop across it; in a band diagram the energy bands of the semiconductor are horizontal (flat).

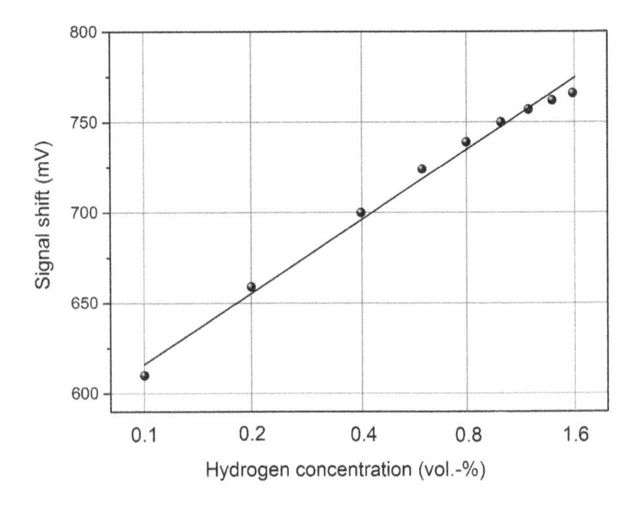

Figure 3.25: Hydrogen sensitivity of a field effect-based hydrogen sensor; linear dependency on logarithmic scale between sensor signal and hydrogen concentration [208].

higher concentrations, where otherwise Pd-metallised capacitors may become saturated [203]. It was shown that sensors with a lower Pt-SiO_2 bonding strength exhibit an improved hydrogen response. Furthermore, the deposition of the metal electrode layer using different methods and conditions results in layers with different properties that impact the sensor response. SiO_2 is often the preferred insulator material in MOS capacitor structures. However, hydrogen detection devices deploying other insulator materials, such as Al_2O_3, Si_3N_4, and Ta_2O_5, have also been investigated, as have alkaline earth titanates and zirconates which allow operation at temperatures up to 250 °C [204]. The surface chemistry of a capacitor's insulator layer has also been shown to influence the capacitor's response to hydrogen. The hydrogen response correlates directly to the concentration of oxygen atoms at the insulator surface and the results support a response mechanism whereby the oxygen atoms of the insulator surface are the hydrogen trapping sites [205].

Traditionally n- or p-type silicon is used as the semiconductor material; however, GaN, a material of high electron mobility, has also been researched. Another alternative is silicon carbide (SiC) which, with its superior chemical resistance and high temperature tolerance, yields MOS capacitor sensors which can operate in harsh conditions and at temperatures up to 700 °C [206].

A MOS capacitor hydrogen sensor, which precludes the need for permanently heating the sensing element, has been achieved by interfacing an additional layer of lanthanum trifluoride (LaF_3) between the metal and the insulator [207]. It resulted in a sensor with an increased signal and a faster response at room temperature. This prototype showed, like many other field effect-based sensors, an exponential dependence of the sensor signal on hydrogen concentration and detected hydrogen in the range of 1 ppm to 10 vol% (see Figure 3.25) [207].

A MOS capacitor hydrogen sensor has been successfully commercialised and is based on a Pd/Ni thin film. It has a hydrogen detection range of 0.001 vo% to 1 vol% with good stability and a high resistance to sulphur poisoning [209]. The capacitor is combined with a Pd/Ni thin film resistor in a single hydrogen detection device which covers a very wide

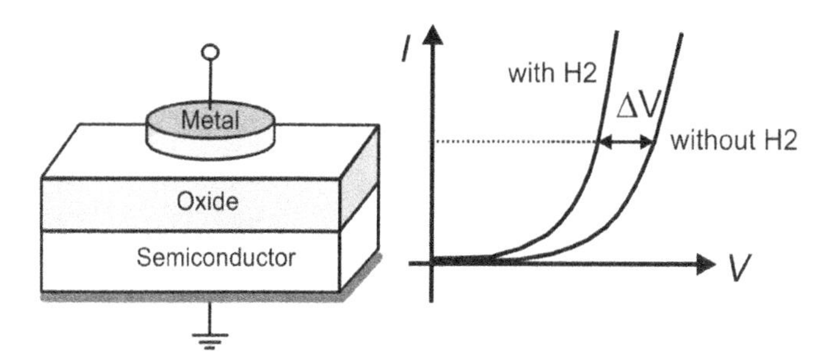

Figure 3.26: Schematic of a MOS diode and *I*-*V* curves demonstrating the effect of hydrogen absorbed from the environment.

hydrogen measuring range from 0.001 vol% up to 100 vol% and has a response time of some seconds.

3.6.2 Metal-Semiconductor (Schottky) Diodes

Schottky diode-type sensors typically comprise a metal in contact with a semiconductor material (MS) or a metal in contact with a very thin insulating (generally an oxide) layer on a semiconductor material (MIS/MOS). The absorption of hydrogen causes a change in the *I*-*V* plot which correlates to the hydrogen gas concentration (see Figure 3.26).

When a metal is brought in contact with a semiconductor, the Fermi level of the semiconductor adjusts to and aligns with the Fermi level of the metal. This adjustment of the Fermi level also occurs if there is a very thin insulating layer, e.g., some atomic layers of SiO_2, between the metal and the semiconductor. The valence and conductive band of the semiconductor bend near the interface and a potential energy barrier, the so called Schottky barrier, is formed. The Schottky barrier, Φ_b, is the difference between the Fermi level of the metal and the semiconductor (n-type) conduction band (see Figure 3.27). Electrons from the semiconductor can diffuse through this barrier to the metal side and a depletion zone is formed in the semiconductor. A forward bias,[4] i.e., positive voltage, pushes electrons from the semiconductor into the depletion zone, the barrier diminishes, and electrons flow into the metal. A reverse bias increases the depletion zone and only a small current of tunnelling electrons is possible. However, an enlarged negative voltage results in a breakthrough and current flow occurs.

The forward current $I(V)$ of a Schottky diode can be expressed by:

$$I(V) = AA^{*}T^{2}\exp\left(\frac{-q\Phi_b}{\mathrm{k_B}T}\right)\exp\left(\frac{qV}{n\mathrm{k_B}T}\left(1-\exp(\frac{-qV}{\mathrm{k_B}T})\right)\right) \tag{3.27}$$

Where V is the voltage, A is the diode area, A^* is the effective Richardson's constant, T is the absolute temperature, $\mathrm{k_B}$ is the Boltzmann constant, q is the electron charge, n is the validity factor, and Φ_b is the Schottky barrier height.

The absorption of hydrogen can cause a change in the *I*-*V* plot of a Schottky diode which correlates to the hydrogen gas concentration (see Figure 3.26). In the presence of hydrogen, hydrogen molecules adsorb onto the catalytic metal's surface. The hydrogen dissociates and

[4]Biasing refers to the process whereby the required voltage or current is established at parts of an electronic circuit to obtain the desired operating conditions.

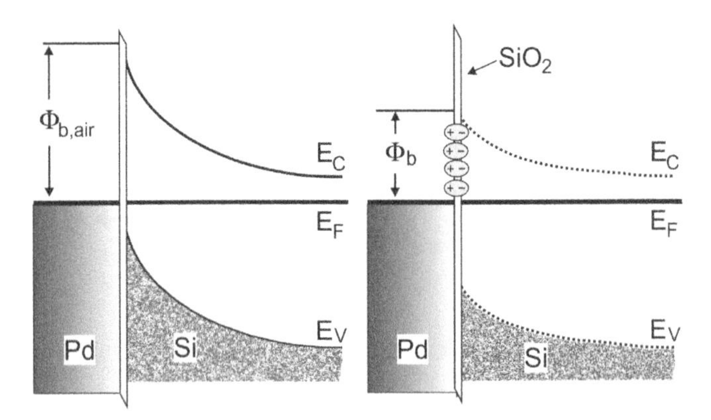

Figure 3.27: Band diagram for a Pd/SiO_2/Si Schottky diode in air on the left-hand side and in the presence of hydrogen on the right-hand side. E_F is the Fermi level, E_V is the valence band, E_C is the conduction band, and Φ_b is the Schottky barrier height.

the atoms diffuse through the metal film to the metal-insulator interface where they are polarised forming a dipolar layer, which changes the Schottky barrier height. This change in the Schottky barrier height causes a shift in the current-voltage I-V characteristics both in the forward and reverse bias. The response to hydrogen can be measured as the change in voltage when the diode is operated at constant bias current.

The first hydrogen sensitive Schottky diodes were reported by Shivaraman et al. [210] and Steele and MacIver [211] in 1976. Since this time, many Schottky diode type sensors for hydrogen detection have been reported in the literature. Various combinations of metal/semiconductor have been used for hydrogen detection. Comparative studies between Schottky diodes using palladium and platinum as the catalytic metals indicate the superior performance of platinum in terms of speed of response and sensitivity to hydrogen. In addition to palladium and platinum other hydrogen sensitive metals and alloys have been proposed including nickel, gold, silver, ruthenium, iridium-palladium, and palladium-gold [102].

In addition to the standard n-doped silicon a wide variety of semiconductor substrates have been reported, such as zinc oxide or cadmium sulphide. The choice of semiconductor is governed by the sensor's operating temperature. Whereas n-doped silicon based sensors are limited to operation below 250 °C, semiconductors with a wider band gap can be operated at higher temperatures. Materials proposed for this purpose are SiC [212], GaN, as well as InP and InAlP to name just a few. High operation temperatures are relevant for application in combustion engines or reformers for hydrogen production.

SiO_2 is the most widely reported insulating material however other materials tested and reported in Schottky diodes include Ga_2O_3, TiO_2, WO_3, and HfO_2. As mentioned previously, a Schottky diode structure may or may not include a thin insulating or oxide layer between the metal and semiconductor materials. A comparison of the performance of MOS structures containing an interfacial oxide layer with the performance of comparable metal-semiconductor structures (i.e., without an oxide layer) highlight that the MOS sensors have a higher sensitivity to hydrogen compared to the MS counterpart. This difference in performance is explained by the elimination of the Fermi-level pinning effect by the insulating oxide layer which otherwise decreases hydrogen detection. A detailed review of Schottky-type sensors is given by Potje-Kamloth [205].

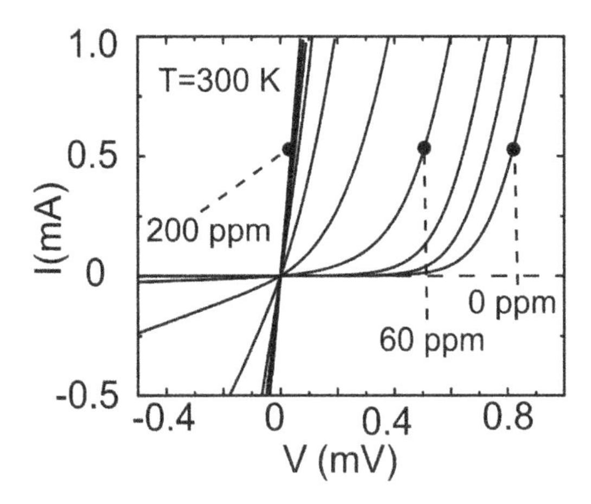

Figure 3.28: *I-V* characteristics of a Pt/TiO2/Ti Schottky diode at various hydrogen concentrations in steps of 20 ppm, measured after an exposure time of 10 min [214].

Schottky diode-type based sensors have a broad measuring range and can detect hydrogen from 0.001 vol% to 4 vol%. Prototype devices have long term stability and a fast response time (< 5 s). However, a strong impact of operation temperature can be expected (see Equation 3.27). Despite the large volume of research done on this type of sensor and the promising performances claimed, only a few commercial Schottky-type sensors are presently available, e.g., based on system Pt/TiO_2 [213] (see Figure 3.28). Nevertheless these sensors boast very low power consumption, room temperature operation, and very fast response times potentially deliverable at a low-cost at high volumes making them also suitable for large volume, price sensitive markets such as hydrogen monitoring in fuel cell vehicles and hydrogen powered homes.

3.6.3 Metal-Oxide-Semiconductor Transistors

Metal-oxide-semiconductor field effect transistor (MOSFET) sensors are another type of field effect device which have been developed for hydrogen detection. Similar to MOS capacitors MOSFETs have a triple layer structure (metal, insulator or oxide layer, and semiconductor substrate).

Traditionally the most common material combination used in MOSFET devices was palladium (or another metal with a catalytic properties for the target gas)–SiO_2–Si triple layer structure.[5] Two regions of the Si substrate are ion-implanted to form a source and a drain. The target gas sensitive metal is deposited on the insulating oxide layer and is known as the gate. The source contact is grounded and used as a reference. A schematic outline of a typical MOSFET device is shown in Figure 3.29. The conductivity between the source and the drain is controlled by a potential applied to the metal electrode. When the applied voltage between the source and gate electrode exceeds the so-called threshold voltage, V_{Th}, an inversion layer (or channel) is formed at the semiconductor surface between the source and the drain. This allows a current to flow between the source and the drain which are now connected by a conductive pathway. The conductivity between the source and the drain can be modulated by varying the potential on the gate electrode.

[5] Instead of an oxide another insulating material, such as Si_3N_4 can be used. In this case the device is called metal-insulator field effect transistor (MISFET).

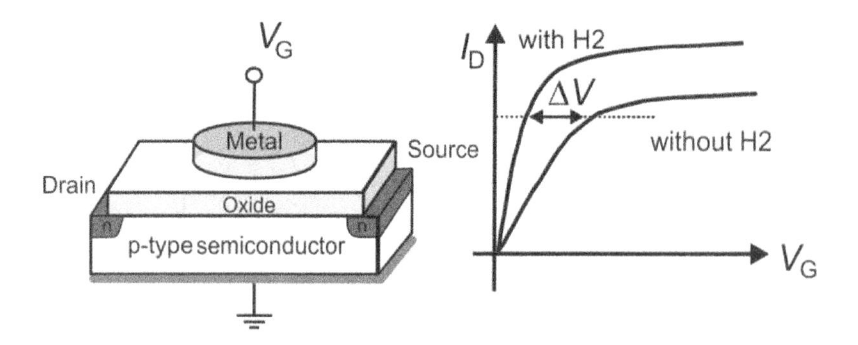

Figure 3.29: Schematic of MOS transistor and its current-voltage (I-V) characteristic. Where I_D is the output current, V_G and V_D are the gate and drain voltages.

The current through the channel between the source and the drain of a MOS transistor is given by:

$$I_D \approx \frac{\mu C_G}{l}[(V_G - V_{Th})V_D] \tag{3.28}$$

Where I_D is the output current, μ is the electron mobility, C_G is the gate capacitance, l is the channel length, and V_G, V_D, and V_{Th} are the gate, drain, and threshold voltages.

MOSFET hydrogen sensors use changes in the current between the source and drain for hydrogen detection. Hydrogen is adsorbed on the surface of the active catalytic metal gate, then dissociates and diffuses to the metal-insulator interface where it forms a dipole layer (see Figure 3.22). This results in a change of the threshold voltage V_{Th} (see Figure 3.29). The voltage drop across the dipole layer, ΔV, can be estimated from the density of adsorbed molecules N', their dipole moment, p, and the dielectric permittivity of the vacuum, ϵ_0, (see Equation 3.29). Assuming an adsorption process at the metal insulator interface, the threshold voltage shift can be correlated to the Langmuir isotherm and described as follows:

$$\Delta V = \frac{N'p}{\epsilon_0} \approx \Delta V_{max} \frac{\mathrm{K}\sqrt{p_{H2}}}{1 + \mathrm{K}\sqrt{p_{H2}}} \tag{3.29}$$

Where ΔV_{max} is the maximum voltage shift, K is the equilibrium constant of adsorption and desorption, and p_{H2} is the hydrogen partial pressure [215, 216].

MOSFET field effect hydrogen sensors based on miniaturised metal-oxide-semiconductor structures were first reported by Lundström and Stilbert in 1975 [200]. Hydrogen MOSFET sensors often use palladium, platinum, or an alloy containing these metals as the catalytic gate material which is deposited as a thin film on the oxide layer. MOSFET hydrogen sensors can operate at room temperature but are generally heated to moderate temperatures of around +150 °C to accelerate the sensor response and reduce the influence of changes in ambient temperature and relative humidity. Fabrication of a MOSFET with an air gap between the gate metal and the insulator layer (suspended gate) allows gas molecules to easily access both surfaces and makes measurements at room temperature feasible [217]. Suspended gate technology is also interesting as it increases the sensitivity of other catalytic metals to hydrogen, such as platinum and iridium, in which hydrogen has a significantly lower solubility than in palladium. Fabrication of suspended gates is however more complicated and expensive compared with conventional gate structures.

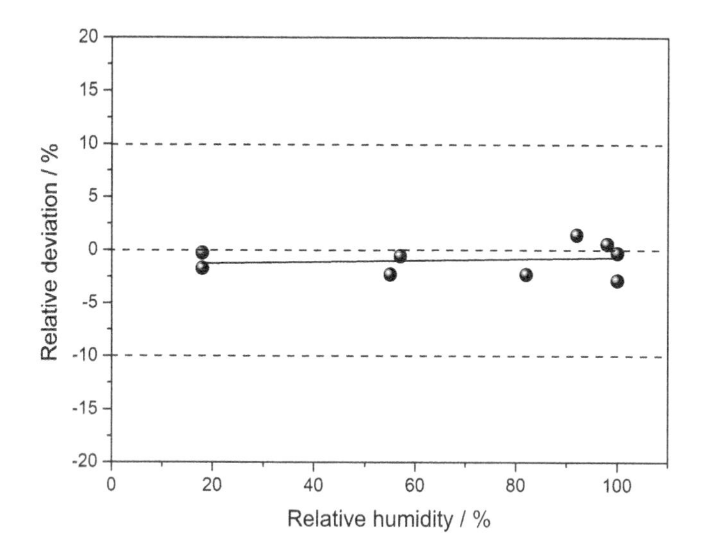

Figure 3.30: Impact of environmental temperature on the sensor signal at 3.5 vol% H_2. The deviation of the sensor signal are smaller than 10 % of reading, less than the requirement from the appropriate standards.

In some applications a hydrogen sensor needs to operate at high ambient temperatures and in a chemically corrosive environment, e.g., for syngas monitoring in an Integrated Gasification Combined Cycle (IGCC) power plant. To withstand such harsh environments sensors based on semiconductors with a wide band gap are required. For high temperature operation, hydrogen transistor sensors based on GaN and SiC have been reported. In addition to their high temperature resistance these sensors also claim very fast intrinsic response time (< 1 s), very low detection limit (0.01 μmol/mol in N_2), and a wide measuring range (up to 50 vol %).

Independent testing of a field-effect based hydrogen sensor revealed its high selectivity and the low impact that changes in environmental conditions (temperature and humidity) have on the response of this type of sensor compared with other types of hydrogen sensor (see Figure 3.30).

Mass fabrication of MOSFET sensors is possible using established microfabrication techniques. Figure 3.31 shows an example of a microfabricated MOSFET mounted on a printed circuit board. Compared with other, more established hydrogen sensing technologies, field effect-based sensors show good potential and further developments will no doubt lead to MOSFET sensors with improved performance capabilities.

While there are a limited number of commercial MOSFET hydrogen sensors currently commercially available independent testing confirmed the superior performance of a hydrogen MOSFET sensor in terms of lower detection limit, accuracy, speed of response, and lack of influence of ambient parameter changes on sensor signal. One commercial device can detect hydrogen in the range of 0.1 ppm to 4 vol % and operates at temperatures between -40 °C to 85 °C and at ambient relative humidities of 5 % to 95 %. Its application in fuel cell vehicles has been successfully tested [218].

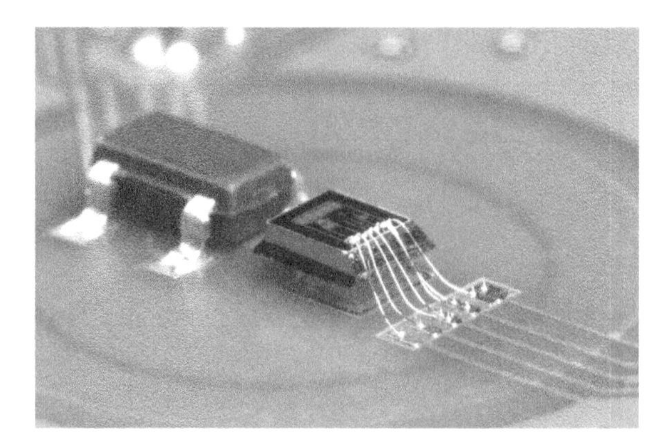

Figure 3.31: Micromachined field effect-based hydrogen sensor.

3.7 ELECTROMECHANICAL HYDROGEN SENSORS

Eveline Weidner, European Commission, Joint Research Centre, Petten

Electromechanical sensors measure physical or chemical phenomena by their effect on the high frequency mechanical vibrations of a sensing element which is typically fabricated from a planar piezoelectric[6] material. The operating mechanism is based on the application of an oscillating electric field to a piezoelectric substrate to generate a mechanical wave along the planar surface or within the substrate. Changes of the wave frequency, phase, or intensity can be detected and transduced into an electrical signal. The physical structures utilized in sensing applications include thin or thick films, membranes, diaphragms, and beams. An overview of the different types of (electro)-mechanical sensors is shown in Figure 3.32.

Electromechanical sensors can be distinguished by the mode of propagation of the acoustic wave. In bulk acoustic wave (BAW) devices, longitudinal or shear acoustic waves are generated in a piezoelectric crystal though the application of an oscillating electric field. The three-dimensional wave travels through the bulk of the crystal. In surface acoustic wave (SAW) devices, the acoustic wave is confined to the surface of the device. These devices are fabricated from materials that can be oscillated at high frequencies (from kHz to GHz).

Another type of electromechanical sensor is based on microcantilevers, which are miniature beam[7] structures, typically produced using microfabrication technology such as that developed in the electronics industry. Microcantilevers operate in two different modes - static and dynamic. For dynamic sensing the cantilever is actuated through the excitation of its structural modes, causing it to oscillate at its resonance frequency. The change in frequency of an oscillating structure due to environmental changes can be used as the basis for chemical sensing. Microcantilevers are commonly fabricated from non-piezoelectric (such as silicon based) substrates and oscillated through an external piezoelectric platform on which the cantilever is mounted. The static operation mode is characterised by a mechanical response (bending or deflection) of the cantilever when exposed to the analyte (for a description see Section 3.11).

---

[6]The piezoelectric effect is the ability of certain materials to generate an electric charge in response to an applied mechanical stress. The word piezoelectric is derived from the Greek piezein, which means to squeeze or press, and piezo, which is Greek for "push."

[7]Beams are load bearing structural elements in which the length is much larger than its width and depth. Beams can be clamped on one side (cantilevers) or on both sides (bridges).

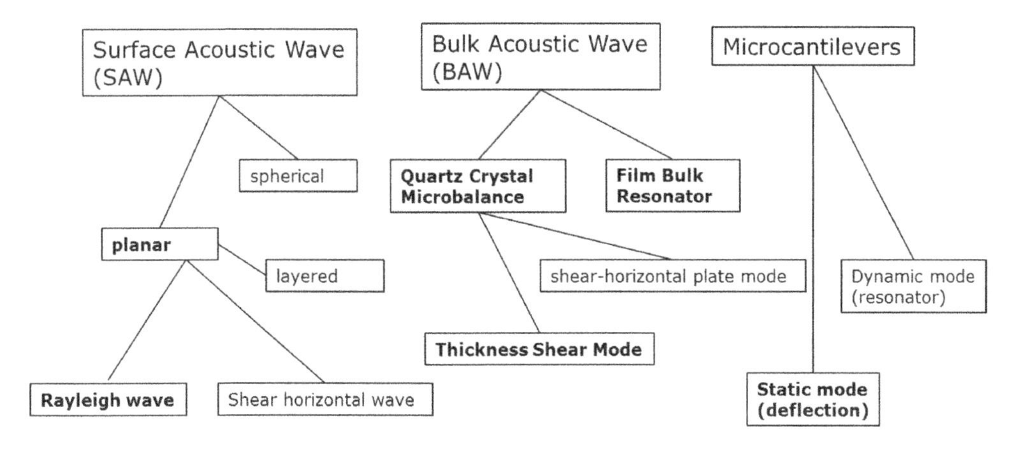

Figure 3.32: Main types of (electro)-mechanical sensors - the main operational modes utilized for hydrogen sensing are marked in bold. Static mode microcantilevers (mechanical) are discussed in Section 3.11.

3.7.1 Basic Working Principle

For gas sensing applications the devices are modified with a thin coating that selectively interacts with or adsorbs the target gas. Platinum, palladium, or palladium alloys are among the most commonly used coating materials for the selective detection of hydrogen. The interaction between the coating and the target gas will change the properties of a propagating wave such as the vibration frequency, phase, or intensity. The quantity measured during exposure to the target gas is commonly correlated to the mass loading associated with the adsorption of the analyte into the coating. This typically causes a shift in the observed frequency. This type of sensor has also been classified as a mass sensitive sensor according to IUPAC (see Table 2.1).

In 1959 Sauerbrey [219] introduced a method for gravimetric measurements based on the change in the frequency of a quartz resonator. The Sauerbrey equation establishes the relationship between the mass loading m and the shift in frequency Δf_m:

$$\Delta f_m = -f_0^2 \frac{2}{A\sqrt{\rho G}} \Delta m \tag{3.30}$$

Where f_0 is the resonant crystal frequency, Δm is the mass change, A is the active area, ρ and G are the density, and the shear modulus of quartz. As can be seen from Equation 3.30, the sensitivity of these devices will increase with the square of the fundamental frequency. Mass loadings of less than 1 ng/cm^2 have been reported [220] for electromechanical gas sensors, enabling lower detection limits in the ppb range. For the detection of hydrogen, however, the relative mass increase is very small, therefore other property changes have been exploited. In addition to changes in mass loading, changes in the elastic properties of the coating influence the vibration frequency. Elastic stiffening will also result in an increase in frequency, whereas elastic softening or swelling of the sensing film will result in a decrease in frequency. The interaction of the analyte with the coating will often change the electrical properties of the coating, and a change in electrical conductivity has an effect on the wave velocity due to the acousto-electric interaction.[8] The magnitude of this effect can be much

[8]The acousto-electric interaction refers to the coupling of the acoustic wave with current carriers, and is caused by an energy transfer between phonons and electrons.

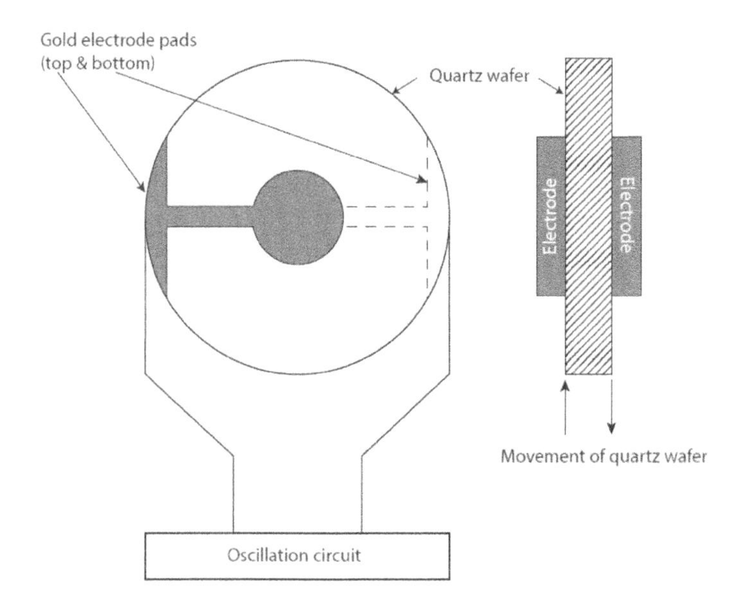

Figure 3.33: Schematic of a quartz crystal microbalance BAW sensor.

higher than that of mass loading. Electromechanical hydrogen sensors have been reported for the SAW, BAW, and microcantilever platforms, with the majority of research being conducted on SAW devices and microcantilevers. To date only few commercial products for hydrogen detection have been developed.

3.7.2 Bulk Acoustic Wave Devices

The most common example of a BAW device is the Quartz Crystal Microbalance (QCM). As depicted in Figure 3.33, a QCM consists of a piezoelectric plate and electrodes deposited on either side. A noble metal based gas sensitive coating is applied onto the piezoelectric substrate. Application of an alternating voltage induces shear waves through the substrate.

Commercial QCMs are well established in gravimetric applications in analytical chemistry or for the measurement of viscosity. QCMs usually operate in the range of 5 MHz to 30 MHz with a minimum detectable mass of around 10 ng/cm^2 [221]. The adsorption or absorption of the target gas into the sensing layer results in an increase in mass, which in turn will decrease the resonant frequency of the quartz crystal. This frequency shift can be directly correlated to the concentration of the target gas. There are, however, only a few reports on BAW devices for hydrogen sensing as compared to surface acoustic wave (SAW) devices. This is because the lower operating frequency of QCM leads to an inherently lower sensitivity relative to SAW devices especially for a transduction mechanism that is based on a change in mass. Sensing hydrogen with BAW devices can be problematic, especially at room temperature, due to the interference from other gases and reaction with pre-adsorbed O_2. The effects of humidity can be avoided by operating at temperatures above 100°C. An increase of the operating frequency (up to 1.1 GHz [220]) and thereby an improved detection of hydrogen has been made possible through the development of film bulk acoustic resonators, for which a thin film of piezo material is sandwiched between two electrodes. A device consisting of an aluminium nitride (AlN) based layered structure, with a 50 nm thick palladium film coated

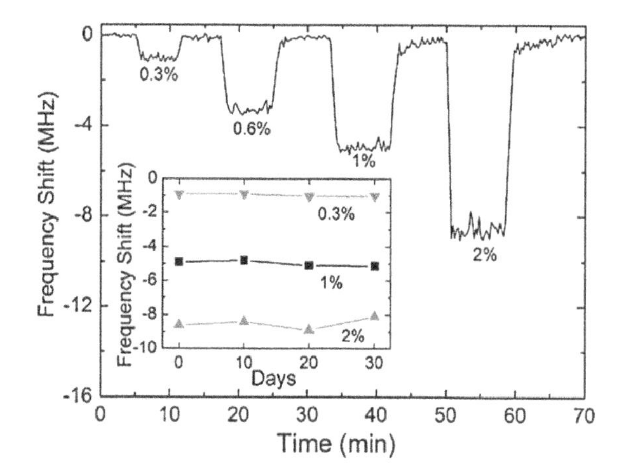

Figure 3.34: Film bulk acoustic resonator sensor - real-time measurement for various hydrogen concentrations from 0.3 vol% to 2 vol%. Reproduced with permission from [222].

as sensitive layer has been reported [222]. Hydrogenation of the palladium layer induced a change in mass density and in the elastic constants, in particular during the transition from α to β phase of palladium. At 1 vol% hydrogen a frequency shift of around -5 MHz was observed (see Figure 3.34). Humidity had a small impact on sensor response. Response and recovery times were 40 s and 70 s, respectively.

3.7.3 Surface Acoustic Wave Devices

Surface Acoustic Wave (SAW) devices are commonly used as temperature, pressure, and acceleration sensors. The response of SAW devices to changes such as mass loading, conductivity or elastic properties led to their development for gas sensing applications. The first SAW gas sensor was reported in 1979 [223]. SAW devices offer a limit of detection more than an order of magnitude lower compared to BAW devices [220].

A SAW device consists of a piezoelectric substrate with an input (transmitting) and output (receiving) interdigitated transducer (IDT) deposited on the substrate (see Figure 3.35). The input IDT generates an acoustic wave from an externally applied AC electrical input signal. For gas sensing applications, a sensitive layer is placed between the input and output transducers. An acoustic wave travels through the sensitized region (also referred to as the delay line) and reaches an output IDT, where it is transduced into an electric signal. The acoustic waves penetrate to a depth of one to two wavelengths at operating frequencies between 100 MHz and 400 MHz. The piezoelectric effect leads to a voltage difference due to the mechanical displacement between the output transducers. The properties of the sensitive layer of the SAW device are changed on interaction with the analyte, which produces measurable changes to the propagating wave, such as a reduction in frequency.

The first hydrogen sensor based on SAW transduction was reported in 1982 [224]. Operating at a frequency of 75 MHz, two delay lines were used, one of which was left uncoated and used as a reference to minimize environmental effects. A palladium layer of 300 nm thickness was coated onto a $LiNBO_3$ substrate. A response to 0.005 vol% hydrogen was observed, albeit with response times on the order of several minutes. For sensors with a palladium sensitive layer, a change in wave velocity can be attributed to the change in

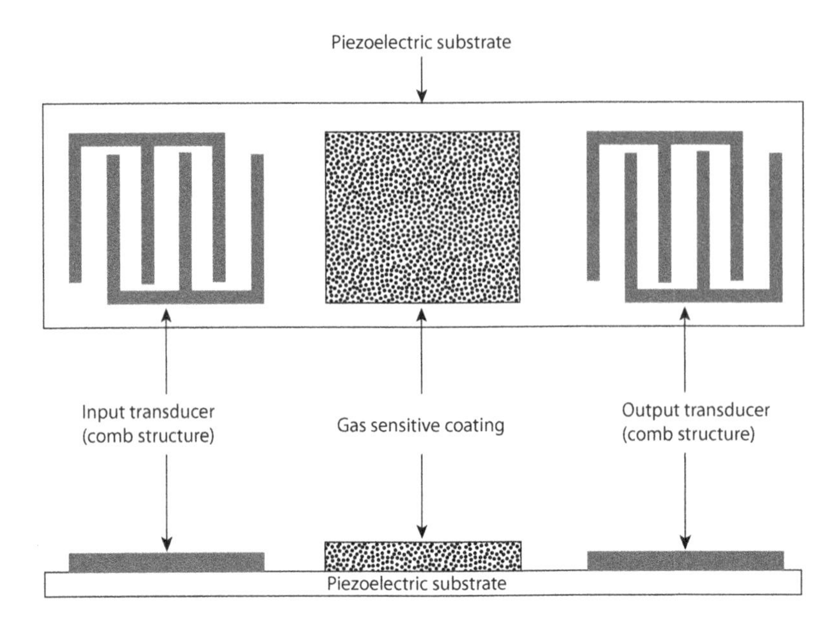

Figure 3.35: Schematic of SAW sensor.

density and elastic constants of palladium due to the hydrogen loading. When the sensing mechanism is based on measuring changes in conductivity, lithium tantalate $LiTaO_3$ is commonly used as substrate material for conductometric sensing. Bi-layered structures were investigated [225] to further exploit the acousto-electric interaction for sensing at temperatures just above room temperature. Other advantages of multi-layer structures include the possibility for temperature compensation by selecting layers with opposite thermal expansion coefficients and protection from the environment. Research into hydrogen selective metal oxide layers [226] and nanostructuring of the surface (e.g., [227]) has led to improved sensitity of the devices. A SAW sensor, based on platinum coated ZnO nanorods, capable of operating at room temperature has been reported to be insensitive to humidity changes [228] due to the use of a dual (reference) delay line. A response time of < 1 s for 3 vol% hydrogen was reported for a ball SAW sensor made out of a piezoelectric crystal sphere. Acoustic waves are generated by RF pulses, which travel around the equator of the sphere. The signal is amplified though averaging over multiple circular paths of the acoustic waves. The fast response was attributed to rapid diffusion of hydrogen in the thin (40 nm) palladium-nickel sensitive layer [229]. The lower detection limit of the sensor was reported as 0.001 vol%.

As the sensitive layer of a SAW device has to be extremely thin (about one acoustic wavelength), achieving uniform layers during production may be challenging. The coating stability of SAW devices is therefore a challenge as it is for any sensor platform with sensitive layer coatings (see also Section 3.11). An advantage of SAW devices is that they can be miniaturized and used in a wireless configuration with an antenna attached to the IDT.

3.7.4 Microcantilever Sensors

Research is being conducted into microcantilever sensors as they can be easily mass produced and have intrinsic high sensitivity coupled with low power consumption. Microcantilevers

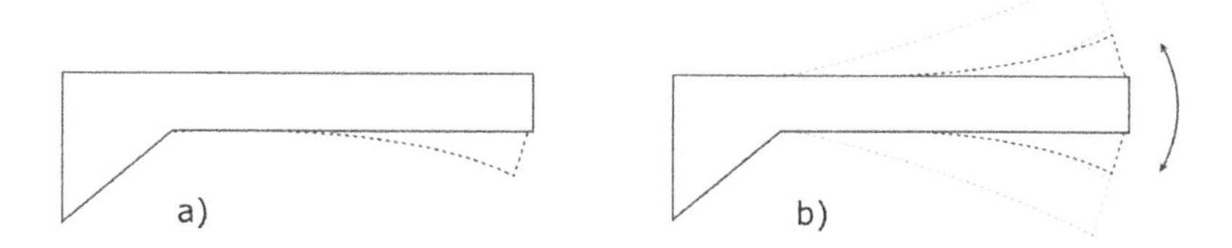

Figure 3.36: Static (a) and dynamic (b) mode of cantilever sensors.

can be considered as microelectromechanical systems (MEMS) sensors as they contain a mechanical moving part responding to the interaction with an analyte. The induced change is a mechanical motion, such as the mechanical resonance frequency of the sensing element or the deflection of the cantilever (see Figure 3.36). The volume of the coating will increase as hydrogen is absorbed into the metal lattice, causing surface stress which in turn will cause a deflection in the cantilever (see Section 3.11) or a shift in the resonance frequency (dynamical mode). Palladium is typically used as the coating.

Microcantilevers were originally developed for surface imaging in atomic force microscopy (AFM). An AFM consists of a cantilever with a sharp tip (probe) at its end that is used to scan the specimen surface at the atomic level. AFM probes are typically made out of silicon or silicon nitride. In 1991 Thundat and co-workers noticed that instabilities in their AFM measurements were due to fluctuations in humidity, which led them to conclude that microcantilevers could be used for chemical sensing [230]. As AFM cantilevers became standard commercial products, their availability has facilitated investigation into their use as transducers for chemical sensing. Using silicon microfabrication techniques, cantilevers can be as thin as a few nanometers with lengths that range from a few micrometers to several hundred micrometers. Typical dimensions are around 100 micrometers long, 20 micrometers wide, and 1 micron thick with spring constants less than 1 N/m.

In the dynamical operating mode the cantilever is excited into resonance. Actuation of the cantilever can be effected by a variety of means, such as an external piezoelectric platform on which the cantilever is mounted. For gas sensing applications microcantilevers can be used as a microbalance where the shift in resonance frequency due to mass loading is measured. Most research is being conducted into the use of the static operation mode of cantilevers for hydrogen sensing. Static mode mechanical sensors may, however, be susceptible to external sources of noise. As the resonant frequency of a structure is robust towards external noise, the signal-to-noise ratio in this mode of operation may be superior to the static mode. In spite of this, few reports can be found concerning dynamic mode hydrogen sensing, as the effect of mass change due to the absorption of hydrogen is low. However, other effects may also shift the cantilever's frequency. The rise in temperature due to the catalytic combustion of hydrogen led to an acoustic resonance frequency shift in a microscale acoustic resonator [231]. The sensor was fabricated on a standard single-mode fibre tip; the vibration of the resonator was excited and detected by optical means. A shift of about 1 kHz resonance frequency upon exposure to 1 vol% hydrogen concentration could be observed and the limit of detection was estimated to be about 0.1 vol%. The stress induced in the sensitive layer through absorption of hydrogen will also lead to a change in resonant frequency.

Hydrogen sensors based on quartz oscillators, such as tuning forks similar to microcantilevers, have been proposed. They are based on the fact that viscosity and molecular weight of gas mixtures containing hydrogen will decrease with the concentration of hydrogen. The momentum transferred from the surrounding gas molecules to a vibrating quartz oscillator causes a change in frequency. Oscillating objects in a gas atmosphere will lose mechanical energy depending on the momentum, and thereby weight, of the gas molecules. As hydrogen is a light molecule, and the viscosity of hydrogen only about half that of air, the loss of mechanical energy of the oscillator will be lower in the presence of hydrogen. The physical properties of the gas are analysed by sensing the energy transfer to an oscillator by the collision with gas molecules. The parameters measured are pressure, viscosity, and molecular weight of binary gas mixtures, or of hydrogen in air. This type of sensor can be operated without the deposition of a sensitive layer [232]. A sensor has been reported to work in a range of up to 100 vol% hydrogen at ambient conditions, with an accuracy of 0.5 vol% and a response time below 1 s [233]. Comparable performance has also been observed for a commercial electromechanical sensor. Severe problems related to the effect of temperature and humidity still need to be overcome. A reference oscillator can be used to compensate for changes in temperature and humidity. A hydrogen selective membrane can be added, as otherwise cross-sensitivity to other light gases such as helium may limit the utility of the sensor.

In summary, electromechanical sensors have the potential to be deployed as reliable hydrogen safety sensors when their susceptibilities to changes in environmental conditions are overcome.

3.8 OPTICAL HYDROGEN SENSORS

Valerio Palmisano and Eveline Weidner, European Commission, Joint Research Centre, Petten

The term optical sensor refers to a broad variety of sensors that measure a change of optical properties of the sensing element or of the external environment in response to a stimulus. The optical properties of some materials change in the presence of hydrogen, and this can form the basis for optical detection of hydrogen. These optical properties are related to the dielectric permittivity of the sensing material, which for hydrogen applications are typically palladium, WO_3, or metal hydrides. In addition to these methods, optical open path sensors measure the optical properties of the external environment (e.g., schlieren method). The dielectric permittivity is a complex function describing the interaction between the material and an electromagnetic field as a function of its frequency: the real part of the dielectric permittivity is related to the (real) refraction index while the imaginary part describes the attenuation of the electromagnetic wave.

Many gas species can be detected directly by absorption spectroscopy techniques, based on the specific interaction of the gas species with the electromagnetic radiation (see Chapter 4). This is difficult in the case of hydrogen gas due to its high transparency to light. As it forms a homonuclear diatomic molecule, hydrogen will not absorb infrared radiation. Furthermore, UV visible light is not commonly used for hydrogen detection since molecular hydrogen is nearly transparent in this region. Thus, hydrogen cannot be easily detected by common spectroscopic methods, such as infrared spectroscopy.

The optical signal may be directly detected by the human eye (e.g., indicator tape), but is typically processed and transformed into an electronic signal. To transform the electromagnetic signal into an electrical signal, the optical sensor makes use of a range of technologies such as photoresistors, photodiodes, and CCD cameras. Fibre optics or integrated optics may be used to generate, transport and process the electromagnetic signal. These sensors are used either as multiplexed sensors where multiple sensing points share a single transmission line of the signal or as fully distributed sensors where sensors intrinsically measure the spatial distribution of the analyte.

Hydrogen has the ability to change the optical properties of some materials or it can modify the environment in a manner that can be monitored by optical means. These effects have been exploited for the development of optical hydrogen sensors. Optical hydrogen sensors can be roughly divided into three groups:

Chemochromic sensors
These sensors are also called colourimetric or gasochromic sensors. They are based on materials that detect the presence of hydrogen by undergoing a visually observable change in colour. The effect is a change of the absorption spectra in the visible range which could also be recorded by a spectrometer. Chemochromic materials are used as the sensing elements and in indicator tapes and paints as a non-electric spot sensing method.

Fibre optic hydrogen sensors
Sensors based on optical fibres allow for a safe and flexible way of detecting hydrogen since the sensing points may be separated from the read-out electronics. This separation can even be over long distances. The fibre optic lines are used either as a means to transmit the electromagnetic signal or as an integral part of the sensing element.

Optical open path sensors
These sensors enable detection of the presence of hydrogen along a beam of light. They are particularly suitable for stand-off detection and wide area monitoring, which may be needed for outdoors or in partially enclosed facilities applications. These sensors do not rely on the direct absorption of light by hydrogen but are based on the change of other optical properties. Open path optical hydrogen sensors are particularly suitable for stand-off detection and wide area monitoring. Methods under investigation include Raman spectroscopy and the background oriented schlieren technique.

Despite the fact that numerous optical methods for hydrogen detection have been proposed and studied, the commercialization of optical hydrogen sensors remains a challenging issue and only a few types are commercially available [122]. Chemochromic sensors and indicators as well as fibre-optic type sensors have been developed into commercial products. There have been attempts at commercialization of other optical sensor types (as described in the following), but these have been met with limited success due to stability issues. However, due to the promising (theoretical) performance capabilities and the associated technological challenges, a wealth of scientific publications describing research and development in this field is available.

3.8.1 Chemochromic Sensors

Reversible chemochromic hydrogen sensors and irreversible indicators have been developed consisting of a thin film coating of a metal oxide. The reduction of the oxide by hydrogen results in a change of the colour of the coating. The colour change can be observed either with the eye, often from a distance, without an electrical signal or electrically by a quantitative measurement of the optical transmittance. In addition to a change in the optical properties, the reduction of metal oxides following exposure to hydrogen often results in a change in electrical resistance, which can also be used for sensing hydrogen (see Section 3.5). Tungsten tri-oxide (WO_3) exhibits colour changes resulting from a change in the oxidation state of the metal ions which are in the normally transparent crystalline oxide.The oxide is coated with a catalytic layer such as of palladium or platinum and will undergo a reversible colouration process upon exposure to hydrogen. The partially reduced oxide turns to blue or even black.

The adsorbed hydrogen atoms intercalate via the spillover process (see also Section 3.5) forming blue tungsten bronze. The chemical reactions can be given as [234]:

$$WO_3 + xH^+ + xe^- \longrightarrow H_xWO_3$$

This results in an increase of absorption in the visible range of light. Many prototype hydrogen sensors based on WO_3 have been reported using either platinum or palladium as the catalyst for hydrogen dissociation. Response and recovery times on the order of 1 s for H_2 concentration greater than 5 % have been reported [235]. Degradation of sensor performance due to the poisoning of the platinum catalyst is a concern for chemochromic sensors. The optimum preparation method of the oxide layer is still under investigation, and the reversibility of the reaction needs to be improved to ensure sensor durability. Test gas indicators for semi-quantitative hydrogen detection are available (e.g., Dräger tubes). Paints, tapes, and coatings as hydrogen indicators are being developed commercially. A paint formulation is reported using nano-particle Pt:WO_3 in a water-based polyvinyl acetate (PVA) paint showing a time constant of 40 s in nitrogen containing 10 vol% H_2 [236]. Hydrogen sensing textile materials are being developed for space applications, based on chemochromic pigment interdispersed within a textile polymer [237]. Optically active materials other than metal oxides have been proposed for use in chemochromic hydrogen sensors including yttrium and lanthanum hydrides [238] and magnesium hydride [239, 240] which also show a remarkable optical response to hydrogen.

3.8.2 Fibre Optic Sensors

Optical hydrogen sensors based on fibre optic systems are the most investigated due to their inherent safety and remote sensing capabilities. The use of optical guides, such as optical fibres, allows flexibility in the location of the reading and controlling electronics remote from the sensing point. Another advantage of note is their inherent safety, similar to other optic hydrogen sensors. Owing to the fact that their raw signal is optical rather than electrical, the optical sensing element will not generate electrical sparks. This facilitates the use of the sensor in potentially explosive environments (as the sensing element itself is highly unlikely to cause ignition). Other advantages are their resistance to corrosion and to electromagnetic interference [241]. Moreover, the use of an optical fibre with sensing material coated at different points along the length allows distributed monitoring of a large area using only one device [234].

Fibre optic hydrogen sensors were first described in 1984 [242]. Since then numerous types of optical fibre hydrogen sensors based on different transduction mechanisms have been developed including: fibre Bragg grating [243], interferometric [244], micromirror [245, 246], evanescent wave optical fibre [247] surface plasmon resonance [248] and colorimetric technology [249]. Optical fibre hydrogen sensors do not yet meet many of the performance specifications required by industry or regulatory bodies. In particular, their response time is often too slow and there is need to regularly calibrate the sensor to compensate for drift and ageing effects. Nevertheless, research is continuing to improve these performance aspects [250].

Palladium and related alloys [251] are among the commonly used materials for the sensing elements of hydrogen fibre optic sensors. Other materials used include switchable mirrors of rare-earth metals, magnesium alloys, and metal oxides. Absorption of hydrogen in palladium is accompanied by changes of both the physical lattice structure, due to expansion, and the electronic structure, due to the filling of the conduction band with electrons from hydrogen.

Figure 3.37: Illustration of the switchable mirror phenomenon: (a) Y is a shiny metal; (b) YH_3 is a transparent insulator/semiconductor. The transparent state was reported to be yellowish [238].

The refractive index of PdH_x decreases with the occupancy, x, of the H atoms in the lattice. Similarly, the reflectance of palladium decreases while the transmittance increases. The actual values of the optical transmission and reflection are strongly related to the geometry of the sensing element and to the wavelength of the optical probe.

In addition to palladium, hydrogen also shows a high solubility in other metals and metallic alloys. The absorption of hydrogen in some metals is associated with a metal-insulator transition. This is the case of magnesium and yttrium, which, in the presence of hydrogen, transform from a reflective metal to a transparent insulator, changing both optical and electrical properties. Each Mg atom can absorb up to two H atoms per atom while each Y atom can absorb up to three H atoms per atom. When the maximum amount of H atoms is dissolved into the lattice the conduction band is filled and the system becomes an insulator. The transition from conductor to insulator also manifests a change in reflectance that can be observed by the human eye [238]. The electrical and optical transitions are reversible by changing the surrounding hydrogen partial pressure, in a process that can occur at room temperature. All the trivalent rare-earth hydrides and some of their alloys exhibit switchable optical and electrical properties and can be referred to as "first generation" switchable mirrors [238](see Figure 3.37). In the transparent state they have characteristic colours: for example, yttrium hydride (YH_3) is yellowish and LaH_3 is red. "Second generation" switchable mirrors are based on alloys of rare-earth metals with magnesium. Magnesium is chosen as the alloying metal due to the large band gap of magnesium hydride which allows switching from the reflective metallic state to a colourless transparent insulating state (as opposed to the coloured rare-earth trihydrides). Therefore a variety of light sources (different coloured LEDs, broadband light sources) can be used in optical signals. "Third generation" switchable mirrors include rare-earth-free Mg-transition metal-hydrides. These switchable mirrors exhibit superior performance as they are more resistant to oxidation than rare earth metal hydrides and therefore improved cycling stability. It should be noted that this type of sensor only indicates the presence of hydrogen above a particular threshold and is not amenable to the precise quantification of hydrogen.

Based on the change of the properties of the electromagnetic signal interacting with the gas sensitive material, fibre optic gas sensors can be grouped as: [252]:

1. Phase-modulated sensors, in which the real part of the dielectric permittivity of the sensing element changes in the presence of the target gas stimulus and corresponds to a change of the phase of the electromagnetic signal.

2. Intensity-modulated sensors, in which the change of the imaginary part of the dielectric permittivity of the sensing element or medium to be tested experiences a change of intensity of the electromagnetic signal in the presence of the target gas.

3. Wavelength-modulated sensors, in which the change of the lattice spacing of a Bragg grating due to the presence of the target gas results in a spectral change of the electromagnetic signal.

4. Polarization-modulated sensors, which show a change in the dichroism of a birefringent crystal due to the presence of the target gas. This induces a change in the polarization of the electromagnetic signal.

Each of these fibre optic technologies has been developed for hydrogen detection, albeit to different extents. In the following sections more detailed information on the research strategies and development of each type, specifically for hydrogen detection, is provided with the exception of the polarization-modulated sensors. Since only a limited amount of research has been reported on this specific platform for hydrogen detection it is not discussed further but further information can be found in (e.g., [253]).

Phase-modulated sensors

Phase-modulated fibre optics sensors make use of interferometric methods to measure the presence of hydrogen by exploiting the high longitudinal coherence of a laser signal. The laser signal is split between two fibres, one of which is coated with palladium. The expansion of the palladium layer upon exposure to hydrogen stretches the fibre in both axial and radial directions, causing a change in its effective optical path length. When the two signals are recombined the optical phase difference results in a change of the interference pattern, which can be related to the hydrogen concentration. Phase-modulated sensors can be made with different geometries, i.e., as Fabry-Perot, Mach-Zehnder, Michelson, and Sagnac interferometers [254]. Phase modulated hydrogen sensors are very sensitive to hydrogen but their measuring range is limited [255]. Thick palladium layers (from a few micrometers to hundreds of micrometers) are necessary in order to measurably stretch the optical fibre. This results in a slow response time of the sensor.

Intensity-modulated sensors

Micromirror type

Micromirror sensors essentially consist of a thin film layer deposited on the cleaved end of an optical fibre. The thin film is made of material which interacts with hydrogen (such as palladium or other metal hydrides) and changes its reflectance and transmittance as a function of the hydrogen uptake.

Different phenomena, attributed to the absorption of hydrogen, contribute to the change in the optical properties, namely the increase of the layer thickness or the change of the dielectric permittivity associated with a change of the electronic structure. For palladium, an increase of the layer thickness is the main cause of the increased reflectance while for switchable mirrors the main effect is a change in the reflective index as the sensing layer goes from a reflective metal to a transparent insulator or a

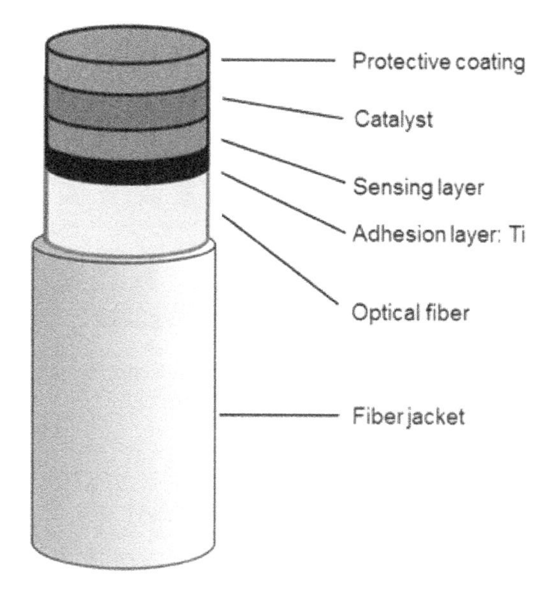

Figure 3.38: Schematic of the sensing point of a micromirror-based fibre optic sensor [256].

dark semiconductor. Durability and response time can be influenced by the choice of alloy composition. Controlling the thickness of the sensing layer allows for optimizing the optical contrast at a given wavelength. It should be noted that for a geometry where the sensing layer surface area (i.e., the amount of the interacting material) is restricted to the cross section of the waveguide, there is a trade-off between the optical contrast and the response time, as a thick layer requires more time for the diffusion of hydrogen into the film. More complex geometries have been explored and include a multi-stack layer (see Figure 3.38).

The main drawbacks of using switchable mirrors (i.e., Y or Mg alloys) for hydrogen sensing are related to the fact that the metal-insulator transition (see Chapter 1) is a sharp transition taking place only when the hydrogen concentration is above the threshold level. Thus, this technique functions more as a switch rather than as a sensor. Moreover, the transition is prone to a strong hysteresis, i.e., a difference is observed between the transition point when the hydrogen partial pressure is increasing and when the hydrogen partial pressure is decreasing. Similar to other metal-hydride based sensors, the transition also has a strong dependence on temperature and pressure. Despite these limitations, micromirror type sensors are the most developed technique among fibre optic hydrogen sensors. A hydrogen sensor based on an palladium coated fibre optic system is commercially available.

Evanescent wave type

Evanescent wave based optical fibre sensors relate the change of the evanescent field of the propagating mode to the presence of hydrogen. A schematic of the configuration of evanescent-wave fibre optic sensors is shown in Figure 3.39. The evanescent field decays exponentially as a function of distance in the space adjacent to the fibre's core. The ratio of the transmitted light intensity in the presence of hydrogen (I_H) relative to the transmitted light in the absence of hydrogen (I_0) depends on several parameters,

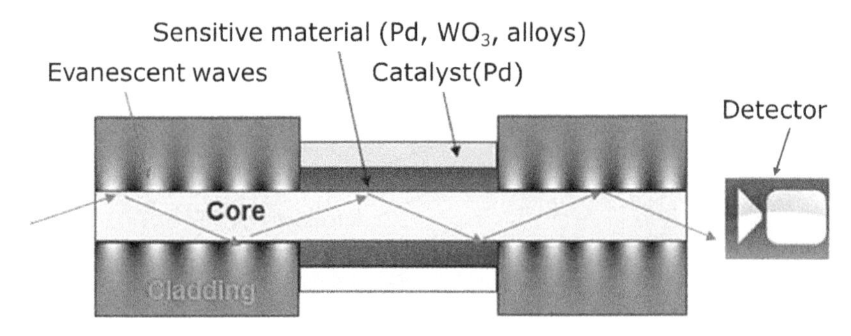

Figure 3.39: Schematic of the configuration of evanescent-wave fibre optic sensors.

according to the formula [257]:

$$I_H/I_0 = exp(-2rl\Delta\alpha) \tag{3.31}$$

Where: l is the interaction length, i.e., the stripped side of the fibre where the material is deposited; $\Delta\alpha$ is the variation in the absorption coefficient of the material due to the absorption of hydrogen; and r is the ratio between the power of the evanescent wave and the total power of the guided light. This dependence on multiple parameters provides great flexibility for optimising sensor performance. Whereas with micromirror type sensors the sensing layer surface is dependent solely on the cross-section of the waveguide, evanescent wave based optical fibre sensors can be optimized by varying the interaction length or the thickness and the composition of the sensing layer. Research has been focused on improving optical contrast and response times: for example, utilizing thin films of WO_3 [234, 258] or palladium and palladium/gold alloy [259] as sensing material.

Evanescent-wave based optic sensors have high sensitivity and fast responses, due to the large sensing area and the thinness of the sensing layer, respectively.

Surface Plasmon Resonance type

Sensors based on surface plasmon resonance have a similar physical structure as the evanescent wave based sensors, where the sensing layer is built on the stripped side of the optical fibre. Whereas the evanescent wave is present at all wavelengths, the surface plasmon effect arises due to the resonant interaction between the incoming wave and the conduction electrons of the sensing layer and, thus, takes place under specific conditions of frequency, direction, and polarization of the incoming wave. The plasmon resonance takes place at the plasma frequency ω_p:

$$\omega_p = \sqrt{\frac{Nq^2}{m\epsilon_0}} \tag{3.32}$$

Where N is the number of electrons; q is the electron charge; m is the effective mass of the electrons of the sensing material; and ϵ_0 is the dielectric permittivity of vacuum. The plasma frequency depends on the direction of the propagating wave. Moreover, because the plasmon resonance is an interaction with the electric field, only the transverse magnetic polarization is able to excite it. Due to the absorption of hydrogen the electronic structure changes and the plasmon resonance shifts. This results in a change of intensity of the incoming light or in a change in wavelength [241].

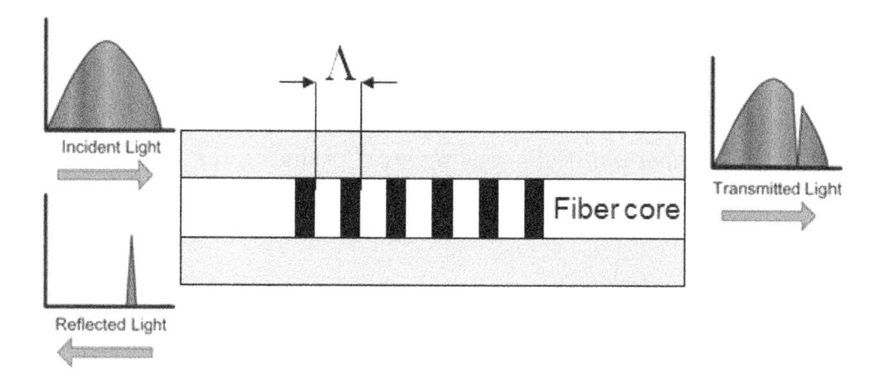

Figure 3.40: Schematic of a fibre Bragg grating with period Λ.

Wavelength modulated sensors

Fibre Bragg grating type

A fibre Bragg grating (FBG) has a repeated structure built along a segment of the core of an optical fibre leading to periodic change of the refractive index (see Figure 3.40). As a consequence of interference effects of the optical waves reflected at the interface between the media with different refractive indexes, part of the light will be reflected back with wavelengths around the Bragg wavelength λ_B (or multiples thereof), which is given by the formula:

$$\lambda_B = 2\eta\Lambda \tag{3.33}$$

Where η is the refractive index in the fibre and Λ is the period of the grating. The reflected wavelength is a function of the periodicity of the Bragg grating, thus an expansion of the grating due to temperature changes (thermal expansion) or stretching by other means will result in a change in the reflected wavelengths. Therefore Bragg gratings can be used for measuring temperature. For chemical sensing, the chemical interaction between the analyte species and the grating will give rise to a change in the period of the grating which induces a change in the reflected wavelengths. FBGs make use of the lattice expansion of metal-hydrides (especially palladium and palladium-based alloys) due to absorption of hydrogen. This effect is particularly interesting for sensing materials (alloys) in the α-phase (see Section 1.1) as the expansion increases continuously with the hydrogen partial pressure, according to Sievert's law which predicts a square root dependence on the partial pressure of hydrogen.

FBGs sensors are strongly influenced by temperature due to the thermal expansion of the grating. Moreover thick sensing layers are required to sufficiently expand the optical fibre along its length, which can lead to sluggish kinetics. The use of substrates with low Young's modulus allows enhancing the reproducibility and the linearity of this sensor type.

Long period fibre Bragg grating (LPG) type

Similarly to FBGs, Long Period Gratings (LPG) exhibit a shift in their resonance wavelength on interaction of the palladium coating with hydrogen. However, while the period of conventional FBGs is of the same order as the optical wavelength, the one of LPGs is significantly longer—typically 0.1 mm to 1 mm. An important difference is that, while the FBGs use fundamental modes that propagate along the core, LPGs are mainly based on the coupling between the cladding[9] modes and the evanescent wave or the surface plasmon resonance at the interface. Typically, LPG type hydrogen sensors make use of thinner sensing layers allowing for faster response times [252]. Moreover a much lower thermal sensitivity was reported, reducing the need for temperature compensation [260].

3.8.3 Optical Open Path Hydrogen Detection

Optical open path sensors detect the presence of hydrogen along a beam of light or along the viewing direction. These sensors are particularly suitable for stand-off detection and wide area monitoring, which may be needed for outdoors or in partially enclosed facilities. Various spectroscopic methods have been investigated for the detection of hydrogen, such as Raman, and laser induced breakdown. Sensors based on these spectroscopic methods can be grouped among the open path sensors, together with the schlieren and shadowgraph techniques.

As open path sensors detect the presence of an analyte along a beam of light, this is a spatially integrating method, i.e., the exact location of the analyte is not measured. However, pulsed laser sources are able to provide spatial resolution by analysing the time of flight of the signal, i.e., the time it takes for the signal light to reach the detector following a laser pulse. Scanning methods enable 2D visualization for wide area monitoring. The receiver and transmitter can either be separated or combined for reflective measurements, enabling stand-off detection.

Shadowgraph and schlieren visualization techniques

This technique was first described by Robert Hooke in 1665, whereby the optical inhomogeneity of a transparent media can be detected through the observation of schlieren (German for "veins" or "streaks") arising from refractive-index gradients. Schlieren can occur in gases, liquids, and solids. They can arise from changes in temperature or changes in other properties that induce density gradients, such as pressure and gas mixture composition. The much lower density of hydrogen as compared to air or nitrogen causes density gradients in the mixing zone of the gases, which can be visualized with both shadowgraph or schlieren techniques. The schlieren method is based on the deflection of a light beam towards a region with a higher index of refraction. The shadowgraph method uses the change in light intensity due to deflection and displacement of the beam.

The two techniques differ in complexity and the type of image formed. A shadowgraph does not need a focused lens and is well suited for large scale observation. It can be used to observe shock waves and turbulent flows. The schlieren method requires a stronger light source than shadowgraphy, but is more sensitive and gives an actual optical image through a lens.

The Background Oriented Schlieren (BOS) method measures the deflection of light as it passes though density gradients present in the gas sample by observing the local displacement of a background pattern. Apart from a patterned background, a high-speed camera is needed to record the changes in the pattern, which can be related to density gradients along the line of sight and thereby to the presence of hydrogen. Apart from visualizing hydrogen flows due to leaks, pressure waves due to explosions can also be observed with background oriented

[9]Optical fibres consist of a light-carrying core encompassed by a cladding with a lower refractive index.

schlieren using high speed cameras. Although quantitative measurements are possible, this method is not sensitive enough to detect hydrogen at concentrations around the LFL [261]. Shadowgraph systems have been implemented using an arc lamp as the light source and a retro-reflective screen as a background surface to visualize gas jet releases, flames, and detonations. This high speed method can be used to detect large scale hydrogen leaks by optical means and may help determine safety distances. Neither of these techniques are hydrogen specific as they are based on density gradients.

Spectroscopic techniques

Although classic direct absorption spectroscopic methods such as IR cannot be applied to detect hydrogen (as mentioned previously, hydrogen does not absorb light in the IR region), emission spectroscopic methods can be used for hydrogen detection. Excitation by a laser can be used to generate a characteristic emission spectrum of electromagnetic radiation of hydrogen. Emission spectroscopic methods are suitable for remote sensing, or stand-off monitoring.

Laser induced break-down spectroscopy

(LIBS): is a fast, non-contact optical method that is used for the analysis of solid, liquid, and gaseous samples. It is used, for example, in materials analysis or forensic studies. As portable systems are available, this technique is well suited for the on-site detection of hazardous materials in the field. A LIBS system consists of a pulsed laser, a sample chamber, a spectrometer, and associated optical systems. A laser generates plasma by dissociation of the sample material, which can be in a gaseous state or on a solid surface. The optical emissions of the plasma, i.e., the atomic emission lines of the elements in the plasma, are analysed. A low detection limit of around 0.002 vol% H_2 can be achieved, both in the gas phase or on a surface [262], However, a main drawback of this technique is that the water molecules in ambient air also give rise to hydrogen emission lines. It may be difficult to distinguish hydrogen leaks from variations in the background due to humidity changes. In addition, the plasma may be a source of ignition, which probably explains why this method is not widely pursued for sensing hydrogen, especially for safety applications.

Raman lidar and Raman based fiber sensors

Raman is a vibrational spectroscopic method for the analysis of solids, liquids, and gases. Raman generates characteristic spectral emission lines through excitation of the sample material. Raman scattering can be used both for distributed sensing based on fibre sensors and for stand-off monitoring with Raman lidar[10] (for an overview of this method refer to, e.g., [263]). Raman spectroscopy is based on the inelastic scattering of monochromatic light, as opposed to IR spectroscopy, which is based on absorption. The scattered light will be observed at a shifted frequency. The magnitude of the shift is unique for a molecule and a given vibrational state and does not depend on the incident light frequency. If the energy of the laser corresponds to an electronic transition of a molecule, resonance Raman scattering with a higher intensity will occur.

Hydrogen has a large Raman scattering cross-section compared to other gases and induces a large wavelength shift, permitting its detection by Raman spectroscopy. Raman scattering can be used for the remote detection of hydrogen either through fibre optics or by analysing the backscattered light of a laser source.

---

[10]Lidar stands for Light Detection And Ranging.

The basic set-up of a Raman lidar consists of a high intensity laser, a receiver and a detector. A pulsed laser source (with pulse widths typically 5 ns to 10 ns) is commonly used to enable distance resolution of the hydrogen leak. Spatial resolution is achieved through measuring the time of flight of the observed backscattered signal. Backscattered light is collected by a telescope, and filters are used to suppress unwanted frequencies. Raman lidar can in principle measure gas concentration as well as temperature and is commonly used for atmospheric sensing.

Raman fibre sensors have been utilised for distributed temperature sensing which enable sensing lengths of more than 30 km. Potential applications for Raman fibre sensors include surveillance of electrical power transmission and monitoring of the integrity of pipelines. The use of Raman scattering for hydrogen sensing was first reported for space applications, as hydrogen leaks are a great concern for this application. Both fibre optic and lidar techniques have been used [264, 265]. A spontaneous Raman sensing system for trace hydrogen gas detection achieved a detection limit below 0.01 vol% [265]. Hydrogen gas concentration from 0.6 vol% to 100 vol% could be measured in outdoor conditions at a distance around 13 m using a compact portable Raman lidar [266]. Hydrogen release rates of 10 L/min to 50 L/min resulted in signals with a signal-to-noise ratio of 50 at 15 m distance and > 5 at distances between 30 m to 50 m [267]. Two-dimensional mapping by scanning the laser enabled observation of the spatial distribution of hydrogen gas.

Potential applications of Raman lidar include outdoor settings such as monitoring of pipelines, refuelling stations, or large-scale hydrogen production facilities. Spatial resolution of a hydrogen plume can be achieved using time of flight analysis methods on the signal. In contrast to other optical wide area monitoring techniques (such as Schlieren) Raman detection does not need a reflective surface. The potential stand-off distance has been demonstrated in the range of 10 m to 50 m [267]. Scanning of an area is possible, but is currently limited to the field of view of the receiver. Reliable quantitative measurements may be challenging, but a comparison of the signal for hydrogen with that of nitrogen may be used to quantify the hydrogen concentration. Disadvantages include a high detection limit and interference from fluorescence. At present a measureable signal can only be obtained from large leaks of 10 L/min H_2. There is also the need for high power lasers. The use of high power lasers is highly regulated and is not feasible for settings in which persons might be endangered. Lasers can also pose a hazard for detecting combustible gases as they can act as an ignition source. Fibre optic Raman systems seem promising, but further development into a commercial product has yet to occur.

3.9 SONIC HYDROGEN DETECTORS

A pressurized gas leak can cause acoustic emissions. The expansion of a gas moving from high pressure to low pressure through an orifice will lead to turbulent flow, which in turn creates acoustic waves in the audible to ultrasonic range. The acoustic waves are typically detected by a microphone. The detection limit is determined by the amount of turbulent gas flow generated by a particular leak. The high frequency, ultrasound emissions are localized around the leak site. Scanning an area "listening" for ultrasound emitted enables detection of a gas leak. Leak detection through the measurement of these acoustic emissions has been established for over a century and has been used for detecting hydrogen leaks; however, it is not specific to hydrogen. A second type of sonic gas detection method is based on measuring

Figure 3.41: Stationary sonic gas leak sensor [269].

the speed of sound in gases. As the speed of sound in gas is a function of pressure and molecular weight of the gas, it can be used to determine the molecular weight of a gas or of a gas mixture.

The methane whistle, or "Schlagwetterpfeife" as it is known in German, is an historic example of a safety gas sensor. A whistle to detect methane (CH_4) for use in coal mines was developed by Haber in 1913. The pitch of the whistle depends on the composition of the gas in the resonating cavity. Due to the lower molecular mass of methane compared to air, the speed of sound will increase in the presence of methane. The difference in the sound produced by the whistle in the presence of methane is, however, difficult to detect by the human ear. Therefore a second reference whistle containing air was used. The sound produced by the two whistles generated an interference effect, revealing subtle differences in pitch due to changes in gas mixture composition. The clearly audible interference beats were directly related to the concentration of methane. The whistle acted as an open-ended resonator, a concept built upon in modern sonic gas detectors. This particular historic safety sensor had to be recalibrated after a few days, which limited its success [268].

Both these sonic gas detection techniques have been developed and applied to hydrogen leak detection and hydrogen concentration measurement, respectively. Details of their use for these purposes are provided in this section.

3.9.1 Leak Detection

Commercial ultrasonic gas leak detectors are used to monitor pressurized pipes and vessels. They detect the noise generated by a leak using a microphone, at distances of up to 15 m. A turbulent flow through a leak occurs when the internal pressure is at least twice the external pressure. The pressure drop, leak size, and gas properties determine the characteristics of

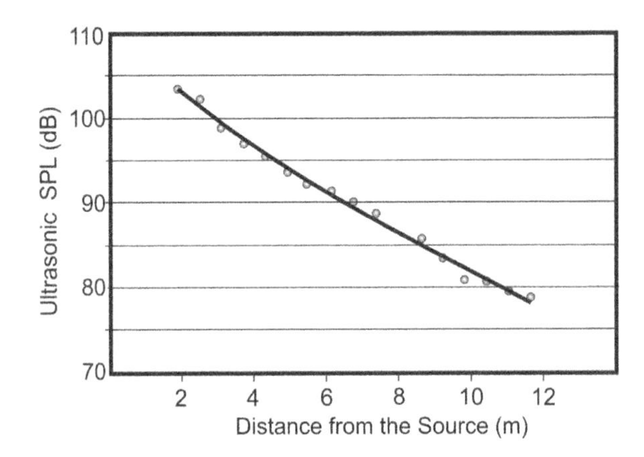

Figure 3.42: Ultrasound pressure level as a function of distance for hydrogen leaks [269].

the acoustic wave. The intensity and frequency of the sound signal is a complex function of the pressure drop and the size and shape of the orifice.

Noise frequencies above 24 kHz to 10 MHz and sound pressure levels of 40 dB to 105 dB are measured to indicate the presence of a leak. The audible range (approximately from 20 Hz to 20 kHz) is removed by band pass filters. Therefore the monitor is not affected by many of the background noises produced in an industrial environment. These devices are typically used to detect leaks of compressed gases ($>$ 10 bar) having a leakage rate above 0.01 kg/s. Ultrasonic gas leak detectors are not affected by leak orientation or concentration gradient of the gas plume. Ultrasonic systems are typically used to pinpoint the source of a leak, but cannot measure gas concentration nor distinguish between different types of gases. This technique can be used for hydrogen leak detection in applications where hydrogen specificity is not required. A commercial stationary ultrasonic gas sensor, using a microphone to detect gas releases, is shown in Figure 3.41. The device operates in the temperature range of -40 °C to +75 °C and relative humidities of up to 100 %. Besides stationary mounted sensors, portable and handheld devices are also available, some of which are combined with gas sensors [270]. Ultrasonic gas leak detectors have a response time in the order of milliseconds; however, an alarm delay time may be introduced to avoid spurious alarms [269].

A great advantage of ultrasonic leak detection is its ability to survey larger areas (also referred to as wide area monitoring). Although the sound pressure level decreases rapidly with distance from the source of the noise, as can be seen in Figure 3.42, acoustic waves can be detected at distances greater than 10 m. Among the disadvantages of this detection method are a limited spatial resolution and a dependency on the leak size and pressure differential.

3.9.2 Concentration Sensors

The measurement of the speed of sound in a gas mixture can be used to detect the presence and concentration of gases including hydrogen. Acoustic gas sensor measurements are based on the dependence of the speed of sound in a gas mixture on the composition of the gas mixture. The speed of sound in gases is a function of gas density, which depends on molecular mass as well as the gas temperature and pressure. This approach has been used to analyse binary and ternary gas mixtures. The speed of sound, ν, in a gas mixture with molecular

Table 3.5: Speed of sound of gases [272]

Gas	**Speed of sound (m/s) at 20 °C**	**Temperature coefficient (m/s·°C)**
Hydrogen	1310	2.2
Dry air	344	0.607
Nitrogen	349	0.85
Oxygen	327	0.57
Methane	442	0.62

mass, M, depends on the adiabatic index, κ ($\kappa = c_p/c_v$), and absolute temperature, T, according to:

$$\nu = \sqrt{\frac{\kappa RT}{M}} \tag{3.34}$$

Where R is the universal gas constant. Measurement of the speed of sound in gases at a known temperature and pressure can be used to determine the molecular weight of the gas. This method is particularly suited to the detection of hydrogen, as the speed of sound of hydrogen is very high compared to air, see Table 3.5.

Different methods are used to determine the speed of sound in a gas. The travelling time method is based on the measurement of the time that an ultrasonic burst needs to travel a defined distance through the gas mixture to be analyzed (see Figure 3.43). The effect of changes in gas composition was analysed with the travelling time method in an experiment where acoustic waves were generated by photo thermal excitation of a black body by pulsed laser [271]. A linear relationship between the time shift and the hydrogen concentration was found, with a lower detection limit of about 0.02 vol%. The response and recovery time were reported to be on the order of seconds. The device was said to operate in a wide temperature range, but the accuracy of the measured concentration depends on the temperature variations of the environment. An even shorter response time of 1 ms was reported [272] for measuring H_2 concentrations down to 0.01 vol%. However, the strong effect of temperature on the sound velocity needs to be compensated. A prototype handheld battery powered sensor has been built. The robustness of this method to poisoning was tested [273] and revealed that continuous exposure to 5 ppm NO_2 did not result in any significant effects.

As mentioned for the methane whistle described previously, the speed of sound can also be measured via the resonant frequency of a resonator. Within a resonator, the wavelength of standing waves are fixed. Therefore the resonance frequencies are proportional to the speed of sound, which is determined by the gas mixture contained in the resonator. This principle of ultrasonic resonance within a cavity has been applied for the detection of hydrogen. Ultrasonic stationary waves can be generated within a cavity, through reflection of the wave by the walls. This will lead to the absorption of the vibrational energy by the gas contained in the cavity. In the case of resonance, the oscillator driving the resonator will stop vibrating. An output voltage proportional to the vibrational amplitude of the resonator can be measured. A lower detection limit of 0.01 vol% hydrogen in air and 0.001 vol% hydrogen in argon was found, with a response time under 1 s [272, 274].

A linear relationship between the resonance frequency and the hydrogen volume fraction has been reported for a magneto electric (ME) drum transducer system [275]. Hydrogen detection in the range of 0.25 vol% to 2.25 vol% in air was realized by monitoring the change

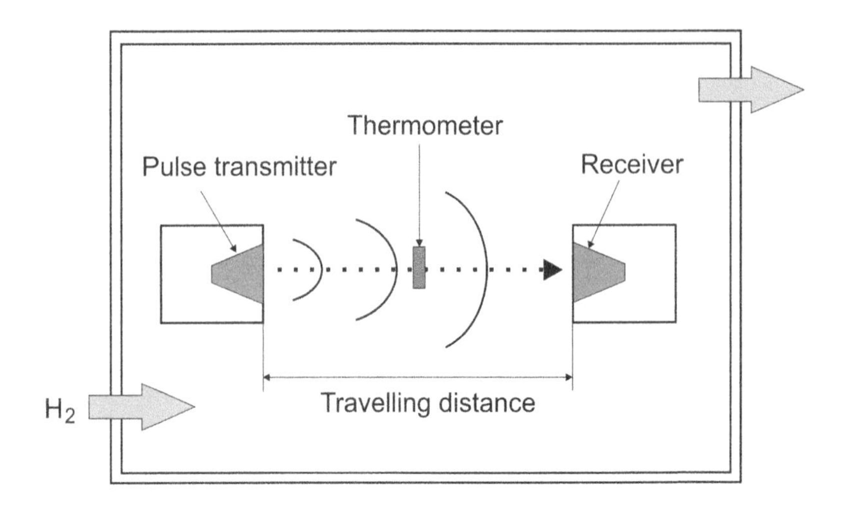

Figure 3.43: Schematic of an ultrasonic measurement system. Used with permission from [272].

of the ME resonance frequency due to changes in the acoustic wave velocity in the resonant cavity.

Acoustic gas sensors can generally achieve high reliability and long-term stability. Depending on the sensor design, they may be suitable for deployment in harsh conditions and their fast response time is very attractive. These performance capabilities may make this technology attractive for hydrogen detection and quantification in some applications. However, compensation of the effects of temperature and humidity might prove difficult. Furthermore it is not possible to distinguish between light gases such as hydrogen, helium, and methane.

3.10 SELECTING A HYDROGEN SENSOR

Equipped with a good knowledge of the foreseen operating conditions and the hydrogen sensor performance requirements of the specific application, the process of selecting a hydrogen sensor begins with an appreciation of the various sensing technologies which are commercially available and a knowledge of the performance capabilities and specificities of the individual sensor products.

3.10.1 Comparison of Hydrogen Sensing Platforms

In this section different platforms for hydrogen sensing are compared based on data published in the literature including information provided by manufacturers in their product specifications as well as from the experience of the authors on the testing and use of hydrogen sensors [85, 102, 276]. Sensing platforms are compared to illustrate similarities and differences in their basic performance parameters. A summary of the data is compiled in Table 3.6. Shortcomings for each sensing platform are listed to highlight areas for potential improvement.

The performance of a hydrogen sensor depends not only on the sensing platform it uses. The techniques and quality of manufacturing technology used to fabricate the sensing element, in addition to the data acquisition and signal processing capabilities have a decisive

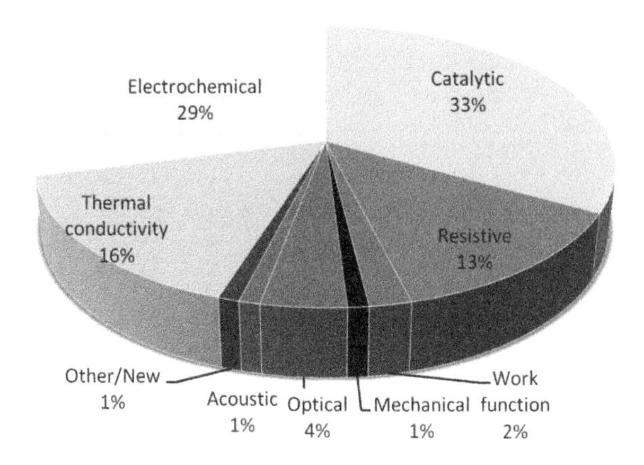

Figure 3.44: Sensor technologies applied in the products listed in the "H2Sense Hydrogen Sensor Database" [277].

influence on a sensor's performance. Proper use of a sensing platform under the correct operating conditions is necessary to ensure optimised sensor performance. For example, catalytic pellistor sensors will not function properly if operated in an environment containing less than the stoichiometric amount of oxygen required for the reaction. Doing so can have a temporary or even permanent effect on the sensor response [123]. In this respect selection of a suitable sensor platform based, not only on the sensor performance requirements of the application but also, on the analysis of the expected operating conditions of the sensor during its lifetime is extremely important. Guidelines for sensor selection are given in Section 6.4.

3.10.2 Commercial Hydrogen Sensor Availability

Any end-user looking for a commercial hydrogen sensor will be spoiled for choice when it comes to selecting a product for their application. Over the past number of years the choice of products has increased enormously. More and more gas sensor manufacturers are developing new or modifying existing products to meet the sensing needs of the growing hydrogen-inclusive economy. Recently the results of a market survey were published within the frame of a European funded research project called H2Sense—Cost-effective and reliable hydrogen sensors for facilitating the safe use of hydrogen.[11] The survey identified more than 90 sensor manufacturers providing more than 400 products—either hydrogen-specific sensors or non-specific sensors with hydrogen sensing capabilities. A database of commercially available hydrogen sensors was created by the consortium for consultation by hydrogen industry scientists and engineers as a convenient tool for sourcing a suitable hydrogen sensor for their specific application [277].

The results of the H2Sense market survey confirm the introduction of many new products onto the market when compared with a similar survey reported in 2010. At this time 53 different hydrogen sensor models from 21 different manufacturers were identified. However, similar to the 2010 survey the H2Sense survey also highlights the prevalence of catalytic,

[11]The H2Sense project was funded by the Fuel Cells and Hydrogen Joint Undertaking (FCH JU), a public private partnership supporting research, technological development and demonstration activities in fuel cell and hydrogen technologies in Europe. The H2Sense consortium comprised European hydrogen sensor manufacturers, sensor end-users and sensor testing bodies. The project included a collaboration and information exchange with a national laboratory of the U.S. Department of Energy (NREL).

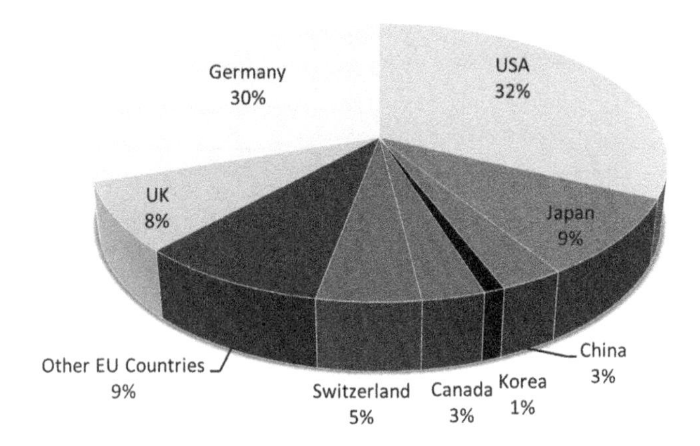

Figure 3.45: Location of hydrogen sensor manufacturers listed in the "H2Sense Hydrogen Sensor Database" [277].

electrochemical, thermal conductivity, and conductometric technologies among the commercial products (see Figure 3.44). Combined, these technologies account for more than 80 % of commercial hydrogen sensors. The survey also highlighted that most sensor manufacturers are located in the United States (32 %) followed closely by Germany (30 %), Japan (9 %), and the UK (8 %) (see Figure 3.45). Furthermore a notable number of hydrogen sensor manufacturers were identified in Switzerland, China, Canada, and other EU countries such as The Netherlands and Italy.

3.10.3 General Guidelines

As has been shown the performance parameters for different classes of hydrogen sensors can differ greatly. Such parameters, including lower detection limit, response time, power consumption, influence of ambient operating conditions, physical size, cross-sensitivity to other species, and ability to operate in oxygen depleted conditions, are critical considerations when selecting an appropriate hydrogen sensor for a specific application. Depending on the application there will be certain performance requirements for the sensor and a cross-check between these requirements, and the sensor specification will aid in the selection of the most suitable sensor. For example hydrogen sensors with a low power consumption should be selected for isolated applications where there is no connection to the grid supply for powering the sensor. In this case power will be supplied by a battery, and sensors with a high power consumption will drain the battery quickly. In applications where the background ambient gas composition can change, e.g., because of use of inert gas purges for system flushing or safety reasons, sensors which are unaffected by changes in ambient gas composition should be used.

Hydrogen sensor selection is not a trivial task and is often hindered by the fact that sensor manufacturers do not always provide a comprehensive list of detection specifications with their product. Furthermore there may be a lack of knowledge concerning the possible operating conditions which the sensor may experience during its working life. The latter point is particularly true for emerging hydrogen applications where experience and precedence are lacking. Detailed guidelines on choosing the most suitable commercial sensor for a specific application are provided in Section 6.4 together with guidance on its correct deployment.

A hydrogen sensor's detection limits and measuring range must be appropriate for the application and compatible with the required alarm levels. For most hydrogen detection devices the first alarm threshold is typically set around 0.4 vol% of hydrogen, while the main alarm is triggered at some concentration in the vicinity of 1 vol% or 2 vol%. Whereas these alarm thresholds may be appropriate for many applications they may be unsuitable for others where an earlier warning threshold may be needed. Similarly rapid detection of hydrogen is often desirable to ensure a timely alert to its presence and to allow sufficient time for engaging mitigative and protective actions, e.g., activation of forced ventilation system, evacuation.

When purchasing a hydrogen sensor for an application it is important to identify clearly the specific performance expectations of the sensor in the context of the application. This will assist in selecting the most suitable product by cross checking against the product specifications. Information on hydrogen sensing technologies provided in this chapter should equip hydrogen sensor end-users with a better understanding of the available sensing principles not only to assist in choosing the right sensor but also to provide better insight into precautions to be taken during its use. Correct use of an appropriate hydrogen sensor as part of an integrated safety system forms an effective risk mitigation strategy by facilitating accurate and fast detection of a potentially explosive atmosphere and activating suitable protective measures.

Table 3.6: Overview of principles used or being developed for hydrogen sensing

Sensing Platform	Operating Principle/Device	Physical Change/ Measurand	Advantages	Limitations
catalytic	pellistor	temperature resistance	robust, stable, long life time, broad measuring range, wide operating temperature range	not hydrogen selective, moderate lower detection limit, high power consumption, requires O_2 to operate, susceptible to poisoning by S and Si compounds
	thermoelectric	thermoelectric voltage	near room temperature operation, low power consumption	slow response time, sensitive to temperature fluctuations, requires O_2 to operate
	pyroelectric	pyroelectric voltage	near room temperature operation, low power consumption, selective to hydrogen	slow response time, sensitive to temperature fluctuations, requires O_2 to operate, not widely commercially available
thermal conductivity	calorimetric	temperature, resistance, voltage	very wide measuring range up to 100 vol%, robust, can operate in the absence of O_2, long term stability, resistant to poisoning, simple construction, fast response time	cross-sensitive to He, moderate lower detection limit
electro-chemical	amperometric	electrical current	limit of detection down to 0.001 vol%, low power consumption, room temperature operation, resistant to poisoning, heating of sensor element not required	narrow operating temperature range, limited lifetime, regular calibration needed, cross-sensitivity to CO, susceptible to humidity variation, ageing
	potentiometric	electromotive force, voltage	operation at high ambient temperature possible	not widely commercially available
conducto-metric	semiconducting metal oxide/ semistor	conductance	high sensitivity, fast response, acceptable lifetime, wide operating temperature range, modest power consumption	poor selectivity, interference from humidity and temperature, high operation temperature, susceptible to ageing and memory effects, requires O_2 to operate

	metal film	resistance	very wide detection range, rapid response, selective, long-term stability, can operate in the absence of O_2	susceptible to signal drift, requires protection from poisons
field effect	MIS capacitor	capacitance	high selectivity, very low limit of detection, fast response, low power consumption, low costs, small size, micromachinable	drift, hysteresis
	Schottky diode	voltage	small size, micromachinable, near room temperature operation	susceptible to drift, not commercially available (only prototype is available)
	MOS field effect transistor,	voltage	very fast response, accurate measurement, low influence from ambient parameters, high sensitivity and selectivity, small size, mass production possible	baseline and hydrogen induced drift, hysteresis
electro-mechanical	Quartz crystal microbalance (QCM)	frequency, time, wave velocity	very high sensitivity, room temperature to 100 °C operation, can operate in the absence of O_2	interference from humidity and temperature, drift
	Surface acoustic wave (SAW)	frequency, time, wave velocity	high sensitivity, room temperature operation, can operate in the absence of O_2	unstable at higher temperature, interference from humidity and temperature
optical	optrode	transmission, reflectance, wave length polarisation, phase shift	no source of ignition in explosive atmospheres, unaffected by electromagnetic interference, wide area monitoring possible, can operate in the absence of O_2	interference from ambient light, drift due to ageing effects, poisoning by SO_2, H_2S
sonic	ultrasonic	speed of sound, noise intensity	robust, operable in harsh conditions, very wide detection range, rapid response, low power consumption, room temperature operation, long-term stability, can operate in the absence of O_2	low selectivity, high lower limit of detection, knowledge of acoustic properties of materials through which the sound travels is necessary

3.11 FUTURE TRENDS IN THE DEVELOPMENT OF HYDROGEN SENSORS

In general, trends in the development of gas sensors, including devices for hydrogen detection, are driven by increasing demands for improved sensor performance and adaptation to specific applications. This includes requirements for lower detection limits, faster response times, broader dynamic measuring range, or enhanced explosion protection. Additional requirements include lower capital costs, reduced installation and operation costs, lower energy consumption, simplified methods for recalibration, and lower frequency for maintenance. Improved communication capabilities are also desirable, such as field bus systems and wireless capabilities.

To meet the increasingly stringent sensor performance requirements, research and development efforts are focused on improving already established sensor technologies to enhance their sensing properties and to develop improved manufacturing technologies. As an example, catalytic properties of sensing layers have been optimised in order to increase their selectivity, long term stability, and tolerance to poisons. Particular applications may call for the wide area monitoring of hydrogen release. This may be needed in large and/or unenclosed spaces, or when the potential release could take place in different locations. New sensitive materials and new sensing and measurement principles are also being investigated. There is a general trend towards miniaturization of electronic components as well as increased system complexity and functionality, which is also reflected in the gas sensing realm [278].

In the following section specific trends in the development of hydrogen sensors are identified including trends in the development of new sensing methods, materials, sensor elements, and sensor design.

3.11.1 Mechanical Microcantilever Sensors

Mechanical sensors based on microcantilevers are subject to much research and development activity. They are fabricated from miniature beams and are typically produced using microfabrication technology. Microcantilevers operate in two different modes—static and dynamic; the latter type is described in Section 3.7. Operating in the static mode, cantilevers will deform due to stresses caused by the interaction of the target analyte with the selective thin-film coating on the surface (see Figure 3.36). The concept of measuring surface stress changes through the bending of a thin plate was first reported in the 1960s [279]. With the bending plate technique, surface stress changes of 1 N/m were reported, but with the advent of micron-sized cantilevers using silicon microfabrication techniques, it became possible to determine surface stress changes as low as 10^{-6} N/m [280].

As described in Section 3.7.4, microcantilevers were originally developed for surface imaging in Atomic Force Microscopy (AFM). An AFM consists of a cantilever with a sharp tip (probe) at its end that is used to scan the specimen surface at the atomic level. AFM probes are typically made out of silicon or silicon nitride.

The first report of a cantilever based hydrogen sensor was published in 1969 [281]. Heated bimetal strips of 100 mm length and 5 mm width, iron strips coated with different palladium-silver alloys as the selective layer, were used. The deflection of the bimetal strip due to exposure to hydrogen was monitored visually with a microscope. The observed deflection ranged from 0.4 mm to 4 mm in response to hydrogen concentrations from 0.005 vol% to 1 vol% hydrogen in nitrogen. Response times of less than 1 min were observed at temperatures greater than 250 °C. This promising approach was exploited several decades later when microcantilevers and the corresponding highly sensitive optical transduction methods became available. Recent research on cantilever hydrogen sensing elements focused on overcoming problems due to:

Limited lifetime and drift of the sensor

Detachment of the hydrogen sensing material from the substrate (also referred to as delamination) has been observed in cantilever hydrogen sensors, limiting their lifetime. Delamination occurs due to the build-up of interfacial stresses between the sensing material and the cantilever substrate, often induced by hydrogenation. Delamination was addressed by using palladium alloys as well as investigating the influence of deposition technique and cantilever surface modification. A deterioration of the palladium layer through plastic (irreversible) deformation and build-up of residual stress was noted by several authors (e.g., [282]). Drift of the sensor signal, which was commonly observed, was linked to residual stress from trapped hydrogen atoms. These can cause a strain field changing the baseline reading of the sensor.

Improvement of response time

Both the bending rate and the magnitude of deflection depend on the hydrogen concentration [283]. The bending rate, however, gives a much faster response, as it was not necessary to wait until the maximum deflection was reached. The response time may also be reduced by optimizing the coating, for example palladium-nickel alloys [284] exhibited a much faster response to hydrogen than pure palladium. A 3 s response time and a 10 s recovery time were reported [285] by using nanoporous palladium thin films, formed through a wet chemical process.

Effect of environmental parameters and interferants

Temperature sensitivity was another potential issue for cantilever hydrogen sensors as strain-induced deformations due to temperature changes can be expected for a bi-material cantilever. The bending due to the unequal thermal expansion of each material in a bi-layer is the operating principle of thermostats and needs to be taken into account in the design of mechanical hydrogen sensors. Another environmental parameter found to affect the sensor performance was the presence of oxygen. Pronounced drift was ascribed to the oxidation of the palladium coating forming a surface metal oxide layer, this phenomenon occurred in cantilever sensing elements with palladium, palladium-nickel, and platinum layers. The formation of oxides on the surface can be prevented by optimizing the deposition technique [286]. The response of the sensor can be affected by other species adsorbed on the surface. Under normal conditions any surface is usually covered by adsorbates such as water vapour. Humidity reduces the activity of the palladium surface, thereby decreasing the signal [286]. The chemically selective coatings were also susceptible to chemical poisoning. Many sulphur compounds irreversibly bind to platinum and palladium surfaces, and this leads to a significant reduction of the sensor response. Protective hydrogen permeable coatings have been incorporated into palladium thin films to protect against poisons and interferants.

3.11.2 Hydrogen Sensing Nanomaterials

Nanomaterials are of interest for sensor development because of their high surface area, possible increased reactivity due to the small particle size, and compatibility with miniaturization of sensing elements and sensor electronics.

Nanoscaled carbon materials

These materials including single and multiwall carbon nanotubes (CNT), as well as two-dimensional graphene or graphene oxide sheets, have outstanding properties due to their high aspect ratio, large surface area with respect to volume, superior chemical and thermal stability, high electrical conductivity, good heat conductance, and excellent mechanical strength. It is expected that gas sensors based on CNT or graphene networked films can provide high sensitivity, fast response, good reversibility, higher resolution, selectivity by functionalization, simplified and stand-alone operation, low power consumption, and low running costs. Graphitic layers, carbon nanotubes, and two-dimensional graphene sheets were tested alone or in combination with other substances, including palladium, for hydrogen sensing mostly in resistive mode [287–289]. Other new nanomaterials are used in resistive or field effect based sensing elements.

Continuous thin metal film sensors

Recent research mainly on palladium thin films has focused on overcoming some of the previously mentioned performance issues encountered by these sensors (see Section 3.5.3), namely slow response times for room temperature operation and a pronounced degradation of the response arising from structural deformations (cracking, blistering, and delamination) of the pure palladium film during its phase transition. Nanostructuring of the sensitive layer by various means has led to improved performance including better adhesion of the metal to the substrate. Nanostructured thin layers have an even greater availability of surface and sub-surface sites for hydrogen adsorption, compared to thin films, which can help in improving the response times. The expected short response times and high sensitivity in large scale manufactured sensing elements still have to be proven. In addition to the manufacturing and positioning of nanowires by lithographically patterning, bottom-up growth or ion-milling are technically challenging.

Discontinuous thin metal film sensors

Whereas most thin palladium-based resistive film hydrogen sensors exhibit a reduced conductivity when hydrogen is absorbed, nanogap ultrathin film palladium film sensors and reversible on-off switches show the opposite behaviour. In these systems hydrogen absorption causes the volume of the palladium to expand, closing nanogaps in the ultrathin film resulting in an increase of conductivity [290–292].

Discontinuous metal nanostructures such as single nanowires or arrays of nanorods have been extensively investigated due to their large surface-to-volume ratio (for an overview see [193, 293]). Palladium and palladium-silver wire arrays have been used to detect hydrogen based on gap closure between the wires. The amplitude of the detection signal is based on the number of connections between the wires due to hydride formation. The detection range depends on the morphology of the wires. By enlarging the distribution in size and shape of the wires and therefore the gaps, the sensor's dynamic measuring range can be extended. A detection range of up to about 60 vol% of hydrogen in nitrogen has been reported for a PdAg alloy wire array [294].

Oxidic nanomaterials

Tin oxide nanorods, zinc oxide, or titanium oxide nano wires, have also been investigated for hydrogen sensing (see Section 3.5) [295, 296]. In a recently developed tin oxide nanorod sensor, a measurable change of resistance was detected at room temperature at a hydrogen concentration of just 0.01 vol%; however,a better sensitivity was obtained at 250 °C. The tin oxide nanorods can be grown by gas phase methods (CVD, thermal evaporation) or liquid phase processes (sol-gel, hydrothermal). Zinc oxide nanorods and nanowires, fabricated using anodized aluminium oxide nanotemplate or deposited by molecular beam epitaxy (MBE), have also been reported. Zinc oxide nanorod arrays exhibit a high sensitivity for hydrogen in a wide concentration range from 5 ppm to 500 ppm, similar to palladium or platinum coated gallium nitride and indium nitride nanowires. Titanium dioxide nanotubes are also used for hydrogen sensing and have been produced by anodic oxidation of titanium in hydrofluoric acid with subsequent annealing at about 500 °C resulting in nanotube pores with a diameter of 70 nm. This material can detect hydrogen in the range of 0.001 vol% to 1 vol% at temperatures between 150 °C and 300 °C.

Porous silicon

Porous silicon is a promising material for gas sensing mainly due to the high surface-to-volume ratio and strong adsorption for gases. It is easy to fabricate by electrochemical anodization using a mixture of hydrogen fluoride (HF) and ethanol in different ratios as the electrolyte solutions. By varying the process parameters the nanoscaled surface morphology can be advantageously controlled. Porous silicon layers have large internal surface areas of 200 m^2/cm^3 to 500 m^2/cm^3, and high activities for surface reactions. Porous silicon is compatible with silicon technology and modification with catalytic palladium or platinum makes it suitable for hydrogen detection [187]. However the distribution of the catalyst over the porous silicon can have a strong influence on the hydrogen sensing parameters.

Polymer thin film hydrogen sensors

Research into nanostructured conducting polymers, based on polyaniline, polypyrrole, or polythiophene for sensing applications has been performed due to their high electrical conductivity and potential for room temperature operation. In contrast to most metal oxide based gas sensors, conducting polymer sensors can be operated at room temperature. Conductivity is attributed to the mobility of charge carriers along the polymer backbone, which can also hop between polymer chains. The electrical conductivity of the polymer depends on its oxidation state and on the degree of protonation, which will influence the concentration of charge carriers. A chemical species may cause resistance changes based on complex redox processes which occur following interaction with the polymer. At present, the use of polymers for hydrogen sensing is still facing several performance issues, among them a low upper detection limit, long response time, and non-stable response. Polymer films may be used to detect hydrogen at ppm level, as recently a detection limit of 5 ppm at room temperature was reported for a palladium-decorated nanoporous mixed polymer nanostructure, with a response time of 90 s and a recovery time of 40 s [297].

Despite their promise of potential low cost, small size, fast response, and high sensitivity, the exploitation of nanostructured metal, metal oxide, polymer, and carbon-based sensing materials for hydrogen detection is hindered by the often costly and elaborate processes that are currently required for their reproducible fabrication. Research has also shown that due to changes in the microstructure, performance properties of these materials are difficult to predict and to stabilise. Research is currently focused on ways of addressing these issues, with the goal of a potential commercialisation of hydrogen sensors based on nanostructured materials.

3.11.3 Advanced Technologies for Hydrogen Sensing

Sensors working near room temperature

Many sensing technologies require the sensing element to be heated to high temperatures for optimum performance, e.g., catalytic pellistor sensors and metal oxide sensors. Resistor heaters are typically used to raise and control the temperature of the sensing element and these heaters contribute significantly to the overall power consumption of the sensor. Accordingly lower sensor power consumption requirements can be achieved by designing sensors to operate near or at room temperature. Effectively this means that the sensing element, where the change of hydrogen gas concentration is transduced into an electrical signal, operates at temperatures not much higher than 30 °C. Room temperature operation can be realized in different ways depending on the type of sensor and sensing material [298]. Examples include:

- Galvanometric hydrogen sensors possessing a liquid or solid electrolyte (e.g., KOH, H_2SO_4, Nafion) and a polymer membrane (Teflon, PTFE), operate at room temperature. However electrochemical sensors have a limited lifetime and consume a certain power for temperature stabilisation and signal display [146].
- In most cases optical sensors operate at room temperature. They can operate without oxygen and in an explosive atmosphere and are hardly influenced by electromagnetic fields. The sensor element can be separated from the electrical readout (see Section 3.8) [299].
- Electromechanical, such as surface acoustic wave (SAW) or cantilever-based sensors operate mostly with interdigitated structures on a piezoelectric substrate near room temperature. The sensor system needs complex electronic tools for signal generation and data aquisition. Also temperature effects have to be compensated.
- Absolute hydrogen concentration can be determined at room temperature by measuring the speed of sound in a test gas; the speed of sound in hydrogen is much faster than in all other atmospheric gases.
- Resistance-based sensors, placed on new materials such as nanocrystalline palladium, nanowires, or polymers like polyaniline, can operate at room temperature.
- Field effect (work function) based sensors work at room temperature; however, the sensors often need periodic heat pulses to refresh their sensitivity [208].

If it is possible to overcome some of the disadvantages associated with sensor operation at room temperature, including slow response and recovery times or saturation effects, then this concept has a promising perspective. However, it should be noted that it is not only the sensor heating element which consumes energy, the sensor electronics which performs the transmission, data processing and display output also consume power.

Advanced manufacturing techniques

Micro hydrogen sensors are typically fabricated from silicon and have dimensions on the micrometer scale. The ability to fabricate hydrogen sensors at this scale offer advantages in terms of performance and cost. Silicon microfabrication offers the potential for mass production of these devices which can drive down their manufacturing costs. Performance parameters such as response time, lower detection limit, and low power consumption can be significantly improved when compared with conventional hydrogen sensors [182, 185]. Microhydrogen sensors have been reported for various detection technologies. Many traditional technologies are conducive to miniaturisation including thermal conductivity sensors, metal-oxide sensors and catalytic pellistors. A new micromachined thermoelectric catalytic sensor, whose working principle is based on the Seebeck effect, shows promising results particularly in terms of lower detection limit [130]. This sensor is capable of detecting hydrogen down to 0.005 vol% with a measuring range up to a few percent. Other micromachined hydrogen sensor types include those based on a thin metal film whose resistance changes in the presence of hydrogen. These conductometeric sensors employ a microhotplate to heat the metal film at low power ratings to increase hydrogen sensitivity and lower the response time to hydrogen. MEMS hydrogen sensors have also been proposed. MEMS sensors are devices which, in addition to being fabricated on a micrometer scale, involve some mechanical motion or vibration in their detection mechanism. Micromachined MEMS cantilevers are an example of such devices.

Combination of hydrogen sensing platforms with temperature and humidity sensing elements

In many hydrogen sensor platforms, the sensor output signal is influenced both by the operating temperature of the sensor and the temperature of the gas in which hydrogen is to be detected. For this reason many contemporary sensors are equipped with a temperature sensor to compensate for the influence of temperature on the sensor response. The humidity of the gas under investigation can also have an impact on some sensors' response to hydrogen. These sensors are commonly protected with a nearly impermeable filter cup to prevent condensation of water on the surface of the sensing element. Another approach is to integrate a humidity sensing element in the hydrogen detection device to compensate for the influence of humidity. For example, the use of a thermal conductivity sensing element in combination with a temperature and a humidity sensor improves sensor performance and reduces uncertainties due to the compensation of the impact of temperature and humidity, e.g, see Figures 3.7 and 3.8.

Combination of two hydrogen sensing platforms

As outlined in Table 3.6 each hydrogen sensing platform has advantages and limitations with regard to its performance capabilities. Clever selection and integration of two or more different sensing platforms into a single sensing device facilitates mutual compensation of performance limitations between the integrated platforms to yield devices with enhanced sensing capabilities. As an example metal oxide semiconductor-based hydrogen sensors have a limited measuring range. The limit of detection is typically around 0.001 vol% and the upper concentration which can be detected is in the range of some percent of hydrogen. The sensor can show saturation effects that may reduce its sensitivity. Furthermore the concentration of oxygen or humidity may have an influence on the sensor response. Thermal conductivity sensors, on the other hand, can detect hydrogen concentrations up to 100 vol% and, advantageously, this sensor platform does not need oxygen for its operation. Therefore, they can be used together as orthogonal (i.e., independent) sensing platforms. The combination of these two different sensing platforms allows evaluation of the signals from both sensors in a conjoint unit and combination using a weighting function. The resulting sensor system offers a selective measurement of hydrogen concentrations in a continuous range up to 100 vol% H_2.

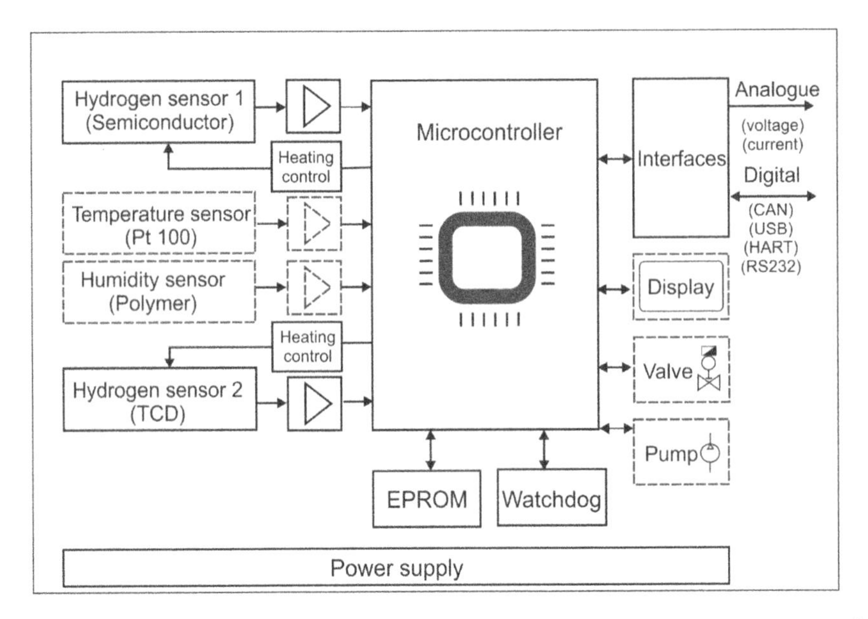

Figure 3.46: Schematic of a hydrogen sensor system with two hydrogen-sensing elements and temperature and relative humidity compensation.

Other commercial products combining two different sensor platforms are known. An example is a commercial sensor module that combines an electrochemical cell and a catalytic pellistor sensing platform in a single device. Electrochemical sensors can be designed for different concentration ranges of hydrogen, typically in the range of 0.001 vol% to 0.2 vol%; however, ranges of up to 4 vol% are also common. Hydrogen sensors based on catalytic pellistors operate in the range of 0.05 vol% up to 4 vol% or even close to 100 vol%. However, it has to be considered that this sensor needs sufficient oxygen for complete combustion of hydrogen. For correct indication of hydrogen values up to the lower flammability level in a gas, at least 2 vol% oxygen are needed. The combined sensing device exploits the advantages of both technologies in terms of measuring range, uncertainty of measurement, and response time. In addition it includes a detection backup for advanced security and reliability.

Signal processing for intelligent hydrogen gas sensors

Intelligent hydrogen sensors use a microcontroller unit[12] to process the signals from different sensing elements. A continuous temperature measurement and corresponding sensor adjustment controlled by a microcontroller ensures almost no temperature dependence of the measurement signal. Also, the integrated microcontroller allows humidity compensation with the result that a reference measurement, which is typical for many thermal conductivity sensors, is not required. Figure 3.46 shows a schematic of the sensor components. Safety-related functions like the recognition and signalling of failure states and error handling procedures, i.e., self-testing features, even during the measurement, are integrated in the intelligent sensor. The sensor output signal can be indicated by a display or provided via an analogue interface as a unit voltage (0 V to 10 V) or current signal (4 mA to 20 mA). In order to supply further types of output signals, the sensor system contains an analogue-to-digital

[12]A microcontroller is an electronic unit which contains a processor core, memory, and programmable input/output peripherals.

converter (ADC) or a frequency-to-digital converter (FDC). Hydrogen measurements can be transferred via different communication protocols or busses, such as Modbus, RS485, CAN and HART (see Section 2.6.2).

3.12 SUMMARY

In this chapter diverse classes of technologies deployed for hydrogen sensing have been described—covering both commercialised technologies and promising technologies still in development. Hydrogen sensors have been used for decades in industrial environments for safety monitoring and process control. Hydrogen sensors commercialised in this period were developed specifically for these markets. In recent years market surveys performed on hydrogen sensors have revealed a rapid expansion in the number and range of products which are commercially available. Several of these products are now marketed also for use in new hydrogen applications such as hydrogen fuelled back-up/emergency power systems, fuel cell powered systems, hydrogen refuelling stations, and hydrogen generation systems. New commercial off-the-shelf hydrogen sensors offer increasingly superior performance metrics which promise to fulfill many of the performance requirements demanded by end-users in existing and emerging hydrogen applications. Examples of the application of hydrogen and other sensors are presented in Chapter 7.

Currently research is being performed to develop innovative hydrogen sensing materials, novel sensing technologies, miniaturisation of hydrogen sensor platforms, and cost-effective sensor manufacturing techniques. The volume and scope of this research reflects the important role hydrogen sensors can and will play to facilitate the detection of unintended hydrogen releases wherever hydrogen is produced, distributed, stored, and used. Hydrogen sensors are a key enabling technology for safety monitoring of hydrogen applications and for alerting the presence of hydrogen before potentially hazardous conditions are reached. Ongoing research is pursuing the development of low-cost hydrogen sensors which are accurate, specific, insensitive to contaminants, resistant to changes in environmental conditions with long term signal stability–important considerations for the wider use of hydrogen sensors in diverse applications.

CHAPTER 4

Sensors for Other Gases

CONTENTS

4.1 OXYGEN SENSING TECHNOLOGIES

4.1.1 Overview

Oxygen determinations are required for a broad range of applications, including process control, health and safety issues, and environmental concerns. For gas-phase applications, the oxygen concentration of interest can range from trace contaminant levels to nearly pure oxygen. For example, hydrogen production via water electrolysis generates hydrogen in the cathode stream, but will also produce a stoichiometric amount of nearly pure oxygen at the anode. Indeed, water electrolysis is one means to produce oxygen for breathing air in confined environments with finite resources, such as in submarines or in the International Space Station [300]. There are applications that require high oxygen levels and purity; oxygen cylinders for welding contain a minimum of 99.5 vol% O_2. At the other extreme, oxygen can be viewed as an impurity with acceptable tolerances in the low ppm range. The maximum oxygen content in hydrogen is rigorously regulated for many applications. The current hydrogen fuel quality standards for PEM fuel cell vehicles restrict the maximum allowable oxygen level to 5 ppm [301, 302]. The allowable oxygen content is regulated for other grades of hydrogen, some of which are presented in Table 4.1. In addition to nearly pure oxygen or for oxygen present at trace concentrations, there is often a need to quantify oxygen concentrations at intermediate levels between these two extremes. The oxygen content in air is typically given as 20.9 vol%. To guard against asphyxiation, the U.S. Occupational Safety and Health Administration (OSHA) mandates that the oxygen content in breathing air must be above 19.5 vol% [303]. The verification that oxygen is present at a safe concentration is mandated for a variety of activities, such as confined space entry. Thus, depending on the application, oxygen measurements can be required from low ppm levels to nearly 100 vol% and at concentrations in between these two extremes. There are some commercial sensors specified for 0 to 100 vol% O_2, but the low-level accuracy or ability to resolve small changes in oxygen concentration in these broad range sensors may be inadequate for some applications. These broad range sensors have a typical lower quantification limit (LQL) not better than approximately 0.05 vol%. To improve performance specifications, some oxygen sensors have been optimized to operate over a smaller concentration range. However, some trace level oxygen determinations remain challenging and may be beyond the current capability of sensor-based analytical methods.

Several major sensor technologies are currently and routinely used for gas phase oxygen measurements. In 2003 Ramamoorthy and co-workers reviewed the major platforms [304]. The most common oxygen sensor types were the galvanic oxygen sensor and the zirconia oxygen sensor. In recent years, optical sensors based on laser diode technology have been commercialized. In addition, there are specialized sensor platforms, such as the paramagnetic oxygen detector. These all offer the ability for on-site, real-time detection of oxygen over a broad range of concentrations and conditions. There are other sensor platforms that respond to changes of oxygen concentration including the thermal conductivity (TC) sensor and high temperature metal oxide (MOX) sensors. The response of these sensors is not selective to oxygen and could arise from any number of other chemical species. The TC or MOX platforms are not usually used for oxygen measurements, and thus will not be discussed here. However, the TC and MOX platforms are extensively used in hydrogen and other applications and are covered in Chapter 3.

4.1.2 Clark Electrode

Perhaps the earliest technology developed for oxygen measurements that can be described as a sensor is the Clark electrode (see Figure 4.1) [305]. It is, however, not generally described as a sensor in journal articles or product literature, but is more often referred to as an analyser for dissolved oxygen. The Clark electrode is based on the reduction of oxygen on a miniature planar platinum or gold electrode, with the oxidation of a silver electrode to form the silver cation (Ag^+) as the counter reaction. Dissolved oxygen is extracted from the test solution through a thin, oxygen-permeable membrane, typically Teflon, covering the electrode assembly. The membrane also serves to isolate a thin potassium chloride electrolyte layer

Table 4.1: Classes of purified hydrogen

	Hydrogen 3.0	**Hydrogen 5.0**	**Hydrogen 6.0**	**Hydrogen 7.0**
	99.9	**99.999**	**99.9999 (liq.)**	**99.99999**
Trace component			in μmol/mol	
O_2	$\leq$10	$\leq$2	$\leq$0.1	$\leq$0.03
H_2O	$\leq$50	$\leq$5	$\leq$0.5	$\leq$0.05
N_2	$\leq$700	$\leq$3	$\leq$0.2	
HC		$\leq$0.5		$\leq$0.03
CO		$\leq$0.5	$\leq$0.1	$\leq$0.03
CO_2		$\leq$0.5	$\leq$0.1	$\leq$0.03

surrounding the electrode assembly from the bulk test solution. Since potassium chloride is often used as the electrolyte, the silver ion will precipitate out of solution as silver chloride (AgCl). The various electrochemical and chemical processes can be represented as:

Working electrode (the cathode) reaction:

$$O_2 + 4\,e^- + 2\,H_2O \longrightarrow 4\,OH^-$$

Counter electrode (the anode) reaction:

$$4\,Ag \longrightarrow 4\,Ag^+ + 4\,e^-$$

Chemical reaction at the counter electrode:

$$4\,Ag^+ + 4\,Cl^- \longrightarrow 4\,AgCl$$

Net cell reaction:

$$O_2 + 4\,Ag + 2\,H_2O + 4\,Cl^- \longrightarrow 4\,AgCl + 4\,OH^-$$

Operationally, it is necessary to apply an external voltage (bias the working electrode) in the range of 0.6 V to 0.8 V through a wire connected from the anode to the cathode to drive the oxygen reduction reaction. The transfer of electrons associated with the reduction of oxygen at the cathode and oxidation of the silver anode can be measured as a current through this wire. The electrochemical current is a direct indication of the rate of oxygen reduction on the platinum electrode. The magnitude of the electrochemical current will be directly proportional to the flux of oxygen molecules to the cathode, which in turn will be controlled by the diffusion of oxygen through the Teflon membrane. Electrochemical reactions that are diffusion controlled will often generate currents that are proportional to the bulk concentration of the analyte (see Section 3.4 where this concept is developed further for the amperometric hydrogen sensor). Although the Clark electrode was developed over 60 years ago, it is still commonly used for the determination of dissolved oxygen, and is commercially available from numerous suppliers. The Clark electrode is an excellent sensor for the determination of oxygen in condensed media, especially aqueous environments. It is not, however, readily amenable for direct gas phase determinations, but there are a variety of commercially mature sensor platforms that perform this function. The Clark electrode is of historical significance for oxygen measurements, and the focus on the remainder of this section will be on gas-phase oxygen determinations, which are most relevant for hydrogen applications.

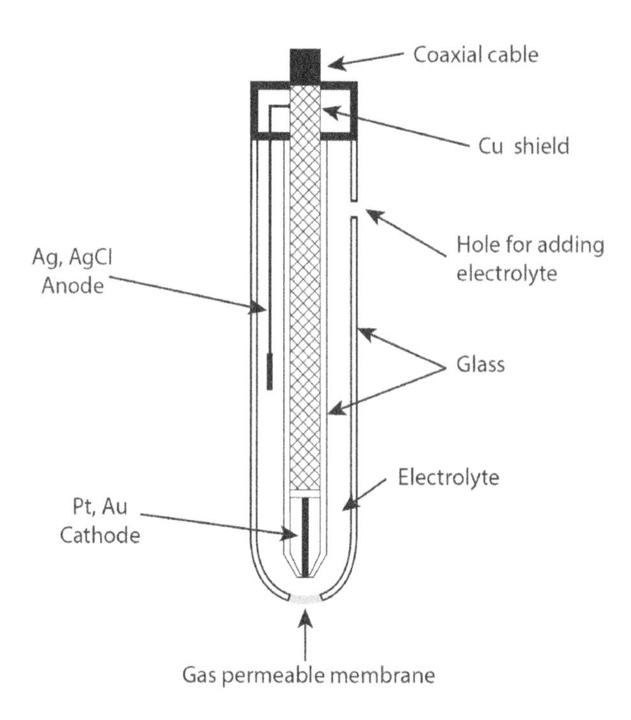

Figure 4.1: Schematic of the Clark electrode.

4.1.3 Galvanic Oxygen Sensor

The galvanic oxygen sensor signal is based on the electrochemical reduction of oxygen:

$$O_2 + 4\,e^- \longrightarrow 2\,O^{2-}$$

Since a current is produced as a result of the electrochemical reaction, this sensor is sometimes called the amperometric oxygen sensor or fuel cell oxygen sensor. It is also often referred to as simply the electrochemical oxygen sensor, but this can be a misleading name since the detection principle for the zirconia oxygen sensor is also based upon an electrochemical mechanism. Stetter and Ling provided an overview of amperometric sensors, covering the design, operation, and typical analytes that can be detected by amperometric gas sensors [144]. The structure of the galvanic oxygen is shown in Figure 4.2. The typical galvanic oxygen sensor uses a membrane working electrode (WE) structure that consists of a thin film of gold or other noble metal affixed to the surface of a hydrophobic, oxygen-permeable membrane. The electrode side of the membrane surface is in contact with an internal aqueous electrolyte. The hydrophobic membrane prevents the aqueous electrolyte from leaking through the membrane structure. The electrolyte is selected to optimize the solubility of cations generated from the oxidation of a sacrificial anode (see below); an acetate electrolyte is often used for a lead anode because of the relatively high solubility of lead acetate salts that are formed during operation of the sensor. In contrast to the Clark electrode which is typically immersed into a liquid test medium, the outer surface of the membrane electrode structure of a galvanic oxygen sensor is in contact with the test gas. Since the working electrode reaction is the reduction of oxygen, the WE is the cathode of the electrochemical cell. Analogous to the amperometric hydrogen sensor (see Section 3.4), the flux of oxygen to the WE generates a current, the magnitude of which is nearly directly proportional to the concentration of oxygen in the test gas sample.

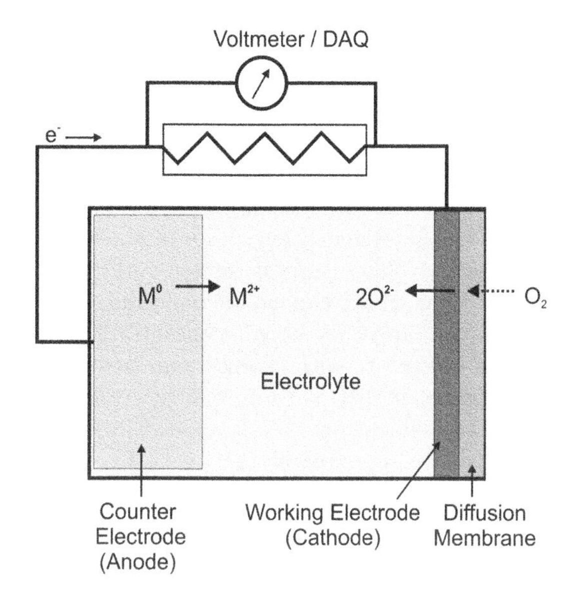

Figure 4.2: The galvanic oxygen sensor [306].

The electrochemical oxygen sensor includes a reactive metal anode electrode such that the counter reaction is the oxidation of the anode ($M^{\circ} \longrightarrow M^{2+} + 2\,e^{-}$). The anode is typically lead (Pb) or zinc (Zn), but other metals can be used. In normal operation the counter electrode (anode) depletes. The extent to which it becomes depleted is directly proportional to the amount of oxygen that reacts at the working electrode. The anode oxidation is typically irreversible, and the availability of anode material is often the controlling factor for the operational lifetime of the galvanic oxygen sensor. At high oxygen concentrations the rate of anode depletion will increase, thereby decreasing the sensor life.

Operationally, the galvanic oxygen sensor is extremely simple to use. Signal measurement requires nothing more than a voltmeter attached to a resistor connecting the anode and the cathode, although many sensor models include circuitry to improve signal characteristics, with, of course, added costs. The resistor transforms the electrochemical current into a voltage. Commercial sensors exist with a vendor-specified range of 0 to 100 vol%. As a low-power device, it is ideal for portable applications, such as in hand-held instruments. Galvanic oxygen sensors typically have a lower detection limit of about 0.05 vol% O_2. Specialized designs, optimized for improved lower detection limits are available, but have a more limited dynamic range. Vendor-specified response and recovery times range between 1 s and 30 s, but empirical response time measurements using validated protocols and apparatus, however, have shown that the t_{90} response times for commercial galvanic oxygen sensors are more in the range between 10 s and 30 s. [307]. The sensor has moderate stability but is prone to long-term drift and noise induced by fluctuations in environmental parameters and ageing. Periodic calibration is necessary to circumvent loss of accuracy associated with systematic drift. Sensor calibration is often performed by measuring the sensor response in ambient air (20.9 vol% O_2) and in an inert gas (e.g., nitrogen or argon with 0 vol% O_2). The liquid electrolyte restricts the operating temperature range of the galvanic oxygen sensor from approximately 0 °C to 40 °C.

The galvanic oxygen sensor is impacted by changes in the surrounding temperature (T) and pressure (p). Changes in ambient humidity will also affect the sensor response, but to a lesser extent than T and p. An internal thermistor is often incorporated into the sensor design to compensate for the temperature dependence. This decreases but does not totally eliminate temperature-induced errors in oxygen measurements, and it may be necessary to empirically verify the impact of temperature-induced variations on the sensor response in the temperature range of interest. The response of galvanic oxygen sensor is also affected by changes in ambient pressure. The galvanic oxygen sensor response is controlled by the oxygen partial pressure (p_{O_2}) and not by volume percent oxygen (vol% O_2). A barometric pressure change from 101 kPa to 98 kPa, which can be encountered during periods of changing weather patterns, will generate an apparent oxygen concentration of 20.5 vol%; this is often a negligible effect. However, larger pressure changes can be encountered if the sensor is transported from one elevation to another. For example, at sea level the pressure is 101 kPa, whereas at 1526 m (5,000 feet), ambient pressure is about 84 kPa; at this elevation a sensor calibrated at sea level would read approximately 17.4 vol% O_2. Large pressure changes can also be encountered in closed pneumatic systems. It should be noted that a release of gas other than oxygen into a closed pneumatic system will change the vol% O_2, but the p_{O_2} will remain unaffected. The output of the galvanic oxygen sensor, which depends on p_{O_2} would remain unchanged. This behaviour and its impact were recently discussed by Buttner et al. [307]. To alleviate the pressure dependency, modified designs that incorporated a capillary diffusion design instead of a membrane diffusion system were developed. Capillary diffusion has a smaller pressure dependency than transverse diffusion through a planar membrane. However, with capillary diffusion the concentration dependency of the sensor response is no longer linear but rather a more complex empirical logarithmic relationship. Changes in humidity can also affect the galvanic oxygen sensor response. The impact of water vapour on the sensor response is ascribed to the presence of a diluent affecting the actual oxygen level. However, the change in ambient water vapour level is typically small, thus imparting a minimal impact on the sensor signal.

The galvanic oxygen sensor typically exhibits good analytical sensitivity around ambient air oxygen levels, accurately differentiating between ambient and typical alarm levels associated with safety concerns. Furthermore it has been reported that electrochemical sensors can detect traces of oxygen (> 1 μmol/mol) in hydrogen [308]. As already stated, a major advantage of the sensor is its ease of operation. This sensor is unique as it requires no power and produces the signal through a spontaneous electrochemical reaction. The sensor is quite small and has no unduly complicated pneumatic requirements. The main disadvantage of the galvanic oxygen sensor is its operational lifetime, which is controlled primarily by the total net amount of oxygen that is reacted on the sensor (which depletes the sacrificial anode). Typical operational lifetime is 1 year, but this varies from 6 months to over 2 years depending upon the model and, of course, whether the sensor is deployed in ambient air versus oxygen-enriched environments. The sensor can also have a short shelf-life if it is stored exposed to oxygen with an electrical contact between the anode and cathode during storage. High concentrations of carbon dioxide have been reported to affect the acid electrolyte used in some designs.

4.1.4 Zirconia Oxide Oxygen Sensor

The zirconia oxygen sensor is one of the most common and commercially successful process control sensors. The sensor was developed for the optimisation of combustion to improve fuel efficiency and to minimize pollution by-products associated with incomplete combustion.

Yttria is added to improve mechanical stability of the zirconia sensor structure, and the resulting structure is referred to as yttria-stablised zirconia (YSZ). The YSZ sensor, often called the Lambda sensor or Lambda probe, is routinely used in automobiles and other combustion systems.

At high temperature (ca. 600 °C to 800 °C), zirconia is a solid-state oxygen ion (O^{2-}) conductor. Oxygen can form a reversible redox couple at the sensing electrode surface that is typically coated with platinum. If the counter electrode is maintained at a known, constant oxygen concentration (e.g., ambient air, 20.9 vol%), a potential is created as given by the Nernst equation:

$$E = \frac{\mathrm{R}T}{z\mathrm{F}} \ln \left(\frac{p_{O_2}}{p_{O_{2-ref}}} \right) \tag{4.1}$$

Where $p_{\mathrm{O}_{2-\mathrm{ref}}}$ is the oxygen partial pressure of the reference side and p_{O_2} is the oxygen partial pressure at the sensing electrode that is exposed to the sample. The other terms include the universal gas constant R, F is the Faraday's constant, z is the number of electrons associated with the redox couple (2 for oxygen oxide), and T is the temperature in kelvin. The zirconia oxygen sensor is another example of an electrochemical oxygen sensor, but is typically operated in a potentiometric mode, thus the signal is a voltage. In addition to high temperature operation, a disadvantage of potentiometric gas sensors, at least for some applications, is that the signal is a logarithmic function of concentration. This usually implies a broad dynamic range, often over several orders of magnitude, but can adversely impact the resolution of the sensor (e.g., the ability of the device to differentiate between small changes in high concentrations of oxygen; this is also called the analytical resolution). The zirconia oxide sensor can be configured for amperometric operation, but this is usually not done for analytical purposes. The YSZ sensor exhibits minimal cross-sensitivity to other gases, but can be poisoned by hydrocarbons, carbon monoxide and lead. The lower detection limit of zirconia oxygen sensors is in the low vol% range, but can operate over a broad range (a large measuring range). However, since the sensor signal is a logarithmic function of concentration, the sensor is unstable in the absence of oxygen. As a high temperature device, fluctuations in ambient temperature do not significantly influence sensor response, but changes in pressure will. Response and recovery time are controlled by both the pneumatic time constant and the time for the sensor to re-establish an equilibrium potential to changes in oxygen concentration. The YSZ sensor has response and recovery times, as reported by manufacturers, on the order of 100 ms to 500 ms.

The sensor is readily available for commercial applications (e.g., as an automotive sensor or as the active sensing element in a combustion gas analyser). Other strong points include robustness and small size. It is frequently used as an in-line analyser. As a high-temperature device the electric power consumption may be high relative to other sensing methods, but this should not be a critical issue for most process-control applications. Control circuitry is a bit more complicated than other electrochemical sensors, but designs are available. Shelf-life is extremely long. Since YSZ oxygen sensors operate at high temperature, some care is required to ensure that no unsafe condition will be present due to thermal hazards. The sensor is less prone to long-term drift than the galvanic sensor. The zirconia sensor is noted for long-term stability with minimal maintenance requirements and should only require routine calibration. Since the sensor is not stable in the absence of oxygen, calibration will typically require the measurement of the sensor response from at least two non-zero oxygen concentrations. Other than anaerobic operation, the influence of gas matrix changes on sensor response should be minimal.

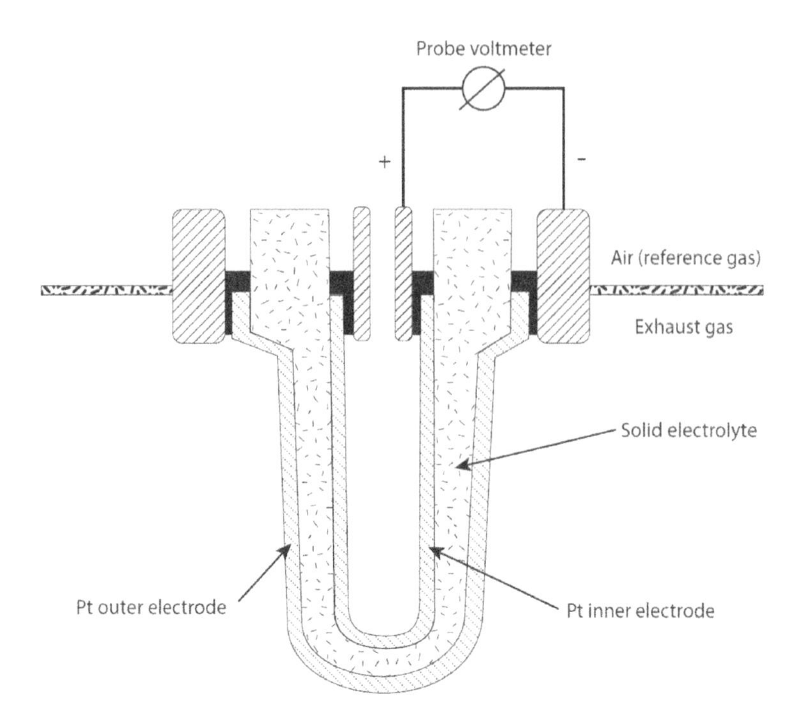

Figure 4.3: The zirconia oxygen sensor.

4.1.5 Conductometric Oxygen Sensors

Conductometric oxygen sensors are based on semiconductive metal oxides and are an alternative to the classical potentiometric zirconia oxygen sensor. This type of sensor can be used in harsh environments and at temperatures of up to several hundred degrees Celsius. The sensor development started in the 1960s with the demand for reducing automotive exhaust pollutants. To minimize pollutants such as carbon monoxide, hydrocarbons, and nitrogen oxides, engines need to operate at the stoichiometric air-to-fuel ratio.

Conductometric gas sensors are based on titania (TiO_2), which have a resistance that is dependent upon the oxygen partial pressure p_{O2}. The resistance of titania sensors increases monotonically in the p_{O2}-range from approximately 10^{-19} Pa to 10^{-3} Pa. This behaviour can be described by the following equation:

$$R_s = \mathrm{A}_0 \times \exp(\frac{E_A}{\mathrm{k_B}T}) \times (p_{O2})^{\beta} \tag{4.2}$$

Where R_s is the resistance of the sensor, A_0 and β denote two constant and E_A is the thermal activation energy of the electrical conductivity.

As seen from Equation 4.2 the resistance of the sensors depends on temperature. Therefore research was performed to develop temperature independent resistors. Doped strontium titanate ($SrTiO_3$), a p-type semi-conductor, was investigated as a material for conductometric oxygen sensors. Both thick-film and thin-film sensors were developed. In addition to $SrTi_{1\text{-}x}Fe_xO_3$, $BaFe_{1\text{-}y}Ta_yO_3$ and $LaFe_{1\text{-}z}Cu_zO_3$ were also found to exhibit a temperature independent response against oxygen.

Temperature-independent conductometric n-type semiconductors were also found, and include gallium oxide (Ga_2O_3) and cerium oxide (ceria). By adding zirconia to ceria, a ceria-zirconia mixed oxide was formed that improved the thermal stability. However, trace levels of both NO_2 and SO_2 permanently affected the sensor response [309].

These sensors do not demand an oxygen-containing reference gas such as that which is needed for the zirconia potentiometric sensor. The titania oxygen sensor can be physically smaller, stronger, and has a faster reaction time than the zirconia type. The sensing element is produced in a hybrid technology using screen-printed thick-films on ceramic tapes that are co-fired. Commercially available sensors have a fast response time of a few ms and have been used in vehicles of different manufacturers. One disadvantage of all conductometric sensors is the dependence of the resistance on geometry and morphology of the sensor films, thus demanding a highly reproducible manufacturing protocol. Changes in device to device sensor characteristics necessitated in-operation recalibration. A less geometry-dependent thermoelectric sensor, which was suggested based on the Seebeck coefficient, has been investigated using materials such as $SrTi_{0.6}Fe_{0.4}O_{3-\delta}$ [310].

Future research will continue to address means to improve the temperature dependence of the material resistance, resistance, long time stability and resistance towards sulphur, NO_x, and reducing atmospheres [311].

4.1.6 Optical Oxygen Sensor

Although optical sensors represent only a small (but growing) share of the oxygen sensor market, this technology is the most common and successful platform for carbon dioxide measurements. Accordingly, a more detailed discussion on the theory of optical absorbance sensors is given in Section 4.2 on carbon dioxide sensors. Optical sensors are available for a variety of other gases, including ammonia, carbon monoxide, hydrogen sulphide, and methane. Several of which are discussed below.

Oxygen has a weak absorbance in the infrared spectrum of the electromagnetic spectrum at around 763 nm. This property is exploited in optical oxygen sensors. An illustration of an optical oxygen sensor is shown in Figure 4.4. Optical oxygen sensors have been developed using a tunable diode laser as a light source. The narrow band-width associated with the light from a tunable diode laser eliminates the need for a band-pass filter that is required for broad-band light sources such as the tungsten lamp. Whereas the galvanic and zirconia oxygen sensors are mature technologies with sensor models available from many different manufacturers, relatively few optical oxygen sensors models are currently commercially available. However, this is changing due to significant improvements in the availability of laser diode technology. Being solid state devices, optical sensors can be deployed in harsher conditions than the galvanic sensors, including broader temperature and pressure ranges. Optical sensors are available for combustion processes. Some designs even include integrated pressure and temperature compensation.

As an optical method based on absorbance, the laser diode oxygen sensor has a good linear range and selectivity. Devices are available with measurement ranges from either 0 to 25 vol% or 0 to 100 vol%. The detection limits and analytical sensitivity are moderate with detection limits around 0.05 vol%. However, this is not a fundamental restriction and much lower detection limits are feasible when longer light paths are used. Response time is fundamentally fast and some manufacturers indicate a 1 s response time. The actual response times are controlled by cell purge times, and thus can often be on the order of 30 s because of the sample cell volume. Drift can be caused by changes in light source intensity and photodetector sensitivity, both of which can be problematic with unattended operation, but the effects can be minimized through the use of a reference light path and signal. Periodic calibration will correct for signal drift. Temperature and pressure which affect gas density

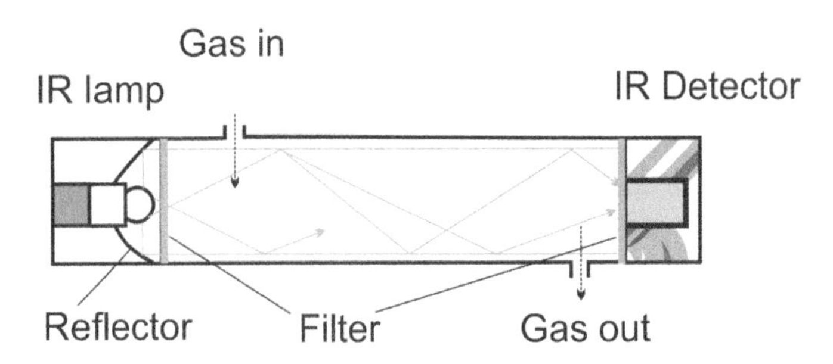

Figure 4.4: Optical oxygen sensor.

will affect signal magnitude, but this can be minimized through compensation. As an optical method, operational life is dictated by the life of the light source and the detector, both of which are solid state devices and as such a long operational lifetime is expected. The sensor should also have an indefinite shelf-life. There is no known chemical interferent for the laser diode oxygen sensor. Unlike many other sensor technologies, optical measurements do not change the composition of the sample (the detector is isolated from the sample by an optical window). Traditionally, optical sensors tended to be sensitive to vibration. However, as solid-state devices, laser diode oxygen sensors should have improved mechanical stability compared to optical sensors using tungsten filaments or other means to generate the light source.

Optical oxygen sensors are also available based on other transduction mechanisms. For example, commercial fibre optic oxygen sensors use the fluorescence of a chemical complex compound to measure the partial pressure of oxygen. The fluorescent compound is fixed in a sol-gel matrix film and glued on the tip of an optical fibre. A pulsed blue LED sends light at 475 nm to the optical fibre and excites the organic compound, which emits light at 600 nm. If the excited chemical compound encounters oxygen molecules, energy is transferred to the oxygen molecules in a non-radiative transfer, decreasing or quenching the fluorescence signal. The degree of quenching correlates to the level of oxygen concentration in the film and to oxygen partial pressure in the environment, respectively. Sensors can measure oxygen up to 25 vol% O_2 or even to 100 vol% O_2 with a lower detection limit (LDL) of around 0.05 vol% O_2.

4.1.7 Paramagnetic Oxygen Detectors

Oxygen is paramagnetic, which means it has a high magnetic susceptibility. Paramagnetic materials are attracted to a magnetic field. This results from the presence of at least one unpaired electron spin in the material's atoms or molecules. Oxygen has two unpaired electrons in its ground state rendering magnetic susceptibility much higher than that for most other gases. Because of its magnetic susceptibility, variations in oxygen concentration in a gas sample will affect the effective strength and direction of applied magnetic fields. This property is exploited as the basis of the paramagnetic oxygen detector. Traditionally paramagnetic oxygen detectors, also known as paramagnetic oxygen analysers, were marketed as bench-top instruments; however, compact versions are now available that can be considered a sensor platform. The most common design for the paramagnetic oxygen detector consists of two nitrogen-filled hermetically sealed spheres linked in a "dumbbell" configuration to a

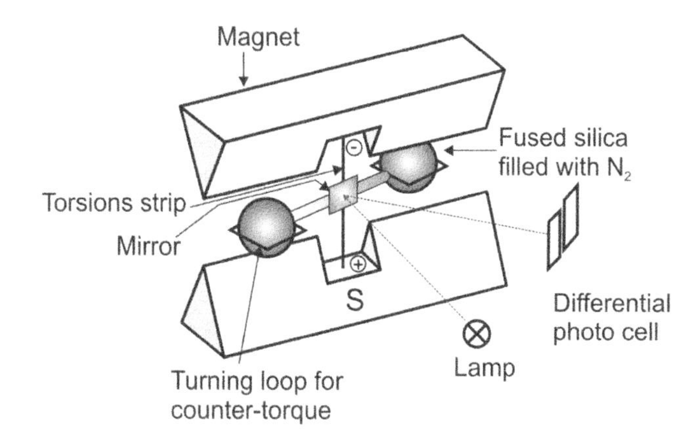

Figure 4.5: Paramagnetic oxygen sensor.

torsion bar, as illustrated in Figure 4.5. This structure is suspended in the magnetic field between two magnets and is free to rotate. A mirror is centrally mounted on the torsion bar. Oxygen attracted into the magnetic field will displace the nitrogen filled spheres causing the torsion bar to rotate. A light source reflected off the mirror onto photocells can be used to detect and quantify the deflection. The magnitude of the deflection is proportional to the oxygen concentration in the surrounding gas, thus changes in oxygen concentration are readily detected and quantified.

The paramagnetic oxygen sensor is sensitive over a broad range of oxygen concentrations, from as low as 0.05 vol% to 100 vol% O_2. However, when used over a broad concentration range, other analytical metrics such as accuracy and analytical resolution will be compromised. Accordingly, the paramagnetic oxygen detector is often tuned to cover only a small portion of this range, for example 90.0 vol% to 99.9 vol% O_2 or 0 to 2.0 vol% O_2. This is not a significant restriction for those applications that require highly accurate and precise oxygen determinations within a narrow range (e.g., oxygen purity determinations). Manufacturers can perform a specific factory calibration from a variety of oxygen concentration range options. This results in an instrument with exceptional accuracy and analytical resolution.

Because the paramagnetic oxygen sensor does not chemically react with oxygen, the system is more stable than other instrument platforms, although periodic calibration is still required. In addition, since a physical property is measured, analysis by the paramagnetic oxygen detector does not change or affect the composition of the sample gas. The paramagnetic oxygen detector is highly selective to oxygen since most major constituents in air have a zero or near-zero magnetic susceptibility. Cross sensitivity coefficients for many gases have been tabulated by many manufactures. Very few interference gases induce a significant response. As with other analytical devices that depend on the "amount" of material in the sample, there will be some environmental effects due to the dependence of density on temperature and ambient pressure, but compensation methods can readily account for these effects, especially in bench-top, fixed-site deployments. Response and recovery times are fast, and are primarily limited by the pneumatic purge time constants (e.g., the time required to fully purge the test chamber with the test gas, including the transport time from the sample source to the detector). Mechanically, the detector is not as robust as other instruments or sensors and is prone to vibration interferences and can be easily damaged.

4.2 SENSORS FOR GASEOUS TRACE COMPONENTS

4.2.1 Overview

Gas detection requirements in hydrogen technologies are not limited to hydrogen and oxygen. The detection and measurement of other, often toxic components are also crucial to protect both human health and devices. Major chemicals of concern are listed in Table 4.2. These gaseous species may have a significant impact even at very low concentrations. Internationally, guidelines for minimal air quality requirements have been published by the World Health Organization, which can serve as the basis for national air quality standards and programmes. Allowable limits for many gas species as well as reference methods for measurements are prescribed in various regulations, e.g., EU Directive 2008/50/EC [312].

Trace levels of many gaseous components can often impede the functionality of devices and facilities in hydrogen technologies. For this reason it is important to supply very pure hydrogen to devices sensitive to impurities. For example, the polymer membrane electrolyte fuel cells (PEM) in fuel cell electric vehicles (FCEVs) are sensitive to very low levels of certain impurities in the hydrogen fuel, including carbon monoxide and sulphur. Irreversible degradation of a PEM fuel cell can occur when these contaminants are present at a level of only 10^{-7} [313]. The maximum allowable level of various components in hydrogen fuel for PEM fuel cell electric vehicles is defined in ISO 14687-2 [302] and are listed in Table 4.3. Almost identical requirements are specified in SAE J2719 [301]. Hydrogen fuel cells also use ambient air to supply oxygen to the cathode, and accordingly, contaminants in air may also affect fuel cell performance.

Table 4.2: Survey on sensing technologies for detection of gaseous trace components in air

Analyte	Limits		Sensor Type			
	WLA[1)]	TLV-TWA[2)]	catalytic	electro-chemical	conducto-metric	optical
CO	30	25		x[3)]	x	x
CO_2	5,000	5,000		x		x
HC[4)]	~500	~300	x		x	x
NH_3	20	25	x	x	x	x
NO	25	25				x
NO_2	1	3		x		x
H_2S	5	10		x		x
SO_2	1	2		x		x
HCS	0.5[5)]			x		

[1)] WLA - Workingplace Limit Value in μmol/mol [314]
[2)] TLV-TWA - threshold limit values - time weighted average (8 h work day and 40 h work week) as specified by U.S. OSHA in μmol/mol
[3)] "x" implies the gas can be detected by this sensor type.
[4)] Hydrocarbons, like methane, propane or octane
[5)] Menthanethiol

This section focuses on sensors for the detection of relevant trace components in air (for air quality) and in hydrogen (for fuel quality). Table 4.2 lists common pollutants together with an identification of sensor types which are commercial available for their detection.

Table 4.3: Maximum content of gaseous constituents in hydrogen for PEM fuel cells applications for road vehicles according to ISO 14687-2 [316]

Non-hydrogen constituents	**Amount /μmol/mol**
Carbon monoxide (CO)	0.2
Carbon dioxide (CO_2)	2
Formaldehyde (HCHO)	0.01
Formic acid (HCOOH)	0.2
Total hydrocarbons (C1 basis)	2
Total halogenated compounds	0.05
Ammonia (NH_3)	0.1
Total sulphur compounds	0.004
Water (H_2O)	5
Oxygen (O_2)	5
Nitrogen (N_2)	100
Helium (He)	300
Argon (Ar)	100

This section will discuss many of these sensor technologies. The discussion is limited to sensors applied to chemicals that are recognized as particularly harmful to fuel cells, to human health, or to those chemicals which are prevalent in air or hydrogen processes. A more thorough review of sensor technology for a broader range of analytes has been recently published as series of books on chemical sensors [5].

ISO 14687-2 specifies suitable analytical methods for detection and quantification of gaseous contaminant components in hydrogen. These methods are not based on sensors but rather on laboratory methods and are often cited as reference methods. An exception is that an electrochemical sensor has been approved for the detection of traces of oxygen in hydrogen [308]. Many gaseous compounds can be determined by gas chromatography using different detectors. Optical methods such as Fourier transform infrared spectroscopy (FTIR) or photo acoustic spectroscopy can be applied [315]. Mass spectrometry has also been applied. A general description of the instrumentation used in these reference methods is described in Section 6.3. Compared to gas sensors, these methods are more complicated and expensive, requiring skilled operators to perform the measurements.

Low-cost sensors for the detection and quantification of air impurities are commercially available for many of the compounds listed in Table 4.2, even at low concentrations. However sensors for impurity detection in hydrogen at the level specified in fuel quality standards are still under development. In this section sensing technologies commonly deployed for the various species listed in Table 4.2 are discussed along with details on the sensor performance and possible gaps in availability of sensors. Many commercial sensors operate according to requirements stipulated in standards like EN 60079-29-1. This standard specifies the acceptable uncertainty of reading for different sensors in the temperature range from -25 °C to 55 °C, at pressures between 800 hPa and 1200 hPa and at relative humidities from 20 % to 90 %. Specialized sensors that operate outside this range (e.g., for high temperature operation) are also available, but will not be extensively discussed in this book.

4.2.2 Carbon Oxides

Carbon Monoxide

Carbon monoxide (CO) is a toxic, colourless and odourless gas which is produced from the incomplete combustion of carbon-based materials. This includes the thermal decomposition of organic materials, which are used commonly as coatings, seals, lubricants, and material of construction. Carbon monoxide is a toxic gas to human beings. To assure breathing air requirements with respect to carbon monoxide are stringent, with threshold limit values (TLV) restricted to 0.0025 vol% (25 ppm). It is also a well-known poison for many catalysts. Whereas high temperature fuel cells typically have a higher tolerance for carbon monoxide as the affinity of CO on a platinum catalyst decreases with temperature, at low temperature (e.g., < 100 °C), carbon monoxide adsorption effectively blocks active sites thereby decreasing the adsorption of oxygen or hydrogen on to the catalyst. This results in a decrease of the fuel cell power output and efficiency [317]. Even very low carbon monoxide levels ($< 2{\times}10^{-7}$) may cause a decrease in cell voltage, especially when the anode gas is recirculated [16]. Therefore only very a small amount of carbon monoxide is tolerable in low-temperature fuel cells and have to be measured, as indicated in Table 4.4. Fortunately, purging with pure hydrogen can often regenerate a catalyst that has been poisoned by carbon monoxide. Thus detection of carbon monoxide is important for both air quality and hydrogen quality.

Carbon monoxide is a combustible gas in the presence of oxygen. Catalytic sensors can detect carbon monoxide in air typically in the range of 0.1 vol% to 12.5 vol% (i.e., up to 100 % LEL). Catalytic gas sensors do not operate in the absence of oxygen and thus are typically deployed in air. The sensors are heated to between 300 °C and 500 °C. These sensors have a warm-up time of less than 5 min and a response time, t_{90}, typically less than 30 s. However the detection limit is high with respect to the threshold values for carbon monoxide as given in Tables 4.2 and 4.3. Also the selectivity of catalytic sensors is poor and other combustible gases, such as hydrogen or methane, can interfere with the measurements. Hydrogen sulphide can act as a poisoning reagent on the catalytic sensor.

Detection apparatus using electrochemical (EC) CO sensors are also commercially available, both as stationary or portable devices. The EC CO sensor typically operates at room temperature. It can quantify carbon monoxide amounts in air in the range from about 0.0001 vol% (1 ppm) to 1 vol%. However, hydrogen or nitrogen oxides may interfere with the measurement. With the exception of hydrogen, most of these interfering gases can be removed with chemical filters. It is generally not feasible to filter out hydrogen, and thus there is a risk of false alarms in the presence of hydrogen. Since carbon monoxide alarms are typically activated in the 0.005 vol% (50 ppm) range, an inconsequential hydrogen level could induce a false carbon monoxide sensor alarm. Due to the influence of temperature on the electrochemical reaction and diffusion processes, the sensor signal will generally increase with temperature. However the signal behaviour is usually well characterized for the specified temperature range of the electrochemical cell and the sensor output can be corrected using an internal temperature sensor.

Conductometric sensors are also available for detection of carbon monoxide and typically measure in the range of 0.005 vol% to 1 vol% in air. These sensors are based on semiconductor metal oxide materials such as tin oxide (SnO_2). These sensors typically operate between 300 °C and 400 °C. The SnO_2 can be pure or doped with Au, In, Cu, Pt, Pd, or in mixtures with other oxides. Other metal oxides investigated for use in carbon monoxide sensors include ZnO, Bi_2O_3, CuO, In_2O_3, Fe_2O_3, TiO_2 [318]. The use of dopants improves carbon monoxide detection limits often at lower operation temperatures, e.g., less than 100 °C. Selectivity of the sensor is also improved through the use of dopants and other metal oxides, such that

Table 4.4: Requirements on CO sensors for fuel cell applications [319]

Parameter	**Reformate (pre-stack)**	**Reformer exhaust**	**Inside reformer**
Measuring range	10^{-6} - 10^{-4}	10^{-4} - 10^{-3}	10^{-3} - 2×10^{-4}
Accuracy		(1 - 10) % of full scale	
Response time		(0.1 - 1) s	
Temperature	< 150 °C	250 °C	250 °C - 800 °C
Pressure		(0.1 - 0.3) MPa	
Gas environment		30 % - 75 % H_2 N_2, CO_2, H_2S, HC, high humidity	

carbon monoxide detection is only slightly influenced by the presence of hydrogen. However, the influence of fluctuations in the ambient temperature and the temperature of operation can still impact the sensor response.

The determination of trace levels of carbon monoxide in hydrogen is also important. To protect PEM fuel cell anodes, hydrogen fuel quality standards mandate that the carbon monoxide level be below 2×10^{-7} (200 ppb). In Table 4.4 performance and operational requirements for carbon monoxide sensors in a variety of fuel cell related applications are summarised.

The detection of carbon monoxide in hydrogen is more difficult than its detection in air as lack of oxygen can impede the performance of many sensor platforms. The required quantification limit is also significantly lower for hydrogen processes relative to typical ambient air requirements. The selective detection of carbon monoxide in the presence of hydrogen by conductometric sensors is particularly difficult, because the detection mechanism of most metal oxide sensors (e.g., SnO_2) and catalytic sensors is based on the catalytic oxidation of gas molecules. In other words, oxygen is required. Furthermore, hydrogen undergoes oxidation more readily than carbon monoxide and consequently most types of ceramic gas sensors show a higher sensitivity to H_2 than to CO. For this reason various solutions for selective detection of carbon monoxide in a hydrogen background are being investigated, driven by the need for carbon monoxide monitoring in reformer and fuel cell gas streams. Although promising, most of these technologies are still in the developmental stage.

An electrochemical sensor capable of detecting carbon monoxide in hydrogen between 0.005 vol% to 0.1 vol% was described by Kirby [320]. The sensor transduction mechanism was based on impeding the oxidation of hydrogen. The sensor had a rapid response time and was claimed to be suitable for use in a fuel cell environment due to its tolerance towards hydrogen, water, and carbon dioxide which are typically present in high concentrations in the gas streams. A similar approach for carbon monoxide detection in hydrogen is the use of miniaturised fuel cell sensors, which indicate the presence of carbon monoxide by the reversible poisoning of the anode catalyst by carbon monoxide. However, drawbacks have been identified including poor detection limits and long response times [321]. The low repeatability of the sensor also makes it difficult to validate the sensor's ability to accurately quantify carbon monoxide levels.

Another electrochemical sensor using a polymer membrane (Nafion) and metal electrodes with different sensitivities to carbon monoxide in hydrogen has been suggested. The measuring principle is also based on the use of a platinum electrode which senses carbon monoxide through the inhibition of the hydrogen oxidation reaction relative to the response of a Pt/Ru electrode which is relatively unaffected by the presence of carbon monoxide. The sensor shows a reproducible response from 0.001 vol% to 0.01 vol% CO at room temperature and from 0.01 vol% to 0.1 vol% CO at 70 °C [322]. A potentiometric sensor based on a samarium doped ceria (CeO_2) electrolyte for carbon monoxide detection in SOFC was reported [323]. The sensor operates between 200° to 300 °C and can detect 0.0001 vol% to 0.4 vol% CO in a gas mixture of 5 vol% hydrogen in nitrogen.

A conductometric copper halide based sensor was reported for detection of carbon monoxide in hydrogen. It has a poor lower detection limit and could detect carbon monoxide only above 0.01 vol%. Furthermore it was observed that humidity influenced the sensor response which would preclude the use of the sensor in numerous hydrogen process applications, especially in reformer processes [324, 325]. Other technologies for carbon monoxide detection in H_2 which have been reported in the literature include a conductometric polyaniline nanofiber-based sensor which was developed for fuel cell application and which was capable of 0.0001 vol% (1 ppm) to 0.01 vol% carbon monoxide in a hydrogen gas background [326]. A field effect transistor based sensor with a platinum-gold alloy gate covered with a thin polymethylmethacylate (PMMA) layer reportedly detects carbon monoxide in hydrogen containing air in the range of 0.01 vol% to 0.001 vol% [327].

None of these sensors reported a range that meets the ISO or SAE standard carbon monoxide requirement for PEM fuel cells. Because of the difficulties to detect carbon monoxide in hydrogen by semiconductor gas sensors and electrochemical sensors, a catalytic reduction of carbon monoxide was explored in a micro-calorimetric sensor configuration. This reverse methane steam reforming reaction uses nickel as a catalyst and is applied in analytical gas chromatography. The reaction of carbon monoxide with hydrogen to form methane is exothermic and can be detected with a pellistor-type sensor. Carbon monoxide can be detected in the range of 0 to 5 vol% with an uncertainty of about 15 %. Carbon dioxide can be reduced to methane in a similar way.

One promising emerging platform for the detection of CO in air as well as in hydrogen is an optical sensor based upon tunable diode laser (TDL) light sources. These sensors are now commercially available for use in air, often with dual cells to allow for the simultaneous measurement of carbon dioxide and monoxide. The TDL sensor is amenable to measure CO in hydrogen as well, but current detection limits are not adequate for verification of fuel quality requirements per ISO 14687 and SAE 2719. Improved detection limits of the CO TDL sensor is, however, an active area of industrial development. A discussion of TDL optical gas sensors is presented for carbon dioxide in the next section.

Carbon Dioxide

Carbon dioxide (CO_2) is produced by combustion of carbon-based fuels and it is a common contaminant in hydrogen produced from hydrocarbon sources. It is also produced metabolically, through fermentation and breathing. The carbon dioxide level is often used as an indicator for indoor air quality. Concentrations exceeding 0.15 vol% are indicative of inadequate ventilation [328] and concentrations above 8 vol% are deadly.

Electrochemical cells are available for the detection of carbon dioxide with a measuring range of approximately 0.2 vol% to 5 vol% CO_2 in air. Carbon dioxide is, however, most commonly detected using optical sensors. In addition, carbon monoxide as well as numerous other gases can be detected by optical methods. Usually infrared methods are used. Infrared spectroscopy is based on detection of energy absorption of molecules in the infrared range between 700 nm to 1,000 μm of the electromagnetic spectrum. Molecules have a characteristic

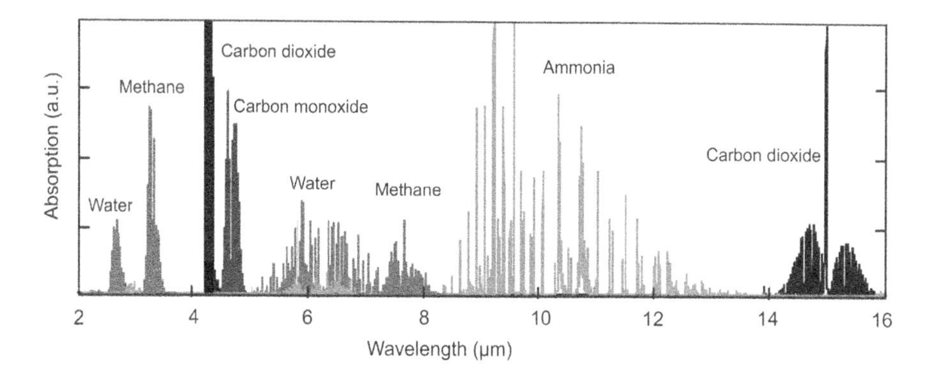

Figure 4.6: Absorption spectrum for different gases in the infrared range.

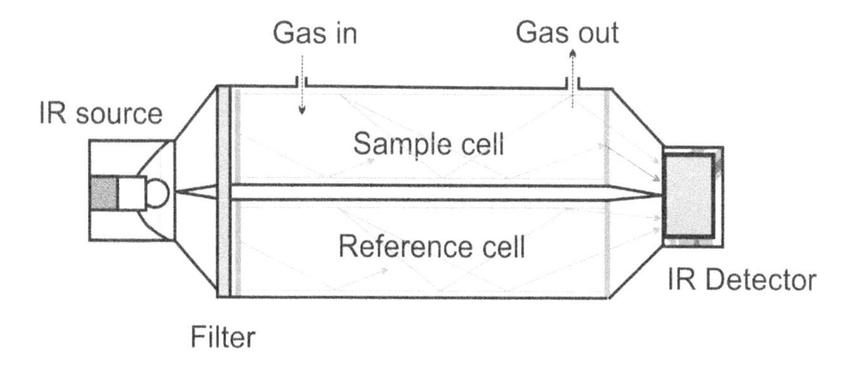

Figure 4.7: Schematic of an infrared sensor.

absorption spectrum (see Figure 4.6 and Table 4.5), which can be exploited for sensing applications. Absorption causes changes in the vibrational states of the molecule (such as stretching, twisting, and bending). The absorption of light is correlated to the amount of the analyte in the gas and can be quantified using the Beer-Lambert law:

$$I = I_0 \exp(-\epsilon_a \cdot c_B \cdot l) \tag{4.3}$$

Where I_0 is the intensity of the incoming light, I is the intensity of the transmitted light, ϵ_a is the molar attenuation coefficient, which depends on wave length λ of the light and temperature T, c_B is molar concentration of the attenuating species in the test medium (gas), and l is the optical length, i.e., the distance the light passes through the gas.

The principle of light absorption in the infrared region is also known as "non-dispersive infrared (NDIR) absorption." NDIR sensors can be constructed with relatively small gas cells. A schematic of an NDIR sensor is shown in Figure 4.7. The IR light source may be a tunable diode laser, a light emitting diode (LED) or a tungsten bulb. The tunable diode laser produces a more stable light with a narrow bandwidth. The wavelength of the light output of the tunable diode laser can be adjusted via modifications in the manufacturing process so as to be optimized for a specific target analyte. For a broad-band light source such as a tungsten lamp, the emitted light has to pass through a bandpass optical filter in order to

Table 4.5: Major absorption bands of various gases commonly used in NDIR sensors

Gas	Absorption [1)] μm	cm^{-1}	Estimated intensity
CO	4.6	2174	medium
	4.7	2128	medium
CO_2	4.22	2370	strong
	4.26	2347	strong
	15	670	strong
CH_4	3.5	3100	medium
	3.3	3000	strong
	3.4	2940	weak
	7.1-8.3	1200-1400	medium
NH_3	3.0	3330	weak
	6.15	1630	medium
	9-12	830-1110	medium
NO	5.2	1920	strong
	5.4	1850	strong
	6.1	1650	medium
	6.2	1600	medium
NO_2	3.4	2940	weak
	6.1	1650	medium
	6.2	1600	medium
H_2O	2.5	3900	strong
	2.7	3700	strong
	5.9	1700	strong
	6.4	1550	strong
H_2S	2.5	3900	strong
	2.6	3800	strong
	3.7	2700	weak
	7.1-9.1	1100-1400	strong
SO_2	7.25	1380	strong
	7.4	1350	strong
	8.6	1160	medium
	8.8	1140	medium

[1)] The energy of electromagnetic radiation in the infrared range is given either as wavelength, e.g., in μm or as frequency in cm^{-1}, ($[\mu m] = 10{,}000/[cm^{-1}]$).

get the required wavelength of "monochromatic light." The light passes through the sample chamber which is filled with the test gas. The intensity of the light is attenuated by absorbing molecules, and the magnitude of the attenuation is related to the amount of analyte present in the test gas (see Equation 4.3). The transmitted light hits a IR radiation-sensitive detector (e.g., pyroelectric detector). IR gas cells often allow multi-reflections of the probe light via properly positioned mirrors. This effectively lengthens the optical path (e.g., the term l in Beers-Lambert law) and hence increases sensitivity. A comparative measurement in a reference cell, filled with a reference gas and having a known absorption, increases accuracy. This allows to compensate the sensor output for changes in sensitivity of the detector and light source intensity.

NDIR CO_2 sensors use a light source with a wavelength output of 4.22 μm or 4.26 μm. The absorption of light at this wavelength is directly proportional to the carbon dioxide concentration in the test cell. NDIR carbon dioxide sensors have a detection limit of around 0.01 vol% (100 ppm). The sensor response time, t_{90}, is typically less than 45 s, and is controlled by the time required to purge the gas cell. Much faster response times are possible with faster cell purge rates. IR sensors are suitable for the detection of carbon dioxide in air, hydrogen, as well as natural gas (methane).

4.2.3 Hydrocarbons and Other Volatile Organic Compounds

Hydrocarbons are molecules that are consisting only of hydrogen and carbon atoms. Some of these compounds can be categorized as volatile organic compounds (VOC), which can include other types of organic chemicals (aldehydes, alcohols, ketones, and other numerous other substituted organic chemicals). Hydrocarbons as well as VOCs are formed in natural processes, such as the metabolism of organisms (e.g., fermentation). They are also formed in oil refining and released during coal mining. Similarly these organic compounds are produced and released from anthropogenic sources, including (incomplete) fuel combustion, chemical processes and vehicle exhausts. VOCs have a decisive influence on indoor air quality and can have an adverse impact on human health. The measurement of VOC is an extensive area of analytical chemistry. In the following section a brief overview on the detection of volatile hydrocarbons and other select VOCs is presented, with an emphasis on sensor-based methods.

The detection of saturated hydrocarbons by electrochemical sensors is limited due to their low electrochemical reactivity without sample treatment (e.g., pyrolysis over a heated filament to form carbon monoxide). Conversely the detection of unsaturated species, such as ethene or propene, can be more readily measured electrochemically as can some oxygen containing organic compounds such as alcohols and aldehydes; however, ketones tend to be unresponsive on electrochemical sensors. Working electrodes formed from high surface area platinum are normally required. The detection range of these sensors is in the range 0.0001 vol% to 0.01 vol%. As combustible chemicals, catalytic sensors can also detect various VOCs, including hydrocarbons. Since catalytic sensors are typically calibrated for the detection of methane (CH_4 – the simplest hydrocarbon molecule), the response towards other hydrocarbons (and VOCs in general) should be adjusted using correction factors for the output signal. The response time of catalytic sensors is typically between 5 s and 10 s, but this can increase when the sensor is protected by a filter. Conductometric sensors can detect 0.001 vol% to 0.1 vol% methane, propane, and butane with an uncertainty of 10 % to 15 %. The sensors are, however, not very selective and will respond to almost any VOC. Furthermore, the sensor output is influenced by changes in humidity.

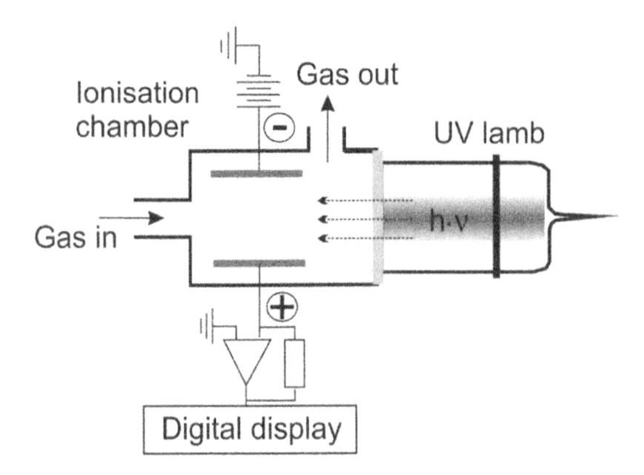

Figure 4.8: Schematic of a photoionization detector (PID).

NDIR sensors are a good sensor platform to detect many volatile organic compounds over a concentration range from about 0.0001 vol% to 1 vol% with an uncertainty below 5 %. Regular recalibration is required to correct for long term drift in the output signal. NDIR sensors for VOCs differ from the CO_2 NDIR sensor mainly by using light with a different wavelength (see Table 4.5), which can be obtained by selecting a suitable TDL light source.

Another useful tool for the detection of hydrocarbons, as well as other VOCs is the photo ionization detector (PID). Gaseous molecules are ionized by high-energy photons emitted from a UV lamp. The presence of the ions affect the current between two electrodes that are held at a fixed voltage (bias). The current is proportional to the amount of analyte in the test gas. A schematic of the PID is shown in Figure 4.8. The energy of the UV light depends on the gas filling the lamp. Lamps using xenon, krypton or argon generate photons having energy of 9.8 eV, 10.6 eV, or 11.7 eV. This energy is sufficient to ionize many types of volatile organic compounds, however, it is insufficient to ionize hydrogen, carbon monoxide, and many saturated hydrocarbons. The PID sensor can be very sensitive and has a detection limit of about 10^{-8} for some VOCs. Portable or stationary mounted PIDs are commercially available for the detection of hydrocarbons, alcohols, aldehydes, and ammonia, in the range of 0.0001 vol% to 1 vol% usually with a t_{90} response time of less than 3 s. The PID can function at ambient temperatures in the range of -20 °C to +50 °C. However, recalibration is frequently required owing primarily to fluctuations in the intensity of the light source. If other gases are present which have an ionization potential above of the energy of the light source, part of the light will be scattered or absorbed, resulting in a lower reading. This is known as the quenching effect. Water vapour is also detected by PIDs and thus can cause false positive readings. Overall, however, the PID is fast and very easy to use and remains a popular detector for indoor air quality and as a quick survey of workspaces.

Whereas numerous sensors are available for detection of trace VOC components in air, sensors for the detection of VOCs in hydrogen are rare. Optical sensors, such as the NDIR platform can still function in hydrogen. Methane in hydrogen can be measured by IR optical methods in the range of 0.001 vol% to 0.01 vol%. Humidity had a small impact on the methane measurement [329]. Photo acoustic spectroscopy (PAS), which is based on IR absorption is a very sensitive detection principle by means of acoustic detection. Miniaturized devices for gas analysis using laser diodes are under development. An electrochemical sensor based on a small polymer electrolyte membrane fuel cell has been suggested to measure VOC impurities in hydrogen [326]. The sensor, which has a low platinum loading, shows a fast response as manifested by a decrease of cell voltage due interferences from H_2S, CO, or

NH_3. This system is non-selective and plagued with a long recovery time, and thus would serve more as a indicator as per the definition provided in Chapter 2 than a sensor.

4.2.4 Ammonia and Nitrogen Oxides

Ammonia

Ammonia (NH_3) is a naturally occurring gas. It is typically present in the atmosphere at low concentrations of around 10^{-9}. It is a colourless and toxic gas but has a characteristic odour. Various sensor technologies are available for its detection. Requirements on the sensor performance are compiled in Table 4.6.

Electrochemical sensors are commercially available to detect ammonia in air in the range of 10^{-7} to 10^{-4}. Amines, hydrogen sulphide, and nitrogen oxides will interfere with the measurement. Catalytic sensors can also be used to detect ammonia. As these sensors are usually calibrated for methane detection, a specific calibration is advisable or a correction factor for the output signal has to be applied. Owing to cross-sensitivity, the use of catalytic sensors for ammonia is only viable if other combustible gases are not present. The catalytic sensor will respond to any combustible gas present, including carbon monoxide, methane, hydrogen, and many solvent vapours. Conductometric metal oxide gas sensors are also commercially available to detect ammonia in the range of 0.001 vol% to 0.1 vol%. However, as discussed previously (see Chapter 3) metal oxide sensors can have a high cross-sensitivity to other chemical species. Nitric oxide (NO) is noted as a problematic interferent on a metal oxide ammonia sensor. The relative selectivity for ammonia and nitric oxide can be adjusted by changing the composition of the metal oxide. For example, it was shown that WO_3 has a higher sensitivity towards ammonia compared to nitric oxide. Adding 5 % MoO_3 by mass to WO_3 eliminated sensitivity to nitric oxide. Furthermore addition of 0.8 % Au by mass increased sensitivity for ammonia with a measure range from 0.0001 vol% to 0.005 vol% (1 ppm to 50 ppm) NH_3 [330]. The response time was about 1 min. The detection is affected by hydrogen sulphide H_2S, but there is little effect on the sensor response to ammonia from hydrogen in the same concentration range [330].

Conductometric polymers such as polypyrrole and polyaniline change their resistivity due to reaction with ammonia. A number of such sensors have been reported for ammonia detection [331]. Some have been commercialized. The lower detection limit is 10^{-6}. One drawback of this type of sensor is that the absorption of ammonia in the polymer is not completely reversible. There is a drop in the sensitivity to ammonia with increasing exposures [332].

A field effect-transistor (FET) platform has been developed as a sensor for ammonia detection in the range from 2×10^{-6} to 30×10^{-6}. FET-based sensors can also indicate NO_2 and other gases [333].

Two optical principles have been used for ammonia sensors. One is an indirect method based on a colour change that occurs when ammonia reacts with a reagent. The second is based on direct optical absorption in an NDIR sensor. The colorimetric method has been applied to a simple and inexpensive system analogous to pH indicator paper. The colorimetric indication is based on the absorption of ammonia in aqueous solution or fixed in a gel target and the formation of coloured substances by the reaction of ammonia with an embedded reagent. Reagents based on the Nessler reaction or Berthelot reaction can be applied for a colorimetric ammonia detection using a spectrometer [332].

Direct optical absorption measurements in the IR range is also feasible. Bench-top spectrometers can measure ammonia down to 10^{-12}. Sensors based on TDL are also emerging in the market place, but with detection only down to around 0.01; however, as with the CO

TDL sensor, the improvement of the detection limit of the ammonia TDL sensor for fuel quality applications is an active area of research. The interference with water and carbon dioxide has to be considered; only at 1.53 μm (6529 cm^{-1}) is a selective ammonia absorption band achievable without interference by these two chemicals [332]. Optical methods can also be applied to detect trace levels of ammonia in hydrogen. A diode laser based quartz-enhanced photo acoustic spectrometer for trace-level detection of methane and ammonia in hydrogen gas has been developed for this purpose. This technique is capable of detecting 2×10^{-6} to 10^{-4} NH_3 in hydrogen at pressures from 144 hPa to 1000 hPa [329].

Nitrogen Oxides

Gas mixtures containing nitrogen oxides, typically abbreviated as NO_x, are often produced during combustion of conventional fuels and hydrogen. Usually NO_x consists of about 90 % nitrogen monoxide, sometimes called nitric oxygen (NO), and 10 % nitrogen dioxide (NO_2). Both of these gases are toxic. NO is a colourless and odourless gas while NO_2 has a characteristic odour and brown colour. Electrochemical sensors are the method of choice for the detection of both of these gases. The measuring range for a NO EC sensor is 10^{-7} to 10^{-4}. The sensor is very selective, however, hydrogen sulphide may be an interferent. The measuring range for electrochemical sensors is typically from 5×10^{-7} to 5×10^{-4} NO_2. Hydrogen sulphide and nitric oxide can act as interferents for NO_2 electrochemical sensors. Several solid electrolyte and semiconductor oxide-based sensing platforms for nitric oxide (NO) detection in exhaust gases have been reported and a conductometric semi-conductive sensors for NO_2 for 10^{-7} to 10^{-6} was commercialized. However these sensors are not widely used [334].

4.2.5 Sulphur Compounds

Sulphur is a common constituent in many fuels. Its presence can span a broad range of concentration, from trace ($< 10^{-6}$) to percent levels. It occurs in reduced form in gases and vapours such as hydrogen sulphide (H_2S) and as sulphur containing organic compounds such as organic sulphides and mercaptans. Sulphur also exists in oxidized form, such as sulphur dioxide (SO_2). Atmospheric sulphur is primarily a product of exhaust gas, and periodically, of volcanic eruptions. Sulphur containing gases and vapours have a characteristic odour and are often quite toxic, as indicated in Table 4.2. Sulphur components are readily and often irreversibly absorbed to the surface of noble metal catalysts and act as a strong poisoning agent. It has been shown that H_2S concentrations of just 0.00025 vol% (2.5 ppm) can induce significant performance degradation of the palladium membrane in PEM fuel cells [335]. While molten carbonate fuel cells (MCFC) and solid oxide fuel cells (SOFC) are resistant to hydrocarbons due to the high operation temperature, sulphides, e.g., in biogas are damaging to both MCFC and SOFC due to poisoning of the anode [336, 337]. Therefore only very low amounts in hydrogen can be tolerated (compare Table 4.3).

Hydrogen Sulphide

Electrochemical sensors are commonly used for the detection of sulphur-containing gases, especially when used for ambient air monitoring. Hydrogen sulphide can be detected by electrochemical sensors in the range of 10^{-7} to 10^{-3}. After a warm-up time of some minutes, the sensor typically has a response time (t_{90}) of less than 30 s. As with many electrochemical sensors, the H_2S EC sensor is influenced by temperature, and thus temperature compensation is often necessary to assure the accuracy of the reading. The sensing element can exhibit

Table 4.6: Requirements on ammonia and sulphur sensors for PEM fuel cell applications

Parameter	**Ammonia (pre-stack)**	**Total sulphur (reformer) exhaust or pre-stack**
Measuring range	$(1 - 10)\times10^{-6}$	$(0.05 - 0.5)\times10^{-6}$
Accuracy		1 % to 10 % full scale
Response time	seconds	< 1 min at 10^{-6} SO_2
Temperature	70 °C - 150 °C	< 400 °C
Pressure	(0.1 - 0.3) MPa	
Gas environment	H_2, CO_2, CO N_2, H_2O HC	(30 - 70) vol% H_2, CO_2, CO N_2, H_2O, traces of NH_3

long term output drift and limited storage life time. Carbon monoxide and carbon dioxide do not interfere with the detection. Electrochemical sensors can also respond to organic sulphur components like methyl mercaptan or tetrahydrothiophene which can be present as an odourant in natural gas in the range of 10^{-6} to 4×10^{-5}.

Photoacoustic spectroscopy (PAS) is a very sensitive optical method for detection of gaseous trace components, such as hydrogen sulphide, ammonia, or carbon dioxide at concentrations below of 10^{-6}. Pulsed and modulated light energy absorbed by an analyte is converted to heat; for gases this generally occurs via molecular collision-induced non-radiative relaxation of excited states. The temperature of the analyte and the surrounding matrix expand. The expansion produces pressure waves that can be detected using a microphone. The detected signal is directly proportional to the intensity of the absorbed light and to the amount of analyte. A disadvantage is that if the background matrix changes, relaxation processes and pathways can change, with a consequent alteration in the signal level. This optical method has been miniaturised to use in sensors [338, 339].

Sulphur Dioxide

Sulphur dioxide (SO_2) in air is most commonly detected by electrochemical sensors and to a lesser extent by optical sensors. EC sensors are, however, more sensitive than optical sensors and have a measuring range from 10^{-7} to 10^{-3}, however EC sensors may be some interference from nitrogen oxides (NO_x). Although sulphur dioxide is not stable in hydrogen, its transformation to hydrogen sulphide is not instantaneous and thus can be a concern in hydrogen, especially for applications the involve hydrogen reactions on catalysts. The presence of sulphur dioxide in PEMFCs can result in degradation of the membrane electrode assembly (MEA), decomposition of the membrane polymer, and a reduction in the cell voltage. It also has a poisoning effect on the electrochemical performance and long term durability of solid oxide fuel cells (SOFC). For this reason stringent impurity levels have been established to ensure high hydrogen purity levels for sensitive applications [319]; these are often stipulated in terms of total sulphur. The maximum allowable sulphur content for PEMFCs is shown in Table 4.3. However, commercial sensors for detection of sulphur components at regulated levels in hydrogen gas are currently unavailable. Specific R&D is on-going to adapt current technology for sulphur detection. Sulphur dioxide sensing elements based on thin metal films have been reported. For example, a rhodium ultra thin film exhibit sensitivity at the 10^{-7} level and has a 400 h life time. A noble metal thin film sensors has reported detection limits of 10^{-8} SO_2, which is close to meeting the PEM fuel quality requirement [319].

4.2.6 Sensors for Inert Gases (N_2, He, Ar)

These components are ubiquitous constituents of air, and also often present at trace levels in natural gas and hydrogen. Miniaturized sensors for the selective detection and quantification of these gases are unknown and will be difficult to develop due to their lack of chemical reactivity. Inert gases can be detected by laboratory methods such as mass spectrometry, gas chromatography, and optical methods. Laser-induced breakdown spectroscopy (LIBS) based optical sensor can detect nitrogen up to 1 vol%, helium and argon up to 0.5 vol%, and oxygen up to 0.04 vol%. The limits of detection are 8×10^{-5} N_2, 9.7×10^{-5} Ar, 2.5×10^{-5} He, and 10^{-5} O_2 [340]. However, this limit of detection is not sufficient to meet the levels defined in the PEM fuel quality standard (Table 4.2). These gases, however, do not impact fuel cell performance, other than acting as a diluent for the hydrogen fuel.

4.3 HUMIDITY SENSING TECHNOLOGIES

4.3.1 Overview

The term "humidity" refers to the amount of water vapour that is contained in a gas. Conversely the term "moisture" refers to the amount of water in a solid or liquid. Water in its gaseous and liquid states is a relevant component not only in air, but also in industrial gases such as hydrogen, oxygen, acetylene, and many others. The ability to measure humidity is essential in many applications including meteorological, industrial, and metrological applications. Humidity has to be quantified over more than eleven orders of magnitude, ranging from about 10^{-9} kg/m^3 to 322 kg/m^3, at different temperatures and pressures. The branch of metrology that deals with quantifying humidity in air and other gases is called "hygrometry," and pertains to vapour amounts up to the critical point of water (374 °C and 22.064 MPa, 322 kg/m^3). As with other gases and vapours, it is relatively difficult to accurately quantify vapour and moisture content compared to parameters such as temperature or pressure. One reason for this is because humidity measurement is based upon temperature measurement, and due to propagation of errors, the accuracy of humidity determinations can never exceed that of temperature measurements. Nevertheless there is a variety of sensor technologies available to quantify humidity [341–343]. Humidity measurements in gas can be divided into three main ranges:

Humidity with respect to ambient climate conditions
For normal altitudes and air pressures of 750 hPa to 1080 hPa the air temperatures are typically between -40 °C and +45 °C and dew-point temperatures are in the range of -40 °C to 40 °C. Indoor climates typically have a narrower temperature range of around of 12 °C to 32 °C and dew-point temperatures between -10 °C and +20 °C.

Trace humidity levels for gas purity considerations
Although it can vary with application, trace humidity can be defined as a humidity level less than a volume fraction of 1×10^{-3}, which corresponds to frost-point temperatures below -20 °C.

Humidity considerations for high temperatures / high pressure applications
These applications pertain to drying processes, heated and compressed gases. Gas

temperatures are typically above 70 °C with dew-point temperatures greater than 40 °C and pressures above atmospheric.

In hydrogen systems, depending on the application, humidity can be a desirable or unwanted component in the hydrogen gas stream. Sensing of water vapour, wanted or otherwise, is of importance in relation to hydrogen technologies for several reasons, including:

Compliance to hydrogen quality requirements
Hydrogen for PEM fuel cell applications in road vehicles must have a water vapour fraction less than 5×10^{-6}. Dew-point measurements are performed to facilitate regulation of the humidity levels in order to avoid unwanted condensation.

Water management in PEM fuel cells
There is an optimum range for water vapour content in fuel cell membranes. High humidity levels could result in condensed water blocking pores in the membrane and hindering the diffusion process of the reactant gases; on the other hand a low humidity level could cause drying out of the fuel cell stack thereby degrading the membrane conductivity [344].

Humidity as an interferent to gas measurement
The determination of hydrogen in air or of trace components in hydrogen can be biased by the presence of or changes in humidity [345, 346]. Ammonia or sulphur dioxide are prone to hydrolysis in the presence of water, thus their accurate detection can be particularly sensitive to humidity.

Condensation on pneumatic and electrical components
Although electronic units are typically encapsulated, water may still diffuse through the protective enclosure and accelerate corrosion. Condensation may induce creeping current or voltage breakdown. Reaction of water vapour with reactive trace components (e.g., acidic gases) in an atmosphere can exacerbate corrosion and degradation [347].

4.3.2 Common Humidity Units

The humidity content of a gas is commonly expressed in a variety of ways including mass, volume, or pressure fraction units, either as a concentration or ratio. The most common units are collected in Table 4.7. All the units to describe water vapour content in air are inter-convertible and there are useful on-line software tools to do the conversion, e.g., [348]. It is noted that the use of abbreviations and symbols used here were developed in part specificalyy for hygrometry and may differ from the symbols otherwise commonly used in other applications. For example, the symbol for water vapour partial pressure is e'. The volume fraction ϕ_v of water vapour can be defined as:

$$\phi_v = \frac{V_v}{V} = \frac{e'}{p} = \frac{n_v}{n} \tag{4.4}$$

Where V_v and V are the volume of the water vapour and the total gas volume, p the gas pressure and n_v and n the number of moles of the water vapour and of the total gas entirety. From the ideal gas law, ϕ also corresponds to partial pressure and molar fraction. The water vapour concentration is called absolute humidity (or water vapour density) and describes the mass of water per volume of wet gas. The most common term used to quantity water content in air is relative humidity, which is the ratio of actual water vapour pressure to its saturation pressure, usually expressed in percent (% rh).[1] In a single component system the

[1]Another common abbreviation is %RH.

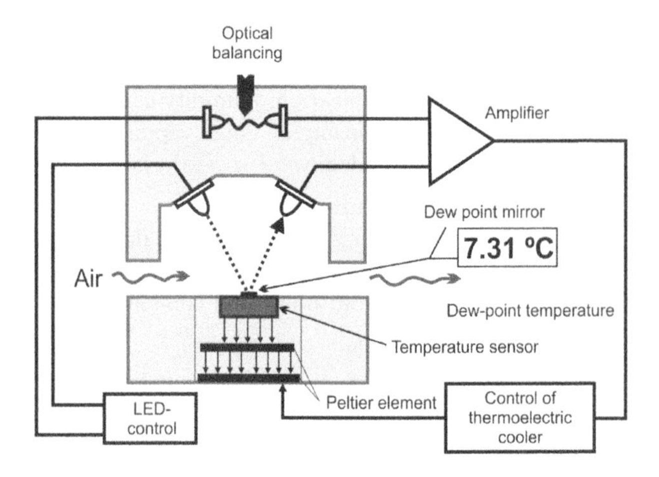

Figure 4.9: Schematic of chilled mirror dew-point hygrometer.

saturation pressure, e_w is the water vapour pressure under saturation conditions above water or ice (e_i), e.g., the water vapour pressure achieved at a thermodynamic equilibrium between gaseous and liquid or solid (ice) phases. It depends on temperature and total pressure and can be calculated by appropriate formulas [349,350]. The temperature of a gas at which the prevailing water vapour pressure is equal to the saturation vapour pressure and at which condensation of water starts is called the dew-point temperature t_d. For temperatures below the freezing point of water which results in the formation of ice instead of liquid condensate, the dew-point temperature is more properly called the frost-point temperature t_f. Dew point or frost point measurements can be performed over a temperature range from -110 °C to 100 °C. This corresponds to a change of water vapour mole fractions from 1.6×10^{-8} to 1. The corresponding water vapour concentration is from 1.1×10^{-9} kg/m^3 to 0.587 kg/m^3, a concentration range that covers nearly 9 orders of magnitude. It is emphasized that the term "dew-point temperature" describes the water vapour content in a gas, and thus should not be equated to the term temperature when it is used as a measure of thermal state of matter.

The following sections focus on those water vapour measurement techniques that are potentially relevant to hydrogen technology. Hygrometers are a family of devices which can be used to measure humidity. Hygrometers contain a humidity sensing element, a signal processing and transfer unit, and a display component. Different methods and devices have been developed to measure humidity in various gases and under different conditions. Many of these techniques have been reviewed elsewhere, [351,352]; however, the focus of this section is on humidity measurement techniques that are potentially most appropriate for use in hydrogen applications. Each technique is described in some detail below and a summary is provided in Table 4.8.

4.3.3 Chilled Mirror Dew Point Hygrometers

The dew-point hygrometer is widely used for precise humidity measurements over a range that encompasses the three regions mentioned in Section 4.3.1. The dew-point or frost-point temperature is detected by isobaric cooling of the gas mixture until the gas at the solid interface becomes saturated and water or ice starts to form on a planar surface of a metal mirror (see Figure 4.9).

Table 4.7: Quantities for water vapour in gases

No.	Quantity	Symbol	Unit	Formula	Comment
1	volume fraction	φ_v	m^3/m^3	$\varphi_v = \frac{V_v}{V} = \frac{e'}{p}$	volume of water vapour in the volume of a wet gas
2	water vapour density (absolute humidity)	d_v	kg^3/m^3	$d_v = \frac{m_v}{V} = 0.0021667 \frac{1}{Z_{mix}} \frac{e'}{T}$	mass of water vapour present in a volume of a humid gas. Z_{mix} is the compression factor of the gas mixture
3	mass fraction (specific humidity)	q_v	kg/kg	$q_v = \frac{m_v}{m_v + m_g} = \frac{M_v}{M_g} \cdot \frac{e'}{p - (1 - M_v/M_g)e'}$	ratio of mass of water vapour to mass of wet gas, M_v, M_g are the molar mass of water and gas
4	mixing ratio	r	kg/kg	$r = \frac{m_v}{m_g} = \frac{M_v}{M_g} \cdot \frac{e'}{p - e'}$	ratio of mass of water vapour to mass of dry gas
5	relative humidity	U	%	$U = \frac{e'}{e'_w(t)} \cdot 100$	percentage of actual water vapour pressure to the saturation vapour pressure under saturation conditions above water (or ice) at equal total pressure and temperature
6	dew-point temperature	t_d	^oC	$e' = e'_w\ (p, t_d)$	temperature at which the prevailing water vapour partial pressure is equal to the saturation vapour pressure and at which condensation starts
7	frost-point temperature	t_f	^oC	$e' = e'_i\ (p, t_f)$	temperature at which the prevailing water vapour partial pressure is equal to the saturation vapour pressure and at which deposition of ice starts

Table 4.8: Humidity measuring methods (selection)

No.	Principle/Type	Humidity range	Application conditions	Uncertainty[a]
1	Chilled-mirror dew-point hygrometer	-95 °C - +95 °C	-40 °C - +105 °C, ≤ 30 MPa	0.1 °C - 1 °C
2	Capacitive polymer sensor	0 % rh - 98 % rh	-60 °C - +200 °C, ≤ MPa	1 %rh - 2 % rh
3	Capacitive alumina sensor	-110 °C - +30 °C	-40 °C - +35 °C, ≤ 45 MPa	2 °C - 5 °C
4	Coulometric method (electrolysis)	-80 °C - +20 °C	0 °C - 60 °C, ≤ 10 MPa	3 % - 10 % of reading
5	Fibre optic sensor	-80 °C - +20 °C	-50 °C - +100 °C, ≤ 25 MPa	3 % - 10 % of reading
6	QMB	-95 °C - +90 °C	ambient	10 % - 20 % of reading
7	Laser absorption spectroscopy	-90 °C - -10 °C	-70 °C - +200 °C ≤ 30 MPa	2 % - 10 % of reading
8	FTIR spectroscopy	-95 °C - -65 °C	-40 °C - +105 °C	2 % - 10 % of reading
9	CRD spectroscopy	-110 °C - -65 °C	ambient	2 % - 10 % of reading
10	Electro-chemical solid state sensor (zirconia)	70 °C - 200 °C	0 °C - 600 °C, ambient	5 % - 10 % of reading

[a]This is only a rough estimation, because uncertainty of humidity measurement depends strongly on water amount and surrounding conditions.

Table 4.9: Enhancement factors f for water vapour in hydrogen

t / °C	p / MPa						
	0.5	5	10	15	20	25	30
50		1.08	1.20	1.31	1.43	1.54	1.65
-20	1.010	1.104	1.212	1.324	1.437	1.552	1.667
-30	1.011	1.117	1.239	1.366	1.494	1.624	1.755
-40	1.013	1.131	1.269	1.412	1.558	1.705	1.853

Typically the bulk of the mirror is made from a material with a high thermal conductivity, e.g., copper. The surface of the material is plated with an inert metal such as silver, rhodium, gold, or platinum to minimise tarnishing and corrosion. An internal Peltier thermoelectric unit cools the mirror and at a certain temperature, determined by the humidity content of the sample gas, dew or ice will form on the surface. Light emitted from a source is reflected at the mirror and received by a detector. The formation of dew or ice on the mirror surface causes a reduction in the reflected light intensity. The change in reflected light intensity is measured by a differential optical detection system which in turn regulates the power to the Peltier unit via a control circuit. The control loop maintains the mirror surfaces exactly at the dew-point temperature which is then accurately measured by an embedded precise platinum resistance thermometer. Accurate temperature measurement is critical since, as already stated in Section 4.3.1, the accuracy of the dew-point measurement cannot exceed the accuracy of the temperature measurement. A broad measuring range capability can be achieved by the addition of a supplementary cooling system for very low humidity levels, or an internal heater for operation at high humidities and elevated ambient temperatures. Dew points can be measured typically over the range -95 °C to +95 °C, which at 100 kPa is equivalent to a water density ranging from about 2.8×10^{-8} kg/m^3 to 0.5 kg/m^3. However, for maximum accuracy it is better to use different devices of this type specifically optimized for the humidity level of interest (e.g., low/trace, medium, or high humidity). Typically the dew point can be measured with a resolution of 0.01 °C. It should be noted that below 0 °C, water can be super-cooled without freezing. Under these conditions, it may be unknown if water or ice has formed on the mirror surface, which can lead to significant measurement uncertainties. As a result many manufacturers of dew-point meters have developed specific strategies for sub-zero operation. One strategy is to cool the mirror surface to well below the frost point to assure frosting of the surface by water vapour. A second strategy is to use an eyepiece to visually inspect for frosting of the mirror surface. The dew-point meter can be used as a reference method for testing and calibration of other humidity sensors and hygrometers. It is noted that due to the interaction between water and gas molecules, the water vapour saturation pressure e'_w of the gas mixture will deviate from the vapour saturation pressure e_w in pure phase. Therefore a correction factor, the so-called enhancement factor f_w, was suggested [353]:

$$e'_w(p,t) = e_w(t) \cdot f_w(p,t) \tag{4.5}$$

The enhancement factor becomes more significant at temperatures and pressures deviating from ambient conditions, see Table 4.9. The enhancement factor has to be considered when converting dew points to other humidity units [354].

It is possible to use a dew-point hygrometer to measure humidity in hydrogen. However, the long-term interaction of hydrogen with metals and other materials should be considered.

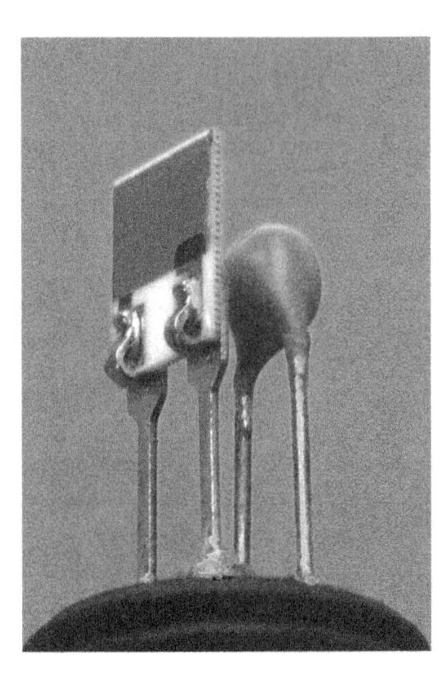

Figure 4.10: Image of a capacitive polymer sensor, showing a polymer thin film on a ceramic substrate on the left and a temperature sensor on the right.

Dew-point hygrometers developed specifically for humidity measurements in PEM fuel cells are available commercially. Dew-point sensors have also been used to ensure that the humidity does not exceed the required levels in hydrogen cooled generators. Humidity measurements in pressurized gases can also be performed with a dew-point hygrometer equipped with a compression-proof chamber. Alternatively, ex-situ measurement can be executed using a by-pass arrangement. Typically in a by-pass arrangement, the pressurized gas is first expanded to ambient pressure and then the dew point can be measured. The humidity in the original pressurized state would then be calculated using the gas laws. It is usually necessary to consider non-ideal behaviour of gaseous components (i.e., including compressibility factor and virial coefficients) and water vapour enhancement factor for accurate calculations.

4.3.4 Capacitive Polymer Sensors

A second humidity sensor platform uses films of polymers. Polymer sensors exploit the large dipole moment of water. When adsorbed into a hygroscopic dielectric polymer, such as polyimide, water will induce a change in permittivity which causes a measurable change in capacitance. The change of capacitance in the kilohertz range of the sensing element can be easily measured (e.g., as an impedance or frequency measurement). This signal is then related to the relative humidity of the surrounding atmosphere. A typical polymer humidity sensing element is shown in Figure 4.10; however, sensing elements are available in different configurations, including different types of parallel-plate or interdigital capacitors on ceramic, glass, or flexible organic substrates. Miniaturised capacitive polymeric sensing elements composed of micro-machined electrodes integrated with a temperature sensor and full control circuitry including electronic amplifier, analogue-to-digital converter with on-board memory and calibration, have been developed. The overall dimension of these sensing elements is a few square mm. Polymer humidity sensors are usually used in the relative humidity range from 10 % rh to 98 % rh. The influence of variations in the ambient temperature on the

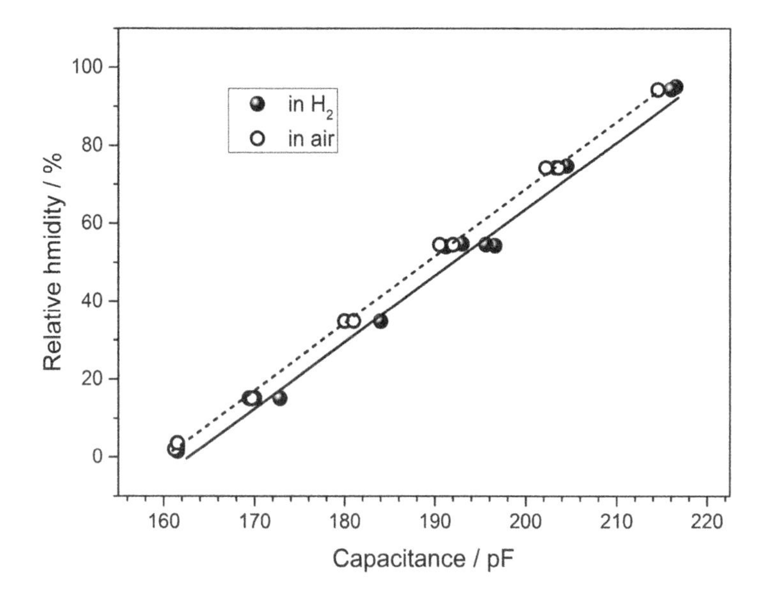

Figure 4.11: Calibration curve of a polymer sensor for humidity measurements in air and hydrogen.

humidity determination is compensated and in some applications the sensor can be heated to maintain a constant but slightly elevated temperature relative to ambient.

Polymer sensors can be used above 100 °C, even up to 200 °C. However the sensor life time decreases when used at high humidity and elevated temperature. Measurement uncertainty also increases with elevated temperature, partly because the electronic units are typically designed for lower temperature operation, but also because of the relative humidity at elevated temperatures is in many cases relatively low, especially compared to the room temperature relative humidity for the same mass of water per unit volume. The measurement of capacitance with higher resolution together with improved signal processing has extended the range of polymer sensors to lower humidity levels. These sensors have been increasingly applied for the determination of water vapour concentrations down to a frost point temperature of -60 °C. To assure accuracy, a temperature sensor should be integrated into sensor platform for this application. This humidity sensor platform is also compatible in compressed gases up to 35 MPa. Capacitive polymer sensors, however, can be plagued with hysteresis, often providing a different indication between increasing and decreasing humidity. It has been reported, however, that this effect is suppressed usually to below 1 % rh. The sensing element is robust and will typically survive condensation; however, the sensing element should be shielded against impurities and chemical precipitations by using a plastic or metal filter. Ammonia, hydrocarbons, and other substances that are absorbed by the polymer or that corrode the electrical leads will impact the sensors accuracy and life time. Polymer sensors can measure humidity in both air and pure hydrogen as shown in Figure 4.11. The sensors were adopted for humidity measurements in PEM fuel cells, specifically to analyse and control the water management [355]. The water vapour content was measured in-situ in the channels of PEMFCs using a miniaturised polymer-based humidity sensor

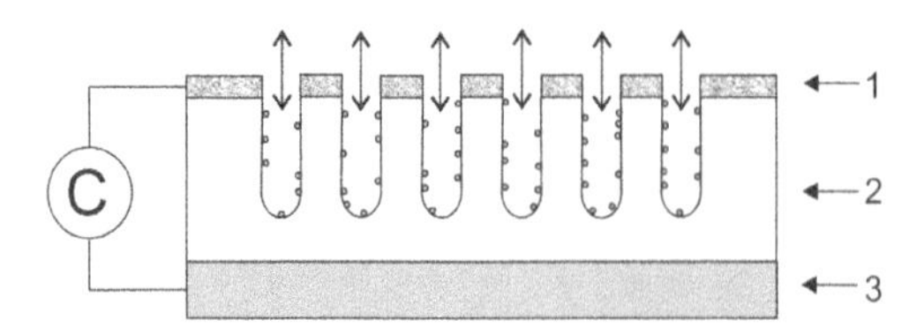

Figure 4.12: Schematic of a metal oxide humidity-sensing element. 1 - water permeable electrode, 2 - porous metal oxide, 3 - back electrode

operating at 80 °C and a pressure of 3 bar both in hydrogen and air. Dew-point temperatures between 60 °C and 79 °C were measured depending on time and operation conditions [356].

4.3.5 Capacitive Alumina Sensors

A third humidity sensor platform is based on changes in capacitance of ceramic substrates upon adsorption of water. Typically, the electrical properties of porous alumina and other metal oxides change following the adsorption of water. The alumina humidity sensing element can be prepared either by an anodic oxidation of aluminium or via a sol-gel process on a metallised substrate. The resulting oxide is then covered by a water permeable thin film metal electrode. A schematic of the sensing element is given in Figure 4.12. The sensor is specified from very low humidity levels (e.g., a frost point of -100 °C) up to a dew point of approximately 30 °C, and an operating temperature range between -40 °C and 35 °C. Higher humidity may result in irreversible damage. Since Al_2O_3 sensing elements tend to show long term drift periodic calibration or replacement is recommended.

4.3.6 Coulometric Sensors

In the coulometric sensor, water vapour is absorbed by a desiccant, usually phosphorous oxide (P_4O_{10}) which also acts as an electrolyte between a pair of noble metal electrodes. The humidity of a gas is related to the amount of absorbed water, which can be quantified by measuring the total number of coulombs (electrical charges) necessary for electrolysis. For that purpose, a voltage which exceeds the decomposition potential of water is applied between the electrodes and the absorbed moisture is electrolysed into hydrogen and oxygen. Two measuring procedures can be used: electrolysis of absorbed water in a defined closed gas volume or in a gas flow of a defined flow rate. The sensor is robust and can be used in reactive gases, including hydrogen; however, hydrocarbons, butadiene and ammonia can interfere with proper operation. An appropriate calibration is required when coulometric sensors are used to measure humidity in hydrogen [357], since the sensor characteristic differs significantly from the behaviour in air, as illustrated in Figure 4.13.

4.3.7 Optical Methods

The basis for optical humidity measurements is the Lambert-Beer Law (see Equation 4.3), which states that the amount of light adsorbed by a chemical is proportional to its concentration. Since chemicals often have very specific and well-defined absorption bands, optical methods can be very selective to the target analyte. For humidity measurements, the characteristic water vapour absorption bands in the infrared (e.g., 1.4 μm and 1.9 μm) or in

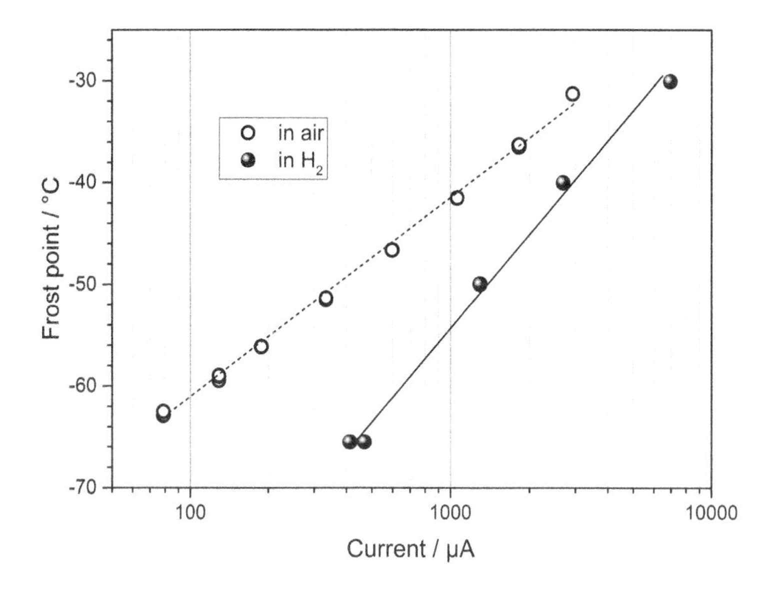

Figure 4.13: Calibration curves for frost point measurements in air and hydrogen.

the ultraviolet (121.6 nm) range are utilised. Optical humidity measurements in hydrogen were discussed as early as 1978 [358]. Typically, either a hollow-cathode lamp or, more recently, tunable diode lasers are used as the light source. A photomultiplier tube or photo diode are used as the detector. Specific measurement methods are Fourier transform infrared spectroscopy (FTIR), tunable diode laser absorption spectroscopy (TDLAS), and Cavity-Ring-Down Spectrometry (CRDS). CRDS is suitable for very trace humidity determinations, down to a frost point of -110 °C. With the possible exception of TDLAS, these methods are not considered sensors but rather are more akin to bench top instrumentation. As such, these methods tend to be experimentally complex and expensive, and therefore are mainly restricted to research areas or the verification of the accuracy of a humidity sensor. Nevertheless, spectroscopic methods are highly sensitive with a fast response time. It is noted that sensors based on tunable diode lasers are becoming more common and economical, and it is likely that a tunable laser diode sensor for humidity will be commercialized. While optical methods are suitable for use with almost any gas, including corrosive ones and hydrogen, absorption bands can shift when changing from air to another gas matrix. This may affect the calibration of the sensor.

Fibre-optic sensing technology provides an alternative optical approach to humidity sensing. It offers several advantages over the use of conventional sensing methods. Remote detection based on fibre optic methods can often be applied in explosive atmospheres, in hot gases up to 450 °C and in the presence of radio frequency (RF) (e.g., microwave) interference. Many fibre-optic humidity sensors are based on humidity-induced changes in light absorption, light scattering, refractive index, fluorescence, or evanescent waves [359]. Other optical humidity sensors are based on a thin film Fabry-Perot interference filter to detect the shift of the interference pattern due to the absorption of water. Fibre optical humidity sensors are commercially available.

4.3.8 Quartz Micro Balances

A crystal oscillator is based on the mechanical resonance of a vibrating piezoelectric material (quartz or ceramics) to create an electrical signal with a very precise frequency. The resonance frequency is affected by the mass of material on the QMB. A coating with an affinity to a specific target analyte is often applied to the QMB surface to enhance selectivity. The application of a hygroscopic surface layer will allow for reversible adsorption of water vapour. The adsorption of water changes the mass of the oscillator and, as a result, its resonance frequency. This change of resonance frequency is proportional to the humidity in the gas. In a measurement cycle the oscillation behaviour is evaluated in humid and dry air as reference. Although the sensing element is inexpensive, the current commercially available instruments used to control and operate the QMB are large and expensive. However, changes in frequency can be precisely measured, which leads to an instrument capable of detecting small quantities of water vapour, with a lower detection limit in the order of 10^{-9}.

4.3.9 Electrochemical Methods

Potentiometric or amperometric sensors with solid electrolytes can be used for indirect humidity measurements of wet gases. At temperatures above 100 °C potentiometric or amperometric sensors with solid electrolytes can be applied. The potentiometric oxygen sensor, e.g., based on a zirconia electrolyte (see Section 4.1), generates a signal due to the difference in oxygen content between the test gas and a reference (typically air). The presence of water vapour in the hot gas reduces the amount of oxygen and results in a decrease of the sensor signal.

4.3.10 Sensor Maintenance

In order to provide precise and reliable measurements humidity sensors should be periodically re-calibrated. The procedures for calibration, and to establish traceability to authoritative standards, are in general the same as described in Chapter 6 for hydrogen sensors. Manufacturer recommendations should be followed for calibration frequency and protocols. Proper calibration ensures measurement accuracy and leads to consistency of measurements among users of different sensors for measurements performed at different times and locations. Primary standards are provided by the national metrological institutes (NMIs). These are so-called humidity generators in which a stream of a gas at an elevated pressure is saturated with respect to liquid or solid phase of water and then expanded to a lower pressure and if necessary mixed with a dry gas stream. The primary standard for calibrating sensors at low humidity is based on gravimetry or the generation of trace water vapour through electrolysis. Precise chilled mirror dew-point hygrometers (see Section 4.3.3) or simplified humidity generators often serve as transfer standards. A simple method often employed to calibrate polymer sensors (and to maintain a fixed humidity level) is the use of saturated aqueous salt solutions [360].

4.3.11 Condensed Water

Condensed water can be problematic. For example, water occurs as a vapour in the gas streams of low temperature fuel cells under normal operating conditions; however, it can also be present as a liquid. Liquid water can be absorbed by the polymer membrane and can form a physical barrier restricting access of the reactant gas to the electrode. Furthermore condensed water can form ice at sub-zero temperatures, which can not only block the polymer membranes, but can even damage it. Appropriate water management in the fuel cell is essential to achieve a high cell performance over a wide range of operating conditions.

Only a limited number of methods for the quantitative determination of moisture (e.g., as condensed droplets or fog) in a gas are available. Because of blockage, many hygrometers do not tolerate condensation. Water in liquid and vapour phase have been semi-quantitatively detected in the different parts of PEM fuel cells by in-situ techniques, like magnetic resonance imaging, neutron radiography, x-ray imaging, fluorescence microscopy, infrared and Raman visualization (different water phases and phase changes), or direct optical visualization of liquid water. These various methods were reviewed in [361]. The optical methods respond to water droplets by light scattering and digital holography. These methods however require a considerable experimental effort and can be applied only in a laboratory. The first occurrence of water droplets (i.e., oversaturation of humidity in vapour) can be indicated by dew-point sensors, e.g., based on a stray field capacitor [347]. However larger amounts of water droplets are incompatible with dew-point meters. Polymer sensors can handle surface condensation without damage, but they will not accurately determine the liquid water concentration. Condensation also passivates the sensor by physically blocking the active surface, which will indicate humidity only after drying of the surface. Coulometric sensors can tolerate and indicate small amounts of water droplets as disturbances. A sensor system for the reliable, in-situ measurements of condensed water has not yet been developed.

CHAPTER 5

Sensors for Temperature, Pressure, Gas Flow, and Fire Detection

CONTENTS

5.1 TEMPERATURE SENSORS

Bernd Fellmuth, Physikalisch-Technische Bundesanstalt PTB, Berlin

5.1.1 Overview

Temperature is probably the most frequently measured and controlled physical process parameter. This is because material properties and efficiency of processes are usually strongly temperature dependent. Temperature monitoring is also critical to ensure safety, as an unexpected and uncontrolled temperature excursion can often lead to hazardous situations. Hydrogen production and applications require temperature measurements over a wide range, from cryogenic temperatures, which are necessary for the storage of liquid hydrogen at −250 °C, to temperatures as high as ca. 1000 °C (high-temperature fuel cells) or even up to 2000 °C (chemical separation from water).

The temperature of a body describes its thermal state, regarded as a measure of its ability to transfer heat to other bodies. Temperature is an intensive quantity, which takes the same value in two systems that are brought into thermal contact and allowed to come to thermal equilibrium. In thermodynamics and statistical thermodynamics, temperature is defined on the basis of basic principles, that is, it is independent of any chosen thermometric working substance [362, 363]. Representative overviews of temperature measurement are given in [364–368].

5.1.2 Temperature Unit and Scales

The unit of the thermodynamic temperature, T, is called kelvin. At present, the base unit kelvin of the International System of Units is defined as the fraction 1/273.16 of the thermodynamic temperature of the triple point of water (TPW). Since temperature appears in fundamental laws of physics in the thermal energy $k_B T$, it is planned to redefine this unit by fixing the value of the associated fundamental constant, the Boltzmann constant k_B [369].

Thermodynamic temperatures are measured by applying primary-thermometry methods, which are performed using thermometers based on well-understood physical systems, for which the equation of state describing the relation between T and other independent quantities can be written down explicitly without unknown or significantly temperature-dependent constants. Examples for such methods include gas thermometry, noise thermometry, and radiation thermometry [362, 363, 369].

Any accurate realisation of the thermodynamic temperature is very time consuming and requires extreme metrological effort. On this ground and to harmonise the differing national temperature scales for practical day-to-day use, the so-called International Temperature Scales were developed by the Consultative Committee of Thermometry and adopted by the General Conference on Weights and Measures [370]. On January 1, 1990, the International Temperature Scale of 1990 (ITS-90) [371] came into force. It is based on temperature fixed points, i.e., phase transitions of pure substances with assigned temperature values, and specified instruments for interpolation between the fixed points. Recommendations for the realisation of the ITS-90 are given in a supplementary information posted on the website of the Bureau International des Poids et Mesures, see [372]. The ITS-90 is disseminated by calibrating customer thermometers within a hierarchical system with the defining fixed points and the interpolation instruments at the top level.

We note that the thermodynamic temperature T, unit K, may also be expressed as a Celsius temperature, t, according to $t/°C = T/K$ - 273.15. The ITS-90 accordingly defines

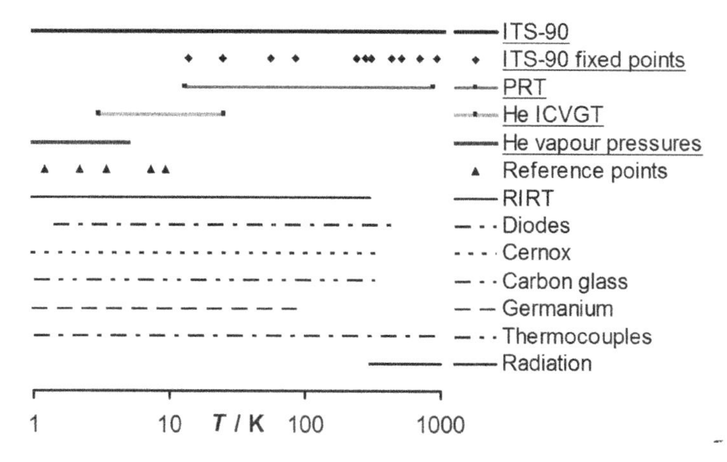

Figure 5.1: Application ranges of some temperature sensors compared with the defining fixed points and the interpolation instruments of the ITS-90. The indicated reference points are the super-to-normal-conducting transitions of aluminium, indium, lead, and niobium as well as the super-to-normal-fluid transition (lambda transition) of ^{4}He. PRT is the abbreviation for platinum resistance thermometer, ICVGT for interpolating constant-volume gas thermometer, and RIRT for Rhodium-Iron Resistance Thermometer (see Section 5.1.4). CERNOX, carbon glass, and germanium are three examples of semiconducting or composite materials used in resistance thermometers that have a high sensitivity at low temperatures, see [375].

both international kelvin temperatures T_{90} and international Celsius temperatures t_{90} by the corresponding relation $t_{90}/°C = T_{90}/K - 273.15$. Both the thermodynamic and the International Temperature Scale have the same units, the kelvin and the degree Celsius. Users sometimes prefer kelvin in the range below 273.15 K and degree Celsius above this point.

5.1.3 Overview of Temperature Sensors

It is not possible to deal comprehensively with practical thermometry in this book because the diversity of both the applications and the available temperature sensors is huge (see Figure 5.1). Users often have conflicting demands concerning the parameters of the sensors and very specialized sensors often have to be used. Furthermore, temperature-measurement errors often have to be reliably estimated separately for each measurement, which requires dedicated experiments under specific experimental conditions. Recently, the following reviews have been published: [364–368]. Additional useful information is contained in [362, 373, 374]. In the temperature range below the TPW, the diversity of sensors is huge because of the complicated temperature dependence of all physical quantities at low temperatures and the limited application range of many sensors. A comprehensive review of cryogenic thermometry is given in [375].

Contact temperature measurements and radiation thermometry are fundamentally different and are covered separately. Table 5.1 contains an overview of typical parameters of sensors for contact thermometry.

The criteria for selecting a temperature sensor vary significantly for different applications, see also Section 6.4. End-users must consider a variety of performance parameters such

Table 5.1: Typical parameters of temperature sensors for contact thermometry

Sensor type	Range K	Sensor Volume cm^{-3}	Measuring power W	Uncertainty $U(T)$ K [1]	Repeatability K	T-difference measurement	Calibration effort
Industrial PRT	20 … 500	< 1	$10^{-6} \ldots 10^{-4}$	10^{-1}	10^{-1}	bad	large
Industrial RIRT	1 … 400	< 1	$10^{-5} \ldots 10^{-4}$	10^{-1}	10^{-1}	poor	large
Thermistor	70 … 400	$10^{-2} \ldots 10^{-1}$	$< 10^{-5}$	10^{-1}	10^{-1}	poor	small
Carbon resistor / Thick film resistor (RuO_2)	0.1 … 300	$10^{-2} \ldots 10^{-1}$	$< 10^{-5}$	10^{-2} [2]	10^{-2} [2]	poor	small
CERNOX resistance	1 … 10	10^{-1}	$< 10^{-5}$	10^{-2} [2]	10^{-2} [2]	poor	small
thermometer	10 … 400	10^{-1}	$<10^{-5}$	10^{-3} [2]	10^{-3} [2]	poor	small
Carbon-glass resistance thermometer	1 … 300	10^{-1}	$< 10^{-5}$	10^{-2} [2]	10^{-2} [2]	poor	small
Diode(Si, GaAs)	1 … 30	10^{-2}	10^{-5}	10^{-2}	10^{-2}	poor	small
	30 … 400	10^{-2}	10^{-5}	10^{-1}	10^{-1}	poor	small
Capacitance thermometer ($SrTiO_3$)	1 … 300	10^{-1}	$< 10^{-5}$	1	1	poor	large
Thermocouple:							
Cu-Constantan	70 … 600	10^{-4} [3]	$< 10^{-4}$	10^{-2} [4]	10^{-2} [4]	very good	small
AuFe-Chromel	2 … 300	10^{-4} [3]	$< 10^{-4}$	10^{-2} [4]	10^{-2} [4]	very good	small

[1] individual calibration
[2] relative uncertainty $U(T)/T$ / repeatability
[3] only the measuring junction, assembling, and thermal anchoring are not considered
[4] relative uncertainty of temperature differences $U(\Delta T)/\Delta T$

as temperature range, sensitivity, signal-to-noise ratio, repeatability, long-term stability, linearity, response time, physical size and mass, thermal resistances, mechanical robustness (e.g., sensitivity to vibrations or mechanical shock, and sensitivity to thermal cycling) as well as deployment parameters such as capital costs, calibration requirements, necessary measuring equipment, and interchangeability.

5.1.4 Resistance Thermometers

Resistance thermometers utilise the temperature dependence of the electrical resistivity of appropriate materials. They can roughly be classified into four groups, which will be treated in this section:

- Platinum resistance thermometers
- Alternative metal resistance thermometers
- Thermistors
- Semiconductor resistance thermometers

Among the various resistance thermometers, platinum resistance thermometers (PRTs) play a special role. This is because the standard ones (SPRTs) are used as interpolation instruments in the ITS-90 over a wide temperature range. SPRTs have rigorous requirements pertaining to the purity and strain-free mounting of the platinum wire. Another group of widely applied resistance sensors are thermistors. Alternative metal resistance thermometers and semiconducting resistance thermometers are used more for special applications, especially at low temperatures.

Compared with thermocouples (see Section 5.1.5), resistance thermometers have often a much better sensitivity and repeatability, a lower and more predictable drift, and a higher signal level. For high-accuracy resistance measurements (uncertainty down to a few parts in 10^8), both AC and DC bridges are commercially available. Examples for high-end ones are the AC bridges from Automatic Systems Laboratories (www.aslltd.co.uk) as well as the DC bridges from Measurements International (www.mintl.com) and Guildline Instruments (www.guildline.com). General information on resistance measurement in thermometry is given for instance in [362, 373]. Some of the sources of error within thermometry bridges and bridge calibration methods are summarized in [376]. Rudtsch et al. (2005) [377], Strouse and Hill (2003) [378], and White (1997) [379] give additional information on typical performance of some resistance bridges. The sources of error in bridges include noise, large scale or "integral" non-linearities, and short-range or "differential" non-linearities.

Platinum Resistance Thermometers

Although it is not possible to develop precise first-principle formulae for the temperature dependence of the resistance of metals [380], the dependence exhibits an extremely stable and well-characterised form described by empirical equations, see below and [371]. Consequently, resistance thermometers are capable of quite reliable and economical temperature measurements. Although nickel and copper can be used in resistance thermometers, pure platinum has proven to be the best material for most purposes. Most national standard specification codes pertaining to resistance thermometers, as well as the IEC-60751 [381], are limited to platinum. Therefore, the most common form of resistance thermometer is based on platinum.

For industrial applications, PRTs must be able to withstand shock and vibration. Furthermore the sheathing must be more robust than for SPRTs. However, industrial PRTs (IPRTs) can have less stringent platinum-wire purity and strain-free-mounting requirements than SPRTs and can still be able to meet the analytical requirements for industrial applications in an economical platform. IPRT calibration requirements can be significantly simplified as well. Establishment, maintenance, and use of water triple point cells are necessary for accurate SPRT calibration and this requires considerable practical expertise. However, the resulting gain in measurement uncertainty is typically not necessary for industrial thermometry and so the ice point can often serve as the reference temperature for IPRTs. The ice point is 0.01 °C lower than the TPW temperature and has a typical uncertainty of about 2 mK. The ice point is realised at the melting temperature of ice at one standard atmosphere at 0 °C.

IEC 60751 defines the temperature-resistance relationship for industrial platinum resistance thermometers (IPRTs) in the temperature range $0\ °C \leq t \leq 850\ °C$ by a simple empirical quadratic equation:

$$R_t = R_0(1 + \mathrm{A}t + \mathrm{B}t^2) \tag{5.1}$$

For the range from -200°C to 0 °C an additional term is used:

$$R_t = R_0(1 + \mathrm{A}t + \mathrm{B}t^2 + \mathrm{C}(t - 100)t^3) \tag{5.2}$$

R_0 is the resistance at the ice point, usually around 100 Ω (Pt100-type). Values for the constants A, B, C, in these two equations are given for ITS-90, e.g., in IEC 60751 or [367]. Using these formulae and constants, the IPRTs can be interchanged and will give temperature readings within the tolerances (δt) at a temperature t defined in IEC 60751 without any calibration. These temperature tolerances are as follows for two classes of IPRTs:

$$\mathrm{ClassA}: \delta t = \pm(0.15 + 0.002 \mid t/°C \mid)°C \tag{5.3}$$

$$\mathrm{ClassB}: \delta t = \pm(0.30 + 0.005 \mid t/°C \mid)°C. \tag{5.4}$$

The tolerances of Class A must not be exceeded in the temperature range $-200\ °C \leq t \leq 650\ °C$, the Class B values are valid in the whole usable temperature range specified by the manufacturer. The 2008 version of IEC 60751 contains additional classes associated with accuracy and more temperature sub-ranges as well as additional sensor constructions. It can be seen from Equations 5.1 and 5.2 that for both classes the first term corresponds to the tolerance in R_0. A single calibration at the ice point to determine R_0 reduces the uncertainty substantially, particularly at temperatures below 100 °C. In general, IPRTs should be considered as replacements for liquid-in-glass thermometers especially where an uncertainty of better than 0.5 °C is required. It is not possible to batch produce IPRTs with tolerances close enough to work at a level of t uncertainty of about 0.1 % or better, so they have to be calibrated individually if greater accuracy is desired. With calibration, individual IPRT may have the capability to be used as laboratory standards with uncertainties approaching 10 mK at temperatures near to room temperature.

Compared to thermocouples, IPRTs offer some advantages in sensitivity, although they have a more limited temperature range. An IPRT of the Pt100 type has an effective sensitivity of about 400 μV/K for a measuring current of 1 mA, which is typically used for IPRT operation. This is a factor about 40 times greater than the sensitivity of a type R or S thermocouple. The main disadvantage of an IPRT is the need of lead compensation, which usually requires three or four leads running to the sensor rather than two wires for thermocouples. However, the advantage of PRTs is achieved after individual calibration, when their measurement capabilities become superior to most other thermometers. For lowest uncertainties the simple formulae (5.1) and (5.2) can be replaced by the relations given in the ITS-90 for SPRTs [371].

As mentioned above, the platinum wire in the sensor of an SPRT is wound in a configuration that is as strain-free as possible. This makes it rather susceptible to vibration and, therefore, IPRT sensors have to be designed differently since vibration and mechanical shock are common in industrial environments. IPRT sensor constructions often use encapsulated Pt-wires wound around a glass or ceramic substrate whose thermal expansion coefficients match that of platinum. Inexpensive IPRT sensors can be manufactured using lithographic techniques. IPRTs of this type are available as thick-film or thin-film sensors; the latter is also available as a surface mounted device (SMD). A comprehensive overview of the different fabrication methods is given in [367]. The direct contact between the platinum sensing element and the supporting material may cause strains, even exceeding the yield strength. The yield strength determines the maximum instability and corresponds to a temperature equivalent of the order of 0.5 K. Thus, strains may cause a hysteresis of several tenths of a kelvin depending on the application range [362, 382].

Alternative Metal Resistance Thermometers

Alternative metal resistance thermometers, especially rhodium-iron resistance thermometers (RIRTs), have been developed [374, 383] for low temperature operation. These devices can function down to about 0.5 K, which is below the operation range of PRTs. Since their resistance versus temperature characteristic is complicated and non-unique, RIRTs and other alternative resistance thermometers have to be calibrated individually at ten to thirty temperatures in the range from 0.5 K to 30 K, depending on the acceptable level of uncertainty. RIRTs are available as precision thermometers designed similar to capsule-type SPRTs, and have a long-term stability that can be on the order of only a few tenths of a millikelvin. RIRTs are also available as industrial thermometers with wire sensors wound on a ceramic substrate or thin-film sensors. The main application range is from 0.5 K to 300 K and the room-temperature resistance is typically between 25 Ω and 100 Ω.

Thermistors

In contrast to metal based resistance thermometers, thermistors are manufactured from semiconducting metal oxides and have negative temperature coefficients of resistance (NTC) [367, 373, 375, 384]. They have relative temperature coefficients of about -0.04 K^{-1} near room temperature, compared with a ten-fold smaller and positive value of +0.004 K^{-1} for metals. They can be obtained commercially with typical room-temperature resistances ranging from 100 Ω to 1 MΩ. Thermistors that can be used within the range from -200 °C to 300 °C are available. Their high sensitivity permits use with inexpensive, direct reading instruments and, as the lead resistance is small compared to the sensor resistance, lead compensation is usually unnecessary. The characteristic of thermistors is very non-linear and is described to a good approximation by the following empirical relationship [385]:

$$1/T = \mathrm{A} + \mathrm{B}\ln(R/R_0) + \mathrm{C}[\ln(R/R_0)]^2 + \mathrm{D}[\ln(R/R_0)]^3 \tag{5.5}$$

Where R_0 is a convenient reference resistance. Measurement uncertainties below 0.1 °C can be easily obtained for limited temperature ranges. Alternatively, the high sensitivity permits the resolution of small temperature changes (< 0.0001 °C).

Semiconductor Thermometers

For low temperatures, a variety of resistance thermometers with negative temperature coefficients are available. Usually, they have a large sensitivity in limited temperature ranges. As for semiconductor resistance thermometers, the most technologically important material for high-precision sensors is germanium. Other widely used sensor materials are carbon, carbon embedded in glass or ceramics, ruthenium dioxide, and zirconium oxynitride (CERNOX sensors), see [362,364,365,367,373,375,384].

5.1.5 Thermocouples

The thermocouple is a device in which the difference between the electro-motive forces (EMFs) in two dissimilar wires gives an indication of the temperature difference between the hot and cold ends of the two wires (thermoelements). This is due to the Seebeck effect. The point of connection of the two dissimilar arms A and B, at temperature T_1, is called measuring junction, and the free ends are referred to as reference junction. The thermoelectric potential difference between the free ends of the thermoelements can be measured with a pair of identical conductors (e.g., copper wires) that are also attached to a voltage detector. It is evident that the Seebeck effect is in no way a junction phenomenon but is, instead, a temperature-gradient phenomenon. For a proper understanding of the behaviour of thermocouples, this cannot be overemphasized. The thermoelectric potential difference between the free ends, both at temperature T_0, is

$$E_{AB}(T_0 \to T_1) = \int_{T_0}^{T_1} [S_A(T) - S_B(T)]dT \qquad (5.6)$$

Where $S_A(T)$ and $S_B(T)$ are the thermopowers (Seebeck coefficients) of the respective conductors. $E_{AB}(T_0 \to T_1)$ is a unique function of T_0 and T_1 only if the thermoelements are homogeneous throughout the temperature gradient. The effect of the presence of an inhomogeneity will be to add a small additional apparent thermopower at the place in the thermoelement, and this will adversely affect measurement accuracy. An unintentional "kink" or bend in a thermoelement will produce such an inhomogeneity. The thermoelectric potential difference of an inhomogeneous thermocouple will be a function of its position as well as the temperature difference between the hot and cold ends. In using a thermocouple to explore the temperature gradient in a furnace, there is thus always the problem of deciding whether or not the change in potential difference, as the thermocouple is moved, indicates the presence of a temperature gradient in the furnace or an inhomogeneity in the thermocouple.

Over 20 different types of thermocouples are extensively used and, of these, 8 have been standardised with their EMF-temperature relationships represented by internationally recognised standard reference functions (IEC 584-1). The metals used in the standardised thermocouple types are given in Table 8.1 with the letters that identify them and the temperature ranges for which tolerances apply (IEC 584-2). The main reasons to use a thermocouple instead of a resistance thermometer are high temperatures, fast response time for thin wires, and high vibration.

The thermocouples of Table 8.1 fall into three groups. The first is the noble-metal group, comprising types B, R, and S, based on platinum and its alloys with rhodium. They are the most accurate of the tabulated thermocouples and may be used at higher temperatures, but they are more expensive and are particularly sensitive to contamination. Their tolerances are higher than those for IPRTs and are in the order of one kelvin.

The second group consists of the two nickel-based thermocouples, types N and K. They are preferred for most applications not requiring the higher temperature limit or accuracy of

Table 5.2: Thermocouple types in common use with their letter designations, typical compositions in % mass, and temperature ranges for which tolerances apply

Type	Positive arm	Thermo element Negative arm	Temperature range °C
B	Pt 30Rh	Pt 6Rh	600 to 1700
R	Pt 13Rh	Pt	0 to 1600
S	Pt 10Rh	Pt	0 to 1600
K	Chromel (Ni 9.5Cr 0.5Si)	Alumel (Ni 5(Si,Mn,Al))	-40 to 1200
N	Nicrosil (Ni 14.2Cr 1.4Si)	Nisil (Ni 4.4Si 0.1Mg)	-40 to 1200
E	Chromel (Ni 9.5Cr 0.5Si)	Constantan (Cu 44Ni)	-200 to 900
J	Fe	Constantan (Cu 44Ni)	-40 to 750
T	Cu	Constantan (Cu 44Ni)	-200 to 350

the rare-metal thermocouples. Usually type N is preferred, but the choice between types N and K may depend especially on whether they are in the bare-wire or the ceramic-insulated, metal-sheathed (MIMS) form. In the MIMS configuration of the thermoelements, their insulation and a sealed sheath are integrated into a flexible cable. In contrast, bare-wire thermocouples, insulated in loose-fitting beads, are exposed to the local atmosphere and are vulnerable to the effects of oxygen, carbon, sulphur, and other chemical interferences. Unsheathed thermocouples, however, may have faster response times.

The third group, comprising the types E, J, and T, is based on the use of the alloy constantan as the negative arm. Constantan has the most negative value of thermopower, so the net thermopowers for these thermocouples are higher than other thermoelement pairs. In general the tolerances are higher than those for the noble-metal group and are in the order of several kelvins.

All the thermocouples discussed above become increasingly insensitive as the temperature drops toward the liquid helium range (around 4 K and below). For that reason, thermocouples are not widely used in cryogenic applications, although Kondo alloys, such as those where the magnetic moment of iron in a gold matrix is strongly temperature dependent, can give sensitivities of about 10 μV/K down to 1 K and have supplanted these thermocouples at low temperatures [375]. Very high resolution of small temperature changes at low temperatures can also be achieved using superconducting quantum interference devices (SQUIDs) [386,387].

5.1.6 Electronic Devices (Diodes and Transistors)

Diodes and transistors have been in use as temperature sensors for more than forty years [364,365,367,375,384,388]. The junction voltage of a semiconductor diode forward-biased at a constant current increases with decreasing temperature. Initially, diodes were attractive because of their wide temperature range and their higher sensitivity and, above about 20 K to 30 K, more nearly linear voltage versus temperature characteristic than most resistance thermometers. More recently, more accurate calibrations have become available on an individual basis and interchangeable devices conforming to the manufacturer's standardized calibration curves can be purchased. A useful review of that area of thermometry has been published [389].

Diodes made from silicon junctions make the most stable and reproducible thermometers. Commercial versions of such devices are described in a variety of papers, see the review [375]. The stability of diodes is limited to about 0.1 K, but better repeatability (perhaps 0.03 K) can be obtained if carefully selected diodes are used. However, in order to achieve this performance, proper measurement techniques must be used. Poorly shielded or improperly grounded measurement systems can introduce AC noise which will create apparent shift in the DC voltage reading across a diode sensor [389].

5.1.7 Piezoelectric Resonators

The piezoelectric properties of quartz have been known for a long time. Many modes of oscillation can be obtained in quartz. By selecting a temperature-dependent mode, various researchers have been able to construct thermometers with reproducible temperature dependent frequency outputs [390, 391].

A commercial version of the quartz thermometer employs probe crystals that are cut so as to exhibit an almost linear frequency temperature response [392]. A second, temperature-independent crystal serves as a stable reference oscillator. In use, the instrument presents a digital display of temperature that is derived from a beat frequency, which itself arises from mixing the probe and reference oscillator signals.

The commercial instrument is sensitive to temperature variations of magnitude 0.01 °C to 10 μ°C, depending upon the sampling time that is chosen. Its application range is at present rather small, -80 °C to 240 °C, but it seems to be possible to extend the range to higher temperatures using new materials as lanthanum gallium silicate (langasite) [393]. The thermometer accuracy is limited by a hysteresis effect that is thought to arise from thermal stresses in the probe-crystal electrical contacts. The hysteresis can amount to 0.02 °C temperature equivalent when the probe is cycled slowly over its full range. Smaller temperature excursions permit more accurate measurements. Manufacturer specifications indicate that, over a 10 °C cycle, the hysteresis will not exceed 0.001 °C.

5.1.8 Phosphor (Luminescence) Thermometry

Phosphor thermometry is an optical method for surface temperature measurement. The method exploits luminescence emitted by phosphor material (not to be confused with the chemical element phosphorus). To excite luminescence, an ultra-violet light source is focussed into an optical fibre to illuminate a small sample of luminescent phosphor at the end of the fibre. The phosphor re-radiates light at a longer wavelength (usually red light) and this is picked up by the fibre. A dichroic mirror and interference filter is used to select the fluorescence signal from the illuminating ultra-violet light and any other stray sources. Some of the most commonly used phosphors are made using a rare-earth, such as Gadolinium or Europium, doped into a ceramic crystal. After excitation, they normally luminesce for several milliseconds. The decay time decreases exponentially with increasing temperature, usually decreasing by an order of magnitude for each 100 °C, making it a sensitive temperature indicator. Because of the rapid change in decay time with temperature and the difficulty in accurate measurement of decay times shorter than a μs, a given phosphor is useful for only a range of several hundred degrees. Different combinations of dopant and host result in different usable temperature ranges [364].

Several commercial instruments utilise luminescence, and a range of phosphors are available. The strong temperature dependence allows good temperature resolution, with some manufacturers claiming less than 0.1 °C, although the wide range of time constants (or intensity) over the usable temperature range for the phosphors leaves these instruments with much poorer linearity than 0.1 °C.

The luminescence methods are just another example of contact probe measurements with the phosphor indicator and fibre optic probe in contact with the test sample. The relative stability of the phosphor materials has allowed some users to adapt this form of thermometry to non-contact surface temperature measurement by coating the test region with the phosphor and using a lens system to illuminate with ultra-violet light and look for luminescence. Systems for the measurement of turbine blades, motors, and generators have been developed. Techniques using CCD cameras and flash-lamps have also been developed for imaging temperature profiles. The progress achieved in phosphor thermometry in the last ten years is described in several papers presented at ITS9, e.g., [394].

5.1.9 Radiation Thermometers

Radiation thermometry is especially useful when contact with the process or test sample is not possible, e.g., if solid products are moving. It is based upon the detection of radiation power that is emitted by every substance in consequence of its temperature. For blackbodies (emissivity equal to one), the emission is described by Planck's law. For total radiation thermometry, this leads to the Stefan-Boltzmann law. Above the silver freezing point (961.78 °C), the ITS-90 is defined via spectral radiation thermometry with measurements performed relative to one of three temperature fixed points, namely the freezing points of silver, gold (1064.18 °C), or copper (1084.62 °C). Detailed guidelines and uncertainty budgets for accurate measurements are given in Chapter 6 of the Guide [395] and the references cited therein.

A wide variety of different designs of radiation thermometers are available for laboratory and industrial use. Practical total radiation thermometers operating at room temperature, are the simplest radiation thermometers. They generally consist of a very basic imaging system which restricts the field of view with an aperture. The detector is usually a thermopile bolometer, which measures the temperature rise of a blackened disk and allows the detection of the whole radiated spectrum. Although they have low sensitivity in comparison to other thermometer types, they are the most sensitive of the radiation thermometers. For example, a 10 % over-estimation of the emissivity will result in a 2.5 % under-estimation of the absolute temperature.

Spectral-band radiation thermometers use a filter to selectively measure only thermal radiation in a narrow band. A relative bandwidth less than a few percent is typical. Usually a silicon-semiconductor photodetector is used. Spectral-band radiation thermometers are used to overcome surface emissivity problems. In this case, the centre wavelength is chosen to be as short as possible whilst still giving enough signal. Typically a 10 % over-estimation of the emissivity will lead to only a (0.5 - 1) % under-estimation of the temperature.

The most common bands are 0.85 μm to 0.9 μm for temperatures above 500 °C, and around 0.65 μm for temperatures above 700 °C. For lower temperatures, longer wavelengths and different detectors are applied. For example, at 3.43 μm many plastics have high absorbance. A radiation thermometer operating at this wavelength will thus see even a very thin plastic film as opaque and having high emissivity, allowing accurate measurements during the manufacturing process at 200 °C to 300 °C.

The thermal radiation emitted by objects around room temperature is comparatively low and exclusively in the infrared. To obtain sufficient signal, most radiation thermometers use a wider band of wavelengths. Atmospheric absorption of infra-red radiation limits the choice of bands to mainly two windows, the 3 μm to 5 μm band and the 8 μm to 14 μm band.

Pyroelectric or thermoelectric detectors are used to detect radiation. The wavelength range is often set by the transmission of the optics materials. Since the filter defining the spectral sensitivity, the optics, and the apertures will also radiate energy in this wavelength region, there must be some form of background temperature compensation within the instrument. Some instruments solve this problem by using a rotating sector disk providing a reference signal of known temperature. There has been considerable progress in the instrumentation for calibration of low-temperature radiation thermometers [396] driven by the European research project TRIRAT (TRaceability in Infrared RAdiation Thermometry) [397].

Process temperature targets do not behave as blackbodies. The total emissivity is the ratio of the total energy radiated by a material to the total energy radiated by a blackbody. The total emissivity of most materials encountered in process applications at various temperatures can be found in tables in literature, e.g., www.monarchserver.com/TableofEmissivity.pdf.

The emissivity is a physical property of the material and is specified for a sample with a highly polished surface. The targets in industrial equipment (e.g., furnaces, fuel cells) have different roughness, geometry, and sizes, and are often covered with oxides and corrosion products. To take into account the effects of installation and operating conditions, a factor called emittance is used instead of emissivity. The total emittance is the ratio of total radiation for the physical and operating conditions of the target to the total radiation from a blackbody at a given temperature. The actual emittance depends upon application conditions and changes to the surface from coatings, corrosion, oxidation, and erosion during a production run.

5.1.10 Digital Thermometers

In a digital thermometer, an electrical output corresponding to the temperature-dependent property of a sensor is digitised and then converted to a temperature display, using a mathematical processor. The entire process from sensor to display is integrated in one unit. Digital thermometers come in a variety of types, varying in resolution, sensor type and price [364]. Compared with resistance bridges, see Section 5.1.4, they may have advantages concerning costs, size, speed, and measuring range.

Modern reversal-switched DC precision thermometers achieve performance levels previously attained only with AC bridges. They employ modern integrated-circuit analog-to-digital converters [398–400].

5.1.11 Static and Dynamic Temperature Measurement Errors

Dynamic temperature-measurement errors originate from the time constant, which is caused by the heat capacity of the sensing element of the thermometer and the thermal resistance between the element and the measuring object [366,368]. The simplest model for the settling response is an exponential decrease of the temperature difference between element and object. But there are situations and probe designs (for instance thermowell assemblies) where there is more than one time constant involved. The most difficult time-constant problems occur in gas-temperature measurements. Because of the extra thermal resistance of the boundary layer, the time constant of a thermometer in a gas may easily be an order of magnitude larger than in a well-stirred calibration bath.

For estimating static temperature-measurement errors, again the thermal resistance between the element and the measuring object must be known, but also the heat flow to the sensing element. The situation for long-stem standard PRTs is described in [401], and for industrial PRTs in [402]. For different types of capsule-type thermometers, parameters and estimates are presented in [403].

5.2 PRESSURE SENSORS

5.2.1 Overview

Pressure is a fundamental thermodynamic property of a gas (see Section 1.1). The kinetic theory of gases, which provides a theoretical understanding of pressure, asserts that small particles (molecules, atoms, or ions) are in random motion and frequently collide with each other and with an adjacent surface. The particles possess a mass, momentum, and kinetic energy. The momentum of a single particle is the product of its mass and velocity, while the kinetic energy is one-half the mass times the square of the velocity. As the gaseous particles collide with the surface they transfer momentum to it thus exerting a force on the surface. The sum of the forces of all the molecules striking the surface divided by the surface area is defined as the pressure. The pressure of a gas is then a measure of the averaged momentum of the moving molecules of a gas or their kinetic energy, respectively:

$$p = \frac{1}{3} \cdot \frac{n \cdot \mathrm{M} \cdot v^2}{V} = \frac{2}{3} \frac{E_{kin}}{V} \tag{5.7}$$

Where v is the averaged velocity of the gas particles, n is the amount of substance (number of particles) of a gas in the volume V, and M is the molar mass.

Pressure is thus the perpendicular component of a force per area exerted on to a surface. It is handled as a scalar quantity and can be considered as a special case of mechanic stress which is a tensor. Pressure is an intensive quantity, i.e., it does not depend on the size of the system (see Section 1.1). The SI unit for pressure is the Pascal (Pa = N/m^2); a related unit is bar (1 bar = 10^5 Pa).

Static pressure is independent of direction in the media and changes slowly, e.g., the atmospheric pressure. In contrast, **dynamic pressure** is relevant for gas flow. Flow requires pressure difference perpendicular to the flow direction. Dynamic pressure is dependent upon gas density and flow rate:

$$p_{dyn} = \frac{1}{2} \rho \cdot v^2 \tag{5.8}$$

Where p_{dyn} is the dynamic pressure, ρ is the fluid density, and v is the fluid velocity. Three kinds of static pressure can be defined:

- Absolute pressure is measured with reference to a vacuum
- Differential pressure is measured with reference to a known pressure
- Gauge pressure is measured with reference to local atmospheric pressure

Gas pressure is measured over more than 20 orders of magnitude (i.e., a range of 10^{-12} Pa to 10^{12} Pa). Various pressure ranges have been assigned generalized definitions:

- Vacuum range: 10^{-10} Pa to 10^4 Pa
- Acoustic pressure range: 2×10^{-5} Pa to 10^2 Pa
- Atmospheric pressure (air): 7×10^4 Pa to 11×10^4 Pa
- Compressed gases: 10^5 to 10^8 Pa
  - Medium Pressure (gas pneumatic systems): typically up to 1.03×10^8 Pa
  - High Pressure (gas pneumatic systems): typically up to 4.14×10^8 Pa
- Shock waves and detonations: up to 10^{12} Pa

Pressure measurements are required for a broad range of hydrogen applications. General examples include:

- The control and monitoring of pressurized hydrogen systems
- Flow measurements through pressure differential
- Leak detection through measured pressure drops

Hydrogen vehicles will have pressurized systems operating at elevated pressures. Commercial fuel cell electric vehicles will have on-board hydrogen storage of up to a working pressure of 70 MPa (700 bar), but because of compression heating, transient elevated pressures will develop following refuelling, these are restricted to 87.5 MPa [68]. Storage of gaseous hydrogen at fuelling facilities may even be at higher working pressures. Furthermore, pressure monitoring and control must be performed over a broad temperature range. For example for fast refuelling of less than 3 minutes pre-cooling of hydrogen to -40 °C may be required to alleviate the impacts of compression heating. SAE Standard J2601 (Fuelling Protocols for Light Duty Gaseous Hydrogen Surface Vehicles) stipulates that the temperature does not exceed 85 °C. Pressure measurements are also necessary to control fuel cell operation. In PEM fuel cells, the pressure sensor shall measure a differential pressure up to 7×10^4 Pa at temperatures from 30 °C to 100 °C with a response time < 1 s and an uncertainty of less than 1 % [319].

In addition to its use in fuel cells, hydrogen can be used as a fuel for combustion in gas turbines, jets, and rocket engines. The characterization of pressure in the combustion process is a key parameter for improving power density and efficiency [404]. This section, however, will focus on pressure sensing technologies and issues specific to pressure measurements in hydrogen containing environments.

5.2.2 Operation Principles

Many pressure measurements are based on a displacement of an object due to the force exerted by the pressure. The exerting force induces an elastic strain typically manifested as an elongation, compression, or deformation. One of the first barometers to measure air pressure was invented by Evangelista Torricelli in 1643. He observed that the height of a mercury column in a glass tube changed daily and concluded that this response was due to changes in the ambient air pressure. Over the centuries various pressure gauges, such as the U-tube manometers, bourdon pressure gauges, or bellow pressure gauges, were developed that convert changes in pressure into a mechanical motion. In addition to sensor, the terms "transducer" (converter of an applied pressure into an electrical signal), "transmitter" (transducer with output amplifier) or "gauge" are used in connection with a pressure-measuring device (see Section 2.1). Pressure gauges, such as the bourdon gauge, are based on a spring or a membrane deformation, induced by pressure changes, that is mechanically linked to an indicator (needle) on a scale. Such mechanical devices are not included in the following discussions because they lack electrical transduction and are therefore not properly classified as sensors (see Section 2.1).

Pressure sensors can be classified by the operating principle of the sensing element. Examples include resistive, piezoresitive, capacitive, magnetic, potentiometric, resonant, and optical sensors. Sensors are designed to measure in absolute, differential, or gauge pressure.[1] The measuring range can be small (e.g., between 800 hPa and 1100 hPa) or can encompass

[1]Therefore the unit "pounds per square inch" (psi) is specified with an suffix "a" or "g" (psia, psig) to indicate that the measured value is an absolute or a gauge pressure.

Table 5.3: Typical measurement range of common commercial pressure sensor platforms

No	Principle/Type	Measuring Range	Uncertainty
1	Resistive	0.1 kPa - 120 GPa	0.1 % - 0.5 % low drift 0.25 % in 6 month
2	Piezoresistive	1 kPa - 100 MPa	0.075 % - 2 % of full scale
3	Capacitive	0.01 kPa - 70 MPa	0.05 % - 0.5 % drift 0.2 % per year
4	Electromagnetic	50 kPa - 70 MPa	0.5 - 1 % of full scale
5	Potentiometric	35 kPa - 70 MPa	0.1 - 1 % of full scale
6	Resonant	0.1 kPa - 42 MPa	0.1 % of full scale drift 0.1 % per 6 month
7	Piezoelectric	0.7 kPa - 500 MPa	1 % of full scale
8	Optical	20 kPa - 200 MPa	0.1 % to 5 %of full scale

several orders of magnitude. The operation principles of pressure sensors are compiled in Table 5.3.

Table 5.3 shows that pressure can be accurately quantified over a wide range. The uncertainty is typically low, e.g., 0.01 % to 1 %. Accuracy is usually given with respect either to the reading value or the full scale. However, pressure sensors can show hysteresis (see Section 2.1), i.e., different indications when pressure is increasing or decreasing. Also signal drift can occur especially in conditions of fluctuating pressures. Pressure sensors that can be used in a temperature range of -70 °C to 120 °C are readily available and specialized sensors are available for application operation up to 700 °C [405, 406].

Principles, features, and limitations of the common pressure sensor platforms are discussed in the following.

Resistive Pressure Sensors

The resistive pressure sensor platform is based upon a sensing element whose electrical resistance changes as a function of pressure. The sensor operates according to a similar principle as the differential strain gauge and consists of a metallic resistor. The strain gauge is bonded onto a diaphragm. When a pressure is applied, the diaphragm and the strain gauge both deform, causing the electrical resistance R of the strain gauge to change. One design of the resistive pressure sensing element is based on the elongation or compression of a metallic conductor placed on a membrane whose resistance changes due to a physical deflection (see Figure 5.2); this changes the length and cross-sectional area of the resistor as well as the resistivity of the resistor ρ, thus resulting in a change in electrical resistance, ΔR. In case of a metallic wire of the length l and the radius r, the resistance R is given by:

$$R = \rho \frac{l}{\pi r^2} \tag{5.9}$$

The relative change of resistance ΔR due to deflection is:

$$\frac{\Delta R}{R} = \frac{1}{R}[\frac{\partial R}{\partial \rho}\Delta\rho + \frac{\partial R}{\partial l}\Delta l + \frac{\partial R}{\partial r}\Delta r] = \frac{\Delta\rho}{\rho} + \frac{\Delta l}{l} - \frac{2\Delta r}{r} = [\frac{\frac{\Delta\rho}{\rho}}{\epsilon} + 1 + 2\nu]\epsilon = GF \cdot \epsilon \tag{5.10}$$

Where, $\epsilon = \Delta l/l$ is the elongation or strain and $\nu = (\Delta d/d)/(\Delta l/l)$ is the Poisson's ratio, a

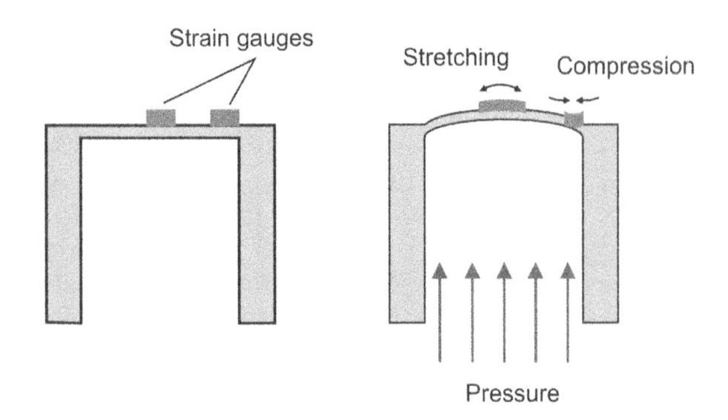

Figure 5.2: Resistive pressure sensor with strain gauges.

material constant which is for most metals in the range of 0.20 to 0.35. "*GF*" is the gauge factor which characterizes the pressure sensitivity of the sensor.

Resistive pressure sensing elements use metals or alloys, such as platinum, constantan (Cu/Ni alloy), manganin (Cu/Mg/Ni alloy), nickel-chromium, or platinum-tungsten, configured into wires, foils, or thin films. Usually a Wheatstone bridge circuit is used with strain gauges. Bridge circuits can be made using 1 to 4 strain gauges, e.g., 2 for compressive strain and 2 for tensile strain, at least one of which is active and changes resistance with strain. The Wheatstone bridge circuit is very sensitive to small changes in resistances, and can readily measure resistance changes on the order of 10^{-3} Ω, and converts the strain-induced resistance change of the gauge to voltage changes which can be directly and more accurately measured with conventional methods, e.g., a voltmeter. The sensing elements are produced using microfabrication technology for thin metallic membranes. Alternatively the resistors can be printed by thick film technology on ceramic substrates and sintered at high temperatures.

This pressure sensor platform is especially amenable for narrow pressure ranges. It can be configured to measure gauge pressure if the low pressure port is left open to the atmosphere or to measure differential pressure if connected to two process pressures. The sensor can also measure absolute pressure if the low pressure side is a sealed vacuum reference. The resistive pressure sensor can also be used to detect very high pressures. This platform is robust and has a good mechanical stability against shock waves and vibration. It can be also used at high temperatures.

However, metal resistive pressure sensors have only a limited sensitivity, expressed by the gauge factor GF of 2 to 6. Piezoresistive pressure sensing platforms, discussed in the next section, do not have this restriction, and thus have a better sensitivity.

Piezoresistive Pressure Sensors

Piezoresistive pressure sensing elements use semiconducting materials, such as silicon that change specific resistivity when a mechanical stress is applied to evoke stretching or compression. A simplified scalar expression for the piezoelectric effect is:

$$\frac{\Delta\rho}{\rho} = \pi_l \cdot \epsilon_l + \pi_t \cdot \epsilon_t \tag{5.11}$$

Where $\Delta\rho$ is the change of specific resistivity ρ, ϵ_l and ϵ_t the longitudinal and transverse strain and π_l and π_t the longitudinal and transverse piezoresistive coefficient [407, 408]. An

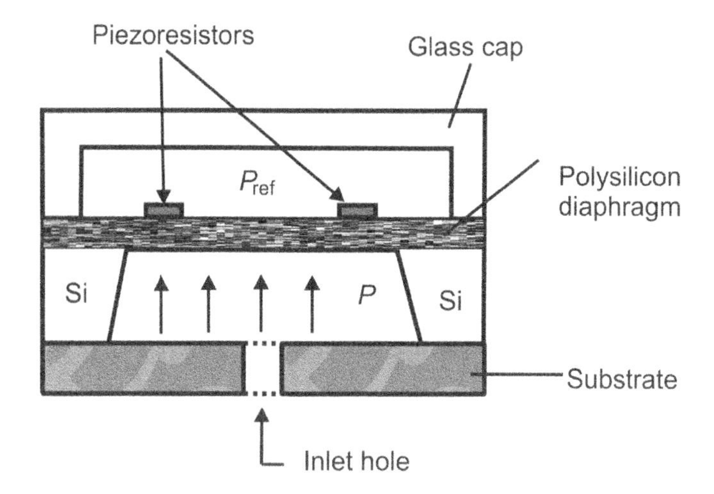

Figure 5.3: Schematic of micro-machined piezoresistive silicon pressure sensor.

illustration of the piezoresistor platform is shown in Figure 5.2. The resistance change of doped silicon due to the piezoresistance effect is much higher than the resistance change observed in metallic resistors (compare Equation 5.11). The gauge factor GF of this type of pressure sensors is in the range of 30 to 200. Piezoresistive pressure sensing elements have sensitivity from 0.5 to 100 mV/kPa and turndown ratio is 1000 : 1.[2]

Thin film sensing elements are produced using micro fabrication technology such as sputtering deposition onto a thin silicon membrane etched in the wafer followed by etching to produce the desired sensing element design. The piezoresistors based on doped silicon are formed either by diffusion or implantation of atoms from the fifth atomic group (e.g., phosphorus which produces an n-type semiconductor) or the third atomic group of the periodic table of chemical elements (e.g., boron which produces a p-type semiconductor). By the use of photolithographic techniques, four resistors are typically placed on the diaphragm. Two of these resistors are positioned on the silicon diaphragm such that they are exposed to a compressive strain and two are positioned where they experience a tensile strain. They are then connected together to form a Wheatstone bridge.

The piezoresistance coefficients are, however, strongly temperature dependent and Si-based piezoresistors are more affected by temperatures than metallic resistive platforms. For example, the temperature coefficient of resistance at 20 °C of constantan or manganin is about 10^{-5} K^{-1} and of platinum typically 3.9×10^{-3} K^{-1} while that of silicon[3] is 7.9×10^{-3} K^{-1}. Without compensation, the temperature dependence of the piezoresistance coefficients will adversely influence measurement accuracy. Nevertheless, the piezoresistive platform is probably the most commonly used technology for general pressure measurements.

Silicon based piezoresistive sensing elements are amenable for absolute, gauge, and differential pressures measurements. For absolute pressure, the sensing elements measure a differential pressure across a thin silicon membrane separating the cell body from a hermetically sealed cavity under a vacuum (see Figure 5.3). This cavity establishes a zero

[2]Turndown is a term that indicates the measuring range of a pressure or flow sensor and is given by the ratio between the largest and smallest value of pressure or flow to be measured with an acceptable accuracy.

[3]These data are an approximation assuming linear behaviour, i.e., see Section 5.1, $B = 0$, $C = 0$ in Equation 5.1.

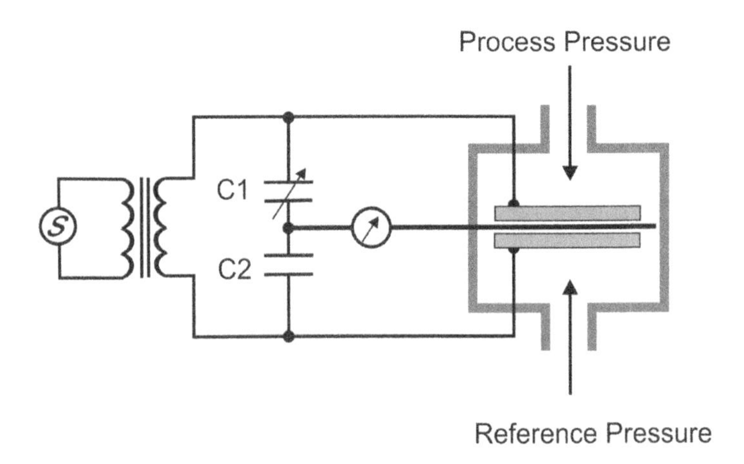

Figure 5.4: Schematic of capacitive pressure sensor.

reference pressure on one side of the membrane. The pressure sensor output is proportional to the difference between the sensed pressure and the reference pressure (vacuum). Another configuration of a pressure sensing element uses a stainless steel membrane on which the poly silicon piezoresistive sensing elements are precipitated (compare Figure 5.2). This structure with stainless steel membrane (e.g., AISI 316L) can be used for hydrogen sensing.

Capacitive Pressure Sensors

Capacitive pressure sensors are based on a parallel plate capacitor design. The capacitance C of a parallel plate capacitor is given by:

$$C = \epsilon_0.\epsilon_r.\frac{A}{d} \tag{5.12}$$

Where ϵ_0 is the permittivity of vacuum, ϵ_r is the relative permittivity of the matter between the electrodes, A, the electrode area and d the gap distance. The change of capacitance due to pressure changes can be described, e.g., by the following relation:

$$C = C_0[1 + ap^\beta] \tag{5.13}$$

Where C_0 is the initial capacitance, a and β are empirical coefficients. In most cases a metal, ceramic (e.g., alumina, quartz), or silicon membrane is placed in a pressure cavity. This results in a variable capacitor because of the pressure-induced physical displacement of the membrane. A schematic of a capacitive pressure sensor is illustrated in Figure 5.4.

Although the sensor can be miniaturized using microfabrication manufacturing methods, it is necessary to address the growing influence of parasitic capacitances associated with miniaturization. Nevertheless, capacitive sensors are widely used and have lower power consumption in comparison to those of piezoresistive type [409]. The turndown ratio of capacitive pressure sensors is 4:1 to 100:1.

Electromagnetic Pressure Sensors

In case of low pressure or small pressure differences the displacement of a membrane would be quite small and would generate a weak signal in the resistive sensor platform.

One possible way to improve sensitivity are inductive sensors which measure the deflection of a diaphragm by means of changes in inductance. The small displacement of the membrane can be transferred to an anchor in a coil and results in a change of magnetic inductance. A variable-reluctance pressure (VRP) sensor measures the deflection of a diaphragm by means of changes in inductance (reluctance). The assembly consists of an E-shaped core and the coil generates the magnetic flux whose field lines travel through the core, the air gap and the membrane. The inductance of the coil-core assembly is determined by the air gap because of its high magnetic resistance. A pressure induced membrane deformation results in a modulation of the inductance, which can be measured with high sensitivity.

Another electromagnetic pressure sensor platform uses changes of magnetic field around a Hall element. A magnetic assembly is placed in close proximity to a bellow. As the bellow expands and contracts due to variations in pressure, the magnetic field changes and an linear output voltage proportional to pressure input can be achieved [410].

Potentiometric Pressure Sensors

The mechanical displacement of a diaphragm, spring, or membrane, such as that in a bourdon or bellow sensing element can be directly transformed into an electrical signal by linking the mechanical deformation with the wiper arm of a precision potentiometer. The displacement of the pressure sensitive assembly results in a movement of the wiper arm across the potentiometer and converts the mechanical deflection into a resistance change, which can by measured by standard means (e.g., a Wheatstone bridge).

Although this is a simple transducing principle, and potentiometric sensors can be made quite small and cheaply, however, this sensor platform is not regularly used in many applications. This is because the sensor typically does not have the range or accuracy of other pressure sensing platforms. The nature of the mechanical linkages connecting the wiper arm to the bellow and temperature effects cause greater uncertainties in this type of measurement relative to other pressure sensor platforms.

Resonant Pressure Sensors

The resonant pressure sensing element is based on changes in resonant frequency of a vibrating element caused by pressure changes. The design of the resonant-wire pressure transducer consists of a wire attached to a static support at one end, and a sensing diaphragm at the other. The sensing diaphragm is subject to a pressure induced displacement. An oscillator induces the wire to vibrate at its resonant frequency. A change in pressure changes the wire tension, which changes the resonant frequency of the wire. The change in frequency can be detected very precisely. This type of transducer can be used for small pressure changes and can be configured for differential, absolute, and gauge pressure measurements. Various designs have been developed, including sensing elements consisting of vibrating wire or vibrating cylinders of quartz. Resonant pressure sensors manufactured using micro-machined silicon provide very stable readings over time.

However, the resonant pressure sensor has some restrictions due to sensitivity to temperature variations and a non-linear output signal. These limitations are typically minimized by using a microprocessor to compensate for non-linearities as well as temperature variations.

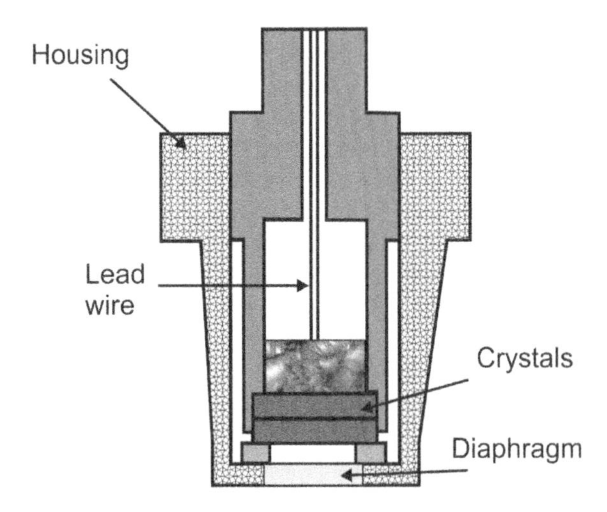

Figure 5.5: Schematic of a piezoelectric pressure sensor.

The sensing element is not as physically robust as other platforms and has some sensitivity to shock and vibration.

Piezoelectric Pressure Sensors

This technology uses the change of electric polarization and the appearance of a voltage of certain materials, such as quartz, in response to elastic deformations arising from applied mechanical stresses, including pressure. If a piezoelectric crystal is exposed to pressure or tensile force in a defined direction, opposite surfaces of the crystal become positively or negatively charged. The induced displacement in the electrically charged lattice elements results in an electric dipole moment which is indicated by the measurable surface charges. The charge quantity Q_x (longitudinal effect) in the x direction is proportional to the applied force F_x and is independent of the size and shape of the element and to the value of the force. Its polarity depends on the force direction.
A simplified expression for this is given as:

$$Q_x = d_{11} \cdot F_x \tag{5.14}$$

Where d_{11} is the piezoelectric coefficient. The electrical voltage created by the surface charges can be measured and amplified for better stability and sensitivity.

Two types of materials may typically be used for piezoelectric pressure sensors—polycrystalline piezoelectric ceramics (e.g., PZT lead zirconate titanate) or single crystal piezoelectric materials (i.e., quartz, tourmaline, lithium niobate, or gallium phosphate). Piezoelectric thin films based on aluminium nitride can be integrated in microelectromechanical systems (MEMS) for pressure sensing [411]. This type of sensor is not influenced by electromagnetic fields and it is robust with good long term stability. However, it is not amenable for continuous use; typically operation is limited to short measurement durations of < 1 s because under static conditions the signal will decrease [412]. For this reason piezoelectric pressure sensors are often operated in pulsed mode.

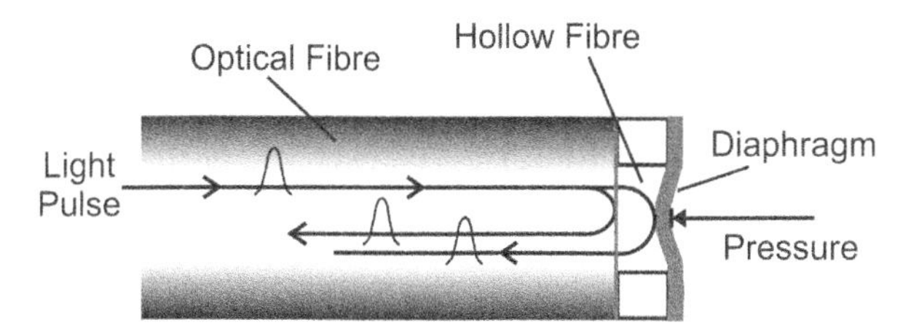

Figure 5.6: Schematic of a Fabry-Perot interferometer pressure sensor.

Optical Pressure Sensors

Fibre Bragg grating pressure sensors

In a fibre optic pressure sensor, the core of the fibre is divided in sections of length Λ (grating period), having different refraction indices. The light travels through the fibre optic and will be reflected and refracted at the section interfaces. The reflected wavelength λ_B, called Bragg wavelength, is:

$$\lambda_B = 2\eta_{eff} \tag{5.15}$$

Where η_{eff} is the effective refractive index. The wavelength shift due to an interacting force can be given approximately by [413]:

$$\frac{\Delta\lambda_B}{\lambda_B} \cong (1 - p_e) \cdot \Delta\epsilon + (\alpha + \xi\Delta T) \tag{5.16}$$

Where $p_e = \frac{\eta^2}{2}[p_{12} - \nu(p_{11} - p_{12})]$ is the effective photo-elastic constant of the fibre material, p_{11} and p_{12} are the strain optic constants, ν is Poisson's ratio, $\Delta\epsilon$ is the strain change, α is the thermal expansion $\alpha = \Delta l/l$ and $\xi = \delta\eta/\delta T$ is the thermo-optical coefficient, respectively. Changes of external pressure induces a wavelength shift, which is then used for pressure sensing.

Fiber Bragg Gratings (FBGs) are so named because the refractive index modulation pattern formed in the fiber acts like a holographic diffractive Bragg grating for light guided in the core of the fibre. Sensors can be used in harsh environmental conditions. However, in addition to pressure, the change in wavelength is also strongly influenced by fluctuations in temperature that is indicated by the second term in Equation 5.16 [414].

Fabry-Perot interferometer

The Fabry-Perot interferometer is based on two partially light transmitting mirrors of high reflectivity which form the basis of an optical resonator. The incoming light is subject to constructive and destructive interference from the reflected light; for example, light reflected 90 ° out of phase with the incoming light of the same intensity will produce a destructive interference. Deflections of the diaphragm, caused by pressure, change the resonance wavelengths. Figure 5.6 illustrates the pressure transducer based on the Fabry-Perot interferometer. Due to constructive and destructive interference, the transmission spectrum contains narrow transmission lines for resonant wavelengths, whereas nonresonance wavelengths are effectively extinguished. Characteristic interference fringes can be observed.

This Fabry-Perot interferometer sensing element can, in principle, be miniaturized and interfaced into an optical fiber. The sensor consists of a thin flexible silica membrane fused on a capillary tube section, which is assembled at the tip of a standard multimode fiber, thus forming a Fabry-Perot air cavity whose length d depends on the applied pressure. The sensor sensitivity (S) characteristic equation is:

$$S = \frac{y}{p} = \frac{r^4}{E \cdot d^3 \cdot \mathrm{A_p}} \tag{5.17}$$

Where p is the applied pressure, y is the centre deflection, r is the diaphragm radius, E is the diaphragm modulus of elasticity, d is the diaphragm thickness, and A_p is a dimensionless stiffness coefficient. A linear spectrum shift in the range of 0.2 to 1 MPa and a sensitivity of 10.07 nm/MPa were reported [415]. Other researchers achieved pressure measurements up to 35 MPa [416, 417]. These optical pressure sensors are insensitive against temperature changes with temperature coefficients of < 0.1 %/K. Optical fibre pressure sensors are immune to electromagnetic interferences and can be employed in applications which require remote pressure monitoring.

Pressure Sensors in Harsh Conditions

Contemporary pressure sensors are manufactured using micro-fabrication techniques, thereby producing at relatively low cost miniaturized, mass produced devices. Silicon (mono and poly crystalline) thin films, bonded metal foils, sputtered thin films are commonly used materials and techniques specifically used to produce micro-fabricated pressure sensors. However harsh environmental conditions, including a combination of high temperatures, high pressure, and aggressive media, preclude the use of silicon technology. For many applications, pressure sensors based on silicon carbide (SiC) or sapphire are the platforms of choice. These materials are stable at high temperature, chemically inert, and amenable for fabrication into many of the membrane platforms discussed above [418]. For protection against contamination and corrosion, a filter can be used to remove reactive chemicals. In many cases, pressure sensors are fragile when exposed to excessive pressures which can often result in irreversible changes. Sensors are usually designed to withstand an overpressure of 50 % to 200 % of the maximum indication.

Reliable operation dictates that both the sensing element (cell body) and sensor packaging must be able to withstand the working environment. Therefore most sensors are available with steel or stainless steel housings. Packaging can also include sealed electrical connections and design features so as to meet explosion-proof and intrinsic safety requirements.

Sensor Technology for Hydrogen Pressure Measurements

Hydrogen poses two major problems for pressure sensors: permeation through seals and embrittlement of the material of construction (see Section 1.1). Both can lead to pressure sensor failure. Thin membranes or thin films, can be particularly prone to adverse interaction with hydrogen, especially under elevated pressure. Hydrogen diffusion into metallic membranes can change the material resistivity which can distort the pressure measurement. If hydrogen diffusion is irreversible, embrittlement can permanently damage the sensor. Permeation of hydrogen through a membrane can also affect the accuracy of a pressure sensor designed to provide the absolute pressure. For example, maintaining a reference vacuum in a cavity formed by the cell body and closed by the silicon membrane is difficult because hydrogen diffuses or permeates through the silicon. Hydrogen can diffuse through other membrane materials as well, including palladium, platinum, and many others. It is in fact difficult to make a hermetic seal for hydrogen using a thin membrane. The amount of hydrogen that diffuses into the reference chamber of an absolute pressure sensor can become significant over time. Hydrogen diffusion will accelerate at slightly elevated temperatures, even temperatures commonly found in automotive applications. As a result, the reference vacuum fills with hydrogen and the uncertainty of the absolute pressure sensing exceeds an acceptable limit, thereby limiting the working lifetime of the sensor. Therefore hydrogen permeation and induced brittleness of material in contact with hydrogen result in increased challenges on pressure sensors to measure compressed hydrogen. Accordingly, hydrogen pressure sensors generally use select materials and designs. For example, a ceramic substrate shields the cavity formed by the silicon membrane and protects the cell body from hydrogen permeation. Other protective measures to prevent embrittlement include increased membrane thickness, doping, or coating the membrane with gold, or through the use of a ceramic diaphragm. Viton seals and 316L stainless steel, which are commonly used for sensor housing, are reasonable materials of choice to guard against hydrogen permeation and embrittlement.

5.2.3 Calibration

Pressure sensors are subject to long term drift and hysteresis. Thus periodic calibration and maintenance of pressure sensors is required. Depending on the sensor platform and application requirements, the calibration can be performed online or offline. Two types of calibration systems are often used for static pressure calibration. One is the deadweight tester (compressed gas calibrator); this is a primary standard. The other means to calibrate a pressure sensor is to use a transfer (secondary) standard. The transfer standard is often a high quality pressure sensor, of the type described in this chapter, calibrated against the primary standard. It is necessary that the transfer standard have a lower uncertainty than the pressure sensor being calibrated. The secondary standard must be calibrated against a primary standard for pressure measurements provided by the national metrological institutes [419].

There are several means available to generate pressure. For example a pump can generate a pressure in a closed system which is then precisely measured by a gauge or sensor, preferably one certified to a traceable device. The sensor to be calibrated will connect to the calibrator and its indication is compared to the reference gauge/sensor under defined ambient conditions. The calibration system should have an uncertainty factor of 10 better than the device to be calibrated. The expanded uncertainty for calibration of a differential pressure sensor and gauges against a national standards in the range of 0.1 kPa to 100 MPa is about $p \times 10^{-5}$ (in Pa). For dynamic pressure mechanically driven calibrators, pressure step change systems and fast valves are employed.

5.3 FLOW SENSORS

5.3.1 Overview

Hydrogen typically has to be transferred from the place of production to storage or usage. At some point in the transfer process, hydrogen will have to flow from one container or location to another. Hydrogen gas flow frequently has to be controlled and monitored. For example, the gas flow rate in PEM fuel cells can be in the range of 30 L/min to 300 L/min at temperatures of about 80 °C and pressures from 1 bar to 3 bar [319]. Precise measurement and control of the flow for other gases, such as oxygen, are important in fuel cells or combustion engines. Hydrogen is considered as a challenging industrial gas with regards to

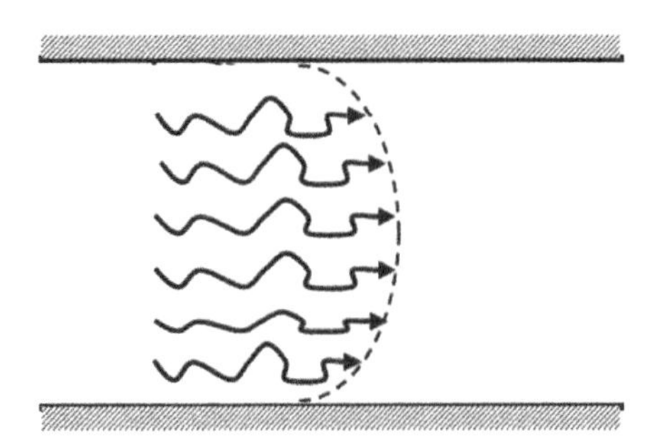

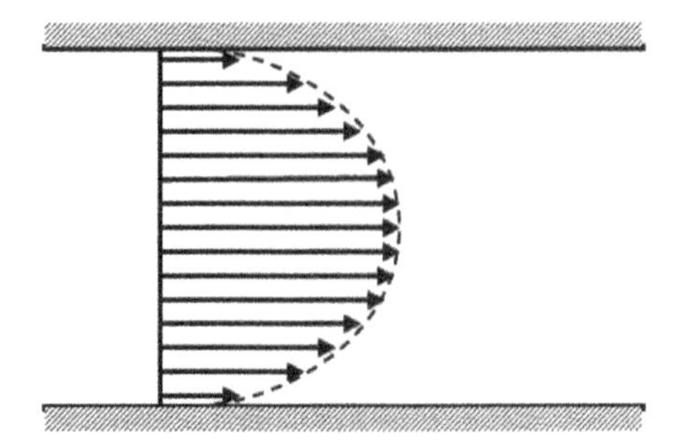

Figure 5.7: Schematic of turbulent (left) and laminar (right) flow.

flow measurements. Accurate hydrogen flow measurement tends to be more difficult than other gas types. One reason is that gases in general are compressible fluids. This means that the volume of a fixed amount of gas depends upon the pressure and temperature that it is subjected to. This is especially challenging for hydrogen due to its low molecular weight and therefore its low operating density, compressibility, and real gas effects. Special precautions are necessary for accurate hydrogen flow measurements by traditional technologies. Gas flow is often expressed in terms of mass per unit time (e.g., grams/min) or a volume per unit time (e.g., L/min).

Two types of flow can be distinguished: turbulent and laminar. Turbulent flow (Latin "turbare" = turn, disturb, confuse) is associated with random often rapid fluctuations in the movement of gases or liquids. This flow pattern is characterized by mostly three-dimensional, apparently random, unsteady movements of the fluid particles which include recirculation and eddies. Laminar flow (Latin "lamina" = board) is a movement of gases or liquids, where no visible turbulence or cross flows occur. In laminar flow, the fluid flows in layers that do not mix with each other. Both types of flow can be characterized by the Reynolds number Re, which is a measure for the ratio of inertial and viscous forces of the fluid. Values of Re $< 10^3$ are typical for laminar flow and Re $> 10^4$ for turbulent flow. Precise measurements and modelling of turbulent flows are difficult.

In a tube of cross-sectional area A the volume of gas moving by Δx in time interval Δt will have a volumetric flow rate given by:

$$Q_v = \frac{\Delta V}{\Delta t} = v \cdot A \tag{5.18}$$

Where v is the flow velocity in m^3/s and ΔV is the volume defined by the product of A and Δx.

By knowing gas density ρ, the mass flow rate is given by:

$$Q_m = \frac{\Delta m}{\Delta t} = \rho \cdot v \cdot A \tag{5.19}$$

Mass flow rate is relevant for the measurement of fuel and air in vehicles in order to optimise engine performance and to ensure efficient fuel economy. Mass is also the metric by which consumers will be charged when refueling their hydrogen vehicle.

The fundamental equation to describe a fluid media is the Bernoulli equation:

$$p + \rho g h + \frac{\rho}{2} v^2 = const. \tag{5.20}$$

Where p is the pressure, ρ is the density, g is the gravitational acceleration, h is the geodetic height, and v is the flow velocity. This equation can be considered as a specific variant of the law of conservation of energy and contains contributions from pressure energy (p), potential energy ($\rho g h$), and kinetic energy ($1/2 \cdot \rho v^2$). The Bernoulli equation describes the underlying principle of several types of flow meters.

The SI units of flow are either cubic meters per second - m^3/s or kilo gram per second - kg/s. Because the volume of a gas often references its pressure, temperature, and compressibility (see Section 1.1), volumetric flow is often referenced to standard conditions (usually STP or NTP, see Section 8.1). The "actual" flow rate through a gauge, e.g., given in actual cubic meters per second - acm/s can be converted to standard cubic meter per second - scm/s by multiplying by the factor:

$$\frac{p_a}{p_s} \frac{T_s}{T_a} \tag{5.21}$$

The subscript a refers to actual conditions; the subscript (s) refers to standard conditions.

Flow meters can be characterised by their measuring range, which is defined by their turndown ratio. For example, a turndown ratio of 1:20 means that the highest measurable flow of acceptable uncertainty is a factor of 20 larger than the lowest flow. Thus, if the upper measuring limit is 2 L/min, the lower measuring limit would be 0.1 L/min. The measuring range of flow meters covers over 15 decades from nL/min to 10^4 m^3/s. The turndown ratio of most flow meters is between 10 and 20.

The uncertainty of flow measurements typically ranges from 0.2 % to 5 % and depends on conditions of application like media, its flow rate, pressure, and temperature. A further parameter of importance for many applications is the time of response to changes of flow velocity.

5.3.2 Operating Principles

There are numerous commercially available meters and sensors based on different operating principles [420, 421]. The selection of a particular flow meter depends on specific conditions and requirements. All flow measurement technologies have advantages and disadvantages. Consideration should be given, but not necessarily limited, to the following specifications when selecting a flow meter for gas flow measurements: accuracy, turndown ratios, pressure drops, process temperatures, additional sensor requirements, and process connections. A mass flow can be a more accurate representation of actual gas consumption in a process than a volumetric flow reading because of possible variations of temperature and pressure associated with the volume gas flow. However, the impact of variations in these parameters can be overcome by pressure and temperature sensors, which then permits the determination of a mass flow from the measured volume flow. Traditionally rotary, turbine, and orifice meters were used in industry. But increasingly industry is relying on ultrasonic technology for large-volume flow measurements. Not all processes will have the same specification requirements as that of hydrogen technology. Thus anemometers, which are devices to measure the wind speed in air, are not discussed in this section, even though the air flow may be relevant for ventilation in hydrogen loaded room atmospheres. In the following, measurements in closed conduits are described. The performance of various flow sensor technologies for hydrogen is compiled in Table 5.4 and discussed further in this chapter. Since hydrogen is a flammable gas, the device should be explosion proof and compatible, e.g., with zone 2 operation (see Section 6.1). This requirement is not necessary for nonflammable mixtures of hydrogen (e.g., less than 4 vol% H_2 in air).

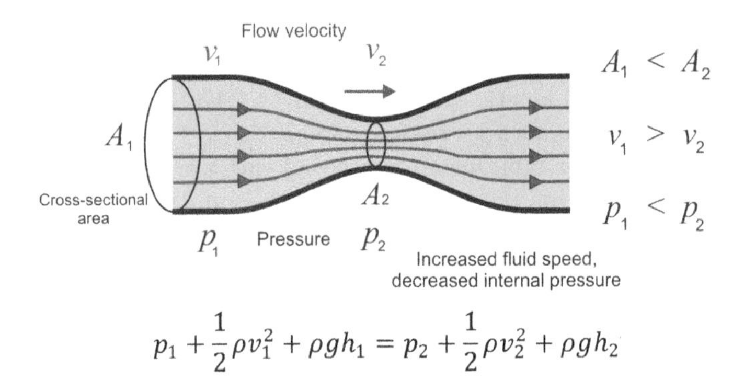

Figure 5.8: Flow behaviour at a tube's constriction.

Differential Pressure Volume Flow Meter

From the Bernoulli equation (see Equation 5.20) it can be deduced that for an incompressible fluid (ρ = constant) the constriction of diameter of a tube results in an increase of flow velocity and decrease of pressure (see Figure 5.8). The pressure change is the basis for many flow meters, e.g., orifice plate, nozzle, venturi, or pitot tube. There is a direct relationship in pipelines between the cross-sectional area, the pressure, and the speed of a flowing fluid. The volume flow is always proportional to the square root of the pressure difference.

An **orifice plate** is a thin plate with a hole in the middle. It is usually placed in a tube in which fluid flows. When the fluid reaches the orifice plate, the fluid is forced to go through the small hole. As it does so, the velocity increases and the pressure decreases. By measuring the pressure difference between the normal tube section on each side of the orifice and within the constricted zone, the volumetric and mass flow rates can be obtained from Bernoulli's equation:

$$Q_m = \rho \cdot Q_V = \mathrm{C} \cdot A\sqrt{2\rho(p_1 - p_2)} \tag{5.22}$$

Where p_1 is the fluid upstream pressure (in Pa), p_2 is the fluid downstream pressure (in Pa), C is orifice flow coefficient (dimensionless), and A is the cross-sectional area of the orifice hole (in m^2, e.g., A_2 in Figure 5.8).

The measuring range of the flow through an orifice plate is limited to a factor of 10. A low uncertainty of approximately 0.2 % can be reached if the influence of temperature and density in the gas flow is considered.

The pitot tube is a pressure measurement device used to measure fluid flow velocity. It gauges the difference between stagnation, the dynamic pressure (p_{dyn}), and static pressure (p_{stat}) in (incompressible) flows:

$$v = \sqrt{\frac{2\rho(p_{dyn} - p_{stat})}{\rho}} \tag{5.23}$$

The **pitot tube** type flow meter is a robust and cheap device for the measurement of velocities larger than 10 m/s. However the flow must be aligned with the probe and be steady; yaw or pitch of more than 50 ° causes serious errors. It cannot measure flow rate fluctuations. A variant of tube with a reduced diameter is the Venturi tube. This device is used in methane or liquid gas driven cars for mixing fuel gas and air. The turndown ratio is in the range of 3:1 and 20:1.

The **sonic nozzle** (also called critical flow nozzle, critical flow venturi, or sonic venturi) limits the maximum flow through a tube and consists of a smooth rounded inlet section converging to a throat of minimum diameter and then diverging to an exit cone. The pressurized gas accelerates its velocity through the throat and its density decreases. The maximum velocity is achieved at the throat, where it reaches the speed of sound for the gas being measured. The sonic nozzle is constructed in such a way that the pressure of the up stream gas exceeds the pressure of the down streaming gas by a factor greater than 1.4. In the "choked" or "sonic" state the flow rate through the nozzle is nearly a linear function of the inlet pressure. Thus by measuring only the inlet pressure and the temperature the flow rate can be calculated. These devices usually operate at pressure up to 10 bar and have an uncertainty in the measured flow rate of about 0.25 % of reading. Flow meters based on the differential pressure principle are robust, and tolerate the presence of trace contaminations. The application in fuel cells can be limited, however, due to low flow rates. On the other hand critical nozzles can be used in technical processes requiring accurate high pressure hydrogen flow measurements [422].

Thermal Mass Flow Meter

The thermal mass flow meter, which is also called the thermal dispersion mass flow meter, uses a positive temperature coefficient (PTC) resistor as the sensing element. The flow velocity influences the heat transfer from the sensing element and therefore also its electrical properties which are then used for flow measurement. **Hot wire probes (anemometers)** use either a very fine wire on the order of several micrometers thick and 1 mm to 2 mm in length) or a thin film of some hundred nanometer thickness which are heated electrically from just above the ambient temperature up to 200 °C to 300 °C. Tungsten, platinum, platinum-iridium alloy, and platinum coated tungsten are popular choices for the heated element, and are typically configured to have a 2 Ω to 3 Ω resistance. The flowing gas has a cooling effect as it passes over the heated element. A relationship can be obtained between the resistance of the heated element and the flow speed. The initial work on hot wire anemometer was done by L. V. King in 1914, who derived the following equation for a heated metallic wire:

$$I^2 \cdot R_s = (T_s - T_f) \cdot (\mathrm{A}' + \mathrm{B}' \cdot v^{0.5}) \tag{5.24}$$

Where I is the current through the sensing element R_s the ohmic resistance of the wire, T_s the sensor temperature, T_f the temperature of the fluid, A', B' are constants, and v is the flow velocity in to the heated element.

There are several methods to evaluate an electrical signal from the sensor. The voltage output from anemometers is typically the result of some sort of circuit within the device trying to maintain a specific parameter (current, voltage, or temperature) constant. In a hot film or hot wire anemometer with temperature difference control, the gas stream flow passes two temperature-sensitive resistors which are part of an electrical bridge circuit. One of the resistors is heated and maintained at a constant overtemperature in relation to an unheated temperature sensor inside the gas flow. The heating power required to maintain the heated reference resistor at its temperature setpoint depends directly on the flow rate and the material properties (thermal conductivity) of the gas. With a known (and constant) gas composition the mass-flow can be determined by electronically evaluating the heater current/mass-flow curve without additional pressure and temperature compensation.

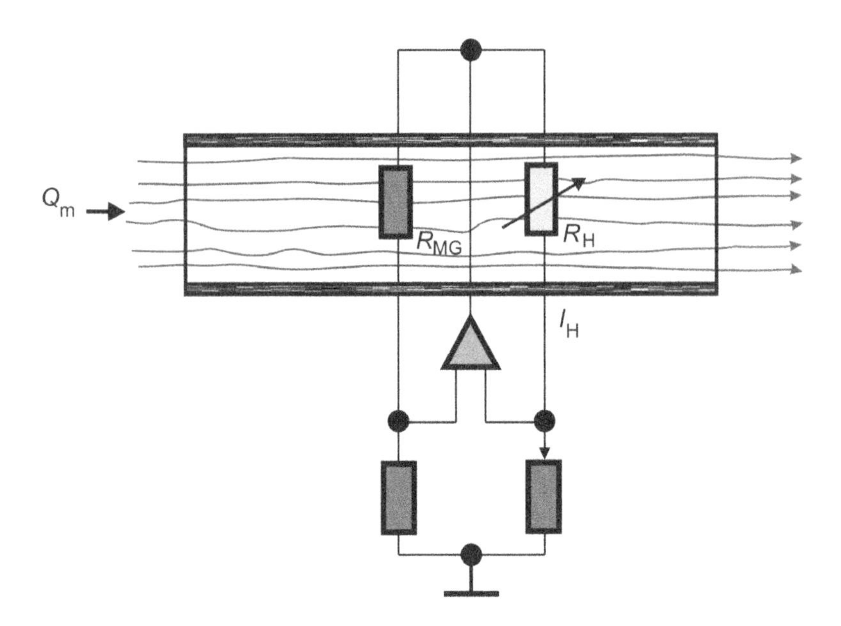

Figure 5.9: Schematic of a thermal mass flow meter.

In a hot film or hot wire anemometer with temperature difference control, the gas flows by two temperature sensitive resistors (R_{MG}, the measuring resistor, and R_H, the heated reference sensor) which are part of a bridge circuit. This is an example of a thermal mass flow meter, and Figure 5.9 illustrates its basic design and operation. By design $R_H < R_{MG}$, R_H is heated by the current I_H. The current I_H is adjusted by an electronic control circuit to produce a constant temperature difference between the heated resistor R_H and the temperature of the gas. The electrical power generated with resistor R_H exactly compensates its loss of heat to the gas flow. I_H represents a measure of the mass flow rate. The output of the electronic circuit can be a voltage (analogue) or a frequency signal (digital).

Thermal mass flow technology is a method of gas flow measurement that does not require correction for changes in process temperature or pressure. This technology also has a benefit of measurement capabilities at low velocities and a larger measuring range than those obtainable with other gas flow measurement devices. The major limitation of thermal type mass flow meters is similar to the limitations of all pitot-type flow gauges, namely that they do not detect the mass flow across the full cross-section of the pipe, but only at the resistive sensing element. Therefore, if the sensing element is installed in a non-representative location across the velocity profile, the resulting reading will be in error. As with many sensors, the sensing element of the thermal flow meter should be protected from condensation and water droplets [319]. However, hot-wire anemometers have extremely fast response times. In addition, the small size of the sensing element allows for fine spatial resolution compared to other measurement methods. Accordingly, they are almost universally employed for the detailed profiling of turbulent flows, or any flow in which rapid velocity fluctuations are of interest [423].

Thermal mass flow meters of different ranges are available to measure gas flow from 1 sccm/min to 2×10^6 m^3/h with an uncertainty of 0.5 % to 2 %. Thermal flow meters are considered as the most cost-effective flow measurement technology.

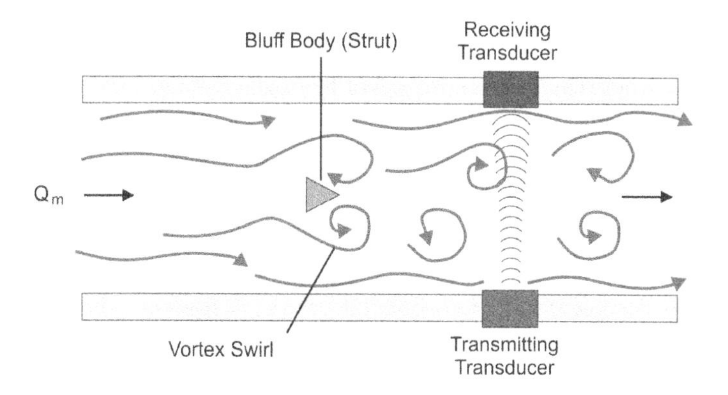

Figure 5.10: Schematic of vortex flow meter.

Vortex Flow Meter

The vortex flow meter is based on the Kármán vortex street, a repeating pattern of swirling vortices. In fluid flow behind an obstacle, e.g., behind a bluff body, vortices are generated. These are regions in a fluid, where the flow is characterised by a spinning rotation about a moving imaginary axis. The principle of the vortex flow meter is illustrated in Figure 5.10. Vortices are a major component of turbulent flow. The vortices flow counter rotating and staggered. Small local pressure differences occur, which can be detected by an appropriate sensor. The mean flow velocity Q_v is

$$Q_v = \frac{f \cdot d}{S_r} \tag{5.25}$$

Where f is the vortex frequency, d the diameter of the bluff body and S_r the Strouhal number (dimensionless).

The pressure sensor counts the number of pressure impulses per unit time, the so-called vortex frequency which is correlated to the flow velocity or volume flow (see Figure 5.10). A temperature and pressure correction is needed for the accurate determination of mass flow.

Coriolis Flow Meter

Coriolis flow meters are based on the Coriolis force which occurs in rotating fluid systems that are perpendicularly orientated to the centrifugal force. The value of this force is proportional to the mass of the moving fluid. A curved tube such as U- or delta-shaped tube will vibrate in the frequency range from 80 s^{-1} to 1000 s^{-1} due to the flowing fluid. The fluid flow generates the vibrations because of the Coriolis force which induces a deflection of the tube. The time and space dependent values of the deflection are detected by motion sensors and are correlated to the mass flow. Coriolis gauges can also determine the density of a fluid. The mass flow of a U-shaped Coriolis flow meter is given as:

$$Q_m = \frac{(k_u - I_u \omega^2)}{2\mathrm{K}d^2} \tag{5.26}$$

Where k_u is the temperature dependent stiffness of the tube, K a shape-dependent factor, d the width, τ the time lag, ω the vibration frequency, and I_u the inertia of the tube. As the inertia of the tube depends on its contents, knowledge of the fluid density is needed for the calculation of an accurate mass flow rate. The devices have no drift and can determine flow rates in the range of 10 g/h to 1,500 Mmt/h with an uncertainty of about 0.2 %. Coriolis flow meters were used in the steam reform process of hydrogen from natural gas [424,425] Coriolis flow meters are also being explored as a means to measure the hydrogen dispensation into fuel cell vehicles for the consumer market.

Mechanical Flow Meters

Mechanical flow meters contain components which are mechanically displaced by the fluid. In many cases the displacement is transformed into a rotary motion. The number of rotations or the speed of rotation is proportional to the gas flow.

A rotary gas meter contains two eight-shaped rotors (also known as impellers or pistons). With each turn, they move a specific quantity of gas through the meter. The rotational movement is counted to determine the flow. Rotary meters are high precision instruments capable of handling high volumes and pressures. The turndown ratio is up to 1:160. This flow meter type is not significantly impacted by fast pressure fluctuations.

Vane Wheel Flow Meter

The vane wheel flow meter is a mechanical flow meter that uses wheels which rotate parallel or perpendicular relative to the flow direction. The turbine flow meter has an axial aligned vane wheel that rotates at a speed proportional to the volume flow velocity of the fluid into which it is placed. The rotational speed is almost independent of density, pressure, and temperature of the gas. Typically, the velocity of air or another gas can be measured in a range of 0.01 m/s to 120 m/s, at pressures up 50 bars. The uncertainty is approximately 1 % for velocities greater than > 1m/s. Turbine flow meters range in size from 5 cm through 40 cm were used to measure the flow of liquid hydrogen and oxygen for rocket propulsion in the U.S. Apollo and Skylab space programs [426]. A capacitance-based measurement approach for measuring the flow velocity and the mean density of liquid hydrogen flows was used for the transducer. The sensor allows for the measurement of liquid hydrogen flow velocity exploiting a spatial frequency-based method for observing gaseous hydrogen bubbles moving through the sensing volume. Additionally, electrodes at both ends of the sensing section were used to determine the mean density of the cryogenic liquid [427].

Volume Air Flow Sensors

The vane air flow sensor (VAF) measures the air flow into a combustion engine using a spring loaded air flap attached to a variable resistor (potentiometer). The displacement of the vane (flap) is in correlation to the airflow. A voltage is applied to the potentiometer and a proportional voltage appears on the output terminal of the potentiometer in proportion to the angle of the flap, or the movement of the flap may directly regulate the amount of fuel injected. This type of flow meter is currently used in many vehicles.

Rotameters

A rotameter consists of a glass, plastics, or metal tube with a specially configured weighted internal "float." Glass is the most common material for the tube when the float position is determined visually. The gas flow is applied from underneath and lifts the float. For a constant flow rate, there is an equilibrium between the downward gravimetric force and the drag force provided by the flowing fluid. The mass flow rate for a given fluid depends on the drag force and float cross-section according to:

$$Q_m = k_f \cdot d_s \sqrt{g \cdot m_s \cdot \rho_m (1 - \frac{\rho_m}{\rho_s})} \tag{5.27}$$

Where, k_f is a flow coefficient, d_s the float diameter at the reading edge, g gravitational constant, m_s is the float mass, ρ_s is the float density, and ρ_m is the fluid density.

A higher flow rate through a given area increases flow speed and drag force, so the float will be pushed upwards. However, as the inside of the rotameter is cone shaped (widens), the area around the float through which the medium flows increases, the flow speed and drag force decrease until there is mechanical equilibrium with the float's weight.

Floats are made in different shapes, with spheres and ellipsoids being the most common. The float may be diagonally grooved so that it rotates axially as the fluid passes. Readings are usually taken at the top of the widest part of the float; the centre for an ellipsoid, or the top for a cylinder. The gas flow is indicated by the position of the float on a scale; however, for industrial applications electronic displays are available. Flow of 1 sccm/min up to 250 m^3/h at gas temperatures up to 130 °C (glass tube) and up to 400 °C (metal tube) can be measured by devices of different dimensions. In order to determine the flow rate of hydrogen by rotameter which usually has a scale calibrated for air, the indication has to be multiplied by gas flow correction factor for hydrogen, which is provided by the manufacturer.

Optical Methods

Optical flow meters use light to determine flow rate. Small particles which are carried in the gas stream pass through two laser beams in the flow path. The laser light is scattered when a particle crosses the first beam. The detecting optics collects scattered light on a photo detector, which then generates a pulse signal. As the same particle crosses the second beam, the detecting optics collect scattered light on a second photo detector, which converts the incoming light into a second electrical pulse. By measuring the time interval between these pulses, the fluid velocity is calculated. The operating principle provides highly accurate flow data, even in challenging environments which may include high temperature, low flow rates, high pressure, high humidity, and mechanical vibration. Optical flow meters do not require periodic calibration after their initial commissioning. Commercially available optical flow meters are capable of measuring flow from 0.1 m/s to faster than 150 m/s in single a device, i.e., a turndown ratio > 1000.

Another important optical method for flow measurements is **Doppler interferometry**. A beam of laser light impinging on a moving particle will be partially scattered with a change in wavelength proportional to the particle's speed (the Doppler effect). A Laser Doppler Velocimeter (LDV) focuses a laser beam into a small volume in a flowing fluid containing small particles (naturally occurring or added). The particles scatter the light with a Doppler shift. Analysis of this shifted wavelength can be used to directly, and with great precision, determine the speed of the particle and thus provide a close approximation of the fluid velocity. However, due to the complexitiy of the optics, this method is applied predominantly in research and development applications [428, 429].

5.3.3 Calibration

Primary standards for traceability of flow measurements are provided by national metrological institutes. The amount of gas that passes through a device in a defined time can be determined by a gravimetric or volumetric measurement to deduce the flow rate. A gravimetric flow system can serve as a primary standard for gas flow in the range of 0.001 L/min to 3 L/min. It determines the mass flow from the decrease of the mass of a pressure bottle placed on an accurate electronic balance. The standard for higher flow rates determines flow by collecting gas in a tank of known volume, while monitoring pressure and temperature over a defined time interval, which is referred to as a PVTt method. The PVTt method uses the real gas law to quantify the relationship between pressure, volume, and temperature. Temporal changes in pressure and temperature can be used to determine the volume or mass that passed through the meter under test over a known time interval. This method allows calibrations in the flow range of 1 L/min to 77000 L/m and pressures from 100 kPa to 7,000 kPa of non-corrosive and nonhazardous gases. The uncertainty of the flow standards for dry air varies from 0.02 % to 0.1 % [430].

Working or transfer gas flow standards are used for calibrations of gas flow meters at facilities other than the national metrological institute using transfer standards calibrated to the primary standard. One meter, the bell prover, can be used as a gas standard to accurately measure flow rate as a displaced volume of gas at a constant temperature. The displaced volume and the time over which the volume is displaced, is measured electronically so that a precise flow can be determined. The uncertainty is 0.2 %.

In the low flow range sonic nozzle based laminar flow molblocs or critical flow venturis or nozzles are used. The working standards shall be periodically calibrated with primary standards. The transfer standards are used to calibrate flow meters with low pressure drop, in dry air, at flows from 0.001 L/min to 70,000 L/min with an uncertainty of 0.15 % [431].

5.4 FIRE AND SMOKE DETECTION

5.4.1 Overview

Gas sensors are often used to indicate the presence of flammable gas mixtures in order to avoid fire, deflagration, or explosion. Other electronic detectors can complement the gas sensor capability by the direct detection of fires and other combustion events once such events have been initiated. In the absence of other burning materials, hydrogen fires will not generate smoke or soot, thereby precluding the use of conventional smoke detectors for early hydrogen fire detection (at least until the hydrogen combustion spreads to other materials). Hydrogen fires have a low emissitivity that burns with a pale blue flame that is nearly invisible to the human eye. Consequently other means are necessary for detecting the presence of a hydrogen flame. One proposed method that was explored to verify the presence of a hydrogen fire involved putting corn straw brooms or dusting particles into a suspected hydrogen fire to generate a flame that emits in the visible spectrum. The "straw broom technique" was supposedly used in the not-too-distant-past to verify the presence (or hopefully the absence) of a hydrogen fire. Fortunately electronic technologies are now commercially available to detect hydrogen fires and are better suited to identify hydrogen flames remotely, quickly, and more safely than the straw broom approach. These electronic technologies include UV and IR optical flame detectors that sense the non-visible electro-magnetic radiation associated with flames.

Hydrogen fires can initiate of other materials which can then be detected by conventional means. Smoke, which is a mixture of airborne gases and solid or liquid particulates, will not be generated when combustion only involves hydrogen but will be produced when the combustion spreads to other (usually carbonaceous) materials [432]. Smoke will often be present prior to the initiation of actual combustion, and thus its presence can be an early warning indicator of a pending fire. The changes in the physical or chemical properties of the surrounding atmosphere are commonly called the signatures of the fire. Although it is best to identify the presence of a hydrogen release before a combustion event has occurred,

Table 5.4: Application range of flow sensors for hydrogen

	Operation principle						
	Differential pressure	**Thermal**	**Vortex**	**Coriolis**	**Vane**	**Rotameter**	**Optical**
turn-down ratio	1:3 to 1:100	1:100	1:7	1:00	1:10	1:10	1:1500
velocity in m/s	25 to 250	0.08 to 200	10.5 to 40		0.01 to 120		0.1 to 150
Flow rate		1 sccm/min to 2,100,000 m^3/h	20 m^3/h to 20,000 m^3/h	10 g/h to 900 t/h	1 to 500 L/min	1 sccm/min to 1,000 m^3/h	
uncertainty	> 0.1 to 2 %	0.5 to 2 %	1 %	0.2 %	1 %	2 %	1 to 5 %
pressure in bar	$\leq$420	$\leq$70	$\leq$250	$\leq$400 to 800	$\leq$50	$\leq$400	20
temperature (of the gas) in °C	-200 to 1,000	-40 to 240	-200 to 450 (-250 to 350)	-50 to 350	-40 to 500	-40 to 130 (glass) -200 to 400 (metal)	-40 to 100

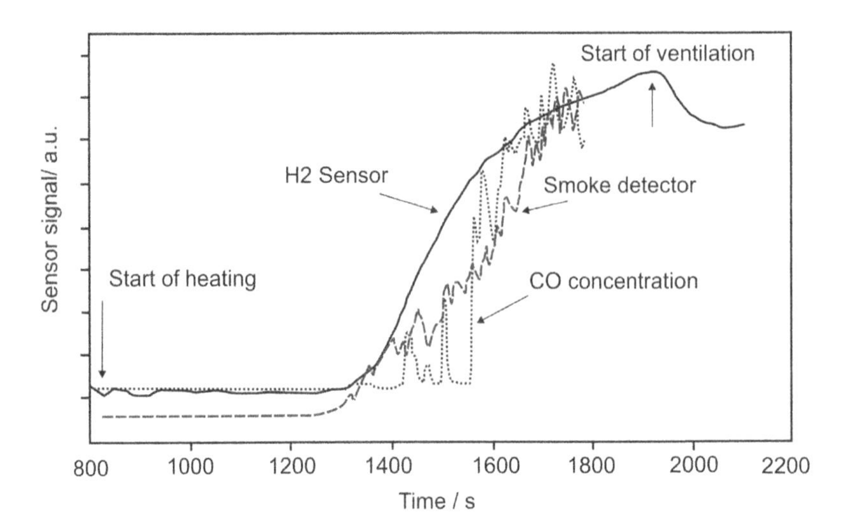

Figure 5.11: Hydrogen and carbon monoxide release during a standardised fire test on beech wood [436].

it is still essential to quickly identify and locate a hydrogen fire so as to initiate corrective measures before it has spread to the surroundings.

In the following section, the different principles used in the electronic detection of fires are described, with an emphasis on hydrogen fires. A compilation of the most commonly used detectors for hydrogen fires is given in Table 5.5.

5.4.2 Gas Sensors for Fire Detection

Gas sensors can be used to detect chemical signatures associated with non-hydrogen incipient fires. These signature compounds include carbon monoxide, carbon dioxide, nitrogen oxides, acrolein and other chemicals that are associated with the burning material. A nascent fire produces trace levels of characteristic gases, that often include carbon monoxide as well as hydrogen [433, 434]. Figure 5.11 shows the detection of the hydrogen and carbon monoxide generated during a fire test on beech wood. The chemical signatures associated with incipient fires are often released prior to actual combustion, and therefore their detection can potentially be used to detect a pending fire prior to the generation of physical signatures such as smoke, soot, and flames. The type and distribution of the chemical signature are related to the specific material undergoing an uncontrolled thermal excursion. Gas sensors can be used to indicate chemical signatures to provide an early warning fire indication faster than other fire detection methods. A combination of a gas sensor, a smoke and heat sensor in one fire detection system is already commercially available [435]. Carbon monoxide, however, is not a signature for a hydrogen fire but may serve as an indicator for the combustion of other materials.

5.4.3 Smoke Detection

Fires will typical generate smoke, whose detection is the basis for the most common residential early warning fire indicators—the smoke detector. Smoke can be detected to indicate the

Table 5.5: Hydrogen fire detectors

	Type of Sensor	**Remark**
	Gas Sensor	
Gas detection	Detection of traces of flammable gas	Requires proper sensor placement to detect gases and vapours.
	Smoke Sensor	
Optical	Detection of light scattering on smoke particles	
Ionisation current	Reduced ionisation current due to radiation absorption by smoke particles	Short range radioactivity may limit domestic applications
	Optical Sensor	
	Very fast response within (3 - 4) ms, however (2 - 3) s delay is incorporated to avoid false alarm	Most suitable principle for hydrogen fires, but false alarms can be triggered by UV sources (e.g., lightning, arc welding, corona);
UV range (180 - 300) nm	Fast response within 150 ms	however, not sensitive to natural background radiation thick smoke, vapours, deposits on the detector's window can prevent detection of light. Detection range up to 50 m
Visible (0.4 - 0.7) μm	Detects the glowing of matter at temperatures above 700 K, e.g., by a CCD camera	Not recommended for pure hydrogen fire due to low emission, needs complex image processing analysis or operator for interpretation
Near IR (0.7 - 1.1) μm	Pyro electric sensor detects IR radiation	Relative cheap, however not applicable for pure hydrogen fire, >3.5 μm water absorption can
Broad band IR (> 1.1 μm)	Thermal imaging camera monitors heat radiation of a flame or hot gases in response time of 3 - 5 seconds.	disturb detection
	Thermal Sensor	
Thermal flame detection	Resistive temperature device (RTD) or thermocouple detects the heat of combustion	Commonly used as flame failure devices to cut off the fuel supply if the flame fails
Linear heat detector	Proprietary cable detects heat conditions anywhere along its length	

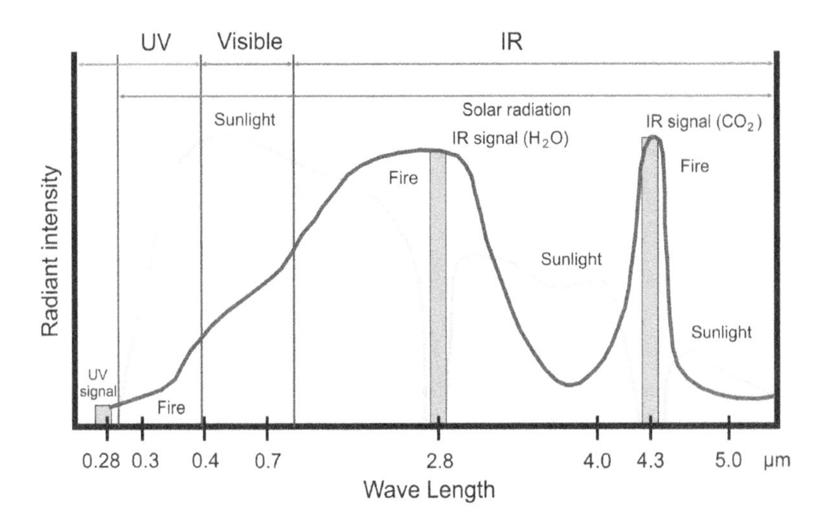

Figure 5.12: Spectra of fire and solar radiation.

presence of fire or nascent fire, in contrast to the direct detection of actual flames. There are two basic types of sensors used in smoke detectors. Ionization type detectors, which are better at detecting the small (less than 1 μm) smoke particles from fast flaming fires, while photoelectric type detectors are better at detecting the relatively large smoke particles from slow smoldering fires. Ionization type smoke detectors are commonly used for residential applications.

The optical or photo-electrical smoke is detector based on light scattering (Tyndall effect). A light beam from a light emitting diode (LED) or a laser diode will not be deflected in pure air. However, light scattering will occur if smoke particles are present in the optical chamber. The scattered light will be detected by a sensitive photo diode, the output of which can then be used to activate an alarm signal.

The ionization smoke detectors use a radioactive source that releases alpha particles (e.g., ^{241}Am) to ionize the air between two metal electrodes. The presence of ionized air produces an electric current when a potential is applied across the two electrodes. Smoke particles between the electrodes absorb part of the alpha radiation and thus fewer air molecules will be ionized. This leads to a drop in the electrical current, which is then used to activate the fire alarm.

5.4.4 Optical Flame Detectors

Optical flame detectors respond to light radiated from a flame. The flame detection falls into three spectral regions: ultraviolet (UV), visible (vis), and infrared (IR). Unfortunately, the spectra of emitted energy from a fire overlap significantly with the light from the sun and other sources, as shown in Figure 5.12. It is thus often necessary to take specific measures to avoid false alarms.

The UV photo tube detectors operate on the Geiger-Mueller principle and consist of a glass bulb filled with a gas and contain two electrodes across which a high AC voltage is applied (see Figure 5.13). Ultra-violet radiation causes one electrode (the cathode) to release electrons. The electrons will be captured by the anode. However as it traverses through the

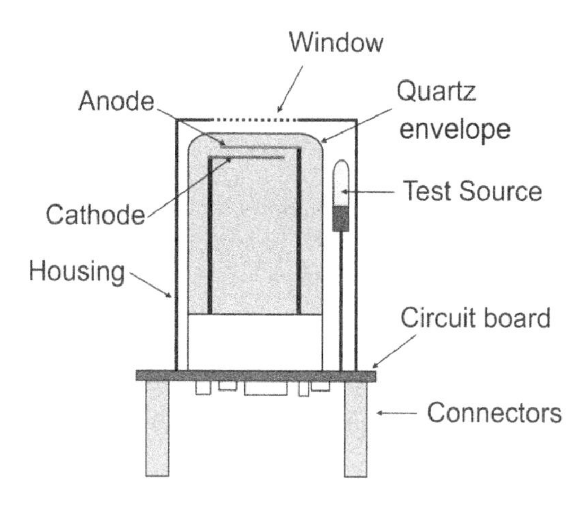

Figure 5.13: Schematic of a UV flame detector.

gas, an ionization process of the gas is initiated leading to the generation of an electronic discharge and hence to the formation of an electric current. The current indicates impinging UV light, which in turn is indicative of a fire. Alternatively, a UV diode can be used to directly detect the UV light associated with a flame. UV systems are extremely sensitive and fast responding, requiring only a few seconds to go into an alarm state in the presence of a fire. UV imaging systems, however, require special optics and are expensive because the intensity of the emitted radiation in the UV region is low and signal amplification is needed. Furthermore this type of detector can be blinded in foggy conditions and is sensitive to arcs, sparks, welding, sun light, and other UV-rich sources. Nevertheless, UV detectors, with their very fast response time and good detection range, are well suited for hydrogen fire detection where the false-alarm sources can be controlled, such as in enclosed rooms.

Infrared systems, which are typically designed for hydrocarbon fires, detect heat radiation and the characteristic absorption band of carbon dioxide at 4.3 μm, and are not very specific to hydrogen fires. However, the specific absorbance bands of water vapour in the range of 2.8 μm can be employed for hydrogen fire recognition. Heated water vapour formed in a hydrogen fire tends to emit radiation over a broad IR range (see Figure 5.12).

Multispectrum IR flame detectors use a combination of IR filters and software analysis to both detect the flame and reduce false alarms. Some broad band IR detectors have been designed specifically to detect the low IR level of hydrogen fires using a unique set of IR filters. A large percentage of the observed radiation emitted from a hydrogen fire originates from water vapour molecules. These special IR flame detectors have a very good detection range with a fast response time to hydrogen flames, and are not prone to false alarms from arcs, sparks, welding, and lightning. In addition, the multispectrum IR detector has complete solar resistance and is insensitive to artificial lights and most "blackbody" radiation, which plague other optical fire detection technologies. The multispectrum IR detectors do have limits, however. For example, their detection range is reduced with the presence of water or ice on the lens. To mitigate this problem, some detectors are manufactured with lens heaters that melt ice and accelerate the evaporation of water.

When used improperly, flame detectors can go into an alarm state without the presence of a hydrogen fire. False alarms can be expensive and can reduce confidence in the technology.

The occurrence of false alarms should be minimized. Users should match the appropriate technologies to the applications they face. It is often preferred and sometimes necessary to use devices with a combination of different detectors with sensitivities in multiple spectral regions to minimize the possibility of false alarm. Some commercial flame detector designs with responses in multiple spectral regions include:

- IR/UV: The signals of an IR and UV detector are combined to avoid false alarms. IR/UV detectors are amenable for both indoors and outdoors applications, and have been used at hydrogen fuelling stations [437].
- Dual IR/IR flame detectors compare the signals in two infrared ranges, e.g., in the 4.3 μm range and an additional sensor at a reference wavelength.
- Triple IR or multi IR detectors compare signals in several specific wavelength bands within the IR spectral region.
- Additional optical flame detection strategies to avoid false alarms include multiple wavelength detection. They can be categorized to their spectral rage as IR/IR/vis, IR/UV/vis, and IR/IR/IR/vis, which are all compatible for hydrogen fire detection.

Furthermore to avoid false alarms, "smart" flame detectors analyse for additional properties associated with a fire including the flickering frequency (2 Hz to 35 Hz) of a fire, radiation intensity thresholds, and mathematical algorithms such as ratios, AND-gate comparisons, correlations, and autocorrelations.

Optical flame detectors can often perform automatic self-testing, and compensation of optical pollution. The devices offer output signals via common electronic interfaces, including RS 232, ModBus, Hart protocol or as "normalized current." Specific and highly sensitive devices for hydrogen fire based on UV and UV/vis and broad band IR detectors are available commercially [438].

Deployment designs for flame detectors should consider that the shape and size of the window, as well as the housing, limits the field of view (cone of vision). At increasing angles of horizontal or vertical alignment with the flame, the sensitivity drops. At 120 ° the detector has only 50 % or less of maximum sensitivity. Also the sensitivity decreases rapidly with the distance from the flame. The relationship between distances and detector performance behaviour can be described by an inverse square law; i.e. doubling the detection distance results in only 1/4 of the radiant energy reaching the detector.

Proposed applications for flame detectors include hydrogen storage facilities, fuel cells, hydrogen vehicle re-fuelling stations, garages, the aerospace industry, battery charging areas, chemical process industry, and refineries. Many applications addressed by NFPA 2 explicitly require flame detectors. The requirements on fire detection and fire alarm systems in general are standardized in EN 54 [439] and application rules for hydrogen fire detection systems are given in [440]. These devices need an approval for explosive areas (zone 1) and shall be certified according IEC 61508 (SIL2).

5.4.5 Thermal Fire Detection

Thermal fire detectors, classified as rate-of-temperature-rise detectors and overheat (fixed temperature) detectors, have been manufactured for many years and are reliable. Thermal detectors need to be located at or very near the site of a fire. Thermal detectors, use a resistance thermometer (RTD), thermocouple, or thermistor to activate an alarm when a certain temperature (e.g., 60 °C), in a room is exceeded. Alternatively a bimetallic is used such that at normal temperatures an electrical contact will be maintained, but at elevated temperatures the electrical contact will break and thereby trigger an alarm. Because a thermal detector will not alarm unless it senses the heat, positioning a thermal detector directly above the probable site of a hydrogen fire is logical. However, the probably location of a fire may not be known with a high degree of certainty. The source of the hydrogen leak might create a fire that is directed away from the detector. The hydrogen fire's low IR radiation may not be enough to set the heat detector into alarm by radiated heat alone. Thermal detectors are helpful, but proper positioning is the biggest challenge.

Glow plugs can also be configured as a thermal fire detection system. They are usually employed in combustion engines in order to ignite the fuel-air mixture. Gas plugs often contain a platinum/iridium coated coil which is heated up to 1000 °C to 1300 °C. Initiating the flaring of hydrogen releases in petroleum processes is a common application. This type of device can also be used for rapid leak detection; however, it is not a true gas detection technique. The glow plugs ignite any combustible mixture present and a heat sensor detects the fire and provides rapid shut down of the process (if necessary). The basic idea is that it is better to burn the hydrogen gas rather than letting a combustible mixture accumulate. For example, a flame burns continuously inside a small chamber of the sensor housing. Flowing flammable gases are incinerated by the flame. A temperature sensor measures the resulting change in flame temperature. This temperature change correlates with the amount of flammable gases and can be displayed, e.g., in % LEL.

CHAPTER 6

Sensor Deployment

CONTENTS

6.1 REGULATIONS, CODES AND STANDARDS ON HYDROGEN SENSOR USE

6.1.1 Overview

Regulations, codes, and standards (RCS) establish requirements for safety and environmental compliance, as well as the interchangeability of products and services. RCS may also prescribe relevant test methods to verify compliance to these requirements. In doing so, RCS are important in ensuring safety, encouraging commercialization, and removing technical barriers to trade. While often treated synonymously or referred to together in North America, there is a clear distinction between a regulation, a code, and a standard.

Regulations and directives, as they are called in the European Union (EU), provide legislative rules and when adopted by an authority, i.e., a body that has legal power and rights locally, nationally, regionally, or internationally, they represent that authority's legally enforceable expectations regarding what is acceptable within the scope of the regulation.

Codes are defined by the NFPA as standards that are an extensive compilation of provisions covering broad subject matter or that are suitable for adoption into law independently of other codes and standards [441]. They provide mandatory practices or procedures for the design, manufacture, installation, maintenance, or utilization of equipment, structures, or products when adopted by a jurisdiction. Although a model code is not a priori legally enforceable, it is typically written in a way that can be adopted into law by an authority, and once adopted become legally binding, similar to a regulation.

Standards are documents that have been established by consensus and approved by a recognized body, commonly termed a standard developing organisation (SDO) [442]. Standards support unification of rules, as well as improve quality assurance and safety of products or services. The development of standards is based upon normative research and experience from academia and industry. Compliance to a standard is voluntary unless explicitly referenced in a regulation or adopted code.

Because of their specific and detailed nature, standards are a very useful tool for engineers and system designers. They are a convenient way of ensuring that the equipment needed to fulfill a certain function is appropriate and has been properly designed and fabricated. Once published, standards are regularly reviewed, revised, and updated in a procedure controlled by the administering SDO. Depending on their area of influence, SDOs may be classed as international, regional, or national. The International Organization for Standardization (ISO) and the International Electrotechnical Commission (IEC) are the most well-known international SDOs. Other SDOs may similarly be classed as international SDOs, such as the Society of Automotive Engineers (SAE), International Telecommunications Union (ITU), and the Institute of Electrical and Electronics Engineers (IEEE), as they also develop standards which are adopted internationally. Regional SDOs have a more localised jurisdiction and acceptance, and include, among others, the European Committee for Standardization (CEN), the European Committee for Electrotechnical Standardization (CENELEC), European Telecommunications Standards Institute (ETSI) and the Pan American Standards Commission (COPANT). National SDOs exist in many countries and they are responsible for standardisation in their own country and often represent the interest of that country and its stakeholders in regional and international standards forums. Examples of national standardisation bodies are the British Standards Institution (BSI) in the UK, the Japanese Industrial Standards Committee (JISC) in Japan, Deutsches Institut für Normung (DIN) in

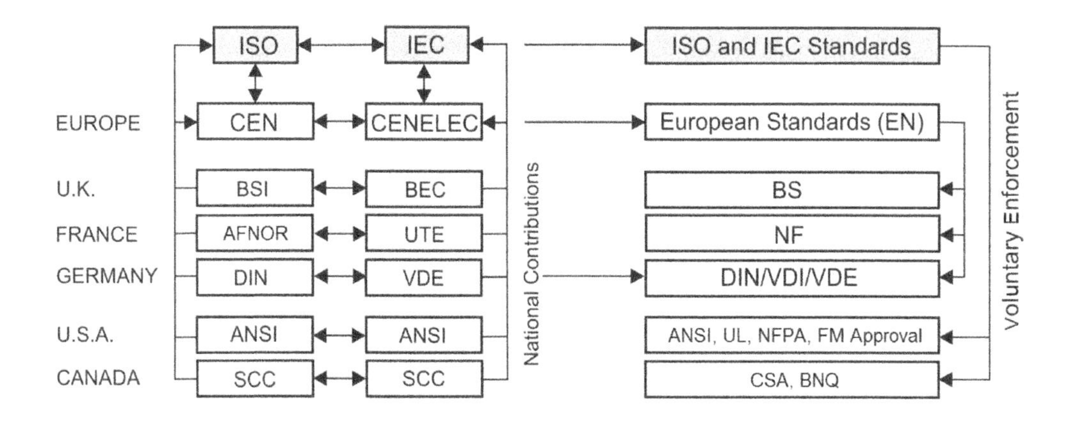

Figure 6.1: Overview on standards bodies.

Germany, Standardization Administration of China (SAC) in China and American National Standards Institute (ANSI) in the U.S. Underwriters Laboratories (UL) and CSA Group may also be considered national SDOs for North America. The distinction between national, regional, and international SDOs is somewhat arbitrary, since standards developed for one area may be recognized or accepted by an authority having jurisdiction (AHJ) in other areas. An overview on standards bodies is given in Figure 6.1.

Standards relevant to sensors for hydrogen technologies exist. Often it is necessary to consider that these sensors will be employed in classified (hazardous) locations, including potentially explosive areas. Similar to other electrical devices, sensors for use in classified locations must be specially designed such that they cannot initiate a fire or explosion. Standards for the safe use of electrical devices (including sensors) in classified environments have been developed. Furthermore codes or regulations require that sensors are certified to these standards when used in classified environments. The distinctions between various types of classified environments are discussed below. In addition, sensor performance requirements and relevant testing procedures are stipulated in appropriate "performance" standards. Standards concerning hydrogen sensors may be classified into the following three categories:

- Performance standards
- Shock and electrical safety standards
- Interface, functional safety, and controls standards

In the following sections a description of the European Union (EU) and North America (particularly the United States - U.S.) RCS relevant to hydrogen sensors is provided. Standards applicable to other gas sensors are also briefly covered.

6.1.2 European Regulations, Codes, and Standards

The rules with regard to the cooperation between European standardisation organisations, national standardisation bodies, member states, and the European Commission are established in the regulation (EU) 1025/2012 [443]. The objectives of this regulation are to ensure the effectiveness and efficiency of standards and standardisation for products and for services in support of European Union legislation and policies.

According to article 114 of the Treaty on the Functioning of the European Union (TFEU), the European Institutions are bound to make legislative provisions to protect health, safety, environment, and consumers. Directives, which are legal acts of the European Union, have been developed for this purpose and some are described below. All EU member states are required to transpose (implement) a directive into national law by a fixed deadline, but are allowed to do so by different means. Typically member states transpose the obligation into their national regulations. Once transposed the directive becomes enshrined as a national regulation at the member state level.

ATEX Directive

ATEX (from the French term **AT**mosphères **EX**plosibles) refers to two European directives covering equipment and protective systems for use in potentially explosive atmospheres, specifically in a working environment. These directives are particularly pertinent to applications using flammable gases, including hydrogen. The directives aim to control or eliminate potential harm to human life and property as a result of explosions. Mitigation of the risk of explosions can be achieved through the use of appropriate equipment to prevent release of explosive substances (gases, vapours, or dusts) and/or by eliminating sources of ignition from the environment.

ATEX directive 94/9/EC[1] also called ATEX 95, ATEX 100a or the equipment directive, concerns (mechanical and electrical) equipment and protective systems intended for use in potentially explosive atmospheres and is valid when this equipment can have its own source of ignition [36]. It stipulates the technical and safety requirements that are to be met and the relevant conformity assessment procedures before such equipment can be placed on the European market. As such, it is relevant to manufacturer of devices, including sensors. The directive has been revised in 2014 and the requirements of the new directive ATEX 2014/34/EU will be mandatory for manufacturers wanting to sell their product in the European community beginning on 20 April 2016.

ATEX directive 99/92/EC also called ATEX 137, ATEX 118a or the workplace directive, specifies the minimum requirements for ensuring the safety and health protection of workers who are potentially at risk from explosive atmospheres[2] in the workplace [37]. Compliance to the requirements established in this directive is the responsibility of the employers and not the manufacturers of equipment.

The ATEX directives are the most relevant directives applicable to hydrogen applications and equipment including sensors. However, depending on the environment of the application, other EU directives may also be relevant to sensors used in hydrogen technologies. These directives include:

- 2004/22/EC on measuring instruments
- 97/23/EC on pressure equipment
- 2004/108/EC on electromagnetic compatibility
- 2006/95/EC on low voltage equipment

[1] The ATEX number refers to the paragraph number in the relevant EU treaty.

[2] Many regulatory agencies define the lower explosion limit (LEL) as equivalent to the lower flammable limit (LFL).

Table 6.1: Classifications of hazardous zones or divisions[3]

Europe and IEC	Definition of zone or division	North America
Zone 0 (gases/ vapours)	A place in which an explosive atmosphere, consisting of a mixture with air and of flammable substances in the form of gas, vapour, or mist, is present continuously or for long periods or frequently (> 1000 h/year[4]).	Class I Division 1 (gases)
Zone 1 (gases/ vapours)	A place in which an explosive atmosphere, consisting of a mixture with air and of flammable substances in the form of gas, vapour, or mist, is likely to occur in normal operation occasionally (10 to 1000 h/year).	Class I Division 1 (gases).
Zone 2 (gases/ vapours)	A place in which an explosive atmosphere, consisting of a mixture with air of flammable substances in the form of gas, vapour or mist is not likely to occur in normal operation, but if it does occur, will persist only for a short period (0.1 to 10 h/year).	Class I Division 2 (gases)
Zone 20 (dusts)	A place in which an explosive atmosphere in the form of a cloud of combustible dust in air is present continuously, or for long periods or frequently (> 1000 h/year).	Class II Division 1 (dusts)
Zone 21 (dusts)	A place in which an explosive atmosphere in the form of a cloud of combustible dust in air is likely to occur in normal operation occasionally (10 to 1000 h/year).	Class II Division 1 (dusts)
Zone 22 (dusts)	A place in which an explosive atmosphere in the form of a cloud of combustible dust in air is not likely to occur but if it does occur, will persist for only a short period (0.1 to 10 h/year).	Class II Division 2 (dusts)

According to ATEX 137, IEC/EN 60079-10-1 and the U.S. equivalent under NFPA 70, the National Electric Code (NEC) [444], it is mandatory to classify hazardous areas. Hazardous areas are classified in zones or divisions according to the frequency and the duration an explosive atmosphere, due to gases, vapours, or dusts, can exist. The classifications described in these documents are summarised in Table 6.1.

According to ATEX 95, equipment (devices, apparatus, machines, etc.) for use in potentially explosive areas are classified into two groups:

Equipment Group I Equipment for use in the mining industry

Equipment Group II Equipment for other hazardous places endangered by explosive atmospheres.

[3]The time span for presence of explosive atmospheres are given according to NEC.

Table 6.2: Gas classification groups and representative examples

European and IEC Classification	North American Classification
IIA (propane)	D (propane)
IIB (ethylene)	C (ethylene)
IIC (acetylene/hydrogen)	A (acetylene)
	B (hydrogen)

Equipment in Group II is subdivided into Categories 1, 2, and 3, correlating respectively to very high, high, or normal safety requirements, depending on where the equipment is intended to be used and whether a potentially explosive atmosphere may be present and for how long (i.e., Zone 0, 1, or 2, see Table 6.1). Equipment certified for use in one area may be used in other less hazardous areas. As an example, equipment compliant with requirements set for Equipment Group II, Category 1 may be used in all Zones as defined in ATEX 137.

The safety requirements for electrical equipment, including sensors, are guided by the nature of the hazardous substances expected in the application's atmosphere. Gases and vapours are classified according to their increasing ignitability into three groups, IIA, IIB, and IIC (see Table 6.2), and into temperature classes from T1 to T6 in order of decreasing auto-ignition temperature. Hydrogen sensors are classified into the Equipment Group IIC and T1 because of hydrogen's high ignitability and its ignition temperature above 450 °C. Data on chemical and engineering properties of about 300 flammable gases and vapours are provided in IEC/EN 60079-20-1 to assist in selecting equipment to be used in hazardous areas [445].

The requirements of the technical Annex II in the ATEX directives are adopted into the international standards of ISO and IEC. Similarly the European Standards Organisation, CENELEC, has elaborated, harmonised European standards that provide solutions for compliance with the legal provisions stipulated in the ATEX directives (including the IEC/EN 60079 series of standards). Compliance with these harmonised standards provides a presumption of conformity with the corresponding requirements of the ATEX directives. General requirements are defined in IEC/EN 60079-7 [446]. The classification for the type of explosive atmosphere, e.g., Zone 0 or Zone 1, is given in the standard IEC/EN 60079-10-1 [447].

Sensor Performance Standards

IEC/EN 60079-29-1 defines the general requirements for construction, testing, and performance of sensors for the detection and measurement of flammable gas or vapour concentrations in air. This standard includes the test methods that apply to portable, transportable, and fixed apparatus. EN 50194 defines the requirements for sensors specifically for use in domestic premises [448].

ISO 26142 [79] pertains specifically to hydrogen sensors. This standard provides performance requirements and test methods for stationary hydrogen detection apparatus. The standard is primarily intended for hydrogen sensors at vehicle refuelling stations, but may also be applied to other stationary installations where the detection of hydrogen is required. This standard covers situations where the user desires the ability to detect hydrogen leaks and monitor hydrogen concentrations relevant to safety. As with other standards (e.g., IEC/EN 60079-29-1 or ISO 15196), this standard uses the terms "lower and upper flammability limit" (LFL, UFL), which can be considered as a more appropriate description of the combustion behavior of hydrogen air mixtures [449].

Requirements on sensors for toxic gases are covered in separate standards. EN 45544, parts 1 to 3 [450], establish the general requirements and test methods for electrical apparatus (sensors) used for the detection of toxic gases and vapours in workplace atmospheres. Standards for specifications and test procedures for carbon monoxide sensors and nitric oxide sensors are defined in EN 50291 [451] and EN 50545 [452], respectively. The performance requirements and test methods for oxygen sensors are specified in EN 50104 [453].

The abovementioned standards have similar structures, and usually contain the following sections:

Scope describes the area or subject to which the standard is relevant

Definition of terms clarification on the meaning of certain terms or words which are used in the standard

General requirements a list of conditions which shall be attained by a product covered by the standard with respect to, for example, construction, labelling and marking, user manual, alarm levels, etc.

Performance requirements level of performance which a product shall meet when tested according to the procedures described in the standard, e.g., influence of environmental parameters, mechanical robustness, measurement accuracy, etc.

Test procedures detailed description of test methods to be used to evaluate compliance of a product with the performance requirements, e.g., description of standard climatic conditions, test gases to be used, description of equipment, etc.

Appendices additional information which may be of a mandatory or informative nature, e.g., detailed description of test equipment

Electrical Safety Standards

When used in hazardous areas it is important that electrical equipment, including sensors, cannot initiate a fire or explosion. Requirements to ensure this are specified in IEC/EN 60079. Part 1 of this standard specifies that the sensor, e.g., of a pellistor-type, shall include a flame proof enclosure (code Ex "d") when used in Zone 1 [454]. Part 7 is applicable for an increased safety design (code Ex "e") when the sensor is used in Zone 1 or 2 [455]. Requirements on sensors for use in Zone 1 or 2 pressurized enclosures (code Ex "p") are prescribed in IEC/EN 60079-2 [456]. IEC/EN 60079-11 specifies the requirements and testing of equipment, including, e.g., electrochemical sensors that are certified as intrinsically safe (code Ex "i"). The intrinsically safe requirement implies that no arc or spark can occur in the sensor that would have sufficient energy to ignite a flammable gas [457]. IEC/EN 60079-25 prescribes the specific requirements on intrinsic safe systems for Zone 0, Zone 1, or Zone 2 (code "ia," "ib," "ic") [458]. The enclosures in oil (Ex "o") and powder (Ex "p") are uncommon for sensors. However equipment protection by encapsulation Ex "m" is a common method to protect critical electronic components [459]. Instead of an Ex "d" detector which is not allowed to be opened in the classified area in Europe, it is becoming more common to combine different protection methods. A common solution would be to use a "demienclosure," which can be openend in an Ex "d" area to replace a sensor without disconnecting it from power. Table 6.3 summarises the usual types of explosion protection for electrical equipment.

Other standards are pertinent to sensors and safety in hydrogen applications:

- The requirements for the protection provided by electrical equipment enclosures against intrusion of solid foreign objects (including body parts such as hands and fingers), dust, and water are included in IEC/EN 60529. It comprises an ingress protection marking, the so-called IP codes [460].
- Safety requirements for electrical equipment for measurement, control, and laboratory use are fixed in IEC/EN 61010-1. The standard is harmonised in Europe under the low voltage directive (LVD) and therefore is mandatory for many gas detection equipment, e.g., for all mains supplied apparatus [461].
- The requirements and tests for electromagnetic compatibility that ensure that the functionality of a device is not disturbed by stray electrical or electromagnetic interferences is standardised in 3 levels:
  - Level 1 are the generic standards of the IEC/EN 61000 series with a large number of standards.
  - Level 2 are the product family standards. For all measuring equipment this is IEC/EN 61326-1 [462].
  - Level 3 are the specific product standards. In Europe the harmonised product standard for gas detection is EN 50270 [463].
- Vibration test protocols for evaluation of the mechanical stability of devices are defined in IEC/EN 60068-2-6 [464].

Table 6.3: Types of explosion protection of electrical apparatus

Code	Type of protection	Diagram	Description
d	flame proof enclosures (Zone 1)		The electrical apparatus or parts of it which can ignite an explosive atmosphere are placed in an enclosure, which can withstand the pressure developed by an internal explosion and which prevents the transmission of the explosion to the explosive atmosphere surrounding the enclosure.
e	increased safety design (Zone 1 or 2)		The occurrence of arcs, sparks, or excessive temperatures which can serve as ignition sources is prevented by additional measures of protection.
p	pressurized enclosure (Zone 1 or 2)	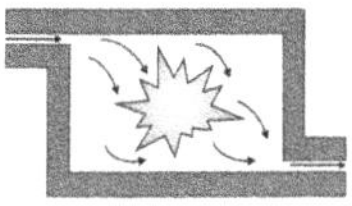	Type of protection in which the enclosure is filled with a protective gas. An overpressure is maintained in such a way that an explosive atmosphere cannot penetrate the enclosure either with or without continuous flow of the protective gas.
i	intrinsic safety (Zone 0, 1 or 2)	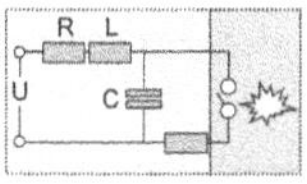 R L U C	The power supply of the electrical apparatus is protected by a barrier for limitation of current and voltage in such a way that a minimum ignition energy or ignition temperature of an explosive atmosphere is not exceeded. This type of protection is further divided in Ex ia for Zone 0 or 1 and Ex ib for Zone 1 or 2.
o	protection by oil immersion (Zone 1 or 2)	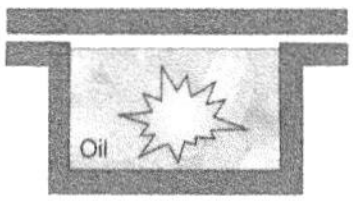 Oil	The electrical apparatus or parts of it are immersed in a protective liquid (oil) in such a way that an explosive atmosphere which may be above or outside of the enclosure cannot be ignited.
q	protection by powder filling (Zone 1 or 2)	 Sand	The enclosure is filled with fine-grained sand. A possible arc occurring within the enclosure of the electrical apparatus is cooled down so that it will not ignite the surrounding atmosphere.
m	encapsulation (Zone 0 or 1)	 Resin	The parts of an electrical apparatus which can ignite an explosive atmosphere are enclosed in a resin sufficiently resistant to environmental influences in such a way that an explosive atmosphere cannot be ignited either by sparking or heating which may occur in the encapsulation.
n	non-incendive, non-sparking in Zone 2		Under normal operation there is no danger of ignition.

Functional Safety Standards

In industrial applications, sensors are often an essential part of a Safety Instrumented System (SIS). A SIS is implemented to prevent or mitigate hazardous events and to ensure functional safety. In this context, functional safety levels ranging from Safety Integrity Levels SIL 1 (lowest) to SIL 3 (highest) are defined as a measure of the effectiveness of the performance of safety systems in terms of probability of failure on demand (PFD). The concept of functional safety and the general requirements for all electrical/electronic/programmable electronic safety-related systems are defined in IEC/EN 61508 [465] and reviewed in [466]. The application of this concept in process industries is stipulated in IEC/EN 61511 [467].

There are two main functional safety standards relevant to devices intended for gas detection and measurement. EN 50271 [468] describes the basic requirements on software and hardware to fulfil the demands for SIL 1. The requirements of this standard include documentation during the process of development of a new sensor as well as minimum requirements on fault recognition via hardware (e.g., watchdog timer) and software (e.g., testing of data storage units such as RAM and ROM). The second standard, EN 50402, provides requirements on the functional safety of fixed gas detection systems for measurements of combustible or toxic gases or vapours or of oxygen [469], and includes specific detailed requirements for each SIL level.

The requirements for gas sensors to reach the different SIL levels can be described as:

- SIL 1: An appropriate design level and ATEX performance certification
- SIL 2: Self-test for hardware, specific software requirements, and documentation, reduced maintenance intervals
- SIL 3: Significantly increased demands on validation, redundancy of hardware, and SIL certified software

The functional safety of hydrogen sensors should meet SIL 2 for hardware and SIL 3 for software. This requires a low probability of failure on demand between 10^{-4} and 10^{-3}, which is equivalent to less than one safety sensor failure in 10^2 or 10^3 years, respectively.

To reach SIL 3 for a sensor system it is necessary that both the software and the central processing unit comply to SIL 3. To increase system reliability duplicate processor systems are used. If one processor fails, the system switches to emergency mode and remains fully operational. Furthermore two redundant sensing elements and redundant relays are included.

Guidance on the selection, installation, use, and maintenance of sensors for the detection of flammable gases is provided in IEC/EN 60079-29-2 [470] and in EN 50244 [471]. Guidance on the functional safety of fixed gas detection systems is established by IEC/EN 60079-29-3 [472]. However these guidelines apply to flammable gases in general and do not take into account the specific properties and behaviour of hydrogen. ISO has produced a technical report, ISO/TR 15916:2004 titled “Basic considerations for the safety of hydrogen systems” [473]. In this book, guidiance on the selection and installation of sensors is briefly discussed in Section 6.4.

A compilation of the regulations, guidelines, and standards relevant to gas sensors in hydrogen technologies is presented in Table 6.4.

Table 6.4: European and IEC standards related to gas sensors

	EU Directives
ATEX 2014/34/EU	Directive 2014/34/EU of the European Parliament and of the Council relating to equipment and protective systems intended for use in potentially explosive atmospheres (valid from 20 April 2016)
ATEX 99/92/EC	Directive relating on minimum requirements for improving the safety and health protection of workers potentially at risk from explosive atmospheres
IEC/EN 60079-0	Explosive atmospheres - Part 0: Equipment - General requirements
IEC/EN 60079-10-1	Explosive atmospheres - Part 10-1: Classification of areas - Explosive gas atmospheres
	Sensor performance standards
IEC/EN 60079-29-1	Explosive atmospheres - Part 29-1: Gas detectors - Performance requirements of detectors for flammable gases.
EN 50194	Electrical apparatus for the detection of combustible gases in domestic premises - Part 1: Test methods and performance requirements
ISO 26142	Hydrogen detection apparatus - Stationary applications
EN 45544-2, 1-3	Workplace atmospheres - Electrical apparatus used for the direct detection and direct concentration measurement of toxic gases and vapours
EN 50291	Electrical apparatus for the detection of carbon monoxide in domestic premises - Part 1: Test methods and performance requirements
EN 50545	Electrical apparatus for the detection and measurement of toxic and combustible gases in car parks and tunnels - Part 1: General performance requirements and test methods for the detection and measurement of carbon monoxide and nitrogen oxides
EN 50104	Electrical apparatus for the detection and measurement of oxygen - Performance requirements and test methods

Continued on next page

Table 6.4 – *Continued from previous page*

	Equipment protection standards
IEC/EN 60529	Degrees of protection provided by enclosures (IP code)
IEC/EN 60079-1	Explosive atmospheres - Part 1: Equipment protection by flame proof enclosures "d"
IEC/EN 60079-7	Explosive atmospheres - Part 7: Equipment protection by increased safety "e"
IEC/EN 60079-2	Explosive Atmospheres - Part 2: Equipment protection by pressurized enclosure "p"
IEC/EN 60079-11	Explosive atmospheres - Part 11: Equipment protection by intrinsic safety "i"
IEC/EN 60079-11	Explosive atmospheres - Part 18: Equipment protection by encapsulation "m"
IEC/EN 60079-25	Explosive atmospheres - Part 25: Intrinsically safe electrical systems
IEC/EN 61000-4	Electromagnetic compatibility (EMC) - Testing and measurement techniques - Electrostatic discharge immunity test
EN 50270	Electromagnetic compatibility - Electrical apparatus for the detection and measurement of combustible gases, toxic gases, or oxygen
IEC/EN 60068-2-6	Environmental testing - Part 2-6: Tests - Test Fc: Vibration (sinusoidal)
SAE J2011	Handbook for Robustness Validation of Automotive Electrical/Electronic Modules
	Functional safety standards
IEC/EN 61508	Functional safety of electrical / electronic / programmable electronic safety-related systems - Part 1: General requirements
EN 50271	Electrical apparatus for the detection and measurement of combustible gases, toxic gases, or oxygen - Requirements and tests for apparatus using software and/or digital technologies
EN 50402	Electrical apparatus for the detection and measurement of combustible or toxic gases or vapours or of oxygen - Requirements on the functional safety of fixed gas detection

Continued on next page

Table 6.4 – *Continued from previous page*

	Standards for selection, installation and use
IEC/EN 60079-29-2	Explosive atmospheres - Part 29-2: Gas detectors - Selection, installation, use, and maintenance of detectors for flammable gases and oxygen
EN 50244	Electrical apparatus for the detection of combustible gases in domestic premises - Guide on the selection, installation, use, and maintenance
EN 45544-2	Part 4, Workplace atmospheres - Electrical apparatus used for the direct detection and direct concentration measurement of toxic gases and vapours
IEC/EN 60079-29-3	Explosive atmospheres - Part 29-3: Gas detectors - Guidance on functional safety of fixed gas detection systems.

6.1.3 U.S. and North American Codes and Standards

Overview

The following discussion provides an overview of the standards and codes relevant to hydrogen sensor use and performance in the United States and North America. This subject was recently reviewed [474].

In the U.S., code requirements are often more prescriptive with explicit requirements, as opposed to performance based requirements that are more prevalent in the EU. In North America and the U.S. in particular, the main documents governing mandatory safety requirements in the emerging hydrogen infrastructure are covered by the International Fire Code (IFC) 2009 Edition [475] or the 2012 Edition [38] and the National Fire Protection Association 2 (NFPA 2) Hydrogen Technologies Code [476]. The IFC and NFPA 2 have both mandated the use of hydrogen safety sensors in select hydrogen operations. The IFC has explicit requirements for hydrogen sensors and flammable detection in hydrogen vehicle repair facilities. Similarly, NFPA-2 explicitly mandates the use of sensors for various hydrogen operations, including dispensing. Thus, the use of sensors will be mandated by enforceable code when the IFC or NFPA-2 is adopted by a local jurisdiction. With regard to code enforcement in the U.S., jurisdictions are local, typically being at the city or county level, but occasionally the authority having jurisdiction (AHJ) may have state-wide authority. Adoption can be either directly or by reference. For example, the 2010 California Fire Code references the 2009 IFC, thus the sensor requirements within the IFC have become codified (e.g., legally binding) in the state of California. The IFC and NFPA are endeavouring to simplify and harmonize their respective documents, in part by eliminating duplicate requirements. Furthermore the next edition of the IFC (IFC 2014) will explicitly reference NFPA 2, thereby mandating compliance to the requirements of NFPA 2 in those jurisdictions that adopt IFC 2014.

The IFC and NFPA 2 are often referred to as model codes. A model code is typically written in a way that can be adopted into law, and once adopted it is legally binding. However, without formal adoption, the requirements of the IFC or NFPA cannot be legally imposed upon stakeholders. Thus, although called the International Fire Code, the IFC would not be enforceable until formal adoption by a jurisdiction. Similarly, NFPA 2 would not be codified until adopted by a jurisdiction. However, the IFC has been adopted by most jurisdictions in the U.S.

Both the IFC and NFPA 2 require the certification of hydrogen sensors ny recognized standards. Specifically section 2311.7.2.1.1 of the IFC 2012 edition explicitly states that the sensors are to be labelled and listed to UL 864 "Control Units and Accessories for Fire Alarm Systems" [477], or UL 2017 "General-Purpose Signaling Devices and Systems" [478], and UL 2075 "Gas and Vapor Detectors and Sensors" [479]. Thus, the use of sensors certified to specific standards is legally enforceable in those jurisdictions that formally adopt the IFC. This is not, however, an absolute requirement, since the AHJ has the option to waive a requirement prescribed in the IFC; this waiving of a requirement can be based on using sensors that were certified to another performance standard deemed acceptable by the AHJ, sensors that were used in other industrial hydrogen operations, or that were otherwise shown to satisfy the safety concerns of the facility managers and AHJs. In other words, the AHJ can approve the use of alternative means to meet the safety requirements. NFPA 2 also requires that gas detection equipment shall be listed (i.e., a product which has been certified to a relevant standard), although specific standards to which it should be listed are not identified in NFPA 2 as they are in the IFC.

Sensors for use in hazardous environments may require certifications such as Class 1, Division 2 certification as prescribed by NFPA 70, the National Electric Codes (see Table 6.1). For example devices certified to ANSI/ISA 12.12.01 (Nonincendive Electrical Equipment for Use in Class I and II, Division 2 and Class III, Divisions 1 and 2 Hazardous (Classified) Locations) or other electrical safety standards accepted by the AHJ, may be used to show compliance to hazardous location operation requirements. In addition to electrical safety standards, there are performance-based standards for gas sensors (e.g., UL 2075, CSA C22.2, No. 152 "Performance of Combustible Gas Detection Instruments," and FM Approval Standard 6310/6320 "Approval Standard for Combustible Gas Detectors"). These standards are not specific for hydrogen sensors, but since hydrogen is a combustible gas, these standards would apply.

The distinction between a code and a standard was described in Section 6.1. In the United States, the American National Standards Institute (ANSI) is the national coordinating body. ANSI coordinates the scope of U.S. standards development so that there are no conflicting requirements for the same product design. Through accreditation of SDOs within the U.S., ANSI ensures that the SDO and their processes meet requirements for transparency, balance, consensus, and due process. ANSI also strives to harmonize domestic and international standards requirements to ensure import and export of safe and quality products into and out of the U.S. Product certification is the process by which the product design undergoes specific test procedures to ensure that the design meets all of the requirements and qualification criteria detailed in a standard. Certification is a formal conformity declaration and is performed by a Nationally Recognized Testing Laboratory (NRTL). The term "certification" is used with respect to a specific standard. "Certification" and "listed" can be and are often used interchangeably, but there are subtle differences. "Product listing" refers to the list published by a NRTL of products certified to a specific standard. Since all listed product designs will be certified and the NRTL will list all products it has certified, the distinction is minor. It is noted that ANSI does not formally endorse the use of the term "listed," but prefers the term "certified" to indicate that a product design has been tested and shown by a NRTL to meet the requirements of a standard [480]. ANSI reserves the term "approved" to refer to a standard that it recognizes and that "approved" should not be used as synonymous with "certified." This concurs with the nomenclature used by many NRTLs (e.g., UL, CSA) that explicitly state that the term "approved" should never be used as synonymous with "certified" when referring to product designs the NRTL has certified. However, the distinction between the use of the terms "certified" and "approved" is not universal. For historic reasons, FM Approvals still use the term "approved" to indicate a product design certification to FM Approvals standards.

The main standards relevant to hydrogen sensors used in North America (U.S. and Canada) are described below. More information on the various standards can be obtained from the specific standards development organization (SDO). Many SDOs develop "Guide Cards" that provide a brief overview of the standard. The Guide Cards (or analogous system for other SDOs) provide stakeholders with an invaluable tool. The Guide Cards assist sensor developers in identifying appropriate standards for their products. For end-users, AHJs, and other regulators, the Guide Cards provide a scope of the technical requirements in the standard.

Sensor Performance Standards

Numerous SDOs have developed standards that relate to the performance requirements of combustible gas sensors, which include hydrogen sensors. The following standards are used within the U.S. to assure performance of the end-product(s):

- UL 2075 "Gas and Vapor Detectors and Sensors" covers specific construction requirements of the end-product that deal with both fire and shock, as well as for reliability of the end-product(s). The testing per UL 2075 is to verify compliance with the construction of the end-product, but also generates test data to support the end-product performance when tested against the gas(es) that the device(s) are intended to detect. This standard has detailed construction requirements such as electrical spacings, maximum temperature rise of components, displays and enclosure material. UL 2075 is intended for end-product certification for the U.S.
- CSA C22.2 No. 152-M1984 "Combustible Gas Detection Instruments" is associated with certification of end-product primarily for use in Canada. The requirements are similar to those of IEC/EN 60079-29-1; however, it requires compliance to additional standards that address construction including CSA C22.2 No. 142, which has general construction details for meters and indications, alarm functions, trouble signals, batteries, control, and adjustment.
- FM 6310/6320 "Approval Standard for Combustible Gas Detectors". As with CSA C22.2, requirements are similar to those of IEC/EN 60079. This standard also only has general construction details, however it does require compliance to FM3810 which addresses additional construction requirements.

There are also international standards on sensor performance, such as ISO 26142 or European standards, including those discussed in Section 6.1.2. While these standards are not extensively applied or accepted in U.S. applications, it is possible that sensors certified to a European or international standard could be accepted by an AHJ.

Shock and Electrical Safety Standards

Numerous SDOs have developed standards that relate to the safe operation of components in classified environments, including combustible gas sensors, which include hydrogen sensors. These standards are developed to assure compliance to NFPA 70 guidelines. The following standards are used or regularly accepted within U.S. jurisdictions for the verification and classification of electrical safety requirements:

- UL 61010-1 "Safety Requirements for Electrical Equipment for Measurement, Control, and Laboratory Use - Part 1: General Requirements" is essentially identical to IEC 61010-1, which specifies general safety requirements for electrical equipment intended for professional, industrial process, and educational use, any of which may incorporate computing devices.

- UL 60079 "Explosive Gas Atmospheres" was adapted from IEC 60079, which specifies requirements for construction, testing, and marking of Group II electrical apparatus with type of protection "n" intended for use in Class I, Zone 2 hazardous (classified) locations as defined by the National Electrical Code (ANSI/NFPA 70).

- CSA C22.2 No. 152-M1984 "Combustible Gas Detection Instruments." Products evaluated to this standard are applied in areas that are or could be a hazardous location due to the presence of an explosive gas atmosphere (which is defined as a mixture with air, under atmospheric conditions, of flammable substances in the form of gas, vapour, or mist in which, after ignition, combustion spreads throughout the unconsumed mixture). This standard covers portable gas detection equipment used by maintenance personnel, first responders, utilities, or any other instances where personnel need to verify the presence or absence of combustible gases. This standard also covers field-installed gas detectors used to monitor an environment, sound an alarm, shut down equipment, or activate ventilation systems designed to remove or dilute the combustible gas atmosphere.

- FM 3600 "Approval Standard for Electrical Equipment for Use in Hazardous Locations" identifies the basis for certification of electrical equipment for use in hazardous (classified) locations. The requirements in the standard are based on consideration of ignition in locations made hazardous by the presence of flammable or combustible materials under normal atmospheric conditions within defined ranges (e.g., -25 °C to +40 °C, oxygen not greater than 21 vol%, and barometric pressure in the range of 0.8 bar to 1.1 bar).

- FM 3615 "Approval Standard for Explosion Proof Electrical Equipment General Requirements" contains the basic requirements for the construction and testing of explosion proof electrical equipment. This standard is expected to be used in conjunction with FM 3600.

- FM 3810 "Approval Standard for Electrical Equipment for Measurement, Control, and Laboratory Use" lists certification requirements for electrical equipment for measurement, control, and laboratory use. Approval requirements include, but are not limited to, performance requirements, marking requirements, examination of manufacturing facilities, audits of quality assurance procedures, and follow-up programs.

Interface Standards

Interface standards provide specifications and requirements associated with the sensor alarm systems and other systems requiring activation (e.g., forced ventilation systems). Several standards have been developed, some of which are explicitly mentioned in the IFC, including:

- UL 864 "Control Units and Accessories for Fire Alarm Systems" is referenced in the IFC 2009 and 2012 editions; thus, for those jurisdictions that adopted the IFC, sensors certified to UL 864 (or UL 2017) or approved by the AHJ must be used. These requirements cover discrete electrical control units and accessories for fire alarm systems to be installed in accordance with the NFPA 70 model code titled "National Electrical Code" and in accordance with the NFPA 72 model code, titled "National Fire Alarm Code." The products covered by this standard are intended to be used in combination with other appliances and devices to form a commercial fire alarm system. These products provide all monitoring, control, and indicating functions of the system.

- UL 2017 "General-Purpose Signaling Devices and Systems" is referenced in the IFC 2009 and 2012 editions; thus, sensors certified to UL 2017 (or UL 864) or approved by the AHJ must be used in those jurisdictions that adopted the IFC. These requirements cover signaling devices intended for emergency or non-emergency use for indoor and/or outdoor locations and, where applicable, installed in accordance with the National Electrical Code (NFPA 70). Emergency-signaling products covered by these requirements are associated with property and/or life safety and are of a non-fire/non-security alarm nature. These products are evaluated with regard to product safety and to appropriateness of signaling. Non-emergency-signaling products covered by these requirements are not associated with property and/or life safety and are only evaluated relative to product safety.
- FM 3010 "Approval Standard for Fire Alarm Signaling Systems" applies to permanently installed fire alarm signaling systems intended for indoor use. The equipment covered by this standard connects to other peripheral equipment to provide indicating and control functions associated with a commercial fire alarm system in accordance with NFPA 72. Examined products may be complete systems of products intended to provide a feature (e.g., auxiliary power supplies or digital alarm communicator transmitters) that would then be used with an approved fire alarm signaling system. This standard provides the operating basis for alarm signaling and does not cover initiating and notification devices as well as the use in damp, wet, or hazardous (classified) locations, where other standards are applied. The equipment covered by this standard may be combined with additional standards for use in other areas.

6.1.4 Specialised Industry and Other Standards

Government agencies and specialized industries often have their own internal standards. For example, the Society for Automotive Engineers (SAE) generates standards for the automotive industry. Although a hydrogen gas sensor standard from this organisation does not exist, one critical document for any electronic module is the SAE Handbook for Robustness Validation of Automotive Electrical/Electronic Modules. This handbook defines requirements and test protocols for semiconductor devices including sensors [481]. However, under development is a technical information report (TIR) SAE 3089 "Characterization of On-board Vehicular Hydrogen Sensors" to provide guidance on the evaluation of on-board hydrogen sensors.

In the aerospace industry, the Federal Aviation Authority (U.S.) and the European Aviation Safety Agency oversee the safety requirements. The underlying standard prescribing safety requirements for all hardware used in aircraft, including sensors, is DO-254 (Design Assurance Guidance for Airbourne Electronic Hardware), developed by Radio Technical Commission for Aeronautics (RTCA) [482]. In some aeronautic hydrogen applications, a hydrogen sensor, if used, must be compliant to the requirements specified in DO-254.

The American Institute of Aeronautics and Astronautics (AIAA) published a Guide to Safety of Hydrogen and Hydrogen Systems (G-095-2004e) [31]. The Guide provides information that designers, builders, and users of hydrogen systems can use to avoid or resolve hydrogen hazards. It is presented for system design, materials selection, operations, storage, and transportation. Space agencies, especially the U.S. NASA have rigorous requirements on the safe use of hydrogen. One critical document was NASA NSS 1740.16 Safety Standard for hydrogen and hydrogen systems, which was cancelled in 2005 and is superseded by AIAA Guide 095-2004.

MIL-STD-810 on Environmental Engineering Considerations and Laboratory Tests is a United States Military Standard. This guide prescribes the design and testing of equipment to the limits that it will be subjected to in deployment. Evaluation criteria include temperature extremes, pressures extremes (associated with terrestrial to high altitude transitions), shock and vibrations, as well as other environmental stresses [483].

Other countries including Russia [484], Japan [485, 486] and China [487] have also developed their standards concerning requirements for the selection, installation, use, and maintenance of detectors for flammable gases and oxygen, and on stationary or portable sensors for combustible gases. These standards may not be harmonized with many of the international standards described above. A comprehensive survey of all national standards is, however, beyond the scope of this book.

6.1.5 Certification

The term "certification" describes a procedure by which a third party gives written assurance that a product, process, or service conforms to specified requirements. The conformity of a sensor with the required specifications fixed in standards or guidelines can be verified by certification. To ensure independent assessment of sensor performance and therefore to ensure greater confidence on the part of the customer, this procedure is performed by competent bodies maintaining a quality management system and being themselves accredited to carry out these activities. NRTLs in North America include Underwriters Laboratories (UL) and the Canadian Standards Association (CSA). Notified bodies for ATEX in the EU include BAM Federal Institute for Materials Research and Testing, TÜV Technischer Überwachungsvereinn, SIRA certification service and Baseefa. Conformity certification consists of the following steps:

1. Test of technical and quality documents, audit at the manufacturer, experimental test of the quality of the method in an independent laboratory
2. Assessment of the results and
3. Awarding of a formal document of certification.

Certification contributes to technical harmonisation, ensuring product quality especially with regard to security and human and environmental safety. It also helps to overcome technical trade barriers in a global market and supports the acceptance of new analytical methods. Certification is important for commercial purposes, but should also be considered in the early stages of sensor research and development, because it indicates the required quality level and the state of development of the technology.

The certification of hydrogen sensors is important in connection with their safe use in potentially explosive atmospheres, where the gas concentration can reach or exceed the LEL and where the presence of ignition sources must therefore be avoided. Sensors may be capable of providing an ignition source which can cause an explosion due to the generation of static electricity and electrical sparks or due to the high operating temperatures of some sensor types. Various aspects of explosion protection must therefore be considered.

In Europe a hydrogen sensor needs to be designated according to equipment directive 94/9/EC (ATEX 95) indicating where it can be used [36]. This designation includes the

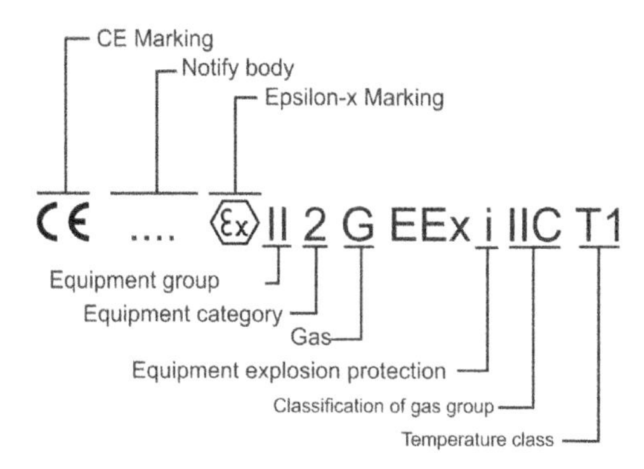

Figure 6.2: Example of marking for safety gas sensors. See Section 6.1.2.

allocation into groups of application areas (mining or non-mining), the category or protection level (very high, high, or normal) and the type of explosive atmosphere (gas or dust) in which they may operate. Gases and vapours are classified according to their ignition properties into three explosion groups (IIA, IIB, IIC) and this information is also contained in the ATEX marking. Hydrogen belongs to the IIC explosion group. In addition, the gas is assigned to a temperature class (T1 - T6) according to its autoignition temperature. Hydrogen belongs to temperature class T1 with an autoignition temperature > 450 °C. The intrinsic safety of a device can be certified, specifically for the purposes indicated by the ATEX marking, by accredited laboratories according to the safety standards series IEC/IEC 60079 series [488] in Europe or JIS M 7626 [489] in Japan and UL 913 [490] or CSA 22.2 in North America [479].

If the equipment passes the necessary tests, it is certified for conformity for a given application. A sensor for deployment in classified areas can only be placed on the market if the device has

- CE marking that attests conformity of a product with the provisions of the ATEX Directive 94/9/EC
- Epsilon-x marking (Hexagon) what is specific marking of explosion protection
- Symbols and letters which characterising the equipment group, zone or division, a type of explosion protection, gas group and temperature classification.

An example is given in Figure 6.2. In addition to this marking of conformity, a certified gas sensor shall also indicate a serial number and the year of construction.

6.2 SENSOR TESTING, CALIBRATION AND VALIDATION

6.2.1 Overview

Sensor testing is best performed under defined conditions for determining specific properties and quantifying performance specifications. These evaluations are necessary to determine the suitability of a promising sensor for an application. Sensor performance can often be assured by using sensors certified to appropriate standards, as described in Section 6.1. In this case, the sensor product design has been shown to meet the performance specifications laid out in the standard. In many cases stakeholders demand an objective and independent assessment confirming that the sensor works as expected and meets the requirements of their specific applications, which may not be covered by a performance standard. This may necessitate additional testing to qualify a sensor for the application. The following discussions provide an overview of the concepts and the principles of sensor testing and evaluation, including sensor calibration. A sensor test or evaluation refers to a procedure that provides for the qualitative or quantitative determinations of characteristics/metrics of a sensor. Calibration is a procedure that establishes a relationship between the quantity values defined in measurement standards (i.e., the concentration of a certified calibration gas mixture) and the corresponding indications of the sensor.

There are two main categories pertaining to sensor testing. The first is the determination of fundamental sensor characteristics, which is usually performed in a laboratory environment under controlled test conditions. Such testing provides data for quantifying sensor specifications (e.g., impact of environmental parameters, dynamic range, cross-sensitivity to potential chemical interferents). This may involve testings for conformity assessment and certification by an accredited test laboratory or a Nationally Recognized Testing Laboratory (NRTL) in North America. A second main category of sensor testing pertains to verification of the sensor performance after its deployment. Calibration of the sensor is the most common field verification assessment performed on a deployed sensor. Calibration pertains to periodic, often mandatory measurements of a sensor response; this usually involves exposure of a sensor to a certified test gas of known composition, and if necessary, an adjustment of the sensor readout to be kept within the manufacturer specified accuracy for the test gas. The adjustment of the sensor readout can be done via adjusting the electronic output signal of the sensor. Calibration of a deployed sensor may be performed either on-site, using field calibration kits, or the sensor may be removed from operation and transported or shipped to a remote qualified laboratory for off-site calibration.

Assurance of compliance of sensor performance metrics to meet end-user requirements can be defined by two terms—verification and validation.

Verification of a sensor is the confirmation that performance properties or legal requirements of a sensor system is fulfilled [73]. For example, sensor performance verification can include confirmation that a target measurement uncertainty can be met. This can be achieved by measurements performed by the manufacturer or by end-users through the use of certified standards e.g., certified test mixtures, as deemed appropriate by the facility authority.

Validation is a confirmation by examination and provision of objective evidence that the particular requirements for a specific end use are fulfilled [491]. It shall give the assurance that a product, service, or system meets the needs of the customer and other stakeholders. Comprehensive sensor testing is performed to meet the specific needs of the intended end-user or customer. In addition to the determination of uncertainty of measurement, further performance aspects of hydrogen sensors must be considered for their technical application. This includes robustness, selectivity, and dynamic behaviour (see Section 2.4). Validation testing is performed in accredited test laboratories shown to be compliant to quality assurance standards such as ISO EN 17025. To be compliant to the ISO EN 17025 standard on general requirements for the competence of testing and calibration laboratories, accredited test laboratories must document both sensor test protocols as well as maintenance protocols. This includes calibration protocols for all test and measurement equipment [492].

6.2.2 Accreditation of Testing and Calibration Laboratories

Qualification tests can be performed in accredited testing and calibration laboratories. The competence of a laboratory for this function has to be demonstrated and documented. Accredited testing laboratories and calibration laboratories must show compliance to the requirements in quality management standards, e.g., in ISO IEC 17025. Accreditation is the formal recognition of the testing laboratory's competence to perform its designated service.

Accreditation of a testing and calibration laboratory testifies to the reliability and wide acceptance of the laboratory's test results. More specifically, it requires a Conformity Assessment Body (CAB) to assess the laboratory's competence to carry out specific tasks and to convey a formal demonstration of this [493].

Laboratories are ISO 17025 accredited by an impartial accreditation body. Accreditation bodies are institutions such as DAkkS,[5] A2LA,[6] UKAS,[7] These institutions joined the "International Accreditation Forum (IAF)" and the "Multilateral Recognition Arrangements (MLA)" and recognizing the equivalence of other members' accreditations to their own. A complete list of the members is available from IAF webside [494]. Accredited testing and compliance laboratories shall have an appropriate and documented quality management system in place that is in compliance with ISO/IEC 17025 and ISO 9001. Among others, the quality management system shall include verification that [492, 495]:

- The personnel is qualified and the responsibilities are defined
- All procedures are regulated, laid down in approved documents, and used in practice
- The test methods are applied in a technically correct manner in a suitable location
- Audits and inter-laboratory tests or proficiency testing are performed

The accreditation procedure includes an on-site review of documents and records to evaluate the conformity with relevant standards and other requirements. Accreditation of the testing and calibration laboratory can be granted by the accrediting body, following a favourable analysis and reporting of findings. Accreditation assures that test and calibration results from different accredited laboratories are comparable and internationally accepted by countries belonging to the MLA.

6.2.3 Traceability and Gas Standards

All test and measurement equipment that impact the accuracy and validity of the test results are to be calibrated traceable to a national standard [492, 496]. In metrology, traceability assures that measurement results can be related to a reference standard preferably to the International System of Units (SI) through an unbroken documented chain of calibrations; it is recognized that each element in that chain will contribute to the overall measurement uncertainty [73]. For example, gas standards which have typical concentration uncertainty of 2 % are commercially available, however they are traceable to primary standards provided by national metrological institutes. Traceability implies that every step of the calibration chain has been documented.

Traceability is a precondition for reliable and defensible test results. However, in most cases certificates of traceable calibrations are only accepted if they are issued by national

[5] Deutsche Akkreditierungsstelle GmbH
[6] American Association for Laboratory Accreditation
[7] United Kingdom Accreditation Service

metrological institutes (NMIs), designated institutes (DIs), or calibration or testing laboratories accredited according to ISO IEC 17025.

The metrological basis for traceable gas analysis relies on the availability of reference gas mixtures of known compositions and sufficient accuracy to serve as primary standards. A primary standard in metrology is a standard which is not subordinated to other standards. Primary standards are defined via SI units, such as mass, length, and time and are used to calibrate other standards referred to as transfer standards. The NMIs or DIs provide primary standards, but lower quality level standards are also available and they have to be traceable to the primary standard. These institutions offer a wide variety of certified gas mixtures, usually compressed in gas cylinders prepared according to procedures prescribed in standards, e.g., ISO 6142 [497]. The German Federal Institute for Materials Research and Testing (BAM) or the U.S. National Institute of Standards and Technology (NIST) offer certified reference gas mixtures consisting of minor components, such as argon, helium, hydrogen, carbon monoxide, carbon dioxide, methane as analytes in air or nitrogen. Certified gas mixtures with concentrations of components greater than 5 μmol/mol are available for a variety of applications. End-users usually acquire gas standards from secondary commercial sources, but these test gas mixtures shall be certified and traceable. In a certified gas standard, the expanded relative uncertainty (k=2) for the minor gas component in a gas mixture is typically in the range of 0.3 % to 1 % of the stated concentration. The use of certified standard gas mixtures [498], calibrated test and measurement instruments, and validated analytical reference methods (see Section 6.3) provide traceability and assures global comparability of results with other testing laboratories.

6.2.4 Testing Protocols

Sensor functionality testing is typically performance based, usually as black-box tests, without regarding the internal construction or working principles of the sensor being tested. The sensor output signal is compared to a reference; this often demands an empirical calibration curve relating electrical outputs to the desired parameter. The sensor output signal may be a direct readout in engineering units (e.g., vol% H_2, °C) or in electrical units (e.g., V, mA) that can be converted to engineering units through a mathematical relationship (e.g., calibration curve). Sensor testing shall be performed according to defined conditions. The whole test procedure, which includes sampling, test conditions, data acquisition protocols, and data work-up is to be documented. Tests to verify compliance with a standard shall follow the instructions of the relevant standards.

In the case that a test is desired under conditions which differ from the requirements prescribed in the standards, e.g., for elevated temperature or pressure, it should be performed according to a specific standard operation procedure (SOP) developed under the auspices of the testing laboratory to verify the compliance with the special requirements; these are often implemented to accommodate the requirements associated with a specific application. The modification of a standard specific test procedure has to be evaluated and validated in advance [499, 500].

Tests can be performed in test laboratories that have the appropriate test facilities and can provide and maintain the required test conditions. Alternatively performance tests, including routine maintenance protocols, can be executed on-site where the sensors are deployed, e.g., for field calibration of the sensors as per manufacturer's specifications.

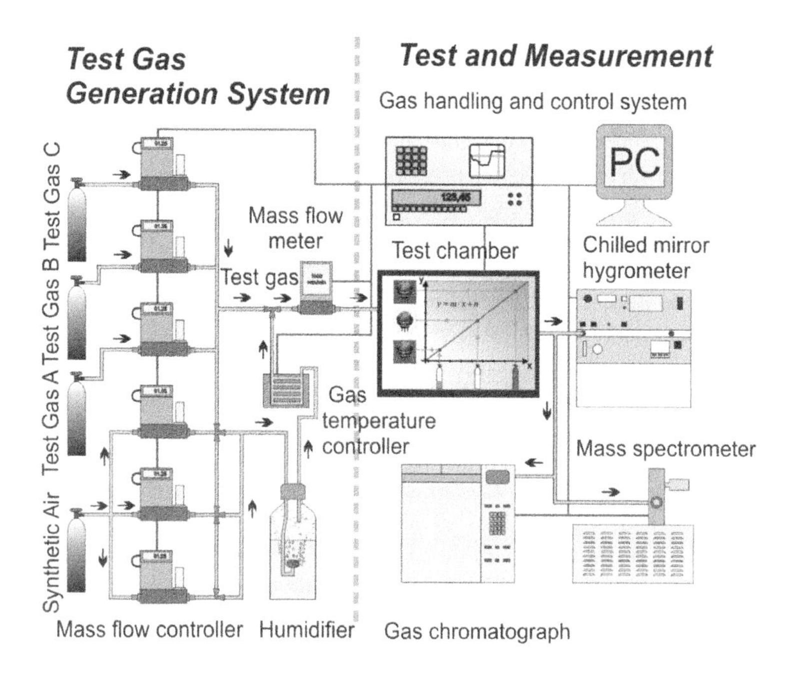

Figure 6.3: Schematic of sensor test facility.

6.2.5 Laboratory Testing of Gas Sensors

The test equipment used for the evaluation of gas sensors consists of several components. Various designs have been developed for sensor testing. Figure 6.3 depicts a schematic of the test facility developed at BAM for this purpose; comparable facilities exist at the European Commission, Joint Research Centre (JRC), Institute for Energy and Transport (IET) and the U.S. Department of Energy, National Renewable Energy Laboratory (NREL). Accredited test and calibration laboratories, NRTLs (see Section 6.1), research institutions, and sensor manufactures also have laboratory facilities for sensor evaluations. Sensor evaluation facilities will differ in the specific designs and testing methodologies they use, e.g., documentation of test protocols and the maintenance and calibration protocols for test and measurement equipment. However, each sensor test facility will have similar general elements, including a gas mixture generation system, a sensor test chamber, regulation system for control of environmental parameters, and a data acquisition system. All equipment having a significant effect on test results shall be calibrated periodically as prescribed by an internally documented program.

Static and Dynamic Test Gas Generation System

Test gas mixtures are necessary for determining the sensor response to different concentrations of the analyte. Several methods are available to obtain the gas mixtures [501]. Test gas mixtures can be prepared in advance or mixed real-time upstream of the sensor test chamber. A common approach for generating a range of test gas concentration is to dilute the pure analyte gas with air or nitrogen in order to generate test gas mixtures of the desired composition. Hydrogen can be supplied in cylinders or produced on-site using commercial hydrogen generators. Certified cylinders of pure hydrogen of various grades (e.g., purity levels) are commercially available from several suppliers. When working with pure hydrogen, special care should be taken to not produce hydrogen-air mixtures in the flammable range (4 vol% to 75 vol% H_2 in air).

When assessing hydrogen safety sensors whose dynamic range is typically up to 4 vol% H_2 it is preferred and considerably safer to use hydrogen cylinders that contain already the appropriate gas concentration. Such cylinders are on the market available and well suited for evaluating sensors for hydrogen safety applications. Dilution of the source gas is possible by blending defined masses or volumes of the gaseous components; for example, controlled mixing of air and a hydrogen/air mixture into an evacuated container is a recognized method to dilute the original hydrogen/air to a desired test concentration (static method) [497, 502]. In this case a homogeneous mixing has to be ensured. Alternatively, a dynamic method of gas generation is often used [503]. This involves the real-time mixing of gaseous components by rigorously controlling the flow rate of each component contributing to the final test gas mixture. An example of a design to achieve dynamic gas mixing is illustrated in the sensor test apparatus in Figure 6.3. Gas flow regulation is best performed using mass flow controllers (MFCs), see Section 5.3. Precision MFCs have an uncertainty of about 1 % of full scale. Rotameters can also be used to control individual gas flow, but with considerable lower accuracy which might preclude the use of rotameters for validation of sensor performance metrics. MFCs operate on a thermal conductivity principle and since different gases have different heat conductivity, it is necessary to implement correction factors when operating an MFC with a gas other than the one for which it was calibrated. Specific protocols for implementing correction factors are provided by the MFC manufacturers. The MFC requires a pressurized gas source (e.g., from a gas cylinder) at the inlet, and as a rule of thumb, the turn-down ratio of a MFC is about a factor of 10 (e.g., a MFC with a range of 0 to 1000 sccm has its best accuracy from 100 sccm to 1000 sccm (see Section 5.3)). Dynamic mixing of a hydrogen source gas with air can easily and accurately produce test gas concentrations of 10 % or below of the original source gas concentration.

Test gases can also be generated using permeation tubes which are capable of adding small, precisely tailored amounts of an analyte to a carrier gas stream. The analyte permeates through a permeable membrane as a gas or vapour into a regulated gas flow [504]. The permeation rate of the analyte through a membrane is a function of temperature but is nearly constant under isothermal conditions. Thus strict control of the temperature of the permeation tube is essential. Permeation ovens and permeation tubes of hundreds of different target analytes are commercially available. Permeation methods can be used to accurately generate trace levels (< 1 μmol/mol) of gaseous components. The permeation system must however be regularly maintained and calibrated to assure accurate gas or vapour generation. Calibration is typically performed by measuring the weight loss of the permeation tube after a long time period of operation (usually between two weeks and six months), based on the assumption that the weight loss is due to constant permeation flux of the material through the tube wall. It is thus necessary that the permeation tube be continuously operated (e.g., maintained at the proper temperature set point and purged with the carrier gas) except during calibration measurements.

Individual gas cylinders can also be purchased for each desired test gas concentration, although not all possible analytes at a required concentration are commercially available as certified gas standards. Nevertheless, commercial gas suppliers offer a wide range of test gases of various grades in compressed gas cylinders. Regardless of grade the test gas mixtures should be prepared by procedures described in standards and certified;

therefore, the analyte concentration is known to a specified accuracy and maximum tolerances for other constituents are specified [505]. Uncertainty values of the gas composition as well as shelf times are provided by the manufacturer. The uncertainty of test gases is usually 2 %, but accuracies of 0.5 % can be obtained (at significantly higher cost). Certified test gas cylinders can be used both for laboratory investigations and for field calibration of deployed hydrogen sensors.

Gas suppliers generally provide dry test gases (i.e., containing < 5 μmol/mol H_2O). The actual maximum allowable moisture content is provided in the product specification sheet. Sensor evaluations often specify that tests must be performed over a certain range of humidity (e.g. relative humidity in the range of 20 % to 90 %). Humidification of dry test gases can be achieved by the addition of liquid water to a gas volume. This can be accomplished by the precise dosing of liquid water into the gas stream using a syringe pump, precision liquid MFCs, or a fine-controlled pump such as those developed for high performance liquid chromatography. The liquid water that is being delivered has to completely evaporate and mix with the gas stream. Another way for dynamic humidification of a gas is to pass part of the gas stream through liquid water. The nearly saturated gas stream is then cooled in a temperature controlled gas container to ensure that the gas stream is completely saturated at this reduced temperature. It should be noted that some gases are readily soluble or will hydrolyse in water, thus care must be taken to verify that this technique is applicable to the test gas of interest. The humidified gas stream is mixed back into the dry gas, resulting in an unsaturated humidified gas; adjusting the ratio of flow rates of the two streams sets the humidity of the test gas to the desired set point. Regardless of the means by which the humidity level of a test gas is controlled, the actual humidity level has to be verified using a calibrated humidity sensor (see Section 4.3).

Sensor Test Chamber

For evaluation purposes, the sensor is typically placed in an enclosure commonly called the sensor test chamber. A sensor test chamber can be constructed from stainless steel, glass, plastics (e.g., acrylic or polycarbonate), or other materials that are inert to the test gas mixtures. The sensor chamber must allow sensor exposure to the test gas but shall be gas tight so as to preclude mixing with outside air. Often several sensors will be tested together in the sensor chamber. This is acceptable as long as the interaction of the sensors with the test gas will not change its composition. This may become an issue when testing multiple pellistor-type sensors (Section 3.2) which can not only deplete the hydrogen concentration but also produce water as a by-product, thereby increasing the humidity of the test gas.

In many facilities the test chambers are designed to measure the sensor response to bulk changes in gas composition as well as to monitor various physical properties, e.g., temperature and pressure. Specialized chambers may be required for specific tests. For example, the determination of sensor response time requires that the test gas change step-wise from one concentration to another; this is not feasible for a test chamber designed for bulk measurements that often have purge times on the order of several minutes. Specialized chamber designs for hydrogen sensor response times measurements have been proposed in ISO 26142, and independently evaluated in [78]. Advanced designs that permit a fast change of test gas composition have been developed [82].

Environmental Conditions

Sensor evaluations have to be performed at different temperatures and pressures. The temperature at which the sensor is tested can be controlled in a thermostated chamber. Climate chambers are commercially available and can often generate a defined humidity, too; however, the impact of the possible temperature gradients in larger volume chambers must be considered. Changes in gas pressure can be achieved by means of compressed gas cylinders, mass flow controllers, pressure controllers, pumps, and corresponding valves. Many sensor test facilities have incorporated the control of environmental parameters within the design of the sensor test chamber. In addition, the parameters of the laboratory air should be recorded. All environmental parameters of the test conditions, such as temperature, pressure and humidity, should be monitored using calibrated sensors.

Independent Gas Analysis

The use of cylinders with certified gas compositions conveys traceability for sensor evaluation of the test gas. Both the static and dynamic methods of generating diluted test gases in the sensor test chamber can generate a test gas of a known composition and uncertainty. However, an independent gas analysis can increase the reliability and reduce the uncertainty of the actual composition of the diluted gas as well as eliminate the impact of systematic errors (e.g., those associated with the uncertainty in the MFC). Independent analyses can be performed on the test gas during the running of sensor evaluations. These analyses are often based on mass spectrometry, gas chromatography, or optical methods, which will have accuracy and precision specification typically better than 5 %. The actual uncertainty and precision is often specified by the manufacturer or empirically determined by the test laboratory. Reference methods for the analysis of gas compositions are discussed in Section 6.3.

Data Acquisition

The collection of so-called "raw data" should include a time synchronised record of sensor output signals (e.g., a time stamp). Reference data including test gas composition, flow, and environmental conditions of temperature, pressure and humidity should also be recorded. The sensor output signals shall be measured with a calibrated precision meter or data acquisition system. The data acquisition rate depends on the test requirements and can vary from 1 ms to 1 min. Further details on data acquisition are presented in Section 2.6.

6.2.6 On-site Sensor Testing

As indicated previously, the assessment of fundamental sensor performance metrics is normally performed in the laboratory under controlled conditions. However, sensors are to be deployed in the field, and sometimes it is necessary that the sensor performance should be verified in the deployed state. Often the sensor must be periodically calibrated in the field. Field calibrations are usually performed using a "calibration kit" consisting of a pressurized gas cylinder with a certified gas mixture and a pneumatic interface system to connect the gas cylinder to the sensor. On-site sensor testing eliminates the need to remove the sensor from its installation and to transport it to a test laboratory for calibration testing. Also, on-site testing assesses sensor performance under the genuine deployment conditions. However, on-site testing typically provides reduced choice of test gases as well as undefined and uncontrollable environmental conditions. Therefore such in-situ tests are performed as intermediate functional checks with a single or a few calibration points. Often the calibration cylinder is of low volume and pressure in order to facilitate transport to the deployment site by minimizing regulations associated with the transport of high pressure cylinders. Compared to laboratory calibrations, field calibration protocols are less encumbered by compliance to data quality standards which are imposed on accredited laboratories. Essentially, field calibrations are performed using protocols consistent with manufacturer recommendations for verifying that the sensor response to a test gas is within the manufacturer's specified accuracy. Manufacturer specification can include test frequency, test gas concentrations, pneumatic interface design, and other recommendations. If the sensor readout is not within the manufacturer's specification of accuracy, an electronic adjustment can often be performed to bring the readout to within the specifications. The sensor (or the sensing element in the sensor unit) should be replaced if the sensor output cannot be adjusted to be within accuracy tolerances. It is recommended that the test gas have a composition uncertainty at least four times better than the sensor uncertainty.

6.2.7 Uncertainty of Measurement

In metrology, accuracy is the closeness of agreement between a measured quantity value and the "actual" value of a measurand. Accuracy consists of two main parameters: trueness (the closeness of the mean of a set of measurement results to the actual value) and precision (repeatability or reproducibility of the measurement) [73, 506]. The actual value can also be referred to as true value, agreed value, or reference value. Trueness is sometimes called accuracy, but this definition is not used in this book. A quantitative measure of accuracy is the uncertainty. Uncertainty in a measurement can be estimated according to the "Guide to the expression of uncertainty in measurement" (GUM) as a cumulative standard deviation resulting from the propagation of uncertainty with contributions of relevant components [507].

The principles of evaluating measurement uncertainty that ensure test outputs compliant with the GUM approach should include the following considerations:

- Identification of what causes the result to change
- Quantification of the uncertainty for each contributing component
- Conversion to standard uncertainties and combination of the standard uncertainties
- Expression of the final result as an expanded uncertainty

Individual contributions to the combined uncertainty are estimated and expressed as standard uncertainties. The calculation of a combined standard uncertainty uses the standard law of propagation of uncertainty. The relationship between the combined standard uncertainty $u_c(y)$ of quantity value of a measurand y and the uncertainties of the independent quantities x_1, x_2, $\cdots$ x_n on which it depends is:

$$u_c(y) = \sqrt{\sum_{i=1}^{n} \left(\frac{\partial f}{\partial x_i}\right)^2 u(x_i)^2} \tag{6.1}$$

Where f is the function expressing the measurand y in terms of x_i, $u(x_i)$ is the standard uncertainty for of i-th component. The partial derivatives $\frac{\partial f}{\partial x_i}$ are called sensitivity coefficients; they describe how the value of y varies with changes in the values of the input values x_i. Wherever possible, sensitivity coefficients are calculated for each component and incorporated in the calculation for the combined standard uncertainty. However, this is not always feasible and therefore the GUM classifies two main types of uncertainties, which are identified as Type A and Type B. Type A results from a statistical analysis of measured quantity values obtained under the defined measurement conditions, as described above. Type B concerns the evaluation of a component of measurement uncertainty determined by means other than a Type A evaluation, e.g., from a calibration certificate or based on reasonable assumptions.

The standard uncertainties $u(x_i)$ result from the scattering of sensor indication, noise, and repeatability. A proper estimate for the combined uncertainty of the sensor indication must also consider impact of other parameters which might influence the sensor reading, including temperature, pressure, gas flow rate, and the composition and homogeneity of the reference gas. It is also important to consider the contribution from the measurement system such as the uncertainty of the data acquisition system. The uncertainty of each of these components contributes to the combined uncertainty, although the individual contribution of a particular component may be relatively insignificant compared to the others.

As far as safety is concerned, and in some other applications, an interval built around the measurement result encompassing a large fraction P of the probability distribution (determined by that result and its $u_c(y)$) should be provided. This requirement is met by reporting the expanded uncertainty U, obtained by multiplying $u_c(y)$ by a coverage factor k [507]:

$$U_c(y) = \mathrm{k} \cdot u(y) \tag{6.2}$$

The fraction p is termed level of confidence of the interval, or coverage probability (if a normal distribution is assumed, $P = 95.45$ % for k = 2).

In a simplified approach, the uncertainty of a sensor response can be characterised by the standard deviation of the response from the value of the reference standard. This deviation has to be evaluated for no less than three different gas concentrations within the measurement range of the sensor [79]. In many applications the uncertainty must not exceed 10 % of the reading determined by a triple measurement [488]. Here the uncertainty of the testing system should be considered. For example, in most cases the gas concentration of certified gas mixtures has an uncertainty of 2 % and is rarely better than 0.5 %. Gas flow meters used for mixing hydrogen and air have a minimum uncertainty of approximately 1 %; reference values can be obtained from the manufacturer or testing laboratory. Therefore it is difficult to obtain accuracy for a calibrated hydrogen sensor better than 1 %.

6.2.8 Evaluation and Calibration of Gas Sensors

During sensor evaluations, including calibrations, the sensor indication is often compared to a reference value under defined conditions. If the sensor indication and the reference value are different, the sensor indication may be corrected:

$$x_{cor} = x_{mes} + a_{cor} \tag{6.3}$$

Where x_{cor} is the corrected value, x_{mes} is the sensor initial indication or measured value, usually the average value of several repeated measurements and a_{cor} is a correction value. This procedure is illustrated in Figure 6.4. The calibration experiment can be repeated for several hydrogen concentrations in air or other gases. The deviation of the sensor indication from the reference value between the compared pairs of values at different concentrations can be estimated using interpolation. However, if reference value and sensor indication are different and the sensor delivers an electrical standard signal (e.g., 4 mA to 20 mA), the calibration curve has to be re-established.

The precision (characterised by repeatability or reproducibility of the measurement) as a part of the uncertainty of measurements has to be also estimated in a calibration experiment using certified gas standards. According to the rules of propagation of uncertainty, the uncertainty of the corrected sensor indication can be given by:

$$u(x_{cor}) = \sqrt{u^2(x_{mes}) + u^2(a_{cor})} \tag{6.4}$$

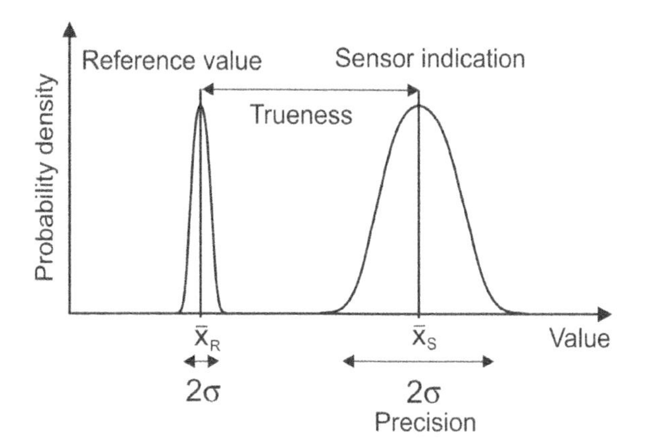

Figure 6.4: Calibration as a comparison of sensor indication with a reference value provided from a certified gas mixture.

Where the uncertainty of the corrected value $u(x_{cor})$ is given by the uncertainty contributions of the sensor $u(x_{mes})$ and the uncertainty of the correction factor $u(a_{cor})$.

The uncertainty of the correction factor $u(x_{cor})$ results from the uncertainty of the calibration facility which cannot be neglected, because it can contribute to the final uncertainty of the sensor indication. From a statistical perspective, the uncertainty of the reference value should be ideally ten times lower, and certainly at least five times lower than the uncertainty of the sensor, otherwise the applicability of the calibration procedure is questionable.

According to the approach of the GUM (see Section 6.2.7), uncertainties of facility parameters including the test gas composition c, the gas flow rate Q_v, the temperature t, the total gas pressure p, and the humidity φ_v (water vapour volume fraction) contribute to the uncertainty of the calibration facility. For an electrical sensor with an output signal, S, the uncertainty of the measurement of the electrical values, e.g., voltage, current, or resistance, also has to be considered. The impact of the uncertainty of the testing facility can be estimated according to the GUM from the contributions of the individual parameters by:

$$
\begin{aligned}
u(S) &= \sqrt{\sum_{i=1}^{n} f_i^2 \cdot u^2(X_i)} \\
&= \sqrt{f_S^2 \cdot u^2(S) + f_c^2 \cdot u^2(c) + f_Q^2 \cdot u^2(Q_v) + f_p^2 \cdot u^2(p) + f_t^2 \cdot u^2(t) + f_\varphi^2 \cdot u^2(\varphi_v)}
\end{aligned}
\tag{6.5}
$$

The weighting factors, $f = \frac{\partial S}{\partial X_i}$, depend on the gas sensor and can be estimated from the sensitivity of the sensor output against the input values. Typically uncertainties of electrical measurements, which combine the uncertainty of the measuring device and the measuring chain, are below 0.1 %. The test gas composition has an uncertainty from 0.5 % to 2 %. The mass flow controllers have uncertainties from 0.5 % to 1 %. The gas temperature can be measured with an uncertainty usually no better than 0.5 % to 1 % and the humidity no better than 1 % to 2 %. Accordingly, for the calibration of hydrogen sensors of an specified uncertainty below 10 %, the uncertainty of the testing facility should be thoroughly analysed and estimated.

6.2.9 Test Report

A list of the items to be included in the test report is provided in ISO 17025 [492]. The test report shall contain a unique identification, such as the names of the customer and the testing laboratory, an identification of the samples, and the employed testing protocol. The test results shall include measurement uncertainties. The test report shall, if relevant, provide a statement of compliance or non-compliance with requirements or specifications, e.g., the deviation of the sensor indication.

6.3 REFERENCE METHODS FOR GAS ANALYSIS

6.3.1 Overview

For the validation of sensors the use of certified reference gas standards of known composition and a validated test protocol are indispensable for establishing traceability of the test results (see Section 6.2). Additionally, analytical reference methods can be used to independently verify test gas composition and accuracy of the sensor response. Reference methods are accepted procedures providing measurement results intended for assessment of the trueness of measured quantity values obtained from other measurement procedures for quantities of the same kind, e.g., in calibration [73]. Reference methods are typically verified by round-robin testing to assure reproducibility among test laboratories and use of calibrated equipment. A reference method is a formally published procedure for specific applications, such as those developed by ASTM.

It is sometimes necessary to perform analytical measurements using a protocol that is not a formal reference method. This may be necessary when an appropriate reference method does not exist or does not meet the end-user requirements, such as for example, the verification of sensor performance metrics under extreme environmental conditions that are outside a range specified in a reference method. Such independent methods are typically developed and documented internally by an organization to independently verify the performance of gas sensors; these customer-tailored methods are not usually as rigorously vetted as formal reference methods but often meet the specific requirements of an application. The development of new methods can be accredited in a flexible scope of testing and calibration laboratories [508]. These independent methods can be used for testing, verification of performance, and calibration of gas sensors, as well as for the verification of the concentration of gas standards. Whether using a published reference method or an independently developed analytical method, it is important that the measurement uncertainty of the method is known and is of a value appropriate for the intended use.

The most common analytical laboratory techniques relevant for gas analysis and for the independent verification of hydrogen sensors are gas chromatography, mass spectroscopy, and optical methods. The following sections discuss these laboratory methods.

6.3.2 Mass Spectrometry

A mass spectrometer draws a gaseous test sample into an ionization chamber which produces charged particles that are then analysed. It can be used for a broad range of analytes. The mass spectrometer is designed to separate ions according to their mass to charge ratio (m/z). Mass spectrometry is amenable to gas phase samples and those samples that can be volatilized. Mass spectroscopy is good for simple mixtures of multiple components with comparable concentrations and distinct molecular ion masses.

The mass spectrometer consists of three main parts, which are the source (ionization chamber), analyser, and the detector. The source generates ions from the constituents of the test samples injected into the mass spectrometer. A variety of methods is available to ionize the samples. The heart of the mass spectrometer is the analyser. The analyser separates the components in the sample according to their respective m/z. Again, there are many different designs and a range of operational principles for the analyser. Generally, electric or magnetic fields are used to transport the ions to the detector. Because this is an electrical interaction, the ions will separate in the analyser based on m/z and not just m. The detector produces the signal from the separated ions.

Depending upon the sensitivity and atomic mass resolving power of the instrument, a mass spectrum of, for example, oxygen, can have 5 main peaks. These arise because oxygen exists as two main isotopes (with atomic mass 15.994 at 99.8 % natural abundance and atomic mass of 17.999 at 0.2 % natural abundance). Thus molecular oxygen can consist of $^{16}O_2$, $^{18}O_2$, and $^{16}O^{18}O$. Ionization would form molecular ions of mass 32, 34, and 36. Ionization could also form atomic oxygen ions of atomic mass 16 and 18. Both process control and research grade mass spectrometers are available. Selectivity is good, but interferences can arise from molecules/fragments of comparable molecular weights; a classic example would be for an atomic weight of 28 ($[N_2]^+ = 28.0061$, $[CO]^+ = 27.9949$, and $[C_2H_4]^+ = 28.0313$). Although resolvable by high resolution mass spectroscopy, such instruments are costly and thus not suitable for routine process control.

6.3.3 Gas Chromatography

Chromatography is a chemical analytical method for the separation and then the analysis of multiple components in a mixture. Several different chromatographic methods have been developed, including liquid chromatography/high performance liquid chromatography which is used for the separation of multiple solutes in a liquid solution and thin-layer chromatography, which is an older method also used to separate components in a liquid solution. The most relevant chromatographic method for gas analysis is gas chromatography (GC). GC is generally applied to samples that are either gaseous or liquids that can be volatilized without decomposition. Thus, GC is usually applied for substances with a boiling point below 400 °C. Thermally unstable or relatively non-volatile substances can be subjected to derivatisation, a chemical reaction to form evaporable compounds. Non-polar silyl groups are often used for this purpose. Alternatively GC can be performed on the thermal degradation (pyrolysis) products of non-volatile test samples for chemical analysis.

The main components of a gas chromatograph include the column, injector, carrier gas, oven, and detector, as illustrated in Figure 6.5. More details on the gas chromatography technique is readily available in the literature, e.g., [509–511]. GC is not usually considered a real-time analytical method. It is performed on discreet samples with analysis times ranging from approximately 1 minute to over one hour.

The heart of a GC is the column, since this is where the separation of components in a sample mixture occurs. A carrier gas flows continuously through the column, and is often referred to as the mobile phase in a GC. The selection of the carrier gas is based on it being inert to the sample such that it will not interfere with the measurement of the target analyte. For example, a common carrier gas is helium, which would be inappropriate for the analysis of helium in hydrogen as per the SAE 2719 fuel quality verification requirements [301]. Once injected, the sample that is to be analysed flows through the column by means of the carrier gas. One critical feature of the column is that the inside is either packed with a solid loading or coated with a viscous, non-volatile liquid. Thus, two main column types are the wide diameter packed column (with an i.d. typically ranging from 2 mm to 5 mm and lengths of up 10 m) and the coated capillary column (with an i.d. typically ranging from 0.1 mm to 0.5 mm and lengths of up to 100 m). Packed columns are usually made from thin-walled stainless steel while fused silica is often used for capillary columns. Both column types are configured in a coil with a diameter of approximately 20 cm. The outside of the capillary column is coated with a polyimide polymer to improve its mechanical stability. The packing/coating within the column is the stationary phase.

In a chromatographic analysis, the most critical parameter for compound identification is the retention time (RT) for each component of a mixture. The RT is the time difference between sample injection and elution from the column. This parameter is controlled primarily by the specific interaction between an individual constituent in the mixture and the stationary phase. Thus, different constituents within a sample will elute from a column with a characteristic RT. The stronger the affinity that a specific chemical has for the stationary phase, the longer its RT; this arises because when the constituent is immobilized on the stationary phase it is not being transported through the column in the mobile phase. A chromatographic analysis will produce a chromatogram that depicts multiple peaks, each of which corresponds to individual components and their amount in the sample mixture. In a gas chromatogram, the RT is the primary means for identification of a constituent within a mixture; this typically requires empirical verification of RTs using independent analysis of standard mixtures of known composition.

Other means of component identification are possible with specialized detectors that provide chemically specific information, as discussed below. Affinity of a component to the stationary phase can be based upon boiling point, vapour pressure, size (e.g., especially with a molecular sieve packing material), or a specific chemical interaction controlling the adsorption of the constituent on to the stationary phase. The affinity of a chemical compound to the stationary phase material is typically temperature dependent, and accordingly, GC columns are mounted in an oven capable of precisely controlling the temperature of the column. Elevated temperature operation will accelerate the chromatographic process, thereby shortening the RT of many compounds. Elevated temperatures will also volatilize many chemicals which have a low vapour pressure at room temperature, and this allows for the chromatographic analysis of many compounds which would normally not elute at room temperature. Thus, GC analyses are performed with rigorously controlled temperature profiles, which start at or near room temperature followed by a ramping up to elevated temperatures. The specific temperature profile is often defined in the reference method. Gas chromatographic results are very repeatable because of the ability of the GC oven to precisely and repeatability control test temperatures and profiles.

The selection of the stationary phase is perhaps the most critical parameter in developing a GC method, and there are many commercial coatings and packing materials available. A typical coating used for gas analysis is based on the polyorganosiloxane class of chemicals (e.g., polydimethyl siloxane, polyphenyldimethylsiloxane). The coatings can have a range of different polarities, and hence different affinities to different compounds, depending on the nature of the organic structure within the organosiloxane polymer. Column manufacturers are an excellent resource to assist end-users on the identification of the best column that should be used for a specific application.

For separation of trace components so called WCOT-columns (wall coated open tubular columns) are used. These are capillaries of fused silica usually of 0.1 mm to 0.5 mm and 10 m to 100 m length. The inner surface of the capillary is coated with a 0.1 μm to 30 μm thick polyorganosiloxane of different polarity. The outside of the capillary is coated with a polyimide polymer for increasing mechanical stability. For permanent gases PLOT-columns (porous layer open tubular columns) with adsorbents such as molecular sieve, alumina, or silica gel can be used as stationary phase.

The injector provides the interface to input a sample into the column. The simplest injection method makes use of a syringe filled with a known amount of the liquid or gaseous sample that is to be analysed by the GC. The contents of the syringe is manually fed directly into the injector and hence on to the column. Sample volume can range from less than 1 μL to several mL. The injector is operated at elevated temperatures to assure volatilization of liquid samples. Liquids have much higher density than gases and it is noted that the volume of a liquid will increase nearly 1000 fold when vaporized from a liquid state to a gaseous state. There are three main types of injectors for liquids to accommodate complex mixtures and include split/splitless injectors, on-column injectors and programmed temperature vaporizing injectors. The later injector allows for partial separation prior to injection on to the column of components of widely varying boiling points within the sample mixture. Gas samples are also often injected using a sample loop, e.g., a tube with a well-defined volume that can typically range from around 10 μL to 10 mL. Standard sample loop designs based on a 6-port valve have been developed. This sample loop design has two pneumatic paths that are isolated when the valve is in its de-powered position; one pneumatic path continuously feed carrier gas to the injector, the other pneumatic path is purged with the test gas. When the valve is powered up, the valve switches positions such that the carrier gas purges the sample loop and injects the sample on to the column. Sample loops based on 6-port valves can be automated, which is particularly beneficial for process control applications.

The final element of a gas chromatograph, as illustrated in Figure 6.5 is the detector. The detector continuously monitors the chemical composition of the carrier gas as it elutes from the column. Detectors are analogous to a sensor in that they are electronic devices which respond to changes in the chemical composition of the carrier gas. Upon elution from the column, the carrier gas, and any constituent in the carrier gas passes over the detector, which then produces an electrical signal in response to the presence of the constituent in the carrier gas. The magnitude of this signal is often proportional to the concentration of the constituent. There are numerous detector types that are commercially available. These can be classified as either non-specific detectors and specialized detectors for the selective and often very sensitive detection of specific classes of chemicals. Two main examples of a non-specific GC detector are the flame ionization detector (FID) and the thermal conductivity detector (TCD). Specialised detectors include electron capture detector (ECD), helium discharge detector (HID), and mass spectrometer (MS).

The FID measures the electrical conductivity of a hydrogen/air flame in an electrical field. As a flammable component within the carrier gas passes through the FID it is combusted and ionized species are formed, which changes the current in the FID electric field. The magnitude of the change in conductivity is proportional to the amount of the analyte. The FID is an excellent detector for most organic substances or other flammable analytes, and has a detection limit on the order of 5×10^{-8} g. However, many gases (e.g., many organic halides, H_2S, COS, CO_2) are not detectable on the FID. The need for a hydrogen flame and its insensitivity to certain chemicals is the major drawback of the FID as a general GC detector.

The second major non-selective GC detector is the TCD. The TCD is based on the principle that the thermal conductivity of a gas is strongly dependent on its composition. Thus, the TCD will detect changes in the thermal conductivity of the eluting carrier gas induced by the presence of components from the injected sample in the carrier gas. Details on the design and operation of the TCD were presented in more detail in Section 3.3, which described the hydrogen TC sensor. The TCD for GC analyses is often designed in a two-cell configuration, with one cell exposed only to the carrier gas and the second to the analyte-containing carrier gas eluting from the column. The signal is the differential electrical response between the reference and the analytic cell. The TCD has comparable sensitivity to the FID, at least for some analytes. The TCD is, however, responsive to a broader range of compounds compared to the FID. One drawback of the TCD is that its sensitivity varies significantly with gases; this variability is controlled primarily by the wide differences in the thermal conductivity of the gas being detected.

There are numerous specialized detectors. These often have extremely low detection limits although often for only a specific class of chemicals. Three will be discussed, along with the use of a mass spectrometer as a GC detector, which is perhaps the most flexible GC detector currently available. One of the most common chemical class-specific specialized GC detectors is the electron capture detector (ECD). The ECD operates by using a beta-source (usually ^{63}Ni, which is a radioactive beta emitter) to ionize compounds containing electronegative elements like the halides. In some regards, the ECD complements the FID in that the FID is insensitive to many of the compounds that the ECD can detect and vice versa. The use of multiple detectors linked in series to simultaneously analyse the same eluting gas is not uncommon in GC configurations. A major drawback of the ECD is the need for radioactive nickel, which is often regulated and may require a special permit to possess.

The pulsed discharge helium ionisation detector (PDHID) generates electrons by ionizing eluting species with photons from a pulsed helium plasma. The released electrons are detected as a current that is collected by applying a bias potential across two electrodes in the discharge region. The detector is very sensitive and can simultaneously detect permanent gases germane to hydrogen purity, including H_2, N_2, CO, CH_4, HCHO, and C_2H_4.

The third specialized detector discussed here is the sulphur chemiluminescence detector (SCD), which is based on the combustion in a hydrogen flame of sulphur compounds to form sulphur monoxide (SO). Sulphur monoxide reacts with ozone producing sulphur dioxide, oxygen, and light. The light intensity is proportional to the amount of the sulphur containing analyte and can be detected by a photo diode. The SCD is extremely sensitive to sulphur containing compounds, but as with the FID, requires a hydrogen flame, which can be a major drawback for some locations and applications.

Perhaps the most powerful general purpose chemical analytical system is the GC interfaced to a Mass Spectrometer (MS). In this configuration the MS is the detector for the GC and provides a unique mass spectra for each eluting constituent, thereby providing not only quantitative information on the amount of a specific constituent in the sample based on the intensity of the MS peaks but also near-unequivocal data on the chemical identity of the eluting chemical. Although powerful, the GC-MS is a large, expensive instrument that can be complicated and expensive to operate and maintain. An overview of the basic theory and operation of the mass spectrometer is provided in Section 6.3.2. A schematic of a gas chromatograph is given in Figure 6.5—more details are available from, for example [509–511].

Gas samples, for example, from a process line can be taken with the aid of collecting tubes, i.e., a gas-tight container of defined volume made from plastics metals, or glass and fitted with a valve. These samples can be pre-treated before injection onto the GC column to remove components which may interfere with the analysis or may damage the GC column. Pre-treatment can be achieved by passing the gas sample through adsorbents like activated charcoal carbon, polymers (Tenax), silica gel, or molecular sieves and desorbed into a smaller gas volume for analysis.

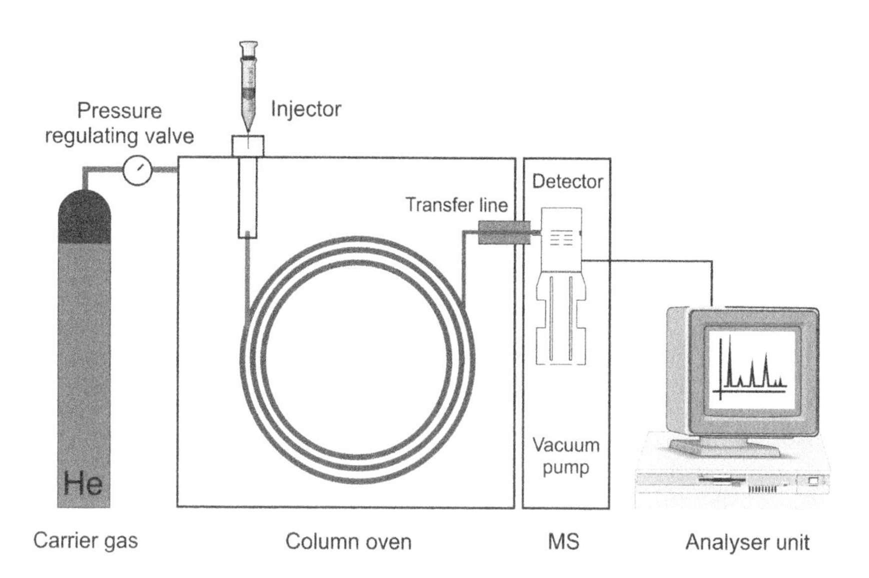

Figure 6.5: Schematic of a gas chromatograph with mass spectrometer detector (GC-MS).

6.3.4 Optical Methods

Analysis of gaseous species by optical methods is based on the attenuation of the intensity of light of different wavelengths as it passes through the sample. The Beer-Lambert law (see Section 4.2), relating the absorption of light to the concentration of the analyte, is most commonly applied for the quantitative determination of gaseous samples. The Lambert-Beer law applies to UV, visible, and infrared absorption by a chemical. Absorption spectroscopy is applied to many possible constituents in hydrogen. Some of the more common optical methods that can be used for these constituents are discussed here.

Fourier Transform Infrared Spectroscopy (FT-IR)

In classical dispersive absorption spectroscopy, the optical spectrum is scanned stepwise by changing the wavelength of light. In contrast, in FTIR, an interferogram is obtained using a Michelson interferometer and a beam of polychromatic collimated infrared light that is divided by a beam splitter into two beams of same intensity. One beam is reflected on a fixed mirror and the other beam is reflected from a moving mirror back to a beam splitter and recombined. A change of optical path length and a phase shift of light occurs that is dependent upon the position of the moving mirror. Recombination of the two beams thus leads to constructive and destructive interferences. The combined beam passes through the sample and on to the detector. The measured signal is an interferogram associated with the modulated signal from the combined light sources and its interaction with the sample gas. The absorption spectrum is obtained by a Fourier transform of the interferogram. The advantages of this method are a higher spectral resolution, lower noise, and shorter measuring time relative to IR measurements performed on a scanning spectrometer. The limit of detection depends on the analyte, but is on the order of 10^{-8} to 10^{-9}. The uncertainty is typically than 5 %. IR measurements can be made using a test cell (e.g., a cuvette) or an open-path configuration.

Laser Absorption Spectroscopy (LAS)

Laser light is used in many optical absorption devices for the analysis of gases. The advantages of lasers, which includes high intensity and narrow bandwidth, can be exploited for better selectivity and detection limits. The use of laser diodes opens prospects for miniaturisation. In tunable diode laser absorption spectroscopy (TDLAS), the frequency of the laser is modulated by variation of temperature or current. The emitted light can be delivered in narrow bands on the order of 10^{-3} cm^{-1} by the use of prism or optical grids and line widths. Thus, a specific absorption band of the analyte can be scanned with high resolution. The use of optical cells with multiple reflections to generate a long optical path results in detection limits as low as 10^{-9}. Also open-path measurements in air or process gases are possible. Gases like CO, CO_2, NO, SO_2, CH_4, and O_2 can be detected in the UV-Vis or IR range of the electromagnetic spectrum and thus can be analyzed by absorption spectroscopy, including LAS.

Cavity Ring-Down Spectroscopy (CRDS)

CRD spectroscopy is a very sensitive absorption technique in which the rate of absorption rather than the magnitude of the absorption of a light pulse is measured. The gas sample is placed inside an optical cavity consisting of two highly reflective mirrors. The internal cavity is illuminated by a short laser pulse, which is reflected back and forth inside the cavity. Every time that the light is reflected, a small fraction of this light is absorbed by the presence of a gas. The intensity of the light pulse decreases exponentially with a time constant dependent upon the concentration of the absorbing gas. Thus, instead of measuring the total intensity of the light exiting the cavity, one determines the decay time associated with the light intensity leaking out of the cavity. In this way the rate of absorption can be obtained; the more the sample absorbs, the shorter the measured decay time [512]. The time dependent intensity decrease of a laser impulse I in a detection cavity resonator is given by:

$$I(t) = I_0 \cdot \exp\left(\frac{-t}{\tau}\right) \tag{6.6}$$

The decay time constant τ is determined from the intensity decrease of the laser impulse proportional to 1/e. It is proportional to the refractive index of the media n, speed of light c, and parameters of the cavity resonator. From the ratio between the time constant τ_{empty} of the empty resonator and time constant τ for the analyte filled cavity, the concentration can be evaluated if the absorption coefficient is known.

$$const. = \frac{1}{cA(\nu)}\left(\frac{1}{\tau(\nu)} - \frac{1}{\tau_{empty}}\right) \tag{6.7}$$

Where A is the absorption cross section and ν is the frequency of laser light. The method is very sensitive due to the multi passing of the light beam in the cavity and independent on scattering of the laser light. A schematic of the method is given in Figure 6.6.

The response time is in the order of 1 min to 3 min. The lowest detection limit depends on the analyte and is in the range of 10^{-6} to 10^{-9} for gases such as CO, CO_2, NH_3, and H_2S.

6.3.5 Analytical Method for Determination of Trace Contaminants in Hydrogen

High sensitivity and selectivity reference methods for the detection of trace components in hydrogen include gas chromatography, mass spectroscopy (MS) and some optical methods [513, 514]. The use of several GC detectors permits analysis of different components

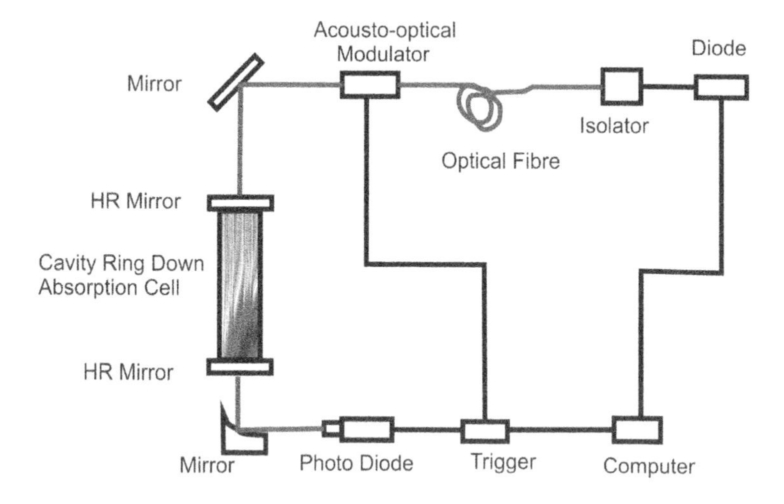

Figure 6.6: Schematic of a cavity ring-down spectrometer.

simultaneously and gives the highest sensitivities. It is often useful to employ enrichment technologies. Table 6.5 compares the highest tolerable amount of trace components in hydrogen and the lower limits of detection (LDL) for methods based on gas chromatography. Methods for the determination of formic acid and halogens are under development.

Advanced quadrupole mass spectrometers coupled with specialised hydrogen separators is an alternative technique to those mentioned in Table 6.5 for highly sensitive analysis of trace components in hydrogen gas [515]. In this method the impure hydrogen flows through a palladium micro channel. Hydrogen diffuses out through the metal while the impurity enriched residual gas can be analysed by mass spectrometry. The LDL is 10^{-12} and the response time is 5 min to 10 min.

Depending on the trace gas to be quantified different analytical methods can be preferred. The method of choice for oxygen, the paramagnetic oxygen sensor, was already described in Section 4.1 while reference methods for water detection were presented in Section 4.3. Optical methods such as Fourier transform spectroscopy (FTIR) may be used to detect gases such as CO, CO_2, NH_3, NO, CH_4, HCHO, HCOOH, and H_2O with high sensitivity. The lower limit of detection is about 10^{-7}. Cavity-Ring-Down-Spectroscopy (CRDS) can detect oxygen in hydrogen even with an LDL of 10^{-10}. It is also being explored to look at CO, CO_2, NH_3, and other contaminants in hydrogen. Near infrared (NIR) gas analysers have been applied to detect carbon monoxide or carbon dioxide in hydrogen, e.g., for monitoring the pressure swing adsorption process (PSA). Analytical reference methods for the determination of different trace components in hydrogen have been developed by ASTM and the current standards are listed in Table 6.6.

This brief overview shows that various established analytical methods are available to detect gaseous trace components in air and in hydrogen gas. The sensitivity is in most cases sufficient for detection of the considerably low concentrations demanded in standards to ensure health and safety protection and long term operation of devices using hydrogen. These analytical systems demand a remarkable experimental effort and qualified personnel are needed for the analysis which most likely have to be performed in a laboratory. However they can be used as reference methods for comparison with gas sensor detection.

Table 6.5: Limit of detection for determination of trace gases in hydrogen by GC according to [513].

Component	**Requirements of ISO/DIS 1487-2** μmol/mol	**LOD** μmol/mol	**GC detector**
Carbon monoxide (CO)	0.2	0.04	FID Meth.[a]
		0.012	PDHID
Carbon dioxide (CO_2)	2	0.015	FID Meth.
		0.012	PDHID
Formaldehyde (HCHO)	0.01	0.2	FID Meth.
Formic acid (HCOOH)	0.2	-	-
Total hydrocarbons	2	0.01	FID
Total halogen compounds	0.05	-	-
Ammonia (NH_3)	0.1	0.5	MS
Total sulphur compounds	0.004	0.001	SCD
Oxygen (O_2)	5	3	TCD
		0.002	PDHID
Nitrogen (N_2)	100	8	TCD
		0.03	PDHID
Helium (He)	300	13	TCD
Argon (Ar)	100	5	TCD
		0.01	PDHID

[a] Meth.—The compound is reduced to methane by a methanizer.
FID (Flame ionization detector), PDHID (pulsed discharge helium ionization detector), MS (mass spectroscopy), SCD (Sulphur Chemiluminescence Detector), TCD (Thermal conductivity detector).

Table 6.6: Standard test methods for determination of trace components in hydrogen

Analyte	Standard	Title/Remark
ASTM D7649-10	CO_2, Ar, N_2, O_2, H_2O	Standard Test Method for Determination of Trace Carbon Dioxide, Argon, Nitrogen, Oxygen, and Water in Hydrogen Fuel by Jet Pulse Injection and GC-MS Analysis. The procedures described in this method were designed to measure carbon dioxide at 0.5 μmole per mole (ppm), argon at 1 ppm, nitrogen at 5 ppm, and oxygen at 2 ppm and water at 4 ppm.
ASTM D7675-11	Total hydrocarbons	Standard Test Method for the Determination of Total Hydrocarbons in Hydrogen by FID Based Total Hydrocarbon Analyzer (THC)
ASTM D7649-10 /WK 37092	Total hydrocarbons	Standard Test Method for the Determination of Total Hydrocarbons in Hydrogen by FID Based Total Hydrocarbon (THC) Analyzer. Sensitivity from 0.1 ppm (μmole/mole) up to 1000 ppm (μmole/mole) concentration are achievable.
ASTM D7550-09	Ammonia, alkaline, alkaline earth	Standard Test Method for Determination of Ammonium, Alkali, and Alkaline Earth Metals in Hydrogen and Other Cell Feed Gases by Ion Chromatography. Sensitivity from low part ppb (μg/l, μg/kg) up to ppm (mg/l, mg/kg) concentration are achievable dependant on the amount of hydrogen or other fuel cell gas sampled.
ASTM D7652-11	Sulfides	Standard Test Method for Determination of Trace Hydrogen Sulfide, Methyl Mercaptan, and Carbonyl Sulfide in Hydrogen Fuel. The detection limit is 0.02 ppbv (nmol/l), based upon the analysis of a 500 mL hydrogen sample.
ASTM D7653-10	NH_3, CO, CO_2, HCHO, HCOOH, CH_4, H_2O,	Standard Test Method for Determination of Trace Gaseous Contaminants in Hydrogen Fuel by Fourier Transform Infrared (FTIR) Spectroscopy

6.3.6 Reference Methods for Detection of Hydrogen

There is no recognized standard method for the detection of hydrogen in air or in other gases. However very low amounts of hydrogen in air can be determined by gas chromatography using molecular sieve columns and a pulsed discharge helium ionisation detector (PDHID). The limit of detection is about 2 pg [516]. Cavity Ring-Down Spectroscopy (CRDS) is also suitable for hydrogen trace detection in gases with a lower detection limit of 5×10^{-9}. However this method demands the previous oxidation of hydrogen to water.

As noted in Section 3.8 hydrogen can be analysed by Raman spectroscopy, although this method does not have a very good quantitation limit. This method has been used, however, for specialised applications, e.g., for investigation of solid hydrogen storage materials and for quantitative analysis of isotopic mixtures of hydrogen, deuterium, and tritium [517].

The absorption of hydrogen in the UV range of the electromagnetic spectrum at about 121.6 nm (called Lyman-alpha line) is used for hydrogen detection. However, because of the absorption of other elements in the UV range, this method is applied only by astronomers for the determination of hydrogen distribution in the universe or by meteorologists for water detection in the upper atmosphere.

6.4 SELECTION AND INSTALLATION OF GAS SENSORS

6.4.1 Overview

With the hundreds of sensors commercially available it can be difficult to choose the best sensor for a particular application. This is particularly true for end-users whose knowledge and past experience may be limited. Selection is made more complex due to the emergence of many new models and types of gas sensors onto the market.

Regardless of the application, any potential user of a gas sensor is faced with many questions including:

- Do I need to install sensors and if so which ones?
- Which sensing platform is the most suitable?
- How many sensors are required?
- Where should the sensors be located?
- Are poisoning or interfering agents likely to be present?
- What are the appropriate alarm thresholds?
- What is the appropriate response time?

These questions are not always easy to answer. Depending on the sensing technologies, commercially available sensors each have their own advantage; however, they also have limitations and drawbacks. The suitability of a given sensor depends largely on the application area in which it is to be used and also on the functionality, i.e., the information which is expected or needed. Thus, in order to choose the optimal sensor, one must first properly define the area of application and the functional requirements expected of the sensor. Furthermore it is essential to install the sensor properly and in the correct location to ensure optimum operation. The aim of this section is to provide general guidelines on the selection and use of gas sensors depending on what the sensor will be used for and under what conditions. Advice on sensor selection and installation for a specific application should be sought from the sensor supplier, sensor installation service, or a specialised engineering consultant. Further information and guidance may be obtained from relevant literature [450, 470, 518, 519].

6.4.2 Selection of Gas Sensors

Identification of the Application Category

The area in which gas sensors are to be applied and the general tasks they have to fulfill is of fundamental importance when choosing a sensor as this determines the sensor performance requirements. This analysis must be performed, even before it is clarified what kind of gas or gases should be detected, and in what concentrations. In terms of sensor functionality the following classification can be made:

Safety and Health Protection

In order to ensure personal safety or equipment protection safety gas sensors can be employed for detection of:

- Flammable gases and vapours (see Chapter 3 and Section 4.2)
- Toxic gases and vapours (see Section 4.2)
- Oxygen (see Section 4.1)

The need to deploy sensors or other analytical measurement equipment is determined by performing a risk analysis. A risk analysis is a systematic procedure in which all hazards, associated with a toxic gas and its release into the environment, are identified and the probability of occurrence of dangerous situations is estimated. Measures should be undertaken to minimise the risks associated with these hazards (see Section 1.2). When using flammable gases an appropriate analysis has to be carried out for the likelihood of occurrence of flammable atmospheres and the probability of ignition. Procedures for the designation and classification of hazardous areas or locations are described in several guidelines, e.g., [520, 521], and are performed according to IEC 60079-10 [447] (see Section 6.1.2). Classified hazardous areas are defined as those areas in which an explosive gas atmosphere is present or likely to be present, in quantities such as to require special precautions for the construction, installation, and use of electrical apparatus. If the concentration of flammable gases can reach explosive limits the deployment of gas sensors or detection systems is mandatory (see Section 1.3). IEC 60079-10 provides guidelines on selection criteria for electrical equipment (e.g. sensors) in hazardous areas depending on the type of flammable gas concerned (which is grouped in relation to its ignition energy), and the expected temperatures [447]. The assessment of the dangerousness of the gas, e.g., hydrogen, following, e.g., the standards IEC 60079-20 [522] in Europe and for North America following standards NFPA 497 [523], NFPA325 [521], API RP 500 [524], and API RP 505 [525] (see Section 6.1.3).

ISO has produced a Technical Report, ISO/TR 15916:2004 on "Basic considerations for the safety of hydrogen systems" [473]. This report provides information on basic safety concerns and risks associated with hydrogen use to those who are unfamiliar with hydrogen technologies. Hydrogen gas and flame detectors are mentioned as devices for mitigation and control of risks. The report recommends the use of hydrogen detectors wherever hydrogen is used. Guidelines on where to install hydrogen gas detectors, considerations in selecting sensors, and indications on appropriate alarm limits are also provided in the report. Provisions for hydrogen fire detection are also recommended for use in areas where leaks, spills, or hazardous accumulations of hydrogen can occur. Considerations for selecting a suitable hydrogen flame detector are provided.

A flammable gas sensor has to fulfill several requirements to ensure its ability to indicate hazardous conditions arising from a facility defect or gas leakage. Clearly, the measuring range of the sensor should be appropriate relative to the lower explosion limit (LEL) of the gas to be detected. In this respect it is important that the sensor is capable of measuring gas concentration levels (i.e., it should not saturate) up to and around the LEL or near the upper explosion limit (UEL). Furthermore the sensor response time should be such that a sufficiently fast response is achievable. Other requirements for operation in potentially flammable areas are described in Section 6.1 and include the requirement that the gas sensor cannot be a source of ignition. While the deployment of sensors is not mandatory in unclassified areas or Zone 0 areas it may nevertheless be desirable to use gas sensors in these areas as they may bring some benefits as described in the paragraphs c) and d) below.

Sensors are also indispensable to avoid poisoning and health impairment when the target gas is toxic. In this case critical concentrations for trace toxic components are given by pre-defined thresholds such as MAC and TLV-TWA thresholds (see Section 4.2). The sensor's lower detection limit should allow detection of the toxic component at concentrations far below its critical threshold.

Oxygen sensing is similarly important to reduce the danger of asphyxiation should the concentration of this vital gas fall below critical levels. A localised decrease in O_2 concentrations may occur in situations where there is fire or where air (and hence oxygen) is displaced due to the release of large volumes of a gas which does not support respiration. For the last two cases, the use of sensors which are certified to the appropriate application is mandatory and also their functional safety should be proven (see Section 6.1).

Process Control

Sensors can be used to control processes where gases are produced or consumed. Gas flow sensors are applied when a certain volume of gas has to be processed. Pressure sensors can be used for monitoring the pressure in a gas stream which can be used for dosage of gas volume (see Section 5.2). Regardless of the type of sensor, process control demands compatibility of the sensor with the control system in which it is deployed and continuous operation over extended periods of time. Accordingly sensors without signal drift and with low maintenance requirements are favourable for this application.

Diagnosis and Analysis

Sensors can be used for single surveillance measurements, e.g., to investigate malfunctions of systems, e.g., of a bio gas fermenter. This can be indicated by a single gas sensor or by a multigas sensor when various gaseous species need to be surveilled. Qualitative sensing, i.e., where simply an indication of the presence of a gas above a certain threshold, is sufficient or a quantitative measurement may be required. For diagnosis and analysis applications sensing is typically required for short periods of time and rapid response is desirable. Depending on the diagnostic requirements the sensor's measuring range may need to be in the low ppm-range or it may need to cover the whole concentration range from 0 to 100 vol%. Such requirements need to be considered when selecting an appropriate sensor. Furthermore the sensor should be portable or at the very minimum transportable to facilitate practical use in the application.

Comfort and Air Quality

Indoor air quality (IAQ) sensors are used to measure indoor parameters such as temperature, humidity, carbon dioxide (CO_2), and volatile organic compounds (VOCs) levels for the purposes of environmental control and comfort. The quality of air in residential environments, as well as in professional environments such as offices and workplaces, can have a significant influence on the comfort and performance of individuals. For example high carbon dioxide levels are known to cause fatigue and drowsiness. For this reason the heating, ventilation, and air conditioning (HVAC) systems in commercial buildings often integrate sensors to measure the concentration of influential species such as carbon dioxide, volatile organic compounds, and humidity. According to the concentration of these species, the building ventilation can be increased or decreased accordingly. Considering the high energy demand of HVAC system operation, the use of sensors for smart environment control in "intelligent" buildings results not only in improved breathing air but also facilitates significant cost savings. However since the consequences of a malfunction of these sensors are relative small, certainly in comparison to safety related applications, the cost of these sensors need to be low. The sensor should also discriminate between low concentrations of the target gas while for many applications small size and low power consumption are favourable.

Sensing Tasks

Functionality

Functionality refers to the basic tasks of gas sensor which can be divided into the following three tasks:

- Indication of the change of gas composition in a mixture
- Identification of toxic or flammable compounds
- Quantification of components at threshold levels

Depending on the sensor's task the requirements for the accuracy of detection or measurement, as well as the indication of results, are different (see Table 6.7 and Table 6.8). Sensors for indicating gas composition change are required to detect small variations in a gas mixture's composition which can arise, for example, due to disturbances in a process. Sensors for identification of toxic or flammable gases typically require high selectivity towards the target gas whereas sensitivity is not so critical. Conversely sensors for quantification of gases at threshold concentrations must be sensitive enough to reliably detect low target gas concentrations. Threshold concentration sensors can be used to indicate increased concentrations of the target gas, possibly arising because of a leakage, and should provide a visual and audible alarm signal when this is relevant for personal safety.

Sensor Performance

Gas sensor performance includes a range of sensor parameters such as measuring range, limit of detection, resolution, response time, recovery time, drift, etc. Prior to selection it is useful to check and consider likely changes in the ambient environmental conditions. For a given application, sensor selection requires comparison of the sensor capabilities and the requirements for that application. Gas sensor measuring ranges are diverse—they can cover gas concentration ranges up to 100 vol% or ranges in the trace concentration region, e.g., from 10^{-6} to 10^{-3}. For applications requiring detection of combustible gases, sensors with measuring ranges up to 100 % of the lower explosion limit (LEL), or a fraction thereof, e.g. 50 % LEL, are required and are commercially available for most combustible gases including hydrogen. The flammable gas concentration is then typically in the range of several volume percent. Sensors with an uncertainty, estimated from the deviation of the sensor indication from a reference value, from 10 % to 30 % are generally appropriate. Conversely, the sensor measuring range required for detection of toxic gases can be in the 10^{-6} (ppm) range. The upper measuring limit of the sensor should be 2 to 5 times higher than the target gas threshold concentration to allow continuous indication of larger concentrations of the toxic gases [70]. Acceptable uncertainty depends on the gas concentration and measuring task and is typically from 20 % to 50 %.

Other sensor performance parameters, such as limit of detection, resolution, warm-up time, response time, recovery time, drift, and behaviour against overload, vary in importance depending on the intended application. A compilation of these and other performance parameters for hydrogen sensors is given in Table 6.8. Another important consideration is the cross sensitivity of a sensor to other gases which may be present in the working environment. Sensors which are sensitive to gases other than the target gas may provide false positive or, more critically, false negative readings.

In the case of particularly harsh working conditions where precise detection of gases may be difficult, other strategies for indirect indication of gas presence, such as pressure or temperature, may be easier to obtain and more cost effective. For example, a pressure sensor can indirectly signal the release of hydrogen into the environment from a pressure drop in a cylinder/tank. A temperature sensor can indicate the catalytic combustion of released hydrogen by a passive auto-catalytic re-combiner in the confinement of a nuclear power plant (see Section 7.6).

Period of Use

Sensors used for continuous monitoring and for control of processes are typically switched on and used continuously or at least for the duration that the process is running. For such applications if sensors with a limited lifetime, due to their sensing principle, such as electrochemical sensors, are used then appropriate sensor maintenance procedures should be implemented. In other applications sensors can be used non-continuously. Portable sensors for indication of toxic gases for health protection is one example of non-continuous use sensors. Such sensors should be able to operate over the whole working day of 8 hours. The frequency of switching on and off may have an influence on the operating life of the sensor.

Mobility

Sensors may be permanently mounted (stationary), transportable, or portable. Accordingly different requirements may arise for power supply, size, and mass of the sensor. Whereas stationary sensors can get power from the grid, portable sensors should have a modest power consumption to allow powering with batteries for reasonable durations. Transportable and portable sensors should be robust against mechanical loads from falling, shock, and vibrations.

Sensor Location

Choosing the correct position for installing permanently mounted sensors is a critical consideration—gas sensors will only be effective if they are positioned proximate to where the gas to be detected may be found. This may be achieved using several individual sensor systems (where sensing element, electronics, and display are integrated in one device) or alternatively by using several basic sensing elements which are networked such that their primary signals are collected and processed by one central electronic control unit.

Sensors for leak detection should be mounted in positions where gas leaks are foreseeable, e.g., close to pipeline connections such as valves. The density of the gas to be detected relative to air will determine whether the sensor should be installed at a height for gases lighter than air (e.g., hydrogen) or whether they should be installed close to the ground for heavy gases (e.g., butane, propane). Sensors for gases with densities comparable to air (e.g., carbon monoxide) should be mounted around head height. When the area to be monitored is large, several sensors may need to be used or wide area monitoring techniques may be deployed.

Sensor Working Environment

A gas sensor can be used in ambient conditions of temperature, pressure, and humidity which can vary in relation to the global location, geographical altitude, and season. The tolerance of sensor performance to changes in these environmental conditions is defined in test standards, such as IEC/EN 690079-29-1; however, the speed and frequency of changes in a sensor's working environment are of particular relevance to precise measurements and long term stability. Furthermore, air movement is of interest because low air circulation can exacerbate condensation. Catalytic and semiconductor sensors working at elevated temperatures tend to be immune to condensation compared to electrochemical and other sensors which work at near room temperature. Dust in the environment surrounding the sensor can clog the sensor filter cap or cover the sensing surface, potentially reducing the sensor signal or slowing the sensor's response time.

A sensor's ambient working environment can also be "harsh." For example, hydrogen sensors, used to measure hydrogen concentration in the exhaust gas stream of PEM fuel cells, are required to operate at temperatures up to 90 °C and relative humidity approaching 100 %, i.e., condensing conditions. The immunity of a sensor's performance to adverse environmental conditions determines whether direct gas concentration measurement is possible or whether a bypass or gas sampling is required.

A further, less obvious consideration for sensor selection concerns the potential for changes in an application's background gas matrix and its possible influence on a sensor's ability to detect the target gas. In particular deviations in oxygen concentration, which can occur during system purging with an inert gas (e.g., nitrogen or argon), may change sensor functionality and even give a permanent malfunction for sensors whose detection principle involves oxygen e.g., catalytic combustion or semiconductor conductometric, sensors (see Chapter 3). In applications where oxygen depletion may occur, oxygen sensors are essential to alert to the danger of asphyxiation (see Section 4.1).

Sensor Use and Maintenance

Ideally, a sensor system should be easy to use with moderate requirements with regard to the qualification of the user. Furthermore sensor signal output and communication protocols are relevant for compatibility, comparability, and integration with other systems for measuring and control, such as those for fire detection and fire suppression.

Regardless of type or function all sensors require maintenance during their operating life including interventions such as periodic inspection, functional tests, sensor recalibration, or hardware replacement including sensing element substitution for sensors with a limited operating lifetime. Maintenance activities should be performed regularly and recommendations for maintenance and calibration time intervals are provided in technical guidelines, such as [518, 519], and are typically in the range of one month (visual check) to 1 year (functional check). Typical checks and tests performed during a sensor scheduled maintenance include:

- Zero reading check in clean air
- Correct sensor response indication to a known concentration of analyte
- Replacement of safety relevant and auxiliary sensor components, such as sensing elements, sockets, cables, and filters
- Plausibility of recorded data (if appropriate)
- Status of battery for portable sensors

In stationary applications, where a sensor should work continuously and reliably for extended periods of time, maintenance intervals of more than one year are desirable. When scheduled, maintenance of sensors should be easily performed by qualified facility personnel with minimal specific training requirements. When necessary, component replacement should be performed at reasonable costs. During routine maintenance procedures (e.g., calibration) or system overhaul (e.g., sensing element replacement and functionality verification), safety features such as explosion protection must not be compromised.

Gas Sensor Specifications

Gas sensor products can deploy different sensing technology platforms. Each platform demonstrates particular trends and limitations with respect to specific performance parameters. Table 6.7 summarises some typical performance capabilities and restrictions for a number of sensing technologies. Some commercial sensor products are available which consist of two or more sensing elements of different technologies. This strategy is used to compensate the mutual limitations of the respective technologies to achieve better overall performance of the sensor product. In addition to these general considerations, a large number of technical parameters are to be taken into account for sensor selection.

Apparent simple specifications of a sensor such as its size (e.g., < 500 cm^3) or mass (e.g., < 100 g) can be relevant for the intended application particularly for transportable or portable devices. Mandatory considerations include consideration of the supplier's declaration of foreseen application field, a user's manual, and the suitability for continuous operation or spot measurements. Table 6.8 compiles sensor performance parameters, requirements from standards of the IEC/EN 60079-29 series and declarations from data sheets of suppliers. Many gas sensors fulfill the spectrum of requirements and in some cases will surpass them.

Table 6.7: Comparison of application areas of different sensing platforms

	Sensing Platform Catalytic combustion	Thermal conductivity	Electro-chemical	Conducto-metric	Work function	Optical IR	PID
Flammable gases	+++	+++	++	++	++	+++	+++
Range	$\leq$ LFL	0.1 - 100 vol%	$\leq$ LFL	$\leq$LFL	0 - 10 vol%	0 - 100 vol%	$\leq$ LFL
Non detectable gases	large mol.	*)	alkanes inert gases	inert gases	-	H_2	H_2, CO
Toxic gases	(+)	-	+++	+++	+++	+	+
Oxygen	-	(+)	+++	-	-	-	-

The symbols +++, ++ to - indicate the degree of suitability of sensor platform application from very suitable to less and impossible.
*) Due to its sensing principle a sufficient difference in the thermal conductivity of the analyte in comparison to the atmosphere is needed.

Table 6.8: Parameters of hydrogen sensors as criteria for sensor selection

Analytical Parameters	**Test conditions**	**Requirements**[8]
Storage	-25 °C and 60 °C for 24 h	No influence on subsequent tests
Warm-up-time	Indication after 5 min	5 % to 10 %
Measurement		20 %
range in air	10^{-4} to 1 vol%	10 %
(or specified	(0.5 - 4) vol%	10 %
for other gases)	(10 - 100) % of LFL)	10 %
	(0 - 100) vol%	45 - 50 % of measuring range
Response time		
t_{50}	standard test gas conc.	< 20 s
t_{90}		< 60 s
Recovery time		
t_{50}		< 20 s
t_{10}		< 60 s
Short term stability	6 × 3 min	10 %
Long term stability (4 weeks)	7 × 8 h	10 %, 20 % to 30 % [9)]
Resolution	test not required	
Environmental Parameters	**Test Conditions**	**Requirements**
Temperature		
Portable/ transportable	-10 °C to 40 °C	10 % to 20 % [9]
Remote sensors	-25 °C to 55 °C	20 %
Separate control unit	5 °C to 55 °C	20%
Fixed sensor with integral sensor	-10 °C to 55 °C	15 % to 30 %
Pressure	(80 - 120) kPa	30 % to 40 % [9)]
Relative humidity	(20 - 90) % @ 40 °C	30 %
Condensation	test not required	
Cross sensitivity / selectivity	CO_2, ethene	
Poisoning	10^{-5} of HMDS	10 %
Gas flow rate	(0 - 6) m/s	10 %
Operational Parameters	**Test Conditions**	**Requirements**
Alarm function (fixed or adjustable)	test	< 120 s

Continued on next page

[8]According to 60079-29-1, deviation of reading from reference, so for detail look there.
[9]Depends on measuring range and type of device (portable or stationary, etc.)

Table 6.8 – *Continued from previous page*

Failure and out of detection indication	No power indication wrong or no signal i. a. no gas flow	indication
Power supply, grid	80 % to 115 % variation of voltage	10 % to 15 %
or battery	8 h to 10 h operating time	5 % to 20 %
Mechanical robustness		
Drop test	3 x fall from:	
portable	1 m height	no loss
transportable ($\leq$ 5 kg)	0.3 m height	of functionality
transportable ($\geq$ 5 kg)	0.1 m height	
vibration	10 Hz to 30 Hz 1 mm excursion 31 Hz to 100 Hz 2 g acceleration	

Deployment Parameters	**Test Conditions**	**Requirements**
Approvals	ISO 24162 EN 60079-29-1 ATEX 94/9/EG EN 50402	
Certification	e.g., ATEX IIG EEx dia IIC T1 or ATEX IIG EEx ib IIC T1	
Functional safety	IEC 61508	
EMV Electrical protection class test according	IP 54, IP 6K7	3 % to 10 %

Selection Algorithm

When choosing a sensor the relative importance of the performance parameters required for the applicaton and the permissible investment and maintenance costs all influence the decisions about which sensor to use. Typically there are a wide range of sensors available on the market from which the user can choose. Mathematical methods have been developed to assist in the selection of a certain product. Using these methods the required performance with respect to various parameters can be considered as selection criteria, each of which can be weighted by scores. The product gaining the highest score is supported for selection by this method. Mathematical methods specifically for the selection of sensors have been developed. A generic mathematical selection method is based on the analytic hierarchy process (AHP), a multi-criteria based decision-making tool [526]. The method has been adapted to sensor selection [527], as this method may prove helpful to some users for selecting a sensor for their application.

6.4.3 Installation of Sensors

Following selection of the most appropriate sensor the end-user must decide where and how to install this sensor to ensure optimal effectiveness. This is particularly, but not exclusively, true for leak detection and safety sensors. A risk analysis should be performed to identify the relevant risks. The risk assessment may include modelling of gas dispersion and should identify the possible flammable or toxic gases which may occur, the source of their release or production, the potential amount of gas which can be released and the volume and relative confinement into which the gas can be released.

A gas sensor should be installed and, if required, integrated into a gas detection safety system, such that it can detect a change in gas concentration or gas leakage and initiate mitigating actions.

There are three basic sensor positioning strategies for stationary applications that can be distinguished; however, it may be necessary to combine or adapt these strategies for specific applications:

- Area monitoring, when potential leaks are spread across large areas, e.g., for room surveillance, which may require a larger number of sensors.
- Spot monitoring, when the location of a possible gas release is known. The sensor can be placed in such a way, that the gas leak can rapidly be detected.
- Perimeter monitoring, an indirect surveillance outside of the room or area where leakage can occur and cross to neighboring unsecured areas.

In the following, considerations and hints for the installation of stationary sensors are presented specifically with regards to indoor surveillance and also for triggering an alarm and taking measures for mitigation.

Gas to be Detected

Correct positioning and installation of a sensor for safety reasons is a complicated task which depends on the properties of the target gas, e.g., its density and diffusivity. For this reason a basic knowledge of the properties of the gas is essential (see Section 1.2). Gases which have a density lower than air and a molar mass less than the average molar mass of air (ca. 29) will rise when released in air. With a density 14 times less than air, hydrogen is extremely buoyant. Conversely, gases or vapours which are denser than air, such as carbon dioxide or liquefied petroleum gas (LPG), will tend to fall when released due to their higher density compared to air. Consequently the positioning of sensors at a height or close to the ground is determined by the target gas' propensity to rise or fall following release. Sensors for gases which are significantly lighter than air (unless they are very cold) should be installed close to the ceiling whereas sensors for heavier gases are more suitably installed close to ground level.

The positive buoyancy and rapid molecular diffusion of gaseous hydrogen means that any release will quickly mix with the surrounding gases and the hydrogen will be rapidly diluted. Therefore the threshold concentration may not be reached at the sensor and a leak may be hard to detect if the sensor is not located correctly. The sensor should be located close to the potential leak site or above this site where hydrogen may accumulate following leakage.

The temperature effect of the liberated gas should also be considered. For example when hydrogen is released as a liquid from a storage vessel its density is initially greater than that of air. This causes it to settle to the ground before heating up, becoming lighter than air and rising. The impact of pressure and gas flow of liberated pressurised gas should also be taken into account for sensor location.

Detection Equipment Specifications and Integration

Only sensors which, according to the specification data of the supplier, are suitable for the detection of the gases or vapours of interest, should be installed. The correct operation of the sensors shall be checked after its installation, as a practical proficiency test for correct indication of critical target gas concentrations. When positioning a sensor, care should be taken not to hinder the regular inspection or maintenance of the sensor. Sensors must also be accessible for re-calibration.

The demands of high functional safety should also be considered if a failure of the sensor could give rise to dangerous situations. In such applications, where failure or indeed removal of an individual sensor could possibly compromise safety, a redundant safety sensor system is required. In this case redundancy, by means of duplication (or triplication) of sensors and control apparatus, is recommended. Depending on a sensor's location and operating environment, the time required to detect a gas leak and for mitigating measures to be enacted, e.g., shutting off of the hydrogen source, depends on factors such as the gas leak rate, room ventilation rate, and the volume of the space being monitored. Accordingly the number and positioning of gas sensors should take these factors into consideration as part of the risk analysis.

The position of the sensor display and control panel should be considered. It should be possible to conveniently and safely read the sensor indication. The sensor control box with readout should be mounted on a wall, where possible, near an exit, or at a height of about 1.5 m to 2 m to allow reasonable access.

The alarm levels of the installed sensor is also a critical consideration. It has to be decided which alarm threshold is appropriate. For example, the alarm indication can be set at different gas concentrations, e.g., at 10 % and 40 % of the lower flammability limit (LFL) of the gas to be detected. The user should also be aware of how the alarm information is subsequently processed, what switching function the sensor shall initiate, and how fast the process interruption shall be executed in case of an alarm. In general, alarm indications should be set to allow sufficient time for personal evacuation and countermeasures to be taken.

In order to define the lowest acceptable alarm threshold, it has to be considered that every sensor has an uncertainty, described in part by the standard deviation, repeatability, or signal noise. As a result, there is a certain probability that incorrect alarms may occur, especially if the alarm threshold is set near (or worse, below) the lower limit of a sensor's measuring range. In safety technology, false positive alarms are regarded as dangerous, because if they occur too often they will be ignored, leading to hazardous situations in the event of a true alarm. A reliable alarm limit, in particular for detection of toxic gases, can be calculated on the basis of sensor data. The lowest alarm limit should not be less than six times the zero-point standard deviation under actual operating conditions. In many cases a sensor can be set to give at least two alarms. The first alarm level, or pre-alarm, prompts mitigation countermeasures such as exhaust ventilation while the main alarm triggers resolute measures such as auto-shutdown, switching off machinery, forced ventilation, activation of fire-fighting procedures, and building evacuation. The time needed to trigger an alarm by the sensor system shall be checked after installation.

For toxic gases, the MAC maximum allowable concentration or TLV-TWA threshold limit values can be used to determine the alarm threshold. For combustible gases 10 % LFL and 20 % LFL for portable gas sensors and 20 % LFL and 40 % LFL for stationary sensor are often recommended as the alarm thresholds. Nevertheless, these values should be validated for each application which may have its own specific requirements. For hydrogen releases alarm thresholds of 25 % LFL and 60 % LFL have been recommended [528]. Actual alarm set points for many applications are often established by codes in the U.S.

Stationary Sensor Location

When sensors are required for safety monitoring and/or control of large facilities or process plants several sensors may be necessary depending on the size of the area to be monitored and the properties of the target gas. A sufficient number of sensors should be installed in order to detect a leakage fast. When choosing the location to install these sensor the following points should be considered:

- Sensors should be located near the potential leak source or in the path of the process gas or vapour. Leakages may occur at pipework flanges, damaged valves or seals, filling nozzles, bellows, or gas cylinder filling equipment. The distance between the sensor and the source should not be too long such that dilution of critical gas concentrations can occur.
- Gas sampling or positioning of the sensor in a bypass is required when harsh ambient conditions (e.g., high temperature or extreme vibration) preclude direct gas measurement.
- Sensor measurements should not be disturbed by environmental conditions. Vibration, mechanical damage, excessive heating, contamination, exposure to dust or water condensation should be avoided. If no alternative locations are possible, other precautions may be taken such as vibration isolation mountings or the use of filters. When such precautions are not implemented sensor lifetime and performance are likely to be compromised.
- Positioning of sensors should consider the air flow in the space being monitored. Air flow arising from passive or active ventilation may promote dilution which may delay or prevent the detection of critical gas concentrations. Sensors should not be located in the stream of the incoming fresh air. Gas surveillance sensors should not be installed near doors or windows due to possible gas dilution effects.
- Gas detectors, used to warn of gas entering a confined space from outside (via air ducts, drains, or underground cable conduits) should be placed near the intake side.

Use of Portable Sensors for Personal Protection

Inspection and functional checks of portable sensors should be performed prior to their use to verify that the sensor is in proper working order. A field calibration should be performed regularly and monitoring of the available remaining battery power is recommended. Possible interferences from user breathing and transpiration, i.e., from humidity, carbon dioxide, or odorants, should be avoided.

6.4.4 Summary

Various regulations, codes, and standards provide a frame for the deployment of sensors for safety applications and to some extent also for process control. Considerations of the requirements in standards is in many cases mandatory and demands financial and personal efforts. Compliance with standards contributes significantly to ensuring that sensors are correctly deployed and that risk mitigation is optimised according to the state of art.

The description of the sensor test methods and protocols, in many cases based on regulations and standards, contributes to transparency in performance assessment of sensors. The use of reference materials (certified gas mixtures) and the comparison of results obtained using precise analytical methods contribute to the traceability of sensor indication and facilitates comparison of the performance between different sensors.

The effectiveness of sensor deployment to facilitate safety depends, to a large extent, on a suitable sensor selection and optimal installation. Nevertheless, this does not preclude the need for regular monitoring and maintenance by safety engineers and end users once the sensors have been installed.

CHAPTER 7

Sensor Applications

CONTENTS

7.1 OVERVIEW

A range of different sensors can be used along the whole process chain of various hydrogen technologies. In this chapter the role of sensors is presented in representative technologies, which include hydrogen production, storage, and distribution as well as processes which consume hydrogen as a fuel or chemical feedstock or generate hydrogen as a by-product. Sensors are used for the following purposes:

- To ensure the health and safety of workers, safeguard facilities, and protect the environment
- To control devices, facilities, processes, and products

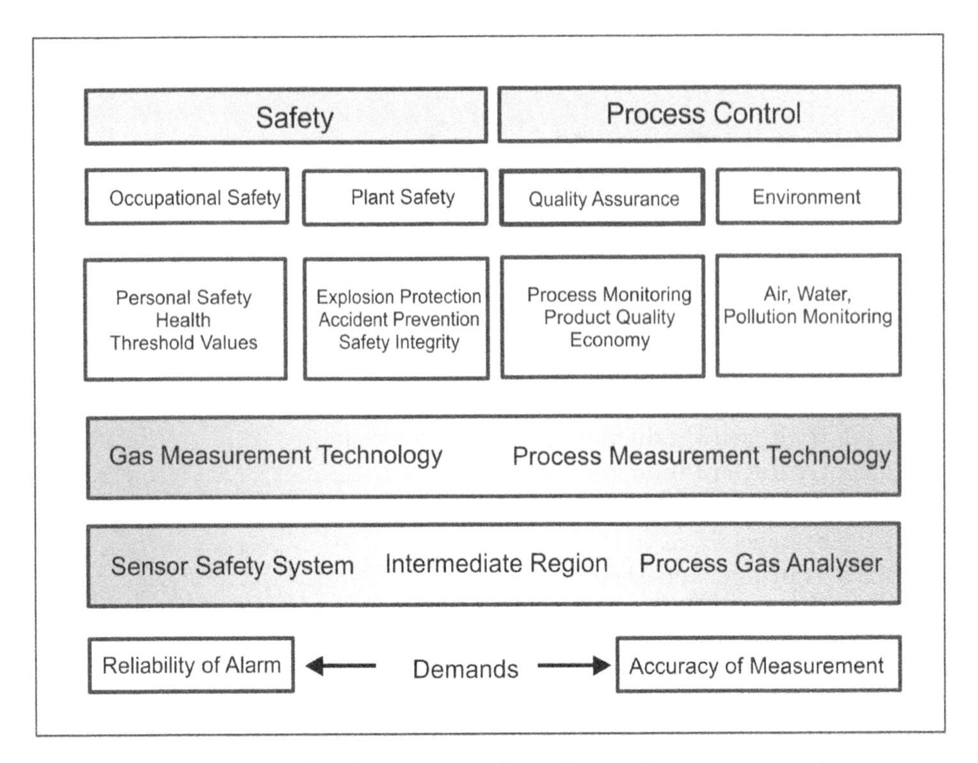

Figure 7.1: Role of gas sensors for safety and process control.

- To detect technical faults, such as unwanted by-products or leaks

This variety of objectives imposes different requirements on sensors. The critical demands on sensor performance can differ with respect to these targets and may also overlap as the schematic in Figure 7.1 illustrates.

Safety sensors are deployed for assuring an acceptable safety level for personnel and for plant or process operations. Indeed, safety sensors are often a critical element of an independent safety instrumented system (SIS), and as discussed in Section 6.1, are necessary for assuring safety integrity levels. Sensors are critical measuring devices for the detection of small leaks or sudden gas releases. Sensors can also function as the actuator for activating an alarm or mitigation measures, e.g., active ventilation or system shut-down. Furthermore, sensors play an active role in fire detection (see Section 5.4). In general, safety sensors operate independent of the process for which they are monitoring. In contrast, process control sensors and measuring devices are used to provide information on the properties and status of processes and products, and thus are integrated into systems being monitored. These sensors work in real time and are important for automation, environmental compliance, process optimisation, reduction of production costs, and understanding the process cycle. Figure 7.2 illustrates different aspects of process control sensors.

The focus of this chapter is on the role and function of hydrogen and other sensors in specific hydrogen technologies. A general overview on relevant hydrogen technologies is provided in Section 1.4. More specifically this chapter includes descriptions of sensor use in the production of hydrogen, hydrogen storage, hydrogen distribution, and the use of hydrogen in stationary and mobile fuel cells. In addition, sensors in processes using hydrogen as a coolant or chemical reagent are also considered as is the use of hydrogen sensors for the detection of technical faults in critical applications.

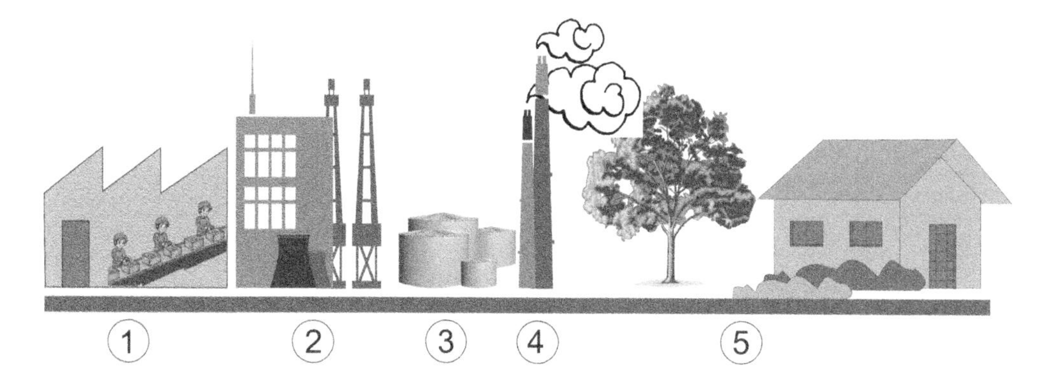

Figure 7.2: Sensors for process control: Monitoring of workplace (1), System state and plant availability (2), Product quality (3), Emissions monitoring and control (4), Environment compliance and control (5).

7.2 SENSORS USED IN HYDROGEN PRODUCTION

7.2.1 Hydrogen Production by Steam Reforming

Steam reforming and partial oxidation are two of the most commonly used technologies for hydrogen production. Steam reforming of methane (SMR) is currently the most cost-effective method for large scale hydrogen production. The basic principles and components of SMR, including the shift converter and the pressure swing absorber (PSA) for purification, are described in Section 1.4. Further details are available elsewhere [529].

Large scale production of hydrogen entails potential hazards that must be addressed in the design and operation of the reformer. All instrumentation and control functions associated with SMR and PSA operation need to obtain the requisite safety integrity level (SIL) (see Section 6.1). An independent safety instrumented system (SIS) is to assure compliance to the relevant SIL level. An SIS often includes an interlock system consisting of a sensor and a control unit that ensures the process is in a safe state or, in the event of a malfunction, ensures that the system is returned to a safe state (including system shut-down).

A fire protection system should be installed in the vicinity of the SMR, such as, in the burner area [530]. For this purpose flame detectors for monitoring radiation in either the UV or IR range should be used (see Section 5.4).

The use of a variety of sensors for both safety functions, e.g., area monitors for external gas releases, and process control in SMR and PSA are well established. For this purpose temperature, pressure, and the amount of combustible and toxic gases, such as H_2, CH_4, CO, CO_2, O_2, NO, NO_2, and H_2S, is routinely measured.

Sensors to detect many of these gases are mandated as safety devices by the International Fire Code [38], NFPA 2 [476], and by the ATEX directives as described in Chapter 6. Specific requirements with regards to hydrogen monitoring are defined in standards including IFC, NFPA, or ISO 60079-29-1 [488]. Additionally, requirements on operational parameters (see Section 2.4) such as robustness against poisons and dust, sensor life time ($>$ 5 years), calibration intervals ($>$ 1 per year), and cost ($<$ 400 EUR or 500 USD per sensor), have been proposed as essential for deployment of sensors in hydrogen production applications [531].

Gas sensors can also be required as part of personal protection equipment (PPE) to be used when working in an SMR plant. Due to the potential hazards from heat, fire, explosion,

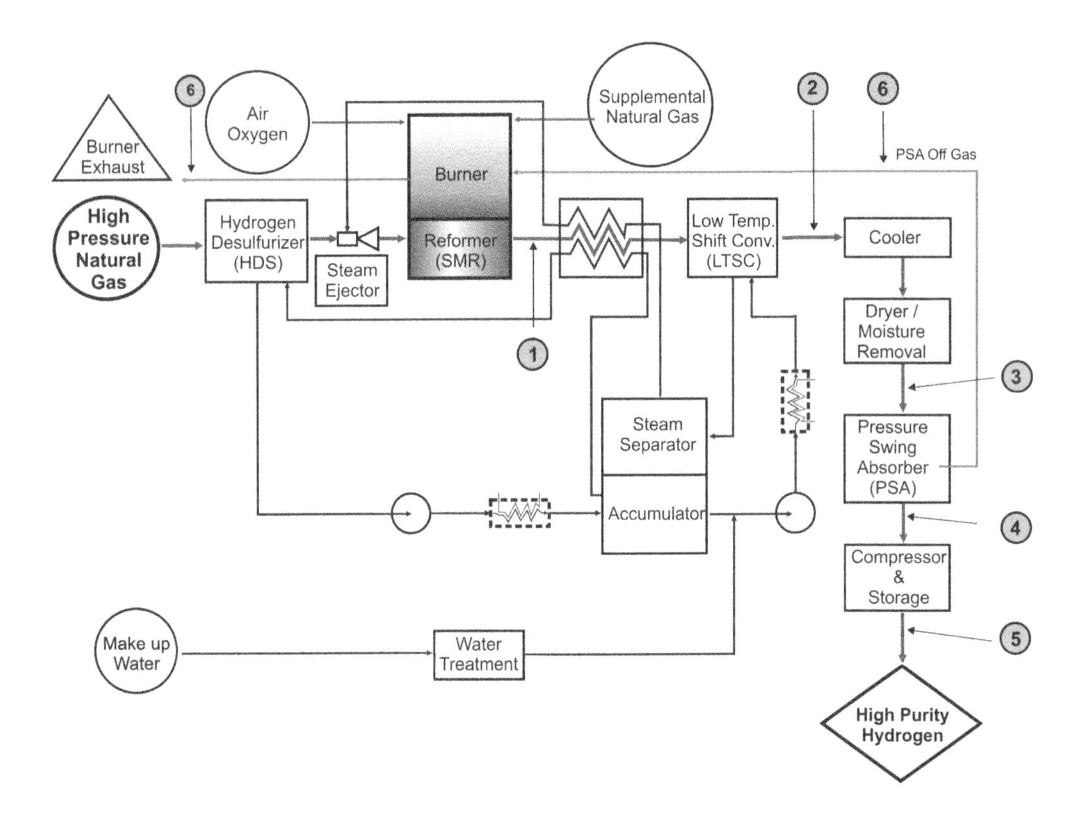

Figure 7.3: Representative locations (1 - 6) for hydrogen measurements in the steam reforming and hydrogen purification process.

toxicity, and asphyxiation, PPE including heat protection clothing and portable personal gas monitors are recommended [532] and often mandated. Leaked hydrogen can be detected using different technologies as discussed in Chapter 3. Stationary and portable handheld sensors for the detection of CO, CO_2, O_2, NO, NO_2, H_2S, and CH_4 are also commercially available (see Section 4.2). To ensure optimum reformer operation a combustion control system is installed to measure and control furnace pressure and temperature. Temperature, pressure, and gas composition of the system are monitored using sensors, and measurements are displayed on the display unit. In addition, gas sensors or analytical devices are installed in different areas of the reformer to monitor fuel levels, gas flow, product composition, and emissions.

The final product, hydrogen, is purified via the PSA process. SMR units will produce a gas stream containing typically 80 % hydrogen, which is then passed through a PSA to produce hydrogen at the required purity level which can be as high as 99.999 %. This purified hydrogen is then fed into other processes, such as the refining of petroleum products. Verification of the hydrogen purity is essential to ensure that the PSA is operating efficiently and to show compliance to various hydrogen purity levels. Figure 7.3 indicates possible places requiring gas composition measurements in the processes associated with SMR hydrogen production and PSA hydrogen purification.

Whereas the process gases (e.g., the natural gas feedstock or products) are typically analysed by gas chromatography, mass spectrometry, or thermal conductivity sensors, the flue gases (i.e., gas exiting to the atmosphere through a stack) are often analysed in-situ using sensors. Flue gases typically include O_2, CO, H_2O, CO_2, and, depending on the particular process, industrial pollutants such as SO_2 and NO_x. Optical sensors, especially those based on laser absorption spectroscopy (LAS) described in Section 6.3, can be used to measure many of the flue and process stream gases [533] except hydrogen. A field effect based sensor (see Section 3.6) based on $Pt/SiO_2/SiC$ is being developed for the detection of hydrogen in the harsh conditions of the flue gas. This sensor operates up to 620 °C and can detect hydrogen in the range from 0.005 vol% to 50 vol% H_2 [534]. Furthermore this sensor reportedly shows no effect on its response to hydrogen in the presence of 40 vol% CO and 5 vol% CH_4.

7.2.2 Hydrogen Production by Electrolysis

The electrolysis of water is another method for the production of hydrogen. This method currently represents only about 4 % of the total worldwide hydrogen production, but this is likely to increase as water electrolysis is a popular on-site hydrogen production method for existing and proposed hydrogen fuelling stations. Moreover, the hydrogen would be viewed as a renewable and environmentally friendly energy carrier provided the required electricity is from renewable sources, e.g., from wind or solar energy. This can be an important consideration for environmentally conscientious consumers. A general description of hydrogen production via water electrolysis is described in Section 1.5 and elsewhere [535].

Hydrogen production via water electrolysis requires sensors for both process gas control and for monitoring of the surrounding atmosphere to ensure safety. The safety and operational requirements associated with hydrogen production through electrolysis is addressed in ISO 22734 "Hydrogen generators using water electrolysis process" [536]. This standard requires a classification of the hydrogen generator enclosure according to IEC EN 60079-10-1 [447]. The standard states that the safety system and the protection equipment shall be flameproof and intrinsically safe (see Section 6.1). Electrolysers are also commonly equipped with a hydrogen leak detector which should be located close to likely points of hydrogen leaks. Passive measures to prevent the accumulation of ignitable gas mixtures are also often implemented and include natural ventilation and gas flow restrictions to limit gas release rates. The hydrogen gas detection system should indicate an alarm as soon as possible after hydrogen is released and should activate safety procedures to avoid hazardous situations from arising. These procedures can include activation of forced ventilation or even purging of the area/system with an inert gas to dilute the released hydrogen. Such safety measures are typically activated when the hydrogen level in air exceeds 0.4 vol% (e.g., 10 % of the hydrogen LFL).

It is also necessary to control the operational parameters of the electrolysers to be within the specified limits of pressure, temperature, current, voltage and hydrogen composition. Sensors monitoring these parameters are installed in commercial electrolysers. Gas sensors are also recommended to monitor the content of oxygen in the hydrogen stream and the content of hydrogen in the oxygen stream. Although it can vary with the process design, the oxygen content in hydrogen is normally in the range of 0.02 vol% to 0.003 vol% while the hydrogen content in oxygen is in the range of 0.1 vol% to 0.2 vol% [537]. Because the oxygen gas stream is often at 100 °C or higher and at a relative humidity of up to 95 %, a by-pass sampling system is used to monitor hydrogen in the oxygen stream. An emergency stop is activated if the typical limits of either 1 % hydrogen in air, 2 % hydrogen in oxygen, or 1.6 % oxygen in hydrogen are exceeded [536]. Hydrogen sensor types used in electrolysers include conductometric, catalytic combustion, field effect-based or thermal conductivity sensors; however, other hydrogen sensor platform types are being investigated for use in the electrolyser gas streams [538].

Hydrogen is also produced as a by-product in the electrolysis of sodium chloride solution in the chloralkali process (see Section 1.4). The hydrogen is not very pure, nevertheless the commercial use of hydrogen generated by this method is being explored because the amount is significant and can offset the cost of chlorine production. The chlorine gas stream contains hydrogen, oxygen, nitrogen, and carbon dioxide which can be removed by filtration, drying, liquefaction and evaporation. Hydrogen levels in chlorine gas must be maintained below the lower explosion limit of 3.5 % H_2 in Cl_2 [539]. Increasing hydrogen levels are indicative of a potential membrane failure in the electrolyser cells. The hydrogen content in chlorine gas can be analysed by gas chromatography. Alternatively, an indirect determination of hydrogen can be performed using thermal conductivity (TC) sensors. For this method hydrogen is converted in a reactor into hydrochloric gas by thermal energy or UV radiation. There are commercially available devices developed for the chlorine industry that measure the thermal conductivity of the gas sample before and after the reactor. The hydrogen content in chlorine gas is then calculated from the differences in the thermal conductivity. This method has been used to determine up to 10 vol% H_2 in chlorine. The TC sensors are robust, corrosion resistant, and have a response time of less than 30 s. Hydrogen concentration can also be calculated from the intensity of optical absorption techniques associated with the formation of gaseous hydrochloric acid.

7.3 SENSORS FOR HYDROGEN INFRASTRUCTURE

7.3.1 Storage

Hydrogen storage technologies can be subdivided into physical storage, where hydrogen is stored in its molecular form either as a compressed gas or liquid and chemical storage, where hydrogen forms a chemical bond with other compounds.

Compressed Hydrogen Storage

High-pressure hydrogen is stored in thick-walled tanks made of metal or composite materials. Metal storage tanks are typically used for bulk stationary hydrogen storage. Stationary gaseous hydrogen is stored at pressures typically around 20 MPa (2,900 psi), but can be up to 100 MPa (14,500 psi). Hydrogen fuel tank systems for vehicle on-board storage are typically carbon fibre reinforced polymer composite tanks with a plastic or metal liner. Vehicle storage tanks have a normal working pressure of either 35 MPa or more commonly 70 MPa. Short-term over-pressures are allowed. Transient overpressure can occur during the refilling of a tank when the heat of compression causes the temperature and pressure in the tank to increase. This transient pressure should remain below the maximum allowable working pressure (MAWP), which is 125 % of the normal working pressure [540]. Composite storage tanks are used on-board vehicles because they weigh significantly less than steel tanks with a comparable pressure rating. The tank systems typically contain temperature and pressure sensors [541]. The vehicle fuel system is also equipped with a pressure relief device (PRD)[1] and shut-off valve. Corrective measures, such as activation of the PRD are initiated if the tank pressure

[1]Pressure relief devices (PRD) are used to protect pressurised vessels or other containment against unacceptable pressure (or temperature) levels. PRDs operate automatically with a full-opening pop action to relieve internal pressure. Some PRD designs re-close when the pressure drops to a value consistent with the requirements prescribed by the applicable governing code or standard [542]. Rupture disks are non-re-closing since the disk is destroyed upon actuation. This type of device may be found in some pressure vessels where a spring-loaded pressure relief device is inappropriate due to the operating conditions or environment.

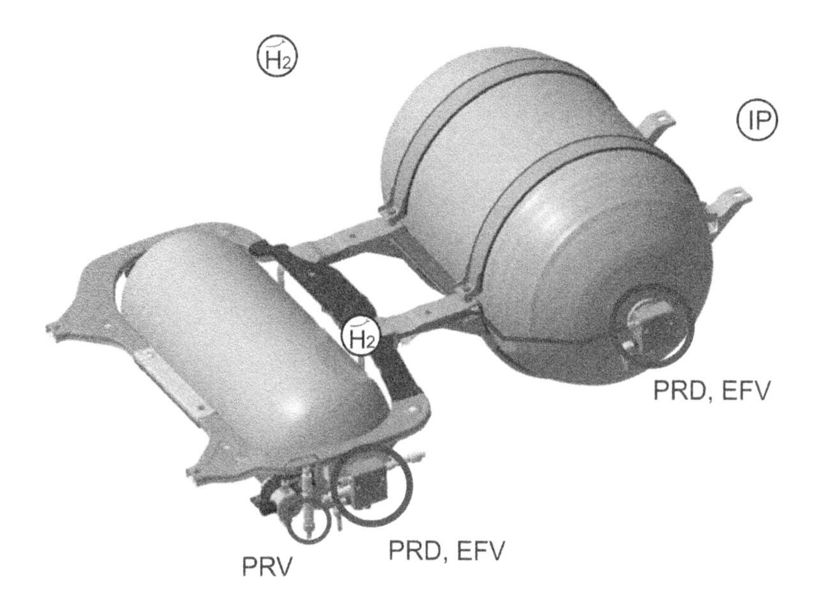

Figure 7.4: Safety devices for fuel cell vehicle Hyundai ix35 with high pressure hydrogen tank including hydrogen sensors, solenoid valves for pressure relief (PRV) and excess flow (EFV), as well as impact sensor (IP) according to [543].

exceeds the Maximum Operating Pressure (MOP), which is defined as 110 % of the MAWP [540]. The additional use of hydrogen sensors in vehicles for leak detection has been suggested although not mandated. An illustration of a high pressure hydrogen tank is shown in Figure 7.4.

Liquid Hydrogen Storage

Cryogenic vessels used to store liquid hydrogen (LH2) are commonly double-walled vessels with a vacuum between the walls to provide thermal insulation. The integrity of this vacuum needs to be monitored and maintained. Temperature and pressure of the vessel need to be strictly monitored and controlled to avoid over-pressurization of the tank. This can more readily occur if the vessel insulation is compromised, which leads to evaporation of the LH2.

Liquid hydrogen tanks have been considered for hydrogen storage in fuelling stations and in fuel cell electric vehicles (FCEV) (see Figure 7.5). The LH2 tank system incorporates several safety components. The fuel tank is equipped with overfill protection using a gauge that measures the level of liquid in the inner tank.

A pressure relief valve is installed for the inner tank. According to ISO 13985 [544], the fuel tank shall be equipped with a pressure maintaining system that is capable of maintaining the fuel tank to within its specified operating pressure range as specified by the manufacturer. The fuel tank shall also be equipped with shut-off valves for the purpose of securing the refuelling lines and the fuel [544]. Leakage due to hydrogen evaporation can be monitored by hydrogen sensors. The hydrogen evaporation rate is system specific and depends upon the particular design of the storage vessel. Although

Figure 7.5: A 114 litre liquid hydrogen highly insulated fuel tank (with permission of Magna Steyr).

often less than 1 % per day, the evaporation losses have proved prohibitive for deployment of this technology in automotive applications. Instead high pressure, gaseous hydrogen has been selected as the method of choice for on-board vehicle storage.

Chemical Hydrogen Storage The storage of hydrogen in solid or liquid chemicals is being developed (see Section 1.5). The advantage of chemical storage is that these systems operate at much lower pressures than the conventional means of gaseous hydrogen storage, thereby improving the fundamental safety of the storage system. At present, no solid storage material fulfills the typical targets set for automotive applications with hydrogen capacity and material stability being the major limitations. This remains an active area of research.

Although there is a relationship between the pressure and the amount of hydrogen contained in a metal hydride, the pressure level cannot easily be used to directly indicate the hydrogen content. Therefore other methods are needed to measure the amount of hydrogen in the storage vessel. An indirect approach is to measure the mass change of the storage system during in or outgoing hydrogen flux. An optical fibre sensor embedded in a solid hydrogen tank has also been suggested as a fuel gauge. The sensor consists of a UV-vis light source (LED), a cylindrical optical wave guide, and a spectrometer detector. Contact of the storage material with the fibre changes the optical properties and causes a change in light intensity which can be correlated to the hydrogen content [545]. A sensor, embedded into a metal hydride storage device, for measuring the hydrogen content according to changes of resistivity and susceptibility of the storage material, ($LaNi_5$), has also been proposed as an alternative fuel gauge [546].

7.3.2 Transport and Distribution

Hydrogen is commonly transported and distributed through pipelines or by road or rail transport using tube or tank trailers. For localized distribution, hydrogen is still more commonly transported by truck in either high pressure cylinders or as a cryogenic liquid.

Figure 7.6: Liquid hydrogen truck trailer (source: United Hydrogen Group).

Gas suppliers have truck fleets to transport compressed hydrogen in steel-tube or composite cylinders. Depending on the volume and pressure level, gas tube trailers can carry up to 400 kg of hydrogen. Liquid hydrogen can be transported in trucks equipped with double-walled tanks with a capacity of up to around 3500 kg. High thermal insulation reduces the hydrogen boil-off rate to below 1 % per day [547]. It is necessary to implement protection measures once hydrogen has been delivered to the end-site, especially during hydrogen transfer operations or when off-loading the transport vessels from the delivery vehicle.

In many large scale production operations, hydrogen is transported to the end-user via pipelines. Often this is to dedicated customers in, for example, the petroleum industry. Presently, the hydrogen pipeline network remains limited to industrial purposes. The German Rhine-Ruhr area and the American state of Louisiana both have pipelines of about 240 km in length. However, a new 965 km hydrogen pipeline supply network is currently under construction in the U.S. (Gulf Coast pipeline system). Globally there are a total of about 3000 km of hydrogen pipeline systems. Pipelines offer the advantage of connecting a centralized hydrogen production site to one or more points of use, enabling lower transport cost. In some cases pipelines made from steel tubes can deliver hydrogen at pressures up to 20 MPa. Hydrogen releases may occur from connections or joints, weld leaks, or even permeation through the walls of the pipeline. Leaks in surface pipelines can be detected, e.g., by colourimetric indicators (see Figure 7.7) or sonic sensors (see Section 3.9). Most of the current and planned hydrogen pipeline network is, however, underground, which restricts access to potential leaks. Special sensors for the detection of hydrogen releases from underground pipelines are commercially available. These sensors are portable and often based on conductometric or thermal conductivity sensing platforms (see Figure 7.8).

Power to gas (often abbreviated P2G) is a technology that converts renewable electrical power to a gaseous fuel, e.g., hydrogen (see Section 1.4.2). The hydrogen produced can be used directly as a fuel or feedstock or distributed by injection into the existing natural gas grid. Because of the high diffusity and reactivity of hydrogen, safety is of course a concern [548]. An additional concern associated with hydrogen distribution in the exiting natural gas pipeline network arises from diffusion of hydrogen into the metal components (such as compressors, pipes, fittings, and valves), which can cause material failures due to embrittlement and crack formation.

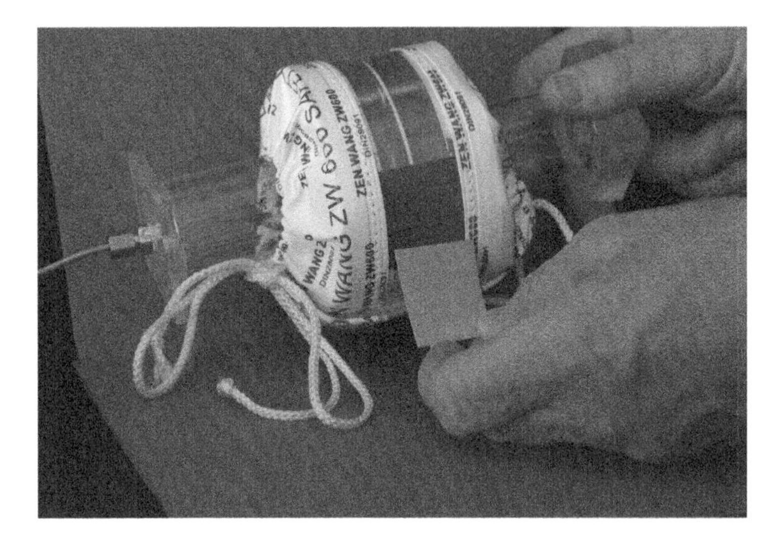

Figure 7.7: Coulorimetric leak indication (source: Element One Inc.)

Figure 7.8: Leak detection in underground pipeline using hydrogen sensor (source: INFICON).

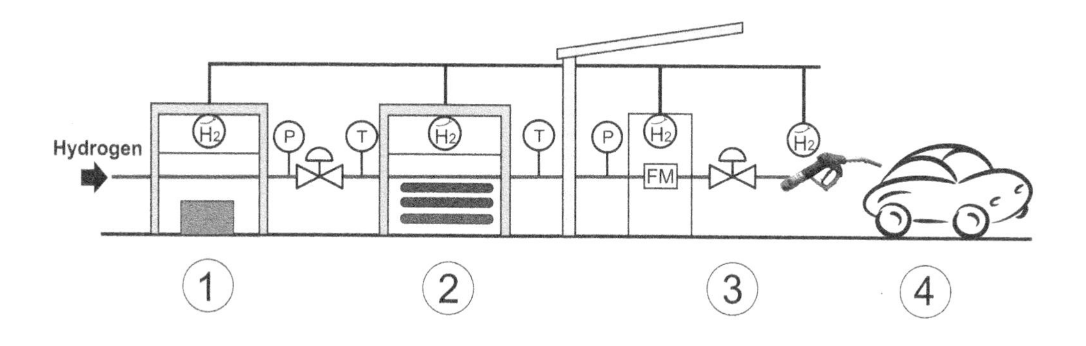

Figure 7.9: Schematic of a hydrogen fuelling station. 1 – Accumulation, 2 – Compressor, 3 – Dispenser, 4 – FCEV (according to [131]).

7.3.3 Hydrogen Fuelling

Hydrogen fuelling operations can be performed indoor or outdoors. In the U.S. monitoring of indoor fuelling stations, such as those that service forklifts in warehouses, requires the deployment of hydrogen sensors. This requirement is mandated by the NFPA 2 [476] (see Section 6.6). The sensor performance requirements are defined in the standards UL 2075 and FM Global Class 6310 and 6320 [479, 549, 550]. Hydrogen sensors can be mounted either on the wall adjacent to the hydrogen dispenser or above the dispenser. They are typically configured for an audible alarm if the hydrogen amount exceeds, for example 10 % or 25 % of the LFL, depending on local requirements. One proposed method to measure hydrogen in indoor or enclosed environments is based on the decreased oxygen level due to displacement from a hydrogen release [551]. Oxygen depletion can be detected by an oxygen sensor. Although the electrochemical oxygen sensor is a good technology for quantifying ambient oxygen levels, the use of this method for hydrogen detection should be used with extreme caution. The large overall uncertainty associated with the indirect measurement of hydrogen precludes the use of this method in many applications [552].

Outdoor fuelling stations that dispense gaseous hydrogen for hydrogen vehicles already exist, and their number is expected to increase to support the pending roll-out of commercial FCEV. Figure 7.9 illustrates the various subsystems of a hydrogen fuelling station. The dispenser is used to transfer hydrogen to the vehicle storage tank. Risks associated with vehicle refuelling include over-filling, over-pressuring, over-heating (greater than 85 °C), and excessive low temperatures (less than -40 °C). Station designs incorporate many different elements to assure safety, including engineering features, such as set back distances and pressure relief devices (PRD), as well as the use of sensors. Dispensers can be outfitted with a hydrogen sensor embedded inside the nozzle housing [553] or fail-safe valves that shut in the case of a system alarm, power loss, or when actuated by a remote emergency shut down switch. A control system manages dispenser operations, monitors for equipment malfunctions, and can even signal the need for hydrogen delivery when reserves run low. The dispenser's control system performs ongoing automatic leak checks, manages fuelling operations, and monitors for potentially dangerous operating conditions. If an alarm triggers or an emergency stop button is pressed, isolation valves close, thereby interrupting hydrogen flow and stopping the fuelling process. Specific requirements for hydrogen leak detection devices for this application are provided in ISO 26142 [79]. Different hydrogen sensor platforms have been applied in

Figure 7.10: Photograph of a hydrogen refuelling station indicating typical locations of hydrogen sensors.

hydrogen fuelling stations including catalytic, field effect-based [554], and thermoelectric hydrogen sensors [131]. Other platform types could be amenable for this application.

7.3.4 Vehicle Repair Facilities

As hydrogen fuelled vehicles are introduced onto the market, vehicle repair facilities will need to be adapted to accommodate these vehicles. Specific safety precautions are necessary to avoid hazardous situations resulting from possible hydrogen releases inside the workshop. Such precautions may include the installation of an active explosion-proof ventilation system on top of the building. Although there are differences in the specific requirements between the EU and the U.S., one common feature is that vehicle repair facilities should be equipped with a hydrogen detection system.

In Europe, a hydrogen warning is activated typically at 0.8 vol% H_2 in air (20 % of LFL) and the main alarm level is 1.6 vol% H_2 (40 % of LFL). When the main alarm is triggered, additional safety measures may be initiated such as automatic opening of windows and activation of the ventilation system to increase the air exchange rate to reduce the concentration of hydrogen. Personnel should be evacuated and explosion-proof lighting switched on. In the U.S. the hydrogen warning state is typically activated at 0.4 vol% H_2 (10 % of the LFL) leading to an audible alarm from the sensor system and activation of forced ventilation. System shut down occurs when 1 vol % H_2 is detected. In the U.S. specific requirements are defined in the International Fire Code IFC (2007 and newer editions). As described in Section 6.1, these requirements are legally enforceable if the jurisdiction has adopted the IFC.

Figure 7.11: Photo of a vehicle maintenance garage [555].

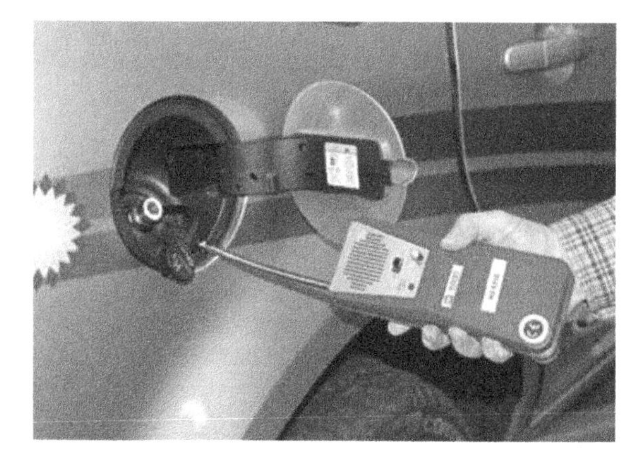

Figure 7.12: Handheld hydrogen sensor for leak tightness check after repair [555].

Hydrogen concentrations during maintenance on hydrogen vehicles in repair facilities, such as those illustrated in Figure 7.11, can be monitored using portable hydrogen or fixed site detectors. Additionally hand held sensors can be used for verification of the fuel system integrity after the vehicle has been subjected to maintenance procedures, such as in Figure 7.12.

7.3.5 Parking Garages

FCEVs will be parked in garages. Hydrogen sensors may need to be deployed to verify that hydrogen levels do not accumulate to dangerous levels. Alternatively, the use of sensors may not be necessary if it can be shown that natural or forced ventilation designs preclude the possibility of a dangerous build-up of hydrogen. However, if sensors are needed, it may be necessary to consider low-power hydrogen sensors, especially for remote facilities where reliable power may not be available and the device may need to operate under battery power for extended times. For any application which requires battery operation, sensors such as those based on a field effect sensing platform may be suitable due to its low power consumption [208]. The sensor should also be resistant to emission products from conventional vehicles.

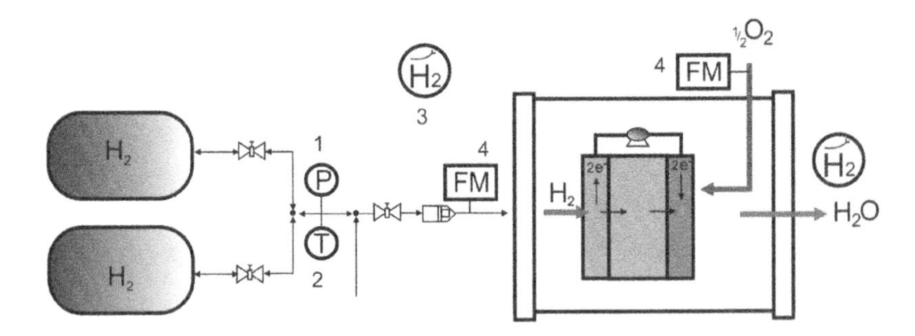

Figure 7.13: Schematic of a PEM FC indicating the location of sensors for pressure (1), temperature (2), hydrogen (3), and gas flow (4).

7.4 SENSORS FOR FUEL CELLS AND INTERNAL COMBUSTION ENGINES

7.4.1 Stationary Fuel Cell Applications

Fuel cell systems of different types are commercially available from 10 mW to more than 1 MW. Of all commercial types the proton exchange membrane fuel cell (PEM FC) and solid oxide fuel cell (SOFC) are among the most widely used. The global fuel cells market is growing steadily. In 2012 about 1,000 new stationary fuel cell systems were shipped and a total number of about 25,000 installations was reached globally [556]. PEM fuel cells are deployed in applications where their low operation temperature and relatively small size and weight are sufficiently advantageous to offset their relatively high cost. Stationary PEM fuel cell systems of 5 kW to 100 kW have been deployed as back-up power for cell towers in the communication industry.

Pressure regulators and gas flow controllers are used for process control of PEM fuel cell systems. In stationary applications hydrogen sensors are often used to monitor the surrounding environment for leaked hydrogen. Hydrogen quality may also have to be monitored; however, affordable sensors capable of reliably working in this environment and which meet the performance requirements are still under development.

Stationary fuel cells combining heat and power capacities of around 2 kW have been deployed in private homes in Japan and Europe. Larger units have been placed in commercial buildings. These systems may need a fuel generator unless the site is supplied directly with hydrogen and auxiliary systems for operation. Solid-oxide fuel cells operate at high temperatures of 750 °C up to 1000 °C and the high-grade residual heat is utilised for space heating and water heating. These systems need to be monitored for operation and safety, e.g., for gas leak detection and fire detection.

Stationary fuel cells may be installed indoors and outdoors. For indoor applications the facility and the fuel cell exhaust gas should be monitored by a hydrogen sensor to protect against fire and explosion. Location of gas detection systems in the container shall be chosen to provide the earliest warning of the combustible gases present. Ventilation is initiated when a warning level of hydrogen (e.g., either 25 % or 40 % of the LFL, depending on the prevailing code) is detected [557]. Indoor placement of a fuel cell may require an automatic shutoff valve interlocked with gas detection for fuel sources located outdoors and serving an indoor fuel cell power system [558]. Sensors based on field effect, semiconductor or thermal conductivity can be applied for this purpose. A sensor system with a combination of semiconductor or field effect-based sensing elements and thermal conductivity sensing platform has the advantage of higher reliability and wider measuring range. However, the sensor must not only be capable of operating reliably in high humidities, since the relative humidity of the fuel cell exhaust gas can be close to 100 %. It should also maintain operability under water condensing conditions. Therefore the sensing element should be protected with a hydrophobic polymer membrane to keep off water droplets or maintained at a temperature to preclude condensation.

Figure 7.13 illustrates a typical PEM FC system and indicates the location of hydrogen, gas flow, temperature, and pressure sensors which can be used to monitor the fuel cell operation.

7.4.2 Mobile Fuel Cell Applications

FCEVs are based on a PEM fuel cell powering an electric motor. The commercial release of light duty FCEVs began in 2014, albeit at a very limited level. Roll-out of additional FCEVs is expected from 2015 and beyond. Fuel cell bus fleets have also been deployed in many cities around the world.

FCEVs including light duty passenger cars and heavy duty busses and trucks are equipped with numerous safety systems. Sensors for measurement of pressure, temperature, and gas flow are used for monitoring and control purposes in these applications. Hydrogen sensors can be used to detect hydrogen in various compartments of the vehicle, such as the passenger cabin, trunks, and engine compartments. It is reported that the Road Transportation Vehicle Law in Japan requires installation of hydrogen sensors in areas where hydrogen accumulation may occur [559].

During the demonstration phase of some FCEVs (e.g., prototype vehicles developed for research purposes and not for general release to the public), up to 10 hydrogen sensors were placed in the vehicle, including multiple sensors in the passenger compartment, the tank system area, the FC system area, the FC oxidant outlet, and the vehicle exhaust. The GM HydroGen4, for example, was equipped with 7 hydrogen sensors – 2 in the engine compartment, 2 in roof liner of the passenger compartment, 2 close to the hydrogen tanks, and 1 in the exhaust (see Figure 7.14). The number of on-board hydrogen sensors in commercial light duty road vehicles will, however, be significantly less. In the GM HydroGen4 the hydrogen sensors were operational when the ignition was switched on and during fuelling. The sensors output was indicated on the dashboard, and if hydrogen was detected above the safety limit an audible alarm signal would be activated and the hydrogen supply to the fuel cell would be stopped. Passengers would be prompted to evacuate the vehicle if the hydrogen concentration remained above the alarm limit for more than one minute. A hydrogen sensor can also be used to monitor the exhaust gas of a PEM fuel cell stack for indication of malfunctions. Several hydrogen sensor platforms for on-board automotive safety applications have been evaluated by vehicle manufacturers, including catalytic combustion sensors, conductometric, thermal conductivity, and field effect sensing platforms, as well as sensors with multiple sensing element types. Although not mandated, hydrogen sensors will be used in the initial release of commercial vehicles (see Figure 7.4) [543]. However, the measurement of hydrogen in FCVs is still a challenge, especially in regard to reliability, accuracy, and durability in a mass-produced device [560].

An additional on-board safety feature is the thermally activated pressure release devices (PRD) on the vehicle fuel system, which in the event of a fire, will open a valve to release the hydrogen into the atmosphere. The design of the thermally activated PRD is based on a metal plug that melts at a designated activation temperature, and opens a vent system. Upon activation, the PRD leads to complete releases of hydrogen in about three minutes for

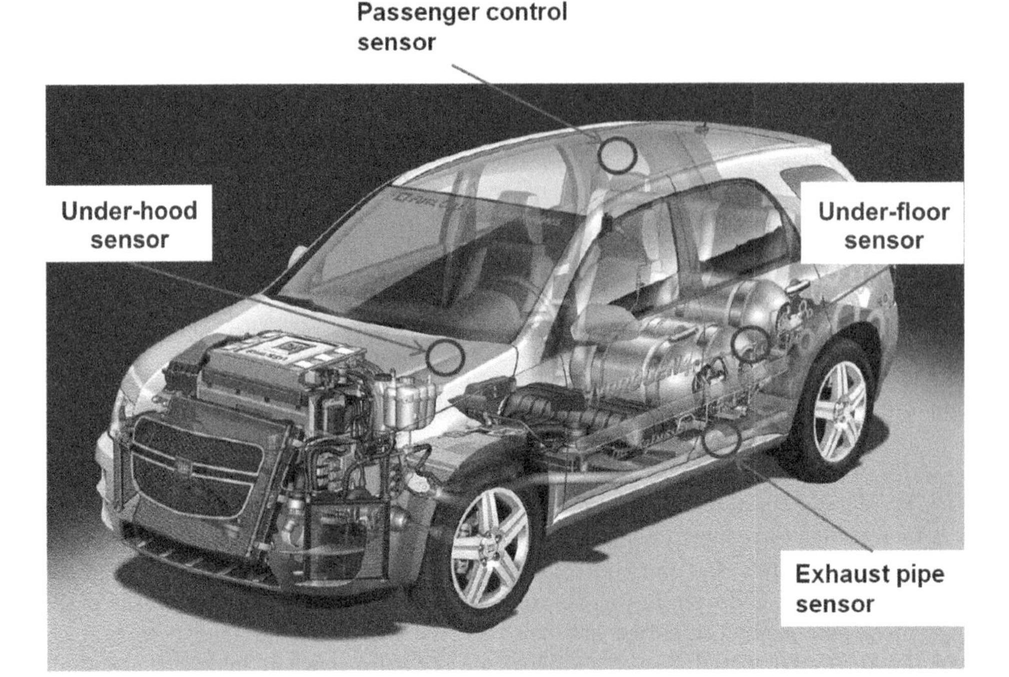

Figure 7.14: Location of hydrogen sensors in GM HydroGen4.

a light duty passenger vehicle. Additional vehicle safety sensors include impact sensors to detect a vehicle collision which will close off the fuel cell system and interrupt the hydrogen supply [561].

Forklifts, also called materials handling vehicles, are an important early market application for the use of mobile fuel cells [562]. Conventional forklifts are powered either by internal combustion engines (ICE) with propane as a fuel or with electric motors powered by on-board batteries. The fuel cell forklift has significant advantages over battery powered systems including longer operation time between fuelling (or recharging in the case of batteries), less down time associated with hydrogen fuelling relative to battery charging (about 3 minutes versus several hours for battery charging), and no loss of power when the fuel supply is low but not depleted. FC forklifts fuelled by hydrogen have different safety requirements. Hydrogen sensors are often used on-board the forklift to monitor for hydrogen releases. Typically this sensor is installed near the fuel tank, which is housed in a protective enclosure. A combustible gas sensor can be used for this application.

PEM fuel cells are also used to supply power in some specialised mobile applications. An example is for submarines such as the type U 212 submarine developed for the German Navy. Since 1985 it has been powered by 9 fuel cells of 34 kW, which have been used for propulsion. The larger type U 214 submarine deploys two 120 kW fuel cells. Both oxygen and hydrogen are stored outside of the hull, oxygen in liquid form, and hydrogen as metal hydride. The safety control system uses electrochemical sensors [563].

The space programme is another specialised application utilising fuel cells for mobile applications. The U.S. space shuttle used three alkaline fuel cells as a power supply. Micro-machined hydrogen leak sensors, consisting of two sensing elements (Pd alloy Schottky diode and Pd alloy resistor) for full range detection from 1 μmol/mol to 100 vol%, were used on the shuttle. Sensors for fire detection and toxic gas detection (e.g., for hydrazine) were also installed in the shuttle and the International Space Station (ISS). Water electrolysis is used to generate hydrogen and breathable oxygen on the ISS. Hydrogen sensors are used to verify that the hydrogen level in the oxygen stream is maintained at a safe level (see Section 7.2.2). Owing to the critical nature of space missions, redundancy using multiple sensors is implemented.

Other applications of fuel cells for mobility have been demonstrated and are under development for commercialization, e.g., for ships, airplanes, railways, bikes, and even Segways. Portable micro fuel cells can be deployed for low-power consumer electronics such as digital camcorders. These systems also utilize low-cost hydrogen sensors that were developed for these applications. The market for portable fuel cells is increasing with about 5,000 transportable and portable fuel cell installations being shipped in 2012 [556].

7.4.3 Combustion Engines and Rocket Propulsion

The use of hydrogen to power internal combustion engine (ICE) for vehicle propulsion is also feasible and was investigated by several vehicle manufacturers including Ford, Mazda, and BMW. Typically a hydrogen ICE is based on a conventional ICE which has been modified to allow the use of hydrogen as a fuel. These engines can use either hydrogen or conventional liquid fuel. For example, the BMW Hydrogen 7 used hydrogen in its internal combustion engine. The hydrogen was supplied from a liquid hydrogen tank. Numerous sensors for pressure, temperature, and hydrogen were used to control the hydrogen system. A gas warning system, with a triple redundant power supply used 5 hydrogen sensors to monitor for leakage in the vehicle's hydrogen system. Sensors were primarily installed in enclosed spaces such as the engine compartment, the passenger cabin, the luggage compartment, and the double-wall of the hydrogen storage tank. The pressure sensors were deployed for the boil-off management system and to avoid excessive over-pressures during fuelling. In a crash, sensors transmit a signal to the tank control unit, thereby shutting-off power to the hydrogen storage tank valves, closing them and interrupting hydrogen supply to the engine [564]. Despite the advantages of exploiting the emerging hydrogen infrastructure in ICE technology, car manufacturers almost exclusively have opted for the use of fuel cells in hydrogen powered vehicles.

In space flight applications, liquid hydrogen and oxygen are used as fuel for the propulsion systems. For example, NASA's space shuttle main engine used a liquid-fuel cryogenic rocket engine. A similar system is planned for the shuttle's successor, the Space Launch System which has two cryogenic tanks for LH2 storage (1,535 m^3) and LO2 storage (55 m^3). The safe use of hydrogen in NASA is regulated by in-house guidelines which demand that sensors are used to detect the presence of hydrogen in all areas where hazardous accumulations could occur [565]. Sensors are being developed and deployed for monitoring all aspects of aerospace systems. Launch and ground test facilities are equipped with catalytic hydrogen sensors, as well as gas chromatographs and mass spectrometers. Fire detection systems are also used (see Section 5.4). A long-range ultrasonic leak detection system has been developed by NASA and private industry to scan for hydrogen leaks in unsafe or difficult-to-approach areas of the space shuttle (see Figure 7.15).

The Ariane 5 rocket of the European Space Agency also uses cryogenic LH2 and LO2 fuel for propulsion. Accordingly the Guiana Space Centre adopts a complex system of protocols to ensure safe operation during launch of the Ariane rocket. This system, which includes gas monitoring, gas leakage detection and control of temperature, relative humidity, and contamination around the rocket, is described in detail elsewhere [566].

Figure 7.15: The NSAS space shuttle, which used a sonic hydrogen detection system for leak detection (source: NASA).

7.5 SENSORS FOR HYDROGEN IN CHEMICAL AND TECHNICAL PROCESSES

Hydrogen is a common industrial chemical used in numerous applications. Historically, hydrogen was used primarily as an industrial chemical feedstock due to its strong reducing capabilities and its high thermal conductivity. The use of hydrogen as an energy carrier is a relatively new and expanding application for hydrogen which had, until recently, accounted for only a minor component of the global demand for hydrogen. In the following section the use of sensors is described for three specific applications where hydrogen is used for purposes other than as a fuel.

7.5.1 Petrochemical Industry

Oil refineries are one of the largest consumers of hydrogen. Hydrogen is used in the petrochemical industry mainly for hydro-cracking and hydro-refining of petroleum. In these applications it is critical to quantify the absolute concentration (partial pressure) of hydrogen to ensure process safety and operational efficiency. The monitoring points for hydrogen concentration in several petrochemical processes are indicated in Figure 7.16. The continuous measurement of hydrogen content in a gas matrix consisting of hydrocarbons, CO, CO_2, H_2, and other gases is a typical refinery application for an in-line hydrogen sensor. In catalytic reforming processes, which are used for the refining of petroleum products, the measurement of hydrogen in the reformer recycle gas ensures the proper efficiency of the catalyst. Additional hydrogen measurements are needed in hydrotreaters and catalytic reformers where hydrogen is used and its purity has to be determined. Currently, traditional process analysers, such as gas chromatographs and gas density analysers are used to measure the hydrogen content. However, in-line process hydrogen sensors can be employed and allow a cost effective direct point-of-use analysers for hydrogen, whose concentration can range from 50 vol% to 100 vol%. Sensors based on thermal conductivity, catalytic combustion, or metal thin film conductometry can be used for this application [209, 567].

7.5.2 Metallurgy

Hydrogen is used for production of metal such as tungsten, molybdenum, or rhenium by reduction of the appropriate oxides. Molten aluminium can contain large amounts of hydrogen due to the reaction of aluminium with water vapour from the ambient atmosphere. The solubility diminishes during solidification and creates a problem in the casting of high-quality aluminium when substantial amounts of hydrogen gas are liberated during solidification. The hydrogen forms pores in the solid aluminium, and these can deteriorate the quality of the cast metalwork piece. Dissolved hydrogen in aluminium and aluminium alloys also adversely impact mechanical properties, such as plasticity, toughness, and stress corrosion resistance. For this reason a precise measurement of hydrogen dissolved in molten aluminium alloys is needed. Many commercially available detection systems are based on a thermal conductivity sensing platform. For example a measuring system for hydrogen in molten aluminium uses a closed sampling loop purged with nitrogen. Hydrogen inter-diffuses until an equilibrium state is reached and the hydrogen/nitrogen gas mixture is then analysed with a thermal conductivity sensor. By this method hydrogen in the range of 0 to 10 $cm^3/100$ g aluminium can be detected with an uncertainty of about 5 % of reading in the temperature range of 650 °C to 1260 °C. The temperature of the melt and the humidity is often measured in parallel to the gas analysis. For this purpose a thermocouple, a platinum RTD thermometer, and a capacitive polymer humidity sensor are used [568]. A new commercial electrochemical sensor operating in a potentiometric mode was used to determine hydrogen content in molten metals at temperatures between 500 °C to 800 °C. The sensor has a solid proton-conducting perovskite type solid electrolyte ($CaZr_{0.9}In_{0.1}O_{3-x}$) and a titanium or zirconium hydride reference electrode. The sensor is placed in a probe which is dipped into the melt and measures hydrogen in the range from 101 to 105 Pa [569]. Alternatively a palladium-nickel thin film technology can be used in a FET-type sensor for hydrogen detection in the range of 0.01 vol% to 1 vol% H_2 [209]. The hydrogen partial gas pressure in molten aluminium can also be determined using a gas pressure sensor.

Many metallurgical processes need to be performed under well-defined gas atmospheres. Gas sensors can monitor the atmosphere of furnaces used in metal treatment and conditioning processes including hardening of steel by nitriding and carburising of steel (martensite formation). The concentration of hydrogen, as well as carbon oxides and oxygen is measured in the working atmospheres and in the exhaust gases. Thermal conductivity based sensing platforms are used for hydrogen measurements, whereas CO and CO_2 are detected by an infrared sensor.

7.5.3 Hydrogen Cooled Turbine Generators

In thermal or nuclear power plants, process steam is produced for propelling a turbine, which in turn drives a generator to produce electricity. Several sensors are used for process control and monitoring, such as sensors for mass flow, temperature, pressure, and to control emissions. Power plant generators and turbine generators above 200 MW output are typically filled with hydrogen [570]. Hydrogen is used to provide a low-drag atmosphere for the armature to rotate in. Hydrogen also assists in the cooling of the entire generator due to its high thermal conductivity. Sensors are installed in, or connected with, and around the generator to measure the hydrogen temperature, pressure, and purity. Sensors are also

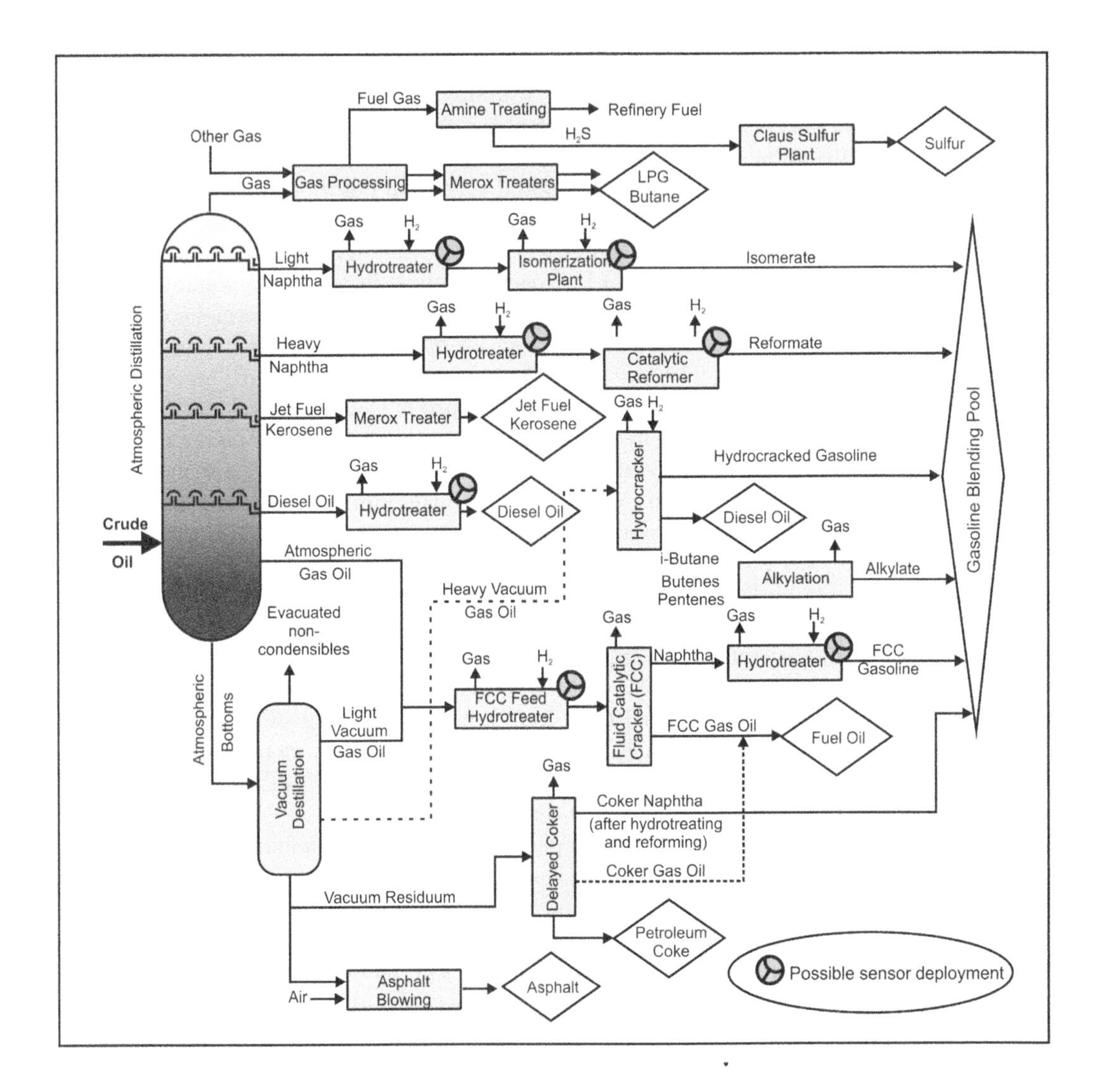

Figure 7.16: Possible monitoring points for in-line hydrogen measurements in a petrochemical refinery. (according to http://commons.wikimedia.org/wiki/File:RefineryFlow.png)

used to monitor and control the filling and draining of the generator with hydrogen. It is important to maintain a high hydrogen purity to reduce frictional losses as well as to avoid the formation of explosive mixtures. Generators usually lose a certain amount of hydrogen during operation. This is typically in the range of some several standard cubic meters per day. However, it is required that uncontrolled losses of more than 12 m^3 per day must be avoided. Therefore sensors monitor the hydrogen filling level of the generator and are used for detection of leakages in feeding tubes and at the generator [568]. Both fixed and portable gas sensors are used for this application. Thermal conductivity sensors are commonly used because of their high reproducibility, low drift, short time of response, and low maintenance requirements. Alternatively, a palladium-nickel thin film configured as a FET-type sensor can be used for hydrogen detection in the range of 0.01 vol% to 1 vol% H_2 and a conductometric sensor for the range 0 vol% to 100 vol% H_2. In Europe, these devices need to be compliant with ATEX requirements.

7.6 SENSORS FOR INDICATION OF TECHNICAL FAULTS

The integrity of pneumatic systems is a critical safety requirement for any hydrogen process. Pneumatic systems consist of tubing or pipes used to transport hydrogen from the storage system to the point of use in a reactor vessel (or fuel cell). Of particular concern would be hydrogen leaks associated with components such as fittings, connections, valves, and seals, which are more prone to leak than an intact pipe. Other components of concern include elements such as in-line pumps, PRDs, compressors, and other gas control systems, which, upon failure may lead to a complete release of hydrogen to the surrounding environment. Unwanted leaks may arise from improper installation (e.g., improper tightening of fittings) or material failure (e.g., embrittlement). Leaks arising from improper installation are most likely on newly built pneumatic systems whereas material failure becomes more likely after the pneumatic system has been in service. Regardless of the mechanism, the sooner hydrogen releases can be detected and acted upon, the lower the potential for injury, property damages and disruption. Detection systems should therefore be applied with the lowest possible detection limit to provide for the earliest possible warning, without the risk of unwarranted "false" alarms.

7.6.1 Leak Detection

The integrity (e.g., gas tightness) of a pneumatic system should be verified by leak detectors prior to commissioning and periodically during operation. This can be done with commercial systems such as the helium leak detector which is based upon probing a pressurized pneumatic system with a helium sensitive sensor (e.g., thermal conductivity-based) or for optimal sensitivity, with a mass spectrometer. Fluorinated hydrocarbon compounds are also used as probe gases for leak detection because of their stability, low toxicity, high vapor pressure, zero background levels, and distinct response on a mas spectrometer. An analogous approach more germane to hydrogen system would be to check for leaks using a gas mixture of approximately 5 vol% hydrogen in nitrogen. The use of hydrogen for leak detection has several significant advantages relative to other gases used for leak detection, such as helium or fluorinated hydrocarbon compounds. Because of its small size and permeability, hydrogen-based leak detection can be performed quickly and with a variety of sensors. Hydrogen sensors can be very sensitive, selective, and low cost, especially when compared to mass spectrometry which is often used for helium leak detection and is often required for fluoro-compound detection. A Pirani gauge equipped with a hydrogen-selective permeable palladium membrane can

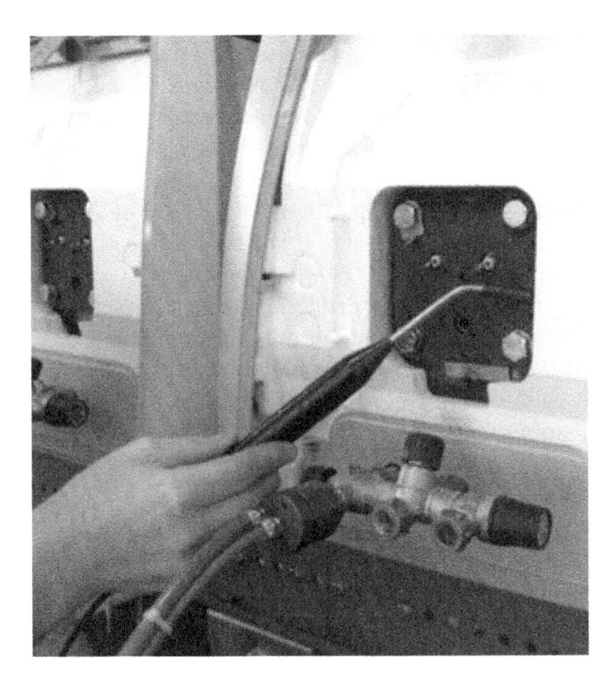

Figure 7.17: Leak detection using hydrogen sensor (source: INFICON).

also be used for hydrogen leak detection. Hydrogen permeates through the membrane and increases the pressure inside the evacuated sensor chamber. This pressure increase is detected by a change in the thermal conductivity. Colorimetric indicators are available which can be applied over fittings, valves, and other potential points of failure; these indicators undergo a dramatic colour change when exposed to hydrogen. Larger leaks can be detected by sonic hydrogen sensors.

7.6.2 Nuclear Power Plants

Nuclear power plants (NPP) generate electricity using steam produced from the heat of fission in the nuclear reactor. Hydrogen can be produced and released in large amounts in the reactor fuel rod container of a nuclear power station as well as in tubes and pipes, especially in Boiled Water Reactor Systems (BWR). Potential hydrogen gas sources under normal operation and under accident conditions are:

- Water radiolysis, i.e., the decomposition of water into H_2 and O_2 due to radiation
- Metal corrosion, e.g., of aluminium and zinc
- Metal oxidation, such as of zirconium, chromium, and iron, by hot steam
- Oxidation of these metals and of the absorber material boron carbide (B_4C) with hot steam
- If present, reactions of the molten core with steam and concrete

The released hydrogen can form a flammable mixture when mixed with air present in the containment. If ignited, different regimes of combustion are possible depending upon factors such as geometry and local conditions. These regimes range from slow deflagration, to fast deflagration, to deflagration-to-detonation transition (DDT), and to detonation. Occurrence of these regimes can lead to high dynamic pressure loads in the containment, which could damage relevant safety systems and even challenge the structural integrity of the containment wall. The consequences of a large hydrogen release and subsequent combustion were clearly demonstrated in the Fukushima accident in 2011.

The International Atomic Energy Agency (IAEA) has published guidelines [571, 572] to increase the safety of NPPs and to facilitate severe accident management. Installed measurement and control instruments are considered as essential elements in normal, abnormal, and emergency operation of NPPs. Modules, sensors, and transmitters measure thousands of variables and processes to provide data to activate pumps, valves, motors, and other electromechanical equipment that control the plant. The instrumentation ensures the plant's safety and efficient production of power [573]. NPPs in most European countries have been or will be retrofitted with systems for the monitoring and control of hydrogen and for filtering the atmosphere in the containment building.

Hydrogen measurement systems have been implemented inside the containment building for analysis of gas outside the containment building. Measurement systems based on catalytic reactions are used inside. For outside measurements a sampling system is often interfaced with a mass spectrometer which provides improved accuracy and range compared with sensors. However, there is a risk of damage to the sampling tube which can influence sensing response. A variety of sensors can be used for measuring hydrogen inside the containment building. A conductometric thick film hydrogen sensor based on a silver-palladium alloy was successfully tested [574]. The sensor was used in an atmosphere having a hydrogen concentration range of 0.5 vol% to 30 vol%, at temperatures from 14 °C to 204 °C and a relative humidity of 0 to 100 %. Stability, repeatability and accuracy of 3 % to 10 % and a response time, (t_{90}), of about 20 s were reported. Palladium-nickel thin films were configurated as a FET-type sensor for hydrogen detection in the range of 0.01 vol% to 1 vol% H_2 and as a conductometric sensor for the range 0 vol% to 100 vol% H_2 in nuclear fuel storage facilities. Sensors based on a thermal conduction platform were used for hydrogen detection in the waste gas or off-gas stream. Additionally, oxygen content was measured by an electrochemical cell [575]. The positioning and the number of the measurement points is of great importance, and it is necessary to have several sensors installed at different locations. Many sensors for the detection of hydrogen were used primarily in the dome of the containment.

Several countermeasures have been developed and implemented in many NPPs in order to avoid and mitigate severe accidents from the hydrogen release. Passive auto-catalytic re-combiners (PAR) facilitate catalytic oxidation of hydrogen in air on the surface of a catalyst. The PAR is typically placed on the ceiling of the containment building [576, 577], and self-initiates when hydrogen is released. The heat generated from hydrogen combustion causes an increase in temperature of the PAR catalyst which in turn can be correlated to the hydrogen concentration; this is analogous to the catalytic hydrogen sensor described in Section 3.2. Other safety measures include igniter flames to burn smaller amounts of hydrogen in air or pre-inertisation, e.g., purging the system with nitrogen or carbon dioxide to bring the oxygen content below 5 vol%. Post-accident corrective measures can include inertisation or pressure venting; however, with venting the release of radioactive substances can be problematic.

Fast metal-cooled reactors use liquid sodium as a coolant. This highly reactive metal burns in the presence of air or forms hydrogen upon reaction with water. Oxygen, hydrogen, and carbon need to be measured for safety reasons in liquid sodium and in the surrounding gas atmosphere. The dissolved hydrogen in the liquid sodium can be detected by an electrochemical sensor down to amounts of about 0.1 μmol/mol. This sensor type, as well as a semiconducting oxide based sensor and a thermal conductivity based sensor were investigated to measure hydrogen amounts between 1 μmol/mol and 100 μmol/mol in argon, which is used in sodium cooled reactors for inertisation [578, 579].

The enclosure of discharged nuclear fuel takes place in casks which are welded closed to contain spent fuel rods for storage and transportation. During the welding process under an inert atmosphere (argon), which takes place over several hours, hydrogen is produced by a radiolysis reaction. The extent of this hydrogen generation has to be monitored for safety reasons. Furthermore, in a particular arc welding method, the so called atomic hydrogen welding (AHW), hydrogen is used to generate high temperatures of more than 3,000 °C via electric arc stimulated hydrogen molecule splitting and their exothermic recombination. Means of hydrogen measurement are required to ensure safety also in this technique.

7.6.3 Battery Room Surveillance

Batteries are increasingly used for the storage of energy in applications where a power supply via the grid is not available or as backup systems to ensure a non-interrupted electrical power supply. Batteries for energy storage are required in remote locations with no access to grid power, e.g., residential homes relying on renewable energy sources (such as wind or solar) for their power requirements. At a larger scale, batteries are also applied to store energy, e.g., from renewable sources, as a peak-shaving reserve. Backup battery systems are used in critical facilities such as hospitals, telecommunications, computing centres, and other technical processes where even a momentary interruption of power supply can have highly detrimental consequences.

Despite competition from other technologies, such as fast responding ultra-capacitors and flywheels, fuel cells, or other battery types, many applications requiring energy storage or back-up power still use lead-acid type batteries [580]. Lead-acid batteries are the oldest rechargeable type battery and have a nominal cell voltage of 2 V. They have lead and lead oxide electrodes immersed in an aqueous solution of sulphuric acid as the electrolyte. There are two main types of lead-acid battery:

Vented type batteries also known as flooded, wet cell, or open type, vented batteries have a large volume of liquid electrolyte flooding the electrodes. Hydrogen and oxygen are produced by electrolysis of the aqueous electrolyte in these batteries during normal operation, including during constant charge mode (when the batteries are sustained at a maximum potential) and particularly during charging mode. The battery is not sealed and a vent allows the release of the gases to prevent pressure building up inside the cell. Water from the electrolyte is consumed in the process and this water needs to be occasionally replenished. In many commercial vented batteries hydrogen venting and water consumption are significantly reduced thanks to the use of external recombinant catalysts [581]. Vented batteries are stacked in open racks and because of the potential hazards associated with hydrogen generation, they are often stored in dedicated "battery rooms." Battery rooms are fitted with an independent ventilation system to remove accumulated hydrogen and to prevent mixing with gases from ventilation systems serving other parts of the same building. Containers are also required to collect leakage of corrosive electrolyte.

Valve regulated lead acid (VRLA) type batteries also known as closed or sealed type, VRLA batteries are sealed and the volume of liquid electrolyte, which is often in gel or absorbed form, is significantly lower than for vented batteries. In the VRLA

Figure 7.18: Hydrogen sensor for safety monitoring in battery charging rooms (with permission of Gunga GmbH).

batteries the separator facilitates the transport of oxygen evolved at the positive plates to the negative plates, where it is electrochemically reduced thus restoring lost water and suppressing hydrogen evolution [580]. Hence, hydrogen and oxygen, which are formed during normal operation, will recombine inside the battery. As indicated by its name, VRLA batteries are equipped with a valve which relieves excessive pressure build-up due to hydrogen generation under fault conditions, including overcharging or overheating. Since they do not release hydrogen under normal conditions VRLAs can be stacked in open racks or in cabinets and their installation does not require an independent ventilation system.

While both battery types are widely used, their installation and monitoring requirements are significantly different. For example the hydrogen out-gassing rate of vented batteries under normal operation is 50-60 times that of VRLA batteries. Depending on the energy storage or back-up requirements of the application, batteries are connected in parallel and/or in series. Battery rooms are typically densely packed with racks of storage batteries and frequently contain supporting electrical equipment such as inverters, switch-gears, and controls. The high density of stored energy and electrical equipment pose risks, e.g., equipment malfunctions, which may result in excessive heat generation leading to fires. Battery rooms housing vented lead acid batteries are classified as hazardous areas due to the venting of combustible gases into the environment. These rooms should be equipped with a safety system that includes an independent ventilation system which is exhausted into the atmosphere outside the building [582]. Accordingly temperature, smoke, fire, and gas detection devices should be used to provide warnings and to initiate mitigating countermeasures (see Section 5.4). In the U.S., vented battery installations must comply with the Occupational Safety and Health

Administration (OSHA) Regulation 29 CFR 1926.441 on Batteries and Battery Charging. The National Fire Protection Association code, NFPA 1, and the International Fire Code (IFC) stipulate specific installation requirements on battery systems depending on the battery technology and the amount of electrolyte. These codes can be enforced by local authorities having jurisdiction (AHJs). The codes require ventilation and smoke detection in rooms containing stationary battery systems. They also require that hydrogen must not be allowed to accumulate in concentrations greater than 1 vol% in air.

While neither NFPA 1 nor the IFC requires the use of hydrogen detectors, in many cases sensors for detecting vented hydrogen are placed on the ceiling above the battery stacks. The hydrogen sensor can indicate an accumulation of hydrogen gas in the battery room and can actuate the ventilation system if hydrogen amounts exceed 1 % so as to dilute and exhaust the flammable gas. If two levels of hydrogen gas alarms are employed, then a higher level, e.g., 2 %, may be used to alert personnel that urgent action is required. Sensors for hydrogen detection based, e.g., on catalytic combustion, electrochemical, field effect- or metal resistor-based sensing elements have been used in battery rooms.

Battery rooms are classed as Zone 2 according to ATEX 95 and so, when sensors are used they shall be of an explosion proof design and certified to IEC/EN 60079-29-1 (see Section 6.1).

7.6.4 Hydrogen in Transformer Oil

Transformers are used to control the voltages from the point of electricity generation (power plant) to the point of use. Transformers step up the voltage of electricity generated at power plants to a higher voltage for distribution via the transmission lines. Transformers are also used to step down the high transmission voltage to deliver typically 4 kV to 69 kV to primary (industrial) customers or 110 V to 220 V to secondary (residential) customers.

High power transformers are filled with a suitable oil that serves both as a coolant and electrical isolating medium. The transformer oil gradually deteriorates due to ageing and moisture ingress. Overload, overheating, or transformer failures can result in chemical reaction and/or the formation of hydrogen which is, mostly dissolved in the oil. The amount of hydrogen and other gases (such as, e.g., ethylene or acetylene) in the oil are monitored together with other parameters including the determination of colour, water content, pH-value, and dielectric properties. Hydrogen gas sensors can be used for monitoring the hydrogen content in the oil. These sensors are usually protected by a hydrogen permeable membrane in order to detain oil from the sensor surface. Thermal conductivity, catalytic, field effect, electrochemical, or metal conductometric sensing elements have been used for this purpose. Sensors can detect hydrogen in oil from 0.005 vol% up to 5 vol% at oil temperatures ranging from -10 °C to +120 °C. Palladium or magnesium-titanium thin films and optical fibre Bragg sensors are also being developed for this application [583]. A method for dissolved gas analysis is described in ASTM D3612 [584].

7.6.5 Biogas Production

The use of biodegradable materials as a renewable energy source is already established and is increasingly being used to replace conventional fossil fuels. The fermentation of plant material, such as corn or manure, yields biogas. Biogas is a mixture consisting primarily of methane in the range of 50 vol% to 75 vol% and carbon dioxide in a range of 25 vol% to 45 vol%. Other minor components include approximately 2 vol% water vapour, $<$ 2 vol% oxygen, $<$ 2 vol% nitrogen, $<$ 1 vol% hydrogen, 0.001 vol% to 2 vol% hydrogen sulphide, along with trace ammonia and carbon monoxide. The composition of the resultant biogas is measured and monitored for the purpose of process control. Early detection of changes is crucial to the evaluation and control of the efficiency of the anaerobic digestion process. If concentration levels of a particular component are too high or too low, this can indicate a drop in the efficiency of the fermentation process. The amount of hydrogen in the biogas produced in an anaerobic digester is an important process parameter. Accumulated hydrogen above a critical value can disturb the normal operation of the methanisation process. For this reason the amount of hydrogen has to be determined in realtime in the produced biogas [585]. Dissolved hydrogen in the liquid phase of the bioreactor can be detected by electrochemical sensors [586, 587]. Due to the flammable nature of methane, the risk of asphyxiation from CO_2 and the risk of poisoning by H_2S, most biogas production plants also feature ambient air monitoring to protect personnel from any hazards linked to the accumulation of these leaked gases. Methane and carbon dioxide are mostly determined by optical or thermal conductivity sensing platforms (see Section 4.2), while oxygen, hydrogen sulphide, and hydrogen can be determined in the range 0.001 vol% to 0.1 vol% by electrochemical cells (see Section 3.4 and 4.2).

7.7 SUMMARY

In this chapter the application and role of sensors in hydrogen technologies were presented. Numerous examples of the use of gas sensors for both safety monitoring and process control were described. A compilation of contemporary applications requiring the use of hydrogen sensors is given in Table 7.1. Table 7.2 contains key near-term applications requiring the use of hydrogen sensors [588]. Different gas sensing platforms employed in various applications can support or, in some cases even, replace traditional analytical methods such as gas chromatography or mass spectrometry. Sensors for the detection of flammable gases are common in applications where hydrogen is produced, stored, and used and may need to be certified as per local requirements. Process control hydrogen sensors are important for monitoring and control purposes and these sensors have to fulfill specific requirements including fast response time and long term stability. The application of sensors in industrial processes which consume large volumes of hydrogen (e.g., petrochemical) was briefly discussed. More detail was provided on sensor use in emerging applications, which are foreseen to increase as the use of hydrogen as a consumer and industrial fuel becomes more widespread. It is clear that each application has its own specificities and requirements with respect to sensor use. In general, information is more widely available for emerging applications compared to established processes where such information is either proprietary or not available. As the demand for hydrogen rises with the increasing uptake and deployment of hydrogen technologies, hydrogen sensors will become ubiquitous wherever hydrogen is produced, stored, or used. Accordingly sensors will play an increasingly important role in ensuring the safety of emerging hydrogen applications and in improving the efficient use of hydrogen as a fuel in a future low-carbon economy.

Table 7.1: Contemporary applications requiring the use of hydrogen sensors

Application	Hydrogen Sensor Use		
	Safety protection	Leak detection	Process control
Hydrogen production			
Steam reforming	x	x	x
Coal gasification	x	x	x
Pressure swing adsorption	x	x	x
Residual product from chemical industry (ethylene production)	x	x	x
Chlorine alkaline electrolysis	x		x
Hydrogen infrastructure			
Storage	x	x	x
Truck transport	x	x	
Pipelines	x	x	
Hydrogen as a fuel			
Hydrogen combustion engines	x	x	
Glass production (fused silica)	x	x	
Gas chromatography	x	x	
Welding	x		
Hydrogen consuming, large scale technical processes			
Petro chemistry: Hydro-cracking, hydro-refining	x		x
Ammonia production	x		
Hydrochloric acid production	x		
Methanol production	x		
Metallurgy	x	x	x
Glass production (float glass)	x	x	
Air separation (de-oxidation)	x		
Industrial and residential safety	x	x	
Indication of technical faults			
Oil-filled high power electrical transformers	x	x	
Hydrogen-cooled turbo-generators	x	x	x
Nuclear plant-produced hydrogen	x	x	
Battery room surveillance	x	x	
Biogas production	x		x
Detection of hot spots in fire security	x	x	

Table 7.2: Emerging applications also requiring the use of hydrogen sensors

Application	**Hydrogen Sensor Use**		
	Safety protection	Leak detection	Process control
Hydrogen production			
Water electrolysis	x	x	x
Very high temperature reactor (VHTR)	x	x	x
Thermo-chemical process	x	x	x
Biological hydrogen production	x	x	
Pyrolysis of bio mass	x	x	
Kvaerner process	x	x	
Hydrogen infrastructure			
Hydrogen refueling stations	x	x	
Vehicle maintenance facilities	x	x	
Fuel cell garages and car parks	x	x	
Industrial and residential safety	x	x	x
Hydrogen as a fuel			
Stationary fuel cells	x	x	
Portable fuel cells	x	x	
Mobile fuel cells, FCEV	x	x	
Hydrogen combustion engines	x	x	x

CHAPTER 8

Practical Aspects of Gas Sensing

CONTENTS

8.1 STANDARD CONDITIONS FOR TEMPERATURE AND PRESSURE

Standard conditions for temperature and pressure are used to define reference conditions for the expression of the volume of gases and the related quantities, e.g., volumetric gas flow. Many properties of gases, such as density, thermal conductivity, and boiling point, depend on temperature and pressure. When specifying values for gas properties, it is necessary to indicate the reference conditions of temperature and pressure. Unfortunately, various different standard conditions are defined which often result in confusion.

The following set of conditions is formally defined as the Standard Temperature and Pressure (STP) for gases according to IUPAC [589]:

- 273.15 K (0 °C or 32 °F)
- 100,000 Pa

These conditions result in a molar volume of 22.711×10^{-3} m^3/mol for ideal gases.

There are other standard conditions that are commonly used, and include the following:

- 273.15 K (0 °C or 32 °F)
- 101,325 Pa

This set of conditions was formerly defined as STP, but the official definition is now at 100,000 Pa. These conditions result in a molar volume of 22.414×10^{-3} m^3/mol for ideal gases.

Further reference conditions for temperature and pressure are also in current use. For example, Normal Temperature and Pressure (NTP) conditions as defined in [35], have the following reference points:

- 293.15 K (20 °C or 68 °F)
- 101,325 Pa (14.7 psia)

The NTP reference conditions are sometimes modified slightly. For example, some gas flow meters are calibrated with gas volumes per unit time referenced to 25 °C (and not 20 °C).

8.2 DESCRIPTION OF THE COMPOSITION OF GAS MIXTURES

The quantity of a gaseous component in a gas mixture, e.g., hydrogen in air, can be expressed in terms of its mass (m), volume (V), pressure (p), or amount of substance, i.e., the number of moles (n). A fraction describes the part of the gaseous compound in relation to the total amount of the mixture, expressed always in the same unit of the compounds. On the other hand a concentration is the amount of a substance divided by the volume of the mixture of the gaseous components at defined pressure and temperature. A ratio gives the amount of a gaseous component relative to one or more other constituents of the mixture expressed always in the same unit. It can be called a mixing ratio, especially when the amounts of several components are mixed together. An overview of important quantities for the amount of a gaseous component in a gas mixture is given in Table 8.1.

Table 8.1: Expressions of the quantity of a gas B in a gas mixture

Measure	**Symbol**	**SI unit**
mass fraction	$w_B = \frac{m_B}{m}$	kg/kg = 1
volume fraction	$\varphi_B = \frac{x_B.V^*{}_{m,B}}{\sum_{i=1}^{i=k} x_i.V^*{}_{m,i}}$	m^3/m^3 =1
molar fraction	$\chi_B = \frac{n_B}{n}$	mol/mol = 1
mass concentration	$\beta_B = \frac{m_B}{V}$	kg/m^3
volume concentration	$\sigma_B = \frac{V_B}{V}$	m^3/m^3 = 1
molar concentration	$c_B = \frac{n_B}{V}$	mol/m^3
mass ratio	$\zeta_B = \frac{m_B}{m-m_B}$	kg/kg = 1
volume ratio	$\psi_B = \frac{V_B}{V-V_B}$	m^3/m^3 = 1
molar ratio	$r_B = \frac{n_B}{n-n_B}$	mol/mol = 1

Where m, V, and n are the total mass, volume, and number of moles or atoms of the k constituents in the gas. The total mass, volume, and number of moles are given by:

$$m = \sum_{i=1}^{k} m_i,\ V = \sum_{i=1}^{k} V_i,\ n = \sum_{i=1}^{k} n_i$$

The sum of all molar fractions of a gas mixture is:

$$\sum_{i=1}^{k} \chi_i = 1. \sum_{i=1}^{k} \chi_i \cdot V^*{}_{m,i}$$

Representing the sum of the volumes of all components as pure substances at the same temperature, i.e., before mixing. The term "amount-of-substance fraction" is preferred instead of the commonly used "molar fraction," because it does not contain the name of the unit "mole." Otherwise mass fraction could be called kilogram fraction.

The quantities of composition exhibit different behaviour concerning the dependence on pressure and temperature, as follows:

- mass fraction and molar fraction are strictly independent of state conditions, i.e., temperature and pressure
- volume fraction and volume concentration depend weakly on state conditions
- mass concentration and molar concentration depend strongly on state conditions

Dalton's law expresses the total pressure p as the sum of the partial pressure of the gas mixture. Therefore, the partial pressure p_B can be used as a further quantity for a gaseous component. For an ideal gas, the partial pressure is proportional to the molar fraction and the pressure of the system $p_B = \chi_B \times p$.

8.3 CONVERSION OF QUANTITIES FOR THE CONTENT OF A GAS IN A GAS MIXTURE

The numerous different units for gas composition in common use can be inter-converted using simple relations. Table 8.2 gives the conversion factors between common units of gas composition. The conversion factors are determined assuming ideal gas behaviour (see Section 1.1). The uncertainty of assuming ideal behaviour is usually less than 1 % for moderate temperatures and pressures (deviation from ideal behaviour is also discussed in Section 1.1). In an ideal gas, volume concentration σ_B, volume fraction φ_B, molar fraction χ_B, and partial pressure fraction are equal:

$$\sigma_B = \varphi_B = \chi_B = \frac{p_B}{p} \tag{8.1}$$

It can be demonstrated that for a volume concentration up to 10 vol% of component B in a gas mixture, e.g., in air, the level of conversion uncertainty due to this approximation is typically 0.5 %, except for conversions involving mass fractions. Conversion factors for real gas behaviour expressed by the compression factor and the mixing factor are given in ISO 14912 [590].

Table 8.2: Conversion factors between quantities of gas composition

Desired quantity	**Given** w_B	**quantity** φ_B	χ_B	β_B	σ_B	c_B
mass fraction, w_B	1	$\frac{M_B}{M_{mix}}$	$\frac{M_B}{M_{mix}}$	$\frac{M_B}{M_{mix}}$	$\frac{M_B}{M_{mix}}$	$\frac{M_B}{M_{mix}}\frac{RT}{p}$
volume fraction, φ_B	$\frac{M_{mix}}{M_B}$	1	1	$\frac{1}{M_B}\frac{RT}{p}$	1	$\frac{RT}{p}$
molar fraction, χ_B	$\frac{M_{mix}}{M_B}$	1	1	$\frac{1}{M_B}\frac{RT}{p}$	1	$\frac{RT}{p}$
mass concentration, β_B	$M\frac{p}{RT}$	$M_B\frac{p}{RT}$	$M_B\frac{RT}{p}$	1	$M_B\frac{p}{RT}$	M_B
volume concentration, σ_B	$\frac{M_{mix}}{M_B}$	1	1	$\frac{1}{M_B}\frac{RT}{p}$	1	$\frac{RT}{p}$
molar concentration, c_B	$\frac{M_{mix}}{M_B}\frac{p}{RT}$	$\frac{p}{RT}$	$\frac{p}{RT}$	$\frac{1}{M_B}$	$\frac{p}{RT}$	1

Notes: M_B and M_{mix} are the molar mass of the component B and the averaged molar mass of the gas mixture.

8.4 PROPERTIES OF HYDROGEN

8.4.1 Gas Law Equations for Non-ideal Behaviour

In addition to the compressibility factor (see Section 1.1), further equations have been developed to more accurately approximate the real gas behaviour of hydrogen and other gases. The van der Waals equation takes into consideration the intermolecular forces and the volume of the gas molecules. The pressure is enlarged by an internal pressure and the molar volume reduced by an incompressible part:

$$p = \frac{\mathrm{R}T}{v_m - \mathrm{b}'} - \frac{\mathrm{a}'}{v_m^2} \tag{8.2}$$

In Table 8.3 the van der Waals constants a' and b' are listed for hydrogen and some other gases.

Redlich and Kwong improved the van der Waals equation to obtain an accurate determination for the compressibility factor [592]:

$$p = \frac{\mathrm{R}T}{v_m - b'} - \frac{a'}{\sqrt{T}v_m(v_m - b')} \tag{8.3}$$

Where $a' = (0.42748R^2T_c^{2.5})/p_c$ is the cohesion pressure, a constant that corrects the pressure due the molecule attraction and $b' = (0.08664RT_c)/p_c$ is a constant that corrects the volume.

The Clausius virial equation gives an approximation to experimental values by:

$$p = \frac{\mathrm{R}T}{V_m}\left(1 + \frac{\mathrm{B}'}{v_m} + \frac{\mathrm{C}'}{v_m^2} + \frac{\mathrm{D}'}{v_m^3} + \cdots\right) \tag{8.4}$$

Table 8.3: Coefficients in van der Waals equation for hydrogen and other gases [591]

Gas	*a'* m^6Pa/mol^2	b'×10^{-5} m^3/mol
H_2	0.02476	2.661
N_2	0.1408	3.913
O_2	0.1378	3.183
CO_2	0.364	4.267
CH_4	0.2283	4.278

Where B', C', and D' are the virial coefficients and are a function of temperature. The values of B' and C' for hydrogen at 273.15 K are 0.01376 m^3/kg mol and 0.404 (m^3/kg mol)^2 [593].

A more extended virial-type equation of state specific for hydrogen was developed based on pressure and temperature dependent terms. This 32-term equation determined the hydrogen density with an expanded uncertainty of 0.2 % over the range of 220 K to 420 K and pressures up to 45 MPa [594]. Nasrafar compared various equations of state for predicting the thermodynamic properties of hydrogen [595].

8.4.2 Hydrogen Physical and Chemical Properties

The properties of hydrogen were already discussed in Section 1.1. In Table 8.4 the most relevant metrics of liquid and gaseous hydrogen are collected where the given data are related to the cited references. A critical review showed that the values for several thermodynamic properties can differ from source to source [596].

Table 8.4: Properties of hydrogen

	Property	Value and unit	Source
1	melting point	-259.05 °C (14.1 K)	[10]
2	boiling point (BP) @ 101325 Pa	20.93 K (-252.22 °C)	[35]
3	triple point:		
	temperature	13.96 K (-259.19 °C)	[35]
	pressure	7.21 kPa	[35]
	density gaseous	0.1298 kg/m^3	[35]
	density liquid	77.21 kg/m^3	[35]
	density solid	86.71 kg/m^3	[35]
4	critical point:		
	temperature	33.19 K (-240.31 °C)	[35]
	pressure	1.315 MPa	[35]
	density	30.12 kg/m^3	[35]

Table 8.4: (*Continued*)

	Property	**Value and unit**	**Source**
5	heat of melting @ MP	59.5 kJ/kg	[7]
	heat of sublimation	379.6 kJ/kg	[10]
	heat of vaporization @NBP	445.5kJ/kg	[7]
	heat of conversion para to ortho	715.8 kJ/kg	[35]
6	liquid phase @BP		
	density liquid	70.96 kg/m^3	[35]
	specific heat capacity c_p	9.8 kJ/kg K	[35]
	specific heat capacity c_v	5.8 kJ/kg K	[35]
	thermal conductivity	0.099 W/m K	[35]
	dynamic viscosity	13.2 μPa s	[35]
	speed of sound	1101 m/s	[35]
7	gas phase @ BP		
	density gaseous	1.331 kg/m^3	[35]
	specific heat capacity c_p	12.2 kJ/kg K	[35]
	specific heat capacity c_v	7.2 kJ/kg K	[35]
	thermal conductivity	0.017 W/m K	[35]
	dynamic viscosity	1.1 μPa s	[35]
	speed of sound	357 m/s	[35]
8	standard conditions (STP) (0 °C and 101325 Pa):		
	density	0.08990 kg/m^3	[10]
	specific heat capacity c_p	14.30 kJ/kg K	[10]
	specific heat capacity c_V	14.30 kJ/kg K	[10]
	(volumetric) caloric value	10.18 kJ/kg K	[10]
	thermal conductivity	0.174 W/m K	[597]
	dynamic viscosity	84 μPa s	[597]
	speed of sound	1290 m/s	
9	normal conditions (NTP) (20 °C and 101,325 Pa):		
	density	0.08376 kg/m^3	[35]
	specific heat capacity c_p	14.33 kJ/kg K	[35]
	specific heat capacity c_v	10.12 kJ/kg K	[35]
	thermal conductivity	0.1838 W/m K	[35]
	dynamic viscosity	89.5 μPa s	[10]
	speed of sound	1294 m/s	[35]
	diffusion coefficient	0.61×10^{-4} m^2/s	[10]

8.5 SENSOR PARAMETERS AND DEFINITIONS

8.5.1 Metrological Parameters

Accuracy
: The closeness of agreement between a test result and the accepted reference value [598]. For chemical sensors, accuracy compares the sensor indication with the nominal (accepted) concentration of the analyte.

Baseline
: Sensor response at the reference stimulus—for chemical sensors the reference stimulus is typically zero concentration of the analyte. This can also be termed as offset.

Cross-sensitivity
: The sensitivity of a sensor generated from stimuli other than the intended use for sensing, e.g., interfering gases (see Selectivity).

Drift
: Gradual change of the sensor output signal under conditions where the input stimulus is constant. Drift in chemical sensors manifests itself as an apparent change in concentration per unit time and is associated with a change in baseline (baseline drift) or degree of response to the desired analyte.

Drift of Zero Point
: Change in the zero point (sensor signal without stimuli) within a specified time period without calibration. This is equivalent to the baseline drift.

Environmental Effect
: The effect of changes in pressure (barometric pressure), temperature, and relative humidity / moisture which induce a change in sensor electrical response. If uncompensated, such changes may be misinterpreted as a change in analyte concentration.

Final Indication
: Indication given by the sensor after stabilization under defined test conditions.

Hysteresis
: The dependence of sensor indication on whether the condition is approached from above or below the value of interest.

Interferent
: Any stimulus or substance, other than the target, that reversibly affects the sensor indication (in contrast, see Poison).

Limit of Quantification (LOQ)
: Smallest value of the quantity which can be quantitatively determined and which can be distinguished from the detection limit. In many cases it is set as 9 times the standard deviation of the background signal added to the background signal.

Linear Range
: The concentration range in which the change in detector (sensor) response is proportional to the change in value of the quantity (e.g., analyte concentration). The output can be represented by the equation: Concentration = (Signalout × Sensitivity).

Lower Detection Limit (LDL)
: The smallest value which can be reliably detected by a sensor; frequently calculated as 3 times the standard deviation of the background signal added to the background signal. Also called limit of detection (LOD). (Some users incorrectly use LDL synonymously with limit of quantification.)

Measuring Range
: Range of values defined by the two extreme values between which a variable can be measured within a specified accuracy. May also be termed Measuring Interval.

Noise
: A random fluctuation in sensor signal.

Offset
: See Baseline.

Operating Range
: Upper and lower temperature, pressure, and humidity environments in which the sensor can be used.

Poison
: A stimulus which produces a permanent or slowly recovering negative impact on the ability of the sensor to perform its designated function (see Interferent).

Uncertainty
: The closeness of agreement between independent test results obtained under stipulated conditions.

Recovery
: See Reversibility.

Recovery Time
: The time for a sensor to adjust to the baseline after exposure to the stimulus. For sensors that reach well-defined steady-state signals it is often calculated as the time to reach a certain fraction of the former response, e.g., 10 % of the previous final indication, e.g., t_{10}.

Response Time
: The time for a sensor signal to adjust during exposure to the stimulus. For sensors that reach well-defined steady-state signals it is calculated as the time to reach a certain fraction of the response, e.g. 90 % of the final indication e.g. t_{90}.

Resolution, Analytical
: The smallest change in a quantity being measured that causes a perceptible change in the corresponding sensor indication.

Reversibility
: The ability of the sensor to recover to baseline following an analyte exposure. May also be termed Recovery.

Saturation
: This term describes a decrease in sensitivity in which a further increase of the stimuli to be sensed does not produce a measurable change of sensor output.

Saturation Stability
: The ability of the sensor to provide quantitative information regarding the stimuli within the specified measuring range of the device following a brief exposure to the stimuli of magnitude in excess of the specified range. For a hydrogen sensor this could be exposure to pure hydrogen. The saturation stability can be defined as the time required for a sensor to be able to measure, within the instrument's specified accuracy, hydrogen at the LFL or user-defined alarm threshold following a brief exposure to pure hydrogen.

Selectivity
: Robustness of a sensor to not respond to stimuli other than the target stimulus.

Sensitivity
: Ratio of the change of the sensor signal to the change in the stimulus (as the first derivative of the calibration function). It is the electronic response of a sensor or instrument to a given amount of an analyte and has units of concentration/electronic-signal. In a linear system the sensitivity is a constant equal to the slope of the calibration curve. (The term sensitivity is frequently used, improperly, as a synonym for LDL.)

Span
: See Measuring Range.

Time of Recovery
: See Recovery Time.

Time of Response
: See Response Time.

Zero Level
: An electronic adjustment point such that the sensor provides a null indication corresponding to a zero value of a quantity to be measured.

8.5.2 Deployment Parameters

Alarm Set Points
: Fixed or adjustable settings of the detector apparatus corresponding to the stimuli levels designated to activate alarms or other output functions. Typical alarm levels for a hydrogen sensor can be described as Warning (typically 10 % to 25 % of the of the LFL) and Alarm (typically set to between 25 % and 100 % of the LFL).

Capital Costs
: The actual cost of the sensor or detector apparatus and the required hardware (i.e., control circuitry).

Control Circuitry
: The required electronic circuitry to control and obtain the electronic output signal from a sensing element. Control circuitry can be trivial (a power source for a chemiresistor), simple (a fixed potential potentiostat with I-V converter for EC sensor), or complex (frequency analysers).

Electronic Interface
: The electrical hardware connecting the sensor to the instrument system and includes both power input requirements and signal output connections.

Installation Costs
: The costs associated with the physical installation of the sensor or detector apparatus, and includes custom pneumatic interfaces, mounts, fixtures, and electrical hardware.

Maturity
: The availability of the sensor technology for current deployment, including the ability to meet market requirements.

Placement
: The location selection of the deployed sensor.

Physical Size
: The basic dimensions of the sensor and of the required circuit board.

Pneumatic Design
: The physical interface of the sensor to the test gas; examples include sampling pumps, pneumatic feedthroughs for accessing gaseous process streams, and safety isolation valves.

Power Requirement
: The voltage and current requirements to operate the sensor and its control circuitry.

Regulations, Codes, and Standards
: Set of agreed rules, often mandatory but sometimes voluntary, pertaining to the use or properties of the sensor. For example, ISO standards are formally viewed as voluntary; the International Fire Code is mandatory for those jurisdictions which have adopted it.

Shelf Life
: The effective time that a sensor can be stored prior to installation.

8.5.3 Operational Parameters

Calibration Requirement
: The protocols and frequency for calibrating the sensor once it has been deployed to assure accuracy and/or compliance to regulations.

Consumables
: The depletable supplementary material requirements for sensor operation. For example, hydrogen in flame ionization detectors for hydrocarbons; calibration and validation gases.

Device-to-Device Repeatability
: The similarity of response characteristics between multiple samples of the same sensor type or model. Compliance to sensor repeatability is mandated by ISO 60079-29-1.

Maintenance
: The required procedures to ensure accuracy.

Matrix Effects
: The effect of background gas on sensor response.

Mechanical Stability
: Durability against mechanical vibrations and shock.

Minimum Analyte Volume
: The smallest amount of the analyte required by a sensor to perform accurate measurements.

Operational Lifetime
: The expected useful life of the sensor under normal operating conditions.

Orientation Effect
: The change in sensor response to a stimulus as the sensor is rotated through various orientations (side, upside down, etc.).

Warm-up Time
: The time required for a sensor to produce output signals of acceptable accuracy upon powering up.

8.5.4 Additional Definitions

Analyte
: The substance in a sample for which a qualitative or quantitative analysis is to be obtained.

Matrix
: The constituents in a sample which are not the object of analysis.

Measurand
: Quantity intended to be measured.

Reliability
: The ability of a sensor to perform a required function under stated conditions for a stated period of time.

Repeatability
: Closeness of agreement of multiple sensor readings repeated under the same conditions of measurement in a specific facility. To quantify repeatability, the spread or dispersion of results may be quoted, e.g., in terms of a standard deviation of a number of readings.

Reproducibility
: Closeness of multiple measurements performed using near-identical equipment and procedures but by different personnel and in different facilities.

Step Function
: Value of a stimulus (or input signal) which changes at a certain time (e.g., t=0) from one steady condition to a new steady condition, as in a step-function change in the concentration of a test gas.

Step Response
: Value of the sensor output signal in response to excitation by a stimulus in form of a step function.

Traceability
: The measuring result is traceable in an unbroken chain of comparisons to a known standard of a national metrological institute.

Trueness
: The closeness of agreement between the average value obtained from a large series of test results and an accepted reference value [598].

Uncertainty of Measurement
: Parameter of a measuring result that characterizes the dispersion of the values attributed to a measured quantity.

Glossary

AAO Anodised aluminium oxide

AC Alternating current

ADC Analogue-to-digital converter

AFC Alkaline fuel cell

AFM Atomic force microscope

AHJ Authority having jurisdiction

AHW Atomic hydrogen welding

AIAA American Institute of Aeronautics and Astronautics

AIST Advanced Industrial Science and Technology of Japan

API RP American Petroleum Institute standard

A2LA American Association for Laboratory Accreditation

ANN Artificial neuronal network

ANSI American National Standards Institute

ANZEx ANZEx: Australian/New Zealand Certification Scheme – Program for the Certification of Equipment for Explosive Atmospheres

ASIC Application-specific integrated circuit

ATEX ATmosphères EXplosibles

ASTM American Society for Testing and Materials

ATR Autothermal reforming

BAM Bundesanstalt für Materialforschung und -prüfung

Baseefa British Approvals Service for Electrical Equipment in Flammable Atmospheres

BAW Bulk acoustic wave (device)

BEV Battery electric vehicle

BOS Background-oriented schlieren

BSI British Standards Institution

BWR Boiled water reactor

CAB Conformity assessment body

CAN Controller area network (bus)

CCD Charge-coupled device

CE Counter electrode

CEN European Committee for Standardization

CENELEC European Committee for Electrochemical Standardization

CFR Code of Federal Regulations (U.S.)

CGS Combustible gas sensor

CHP Combined heat and power units

CNT Carbon nanotube

COPANT Pan American Standards Commission

COTS Commercial off-the-shelf

CPU Central processing unit

CRDS Cavity ring-down spectroscopy

CSA Canadian Standards Association

CVD Chemical vapour deposition

DAkkS Deutsche Akkreditierungsstelle

DAQ Data acquisition system

DC Direct current

DDT Deflagration to detonation transition

DIN Deutsches Institut für Normung

DoE or U.S. DOE U.S. Department of Energy

EC Electrochemical (sensor)

ECD Electron capture detector

EMF Electromotive force

EN European Standard

EU European Union

FBG Fiber Bragg grating

FCEV Fuel cell electric vehicle

FEP Fluorinated ethylene propylene

FET Field effect transistor

FID Flame ionisation detector

FTIR Fourier transform infrared spectroscopy

GC Gas chromatography

GDE Gas diffusion electrode

GF Gauge factor

GUM Guide to the Expression of Uncertainty in Measurement

HART Highway addressable remote transducer

HMDO Hexamethyldisiloxane

HVAC Heating, ventilation, and air conditioning

IAEA International Atomic Energy Agency

IAF International Accreditation Forum

ICVGT Interpolating constant-volume gas thermometer

IDT Interdigital transducer

IEC International Electrotechnical Commission

IEEE Institute of Electrical and Electronics Engineers

IET Institute for Energy and Transport of JRC

IFC International Fire Code

IGCC Integrated gasification combined cycle

IPRT Industrial platinum resistance thermometers

IR Infrared (part of the electromagnetic spectrum)

ISO International Organization for Standardization

ITO Indium tin oxide

ITS-90 International temperature scale 1990

ITU International Telecommunication Union

IUPAC International Union of Pure and Applied Chemistry

JCGM Joint Committee for Guides in Metrology

JIS Japan Industrial Standard

JISC Japanese Industrial Standards Committee

JRC Joint Research Centre

LAS Laser absorption spectroscopy

LDL Lower detection limit

LED Light emitting diode

LEL Lower explosive limit

LFL Lower flammability limit

LH2 Liquid hydrogen

LIBS Laser-induced breakdown spectroscopy

LOD Limit of detection

LO2 Liquid oxygen

LPG Liquefied petroleum gas

LPG Long period grating

LOQ Limit of quantification

LOX Liquid oxygen

LVD Low voltage directive (EU)

MAC Maximum allowable concentration

MAWP Maximum allowable working pressure

MBE Molecular beam epitaxy

MCFC Molten carbonate fuel cell

MEA Membrane electrode assembly

MFC Mass flow controllers

MEMS Micro-electromechanical system

MIL-STD Military standard

MIMS Mineral-insulated metal-sheathed (thermoelement)

MIS Metal-insulator-semiconductor (structure)

MOS Metal oxide semiconductor

MOSFET Metal-oxide-semiconductor field-effect transistor

MOX Metal oxide (sensor)

MS Mass spectrometry

MUX Multiplexer

NAFION Trade name, proton-conducting polymer

NASA National Aeronautics and Space Administration (U.S.)

NASICON $Na_{1+x}Zr_2Si_xP_{3-x}O_{12}$, $(0 < x < 3)$ an inorganic solid electrolyte

NBP Normal boiling point

NEC National Electric Code (U.S.)

NDIR Non-dispersive infrared (sensor)

NFPA National Fire Protection Association (U.S.)

NIST National Institute of Standards and Technology (U.S.)

NMI National Metrology Institute

NPP Nuclear power plant

NREL National Renewable Energy Laboratory (U.S.)

NRTL Nationally Recognized Testing Laboratory (U.S.)

Nm^3 Normal cubic meter

NTP Normal temperature and pressure (conditions)

NTC Negative temperature coefficients of resistance

OEM Original equipment manufacturer

OpAmp Operational amplifier

OSHA Occupational Safety and Health Administration

PA Polyamide

PAFC Phosphoric acid fuel cell

PAR Passive autocatalytic recombiner

PAS Photoacoustic spectroscopy

PC Personal computer

PCA Principal component analysis

PCB Printed circuit board

PEL Permissible exposure limit

PEM FC Polymer electrolyte membrane fuel cell or proton exchange membrane

PFD Probability of failure on demand

PDHID Pulsed discharge helium ionisation detector

PID Photoionisation detector

PLS Partial least-squares (regression method)

POX Partial oxidation

PPE Personal protection equipment

PRD Pressure relief device

PSA Pressure swing adsorption

PSD Position sensitive device

PRT Platinum resistance thermometers

PTFE Polytetrafluoroethylene

PTC Positive temperature coefficient of resistance

PVA Polyvinyl acetate

PVD Physical vapour deposition

PZT Lead zirconate titanate

QCM Quartz crystal microbalance

RAM Random access memory

RCS Regulations, Codes, and Standards (for NREL)

R&D Research and development

RE Reference electrode

RES Renewable energy sources directive (EU)

RF Radio frequency

rh, RH Relative humidity

RIRT Rhodium-iron resistance thermometer

RLC Resistor-inductor-capacitor (circuit)

RS-232 Standard for serial communication transmission of data

RS-485 Standard defining the electrical characteristics of drivers and receivers for use in balanced digital multipoint systems

RT Retention time

RTD Resistive temperature detector

SAC Standardization Administration of China

SAE Society of Automotive Engineers

SAW Surface acoustic wave (sensor)

SCD Sulphur chemiluminescence detection

SDO Standards developing organisation

SI International System of Units

SIL Safety integrity levels

SIRA Scientific Instrument Research Association (British notification body)

SIS Safety instrumented system

SISO Single-input single-output systems

SMD Surface-mount device

SMR Steam methane reformer

SOEC Solid oxide electrolyser

SOFC Solid oxide fuel cell

STP Standard temperature conditions

STPRT Standard platinum resistance thermometers

SQUID Superconducting quantum interference device

TC Thermal conductivity (sensor)

TCD Thermal conductivity detector

TDL Tunable diode laser

TDLAS Tunable diode laser absorption spectroscopy

TGS Taguchi gas sensor

TLV-TWA Threshold limit value time–weighted average

TO Transistor outline

TPW Triple point of water

TRIRAT TRaceability in Infrared RAdiation Thermometry

UEL Upper explosive limit

UFL Upper flammability limit

UL Underwriters Laboratories

UKAS United Kingdom Accreditation Service

UPS Uninterruptible power supply

USB Universal serial bus

UV Ultra violet (part of the electromagnetic spectrum)

VAF Volume air flow (sensor)

VOC Volatile organic compound

VRLA Valve regulated lead acid (battery)

VRP Variable-reluctance pressure (sensor)

WAM Wide area monitoring

WE Working electrode

YSZ Yttria stabilised zirconia

List of Symbols

A, B, C, D
: empirical constants

A cross-sectional area

A active area

A* Richardson's constant

A_0 pre-exponential factor in Arrhenius equation

A_P stiffness coefficient

a', b'
: constants of van der Waal equation

a_{cor} correction value

a_B activity of the component B

$a(t)$ response function

C_D capacitance of the depletion layer

C_G gate capacitance

C_O capacitance of the oxide layer

C_1,C_2,C_3,C_4,C_5
: constants for calculation of c_p

c speed of light constant

c analyte concentration

c_B molar concentration (amount-of-substance concentration)

c_p specific heat capacity at constant pressure

c_V specific heat capacity at constant volume

D damping constant

D diffusion coefficient

d thickness

d_s float diameter

d_v water vapour density

dT/dx
: temperature gradient in the direction of heat flow in K/m

d_v water vapour density (absolute humidity)

E modulus of elasticity

E energy

E_i^o standard potential

E measured potential

e water vapour partial pressure on the pure phase

e' water vapour partial pressure in air

e_w water vapour saturation pressure on the pure phase

e'_w water vapour saturation pressure in air

F Faraday's constant, approximately 96500 coulombs and is the amount of electrical charge in 1 mole of electrons

F force

$\mathcal{F}(s)$ frequency dependent functions

f enhancement factor

f_i function describing the estimate of the measurand y in terms of x_i, sensitivity coefficient

$f(t)$ function of time

f_0 resonant crystal frequency

G shear modulus

G electrical conductance in Siemens, S

G_0 conductance in the absence of the test gas

$\mathcal{G}(s)$ Function of the complex frequency s

g gravitational acceleration

H enthalpy

I intensity of the emitted light

I_H intensity of the transmitted light in the presence of hydrogen

I_0 intensity of the transmitted light in the absence of hydrogen

I_0 intensity of the incoming light

$I(t)$ intensity, dependent on time

I electrical current, flux of charges

I_D output current

K, K'
: constant

k coverage factor

k_B Boltzmann constant $1.3806488 \times 10^{-23}$ J/K

k_f flow coefficient

k_u temperature dependent stiffness

l length
$\mathcal{L}\{X(t)\}$
Laplace transforms of the input $X(t)$
$\mathcal{L}\{Y(t)\}$
Laplace transforms of the output $Y(t)$ signals
M molar mass, m/n in kg/kmol
m mass
m effective mass of the electrons of the sensing material
m mass loading
m_O diffusion related constant
N number of particles, e.g., electrons
N' density of adsorbed molecules
N_A = 6.022 1023 L/mol the Avogadro constant
N_O^* analyte concentration at the outer surface of the gas permeable membrane
n amount of substance in mole (n = N / N_A)
n density of particles, e.g., electrons
n validity factor
P probability
p total gas pressure in Pa
p_B partial pressure of gas B in Pa
p_{crit} critical pressure
p_{dyn} dynamic pressure in Pa
p dipole moment
pH The negative log of the activity of the hydrogen ion in an aqueous solution.
Q total charge
Q_v volumetric flow rate
Q_m mass flow rate
q elementary charge
q conductive heat transfer rate in W
q_v mass fraction of water vapour
R universal gas constant = 8314.472 J/kmol K
R electrical resistance
Re Reynolds number
r radius
r ratio between the power of the evanescent wave and the total power of the guided light
r_B amount-of-substance ratio
r mixing ration of water vapour and dry air
S sensor signal
s entropy
s complex frequency
T thermodynamic temperature in kelvin
T_{90} temperature according to ITS-90 in kelvin (K)
T_{crit} critical temperature in kelvin
T_{hot} temperature of the hot thermoelectric region
T_{cold} temperature of the cold thermoelectic region
t Celsius temperature in grade Celsius, °C, t/°C = T/K - 273.15
t_{90} temperature according to ITS-90 in grade Celsius, t_{90}/°C= T90/K - 273.15
t_d dew point temperature in °C
t_f frost point temperature in °C
t time
t_a adjustment time
t_d dead time
t_u delay time
t_{90} response time, 90 % of final indication
t_{10} recovery time, 10 % of previous indication
U relative humidity in % rh
$U_c(y)$
expanded uncertainty of y
$u(a_{cor})$
uncertainty of the calibration correction factor
$u(x_i)$
standard uncertainty of x_i
u_c(y)
combined standard uncertainty
$u(x_{cor})$
uncertainty of corrected value in result of calibration
$u(x_{cor})$
uncertainty contribution of the sensor u_{mess}
$u(S)$
uncertainty of the sensor signal
$u_{(\text{meas-bridge})}$
uncertainties of the electrical measuring instrument
$u_{(\text{meas-chain})}$
uncertainties of the electrical measurement chain
V volume in m^3
V thermoelectric voltage
V_G gate voltage

V_D drain voltage
V_{Th} threshold voltage
V_O* oxygen vacancy
V_{crit} critical specific volume
V_m molare volume (V/M) in kg/kmol
v specific volume (V/m)
v flow velocity in m^3/s
w_B mass fraction
X sensor stimulus
x amount of a sensor stimulus
x distance from the electrode through the membrane to the outer surface
x_{cor} corrected value in result of calibration
x_i parameter (e.g., gas concentration)
x_{mes} sensor initial indication or measured value
$x(t)$ timedependent input signal
x_d thickness of the depletion layer
x_O thickness of the oxide layer
Y sensor output signal
y variable (e.g., sensor signal) depending on x_i
$y(t)$ timedependent output signal
z number of electrons involved in electrochemical reactions
Z compressibility factor of gases
α Seebeck coefficient
α thermal expansion coefficient
β constant, exponent
β_B mass concentration
γ_B activity coefficient
$\Delta\alpha$ variation in the optical absorption coefficient of a material due to the absorption of hydrogen
$\Delta_c H^\circ$ standard heat of combustion in kJ mol^{-1}
Δf_m shift in frequency
Δm mass change
$\Delta_R H$ reaction enthalpy
ΔV voltage drop
ϵ strain
ϵ_a molar attenuation coefficient
ϵ_{Ox} permittivity of the oxide
ϵ_S permittivity of the semiconductor
ϵ_0 dielectric permittivity of vacuum
ζ_B mass ratio
η refractive index
η_{eff} effective refractive index
Θ fractional surface coverage by atoms or molecules
κ adiabatic index, $\frac{c_p}{c_v}$
Λ period of the grating
λ_B Bragg wavelength
λ wave length of light
λ thermal conductivity, typically expressed in $W/m{\cdot}K$
μ_n electron mobility
p density of electron holes
μ_p the mobility of electron holes
ν Poisson's ratio
ν speed of sound in a gas mixture in m/s
ν frequency
ρ electrical resistivity
ρ density in kg/m^3
σ specific conductivity in S/m
σ standards deviation
σ_B volume concentration in kg/m^3
τ time constant
Φ_F Fermi potential
Φ_b Schottky barrier height
φ_B volume fraction
φ_v volume fraction of water vapour
χ_B molar fraction
ψ_B volume ratio
Ω Ohm = V/A
ω vibration frequency
ω_p plasma frequency

Bibliography

[1] Wurster R., M. Zerta, C. Stiller, and J. Wolf. *Joint Energy Infrastructure 21 - Role of Hydrogen in Addressing the Challenges in the new Global Energy System.* Brussels: European Hydrogen Association, 2009.

[2] Deutsche Energie-Agentur dena, Strategieplattform Power to Gas, http://www.powertogas.info (Accessed 26.01.2015), 2013.

[3] Gupta R. B. *Hydrogen Fuel: Production, Transport, and Storage.* CRC Press, 2008.

[4] Rigas F. and P. Amyotte. *Hydrogen Safety.* CRC Press, 2013.

[5] Korotcenkov G. *Chemical Sensors, vol 1-6.* Momentum Press, 2010.

[6] Tränkler R. and L. Reindl. *Sensortechnik.* Springer, Berlin Heidelberg, 2014.

[7] Lide D. R. (ed.). *CRC Handbook of Chemistry and Physics 89th edition.* CRC Press, 2009.

[8] Holleman A. F. and E. Wiberg. *Inorganic Chemistry, 34th edition.* Academic Press, 2001.

[9] Hydrogen, http://webbook.nist.gov/cgi/cbook.cgi?ID=C1333740&Mask=1 Thermo-Gas.

[10] HySafe, Biennal report on hydrogen safety, Ch. 1 hydrogen fundamentals, 2006.

[11] Cohen E. R., T. Cvitas, J. G. Frey, B. Holmström, K. Kuchitsu, R. Marquardt, I. Mills, F. Pavese, M. Quack, J. Stohner, H. L. Strauss, M. Takami, and A. J. Thor. *Quantities, Units and Symbols in Physical Chemistry.* IUPAC Green Book, 3rd Edition, 2nd Printing, IUPAC & RSC Publishing, Cambridge, 2008.

[12] Molkov M. *Fundamentals of Hydrogen Safety Engineering.* Ventus Publishing ApS, 2012.

[13] Perry R. H., D. W. Green, and J. O. H. Maloney. *Perry's chemical engineers' handbook. 7th edition.* McGraw Hill Company, New York, 1999.

[14] Alcock J. L., Shirvill, L. C., and Cracknell, R. F. Compilation of existing safety data on hydrogen and comparative fuels, Deliverable Report on European FP5 project EIHP2, http://www.eihp.org/public/documents/Compilation Existing Safety Data on H2 and Comparative Fuels.pdf. accessed 25.01.2015, 2001.

[15] Pischinger R., M. Klell, and T. Sams. *Thermodynamik der Verbrennungskraftmaschine.* Springer, Wien New York, 2002.

[16] Christmann K. Interaction of hydrogen with solid surfaces. *Surf. Sci. Rep.*, 9:1–163, 1988.

[17] Bond G. C. *Metal-Catalysed Reactions of Hydrocarbons, Ch. 3 Chemisorption and Reactions of Hydrogen.* Springer, 2005.

[18] Jewell L. L. and B. H. Davis. Review of absorption and adsorption in the hydrogen-palladium system. *Appl Catal A-gen*, 310:1–15, 2006.

[19] Johansson M., E. Skulason, G. Nielsen, S. Murphy, R. M. Nielsen, and I. Chorkendorff. Hydrogen adsorption on palladium and palladium hydride at 1 bar. *Surf Sci*, 604:718–729, 2010.

[20] Zyubin A. S., T. S. Zyubina, Yu. A. Dobrovol'skii, and V. M. Volokhov. Behavior of molecular hydrogen on the platinum crystal surface: Quantum chemical modeling. *Russ J Inorg Chem+*, 57:1460–1469, 2012.

[21] Appel C., J. Mantzabas, R. Schären, R. Bombach, and A. Inauen. Catalytic combustion of hydrogen-air mixtures over platinum: Validation of hetero/homogeneous chemical reaction schemes. *Clean Air*, 5:21–44, 2004.

[22] Hellsing B., B. Kasemo, and V. P. Zdanov. Kinetics of the hydrogen-oxygen reaction on platinum. *J. of Catalysis*, 132:201–228, 1991.

[23] Fukai Y. *The metal-hydrogen system.* Springer, Berlin, 2005.

[24] Manchester F. D., A. San-Martin, and J. M. Pitre. *Phase Diagrams of Binary Hydrogen Alloys.* ASM International, Materials Park. OH, 2000.

[25] Flanagan T. B. and W. A. Oates. The Palladium-Hydrogen System. *Annu Rev Mater Sci*, 21:269–304, 1991.

[26] Barnoush A. *Hydrogen embrittlement, revisited by in situ electrochemical nanoindentation.* Publisher Shaker, 2009.

[27] Gangloff R. P. and B. P. Somerday. *Gaseous hydrogen embrittlement of materials in energy technologies: The problem, its characterisation and effects on the particular alloy class.* Woodhead Publ. LtD, 2012.

[28] Guidelines for Chemical Process Quantitative Risk Analysis, 2nd ed. American Institute of Chemical Engineers, 2000.

[29] http://www.h2incidents.org/Hydrogen Lessons Learned from incidents and near-misses, Pacific Northwest National Laboratory with funding from the U.S. Department of Energy.

[30] Galassi M. C., E. Papanikolaou, D. Baraldi, E. Funnemark, E. Haland, A. Engebo, G. P. Haugom, T. Jordan, and A. V. Tchouvelev. HIAD - hydrogen incident and accident database. *Int J Hydrogen Energ*, 37:17351–17357, 2012.

[31] The American Institute of Aeronautics and Astronautics (AIAA) Guide to Safety of Hydrogen and Hydrogen Systems (G-095-2004e). ISBN: 978-1-56347-676-1.

[32] EN 1839:2012. Determination of explosion limits of gases and vapours.

[33] ASTM E 681:2015. Standard Test Method for Concentration Limits of Flammability of Chemicals (Vapors and Gases).

[34] Schröder V. and K. Holtappels. Explosion characteristics of hydrogen-air mixtures at elevated pressures. *Proc. of the First Intern. Conf. on Hydrogen Safety. Pisa, Italy. 8-10 September 2005*, 2005.

[35] ISO15916:2004. Basic considerations for the safety of hydrogen systems.

[36] Directive 94/9/EC of the European Parliament and of the Council of 23 March 1994 on the approximation of the laws of the Member States concerning equipment and protective systems intended for use in potentially explosive atmospheres (OJ L 100, 19.4.1994).

[37] Directive 1999/92/EC of the EU Parliament and the council of 16 December 1999 on minimum requirements for improving the safety and health protection of workers potentially at risk from explosive atmospheres.

[38] International Code Council, Inc., 2012 International Fire Code, Country Club Hills, IL: International Code Council, Inc., 2012.

[39] Carter D., and J. Wing., Market Data: Hydrogen Infrastructure. Fuel Cells, Power-to-Gas, and Laboratory Applications: Global Market Analysis and Forecast, marketsandmarkets report EP 1708, Hydrogen Generation Market, published Dec. 2011, www.marketsandmarkets.com.

[40] IEA Energy Technology Essentials: Hydrogen Production & Distribution, OECD/IEA 2007.

[41] Turner J. A. Hydrogen production: Overview. 30.10.2013, NREL. Accessed 30.01.2014.

[42] *Ullmann's Encyclopedia of Industrial Chemistry, 7th ed. 2011, vol. 18 p. 249-392 and vol 13A p. 310ff.* John Wiley & Sons, Inc., 2011.

[43] Kerry F. G. *Industrial Gas Handbook: Gas Separation and Purification.* CRC Press, 2007.

[44] Neumann P. and F. von Linde. Opportunities for an economical hydrogen supply. *inform*, 15(18):313–315, 2003.

[45] http://www.methanol.org/Methanol-Basics/Resources/MMSA-Global-Methanol-Supply-and-Demand.aspx. Accessed 09.11.2015.

[46] Sharaf O. Z. and M. F. Orhan. An overview of fuel cell technology: Fundamentals and applications. *Renew Sust Energ Rev*, 32(0):810–853, 2014.

[47] US DOE Fuel Cell Technologies Programme. http://d39b2lgzxqyc8i.cloudfront.net/wp-content/uploads/fuel-cell-comparison-chart.jpg. Accessed 25.02.2015.

[48] AFC Energy Signs Landmark Agreement for Stationary Fuel Cell Systems in South Korea. http://www.businesswire.com/news/home/20140716006452/en/AFC-Energy-Signs-Landmark-Agreement-Stationary-Fuel. Accessed 25.02.2015.

[49] Carter D. and J. Wing. The Fuel Cell Industry Review 2013. *Fuel Cell Today,* p. 50, 2013.

[50] H2 Mobility initiative: Leading industrial companies agree on an action plan for the construction of a hydrogen refuelling network in Germany. http://media.daimler.com/dcmedia/0-921-656547-1-1636552-1-0-0-0-0-0-0-0-0-0-0-0-0-0.html, 2013. Accessed 09.11.2015.

[51] Energy Department Launches Public-Private Partnership to Deploy Hydrogen Infrastructure. http://energy.gov/articles/energy-department-launches-public-private-partnership-deploy-hydrogen-infrastructure. Accessed 25.02.2015.

[52] California Fuel Cell Partnership. http://www.fuelcellpartnership.org/. Accessed 25.02.2015.

[53] Scandinavian Hydrogen Highway Partnership. http://www.scandinavianhydrogen.org/. Accessed 25.02.2015.

[54] Clean Hydrogen in European Cities. http://chic-project.eu/. Accessed 25.02.2015.

[55] Ramsden T. An Evaluation of the Total Cost of Ownership of Fuel Cell-Powered Material Handling Equipment. Technical Report NREL/TP-5600-56408, NREL, 2013.

[56] Dincer I. and C. Acar. Review and evaluation of hydrogen production methods for better sustainability. *Int J Hydrogen Energ*, 40(34):11094-11111, 2015.

[57] Gahleitner G. Hydrogen from renewable electricity: An international review of power-to-gas pilot plants for stationary applications. *Int J Hydrogen Energ*, 38(5):2039–2061, 2013.

[58] NRELwindH2, http://www.nrel.gov/hydrogen/proj_wind_hydrogen.html. Accessed 09.11.2015.

[59] Schiller M. and E. Anderson. Five considerations for large scale hydrogen electrolyzer development. *Gas for Energy*, 1:44–47, 2014.

[60] Koumi Ngoh S. and D. Njomo. An overview of hydrogen gas production from solar energy. *Renew Sust Energ Rev*, 16(9):6782–6792, 2012.

[61] Hore-Lacy I., Hydrogen production from nuclear power. http://www.eoearth.org/view/article/153622, 2009. Accessed 22.03.2015.

[62] Lee D. H. Toward the clean production of hydrogen: Competition among renewable energy sources and nuclear power. *Int J Hydrogen Energ*, 37:15726–15735, 2012.

[63] Alves H. J., C. Jr. Bley, R. R. Niklevicz, E. Pires Frigo, M. Sato Frigo, and C. H. Coimbra-Araújo. Overview of hydrogen production technologies from biogas and the applications in fuel cells. *Int J Hydrogen Energ*, 38(13):5215–5225, 2013.

[64] Kawasaki Heavy Industries Obtains AiP for Liquefied Hydrogen Carrier Cargo Containment System. https://www.khi.co.jp/english/news/detail/20140106-1e.html. Accessed 25.02.2015.

[65] Okada Y. and M. Shimura. Development of large-scale H2 storage and transportation technology with Liquid Organic Hydrogen Carrier (LOHC). In *Joint GCC-Japan Environment Symposium*, 2013.

[66] McPhy Hydrogen Storage. http://www.mcphy.com/en/technologies/solid-hydrogen-storage/. Accessed 25.02.2015.

[67] ISO TS 15869:2009. Gaseous Hydrogen and Hydrogen Blends - Land Vehicle Fuel Tanks.

[68] SAE TIR J2601. Fueling Protocols for Light Duty Gaseous Hydrogen Surface Vehicles.

[69] Fraden D. J. *Handbook of Modern Sensors: Physics, Designs, and Applications.* Springer, 2010.

[70] EN 45544-4:2000. Workplace atmospheres - Electrical apparatus used for the direct detection and direct concentration measurement of toxic gases and vapours - Part 4: Guide for selection, installation, use and maintenance.

[71] EN 60079-29-1:2008. Performance requirements of detectors for flammable gases.

[72] Stetter J. R., W. R. Penrose, and S. Yao. Chemical sensors, electrochemical sensors, and ecs. *J Electrochem Soc*, 150:S11–S15, 2003.

[73] International vocabulary of metrology - basic and general concepts and associated terms (vim) 3rd ed. Technical report, Joint Committee for Guides in metrology (JCGM), 2012.

[74] Hulanicki A., S. Glab, and F. Ingman. Chemical sensors: definitions and classification. *Pure Appl. Chem.*, 63(9):1247–1250, 1991.

[75] Sinclair I. *Sensors and Transducers.* Butterworth Heinemann, 2001.

[76] Chapman P. W. *Smart Sensors.* Instrument Society of America, 1996.

[77] Stetter J. R. and W. R. Penrose. *Understanding chemical sensors and chemical sensor arrays (electronic noses): past, present, and future, Sensors Update, vol. 10, Chapter 2.3, pp. 189-229.* Wiley-VCH, Weinheim, 2010.

[78] Boon-Brett L., P. Moretto G. Black, and J. Bousek. A comparison of test methods for the measurement of hydrogen sensor response and recovery times. *Int J Hydrogen Energ*, 35:652–7663, 2013.

[79] ISO 26142:2010. Hydrogen detection apparatus - Stationary applications.

[80] Burns R. S. *Advanced Control Engineering.* Butterworth Heinemann, Oxford, 2001.

[81] Northrop R. B. *Introduction to Instrumentation and Measurements, 3rd ed.* CRC Press, Taylor & Francis, 2014.

[82] Hübert T. and U. Banach. Response time of hydrogen sensors. *ICHS Proceedings 2013, ed. by M. N. Carcassi, ID 124*, pages 1–8, 2013.

[83] Killian C. *Modern Control Technology.* Thomson Delmar Learning, 2005.

[84] Watson J., K. Ihokura, and G. S. V. Colest. Chemical Sensors - Definitions and Classification. *Meas Sci Technol*, 4:711–719, 1993.

[85] Buttner W., and R. Burgess, M. Post, and C. Rivkin. Summary and Findings from the NREL/DOE Hydrogen Sensor Workshop Technical Report NREL/TP-5600-55645, 2011.

[86] Shrivastava A. and V. B. Gupta. Methods for the determination of limit of detection and limit of quantitation of the analytical methods. *Chronicles of Young Scientists*, 2:21–25, 2011.

[87] Palmisano V., L. Boon-Brett, C. Bonato, F. Harskamp, W. J. Buttner, M. B. Post, R. Burgess, and C. Rivkin. Evaluation of selectivity of commercial hydrogen sensors. *Int J Hydrogen Energ*, 39:20491–20496, 2014.

[88] Gupta T. K. *Handbook of Thick- and Thin-Film Hybrid Microelectronics.* John Wiley & Sons, Inc. Hoboken, N.Y., 2003.

[89] Xu Y. and X-T. Yan. *Chemical Vapour Deposition, An Integrated Engineering Design for Advanced Materials.* Springer, 2010.

[90] Mattox D. M. *Handbook of Physical Vapour Deposition (PVD) Processing. 2nd ed.* William Andrew, 2010.

[91] Panzini M. I. (ed.). *Thick Films: Properties, Technology and Applications (Materials Science and Technologies: Electrical Engineering Developments).* Nova Science Pub Inc, 2011.

[92] Capper P. and M. Mauk (ed.). *Liquid Phase Epitaxy of Electronic, Optical and Optoelectronic Materials.* John Wiley & Sons, Ltd, 2007.

[93] Jonschker G. *Sol-Gel-Technology in Praxis.* Vincentz Network GmbH & Co KG, Hannover, 2014.

[94] Taylor H. R. *Data acquisition for sensor systems.* Chapman & Hall, 2010.

[95] Zurawski R. (ed.). *Industrial Communication Technology Handbook, 2nd ed.*, CRC Press, 2014.

[96] IEC 61131-3. Programmable controllers - Part 3: Programming languages.

[97] Hirlemann A. and R. Gutierrez-Osuna. High-order Chemical Sensing. *Chem Rev*, 108(1):563–613, 2008.

[98] Varmuza K. and P. Filzmoser. *Introduction to Multivariate Statistical Analysis in Chemometrics.* CRC Press Taylor & Francis, 2009.

[99] Korotcenkov G. *Handbook of Gas Sensor Materials: Properties, Advantages and Shortcomings for Applications Volume 1: Conventional Approaches.* Integrated Analytical Systems. Springer London, Limited, 2013.

[100] Umar A. and Y. B. Hahn. *Metal Oxide Nanostructures and Their Applications: Growth and properties ; vol. 2, Growth and properties ; vol. 3, Applications ; vol. 4, Applications; vol. 5, Zn0 nanostructures and nanodevices.* Metal Oxide Nanostructures and Their Applications. CA, 2010.

[101] Gründler P. *Chemical Sensors: An Introduction for Scientists and Engineers.* Springer, 2007.

[102] Hübert T., L. Boon-Brett, G. Black, and U. Banach. Hydrogen sensors - A review. *Sensor Actuat B-chem*, 157(2):329–352, 2011.

[103] Baker A. R. Improvements in or relating to Electrically Heatable Filaments, 01 1958.

[104] Sonley J. M. Gas sensing element and method for its production, September 19 1979. EP Patent App. EP19,790,300,345.

[105] Boon-Brett L. Microstructural analysis of noble metal-based commercial hydrogen safety sensors to investigate why their performance deteriorates. *University of Ulster, Dissertation*, 2011.

[106] Jones E., The pellistor catalytic gas detection, in: P. Moseley, B. Tofield (Eds.), Solid State Gas Sensors, Adam Hilger, Bristol, 1987, Chapter 2, pp. 17-31.

[107] Unwin I. D. Mine monitoring for safety and health. Technical report, National Coal Board, 2007.

[108] Jones T. A. and P. T. Walsh. Flammable Gas Detection - The role of the platinum metals. *Platinum Metals Rev*, 32(2):50–60, 1988.

[109] Jones M. G. and T. G. Nevell. The detection of hydrogen using catalytic flammable gas sensors. *Sensor Actuator*, 16(3):215–224, 1989.

[110] Firth J. G. and T. A. Jones. The principles of the detection of flammable atmospheres by catalytic devices. *Combustion and Flame*, 20(3):303–311, 1973.

[111] Katti V. R., A. K. Debnath, S. C. Gadkari, S. K. Gupta, and V. C. Sahni. Passivated thick film catalytic type H_2 sensor operating at low temperature. *Sensor Actuat B-chem*, 84(2-3):219–225, 2002.

[112] Gentry S. J. and P. T. Walsh. The influence of high methane concentrations on the stability of catalytic flammable-gas sensing elements. *Sensor Actuator*, 5(3):229–238, 1984.

[113] Lee E.-B., I.-S. Hwang, J.-H. Cha, H.-J. Lee, W.-B. Lee, J.J. Pak, J.-H. Lee, and B.-K. Ju. Micromachined catalytic combustible hydrogen gas sensor. *Sensor Actuat B-chem*, 153(2):392–397, 2011.

[114] Trautweiler S., N. Moser, H. Delprat, S. Lepoutre, and C. Alepee. Disruptive MEMS Technology Replaces Conventional Bead Pellistor Device, Proc. of the 14th Intern. Meeting on Chemical Sensors - IMCS 2012. pages 298–301, 2012.

[115] Gall M. The Si planar pellistor: a low-power pellistor sensor in Si thin-film technology. *Sensor Actuat B-chem*, 4(3-4):533–538, 1991.

[116] Zanini M., J. H. Visser, L. Rimai, R. E. Soltis, A. Kovalchuk, D. W. Hoffman, E. M. Logothetis, U. Bonne, L. Brewer, O. W. Bynum, and M. A. Richard. Fabrication and properties of a Si-based high-sensitivity microcalorimetric gas sensor. *Sensor Actuat A:-phys*, 48(3):187–192, 1995.

[117] Visser J. H. and M. Zanini-Fisher. High-sensitivity, silicon-based, microcalorimetric gas sensor, 09 1995. US Patent 5451371.

[118] Guerin S. and P.N. Bartlett. Pellistor, May 28 2002. US Patent 6,395,230.

[119] Han C.-H., D.-W. Hong, Il-J. Kim, J. Gwak, S.-D. Han, and K. C. Singh. Synthesis of Pd or Pt/titanate nanotube and its application to catalytic type hydrogen gas sensor. *Sensor Actuat B-chem*, 128(1):320–325, 2007.

[120] Han C.-H., D.-U. Hong, J. Gwak, and S.-D. Han. A planar catalytic combustion sensor using nano-crystalline F-doped SnO_2 as a supporting material for hydrogen detection. *Korean J Chem Eng*, 24(6):927–931, 2007.

[121] H2Sense Consortium. H2Sense - Cost-effective and reliable hydrogen sensors for facilitating the safe use of hydrogen, 2013.

[122] Boon-Brett L., J. Bousek, G. Black, P. Moretto, P. Castello, T. Hübert, and U. Banach. Identifying performance gaps in hydrogen safety sensor technology for automotive and stationary applications. *Int J Hydrogen Energ*, 35(1):373–384, 2010.

[123] Buttner W. J., R. Burgess, C. Rivkin, M. B. Post, L. Boon-Brett, G. Black, F. Harskamp, and P. Moretto. Inter-laboratory assessment of hydrogen safety sensors performance under anaerobic conditions. *Int J Hydrogen Energ*, 37(22):17540–17548, 2012. HySafe 1.

[124] McAleer J. F., P. T. Moseley, P. Bourke, J. O. W. Norris, and R. Stephan. Tin dioxide gas sensors: Use of the Seebeck effect. *Sensor Actuator*, 8(3):251–257, 1985.

[125] Shin W., K. Imai, N. Izu, and N. Murayama. Thermoelectric Thick-Film Hydrogen Gas Sensor Operating at Room Temperature. *Jpn J Appl Phys*, 40(11B):L1232, 2001.

[126] Matsumiya M., F. Qiu, W. Shin, N. Izu, N. Murayama, and S. Kanzaki. Thin-film Li-doped NiO for thermoelectric hydrogen gas sensor. *Thin Solid Films*, 419(1-2):213–217, 2002.

[127] Shin W., M. Matsumiya, F. Qiu, N. Izu, and N. Murayama. Li- and Na-Doped NiO Thick Film for Thermoelectric Hydrogen Sensor. *J Ceram Soc Jpn*, 110(1287):995–998, 2002.

[128] Tajima K., F. Qiu, W. Shin, N. Izu, I. Matsubara, and N. Murayama. Micromechanical fabrication of low-power thermoelectric hydrogen sensor. *Sensor Actuat B-chem*, 108(1-2):973–978, 2005.

[129] Nishibori M., W. Shin, L. Houlet, K. Tajima, N. Izu, T. Itoh, N. Murayama, and I. Matsubara. New structural design of micro-thermoelectric sensor for wide range hydrogen detection. *J Ceram Soc Jpn*, 114:853–856, 2006.

[130] Shin W., M. Nishibori, L. F. Houlet, T. Itoh, N. Izu, and I. Matsubara. Fabrication of thermoelectric gas sensors on micro-hotplates. *Sensor Actuat B-chem*, 139(2):340–345, 2009.

[131] Nishibori M., W. Shin, N. Izu, T. Itoh, I. Matsubara, S. Yasuda, and S. Ohtani. Robust hydrogen detection system with a thermoelectric hydrogen sensor for hydrogen station application. *Int J Hydrogen Energ*, 34:2834–2841, 2009.

[132] NAST Co., Japan, E-mail: nagoya.sensor@gmail.com.

[133] Hall J. P., R. W. Whatmore, and F. W. Ainge. A black platinum catalyst/pyroelectric gas sensor. *Ferroelectrics*, 54(1):211–214, 1984.

[134] Schreiter M., R. Gabl, J. Lerchner, C. Hohlfeld, A. Delan, G. Wolf, A. Bluher, B. Katzschner, M. Mertig, and W. Pompe. Functionalized pyroelectric sensors for gas detection. *Sensor Actuat B-chem*, 119(1):255–261, 2006.

[135] Daynes H. A. Gas Analysis by Measurement of Thermal Conductivity. *ZAMM - Journal of Applied Mathematics and Mechanics / Zeitschrift für Angewandte Mathematik und Mechanik*, 13(4):328–328, 1933.

[136] Lopez M., J. Hobby, B. Alizadeh, and R. P. Kovacich. Thermal conductivity sensor, July 13 2010. US Patent 7,753,582.

[137] MAN Company. Vorrichtung zur fortlaufenden Bestimmung des Wasserstoffgehaltes in Gasgemische, 1904.

[138] Jessop G. Katharometers. *J Sci Instrum*, 43(11):777–782, 1966.

[139] Simon I. and M. Arndt. Thermal and gas-sensing properties of a micromachined thermal conductivity sensor for the detection of hydrogen in automotive applications. *Sensor Actuat A-phys*, 97-98:104–108, 2002.

[140] Daamen R., A. Humbert, and P. Bancken. Integrated circuit comprising a thermal conductivity based gas sensor, April 17, 2014. US Patent App. 14/047,137.

[141] Xensor Integration B.V., URL: http://www.xensor.nl/), timestamp = 2015.04.20.

[142] Post M. B., R. Burgess, C. Rivkin, W. Buttner, K. O'Malley, and A. Ruiz. Onboard Hydrogen/Helium Sensors in Support of the Global Technical Regulation: An Assessment of Performance in Fuel Cell Electric Vehicle Crash Tests, 2012. EP Patent App. EP19,900,301,086.

[143] Perry R. H., D. W. Green, and J. O. H. Maloney. *Perry's Chemical Engineers' Handbook*. Number v. 1-3 in Chemical Engineering Series. McGraw-Hill Professional Publishing, 1997.

[144] Stetter J. R. and J. Li. Amperometric Gas Sensors - A Review. *Chem Rev*, 108:352–366, 2008.

[145] Guth G., W. Vonau, and J. Zosel. Recent developments in electrochemical sensor application and technology - a review. *Meas Sci Technol*, 20(4):042002 (14pp), 2009.

[146] Korotecenkov G., S. D. Han, and J. R. Stetter. Review of Electrochemical Hydrogen Sensors. *Chem Rev*, 109:1402–1403, 2009.

[147] Schelter M., J. Zosel, W. Oelósner, and M. Mertig. A novel method for measuring dissolved gases in liquids. *Sensor Actuat B-chem*, 193:113–120, 2014.

[148] Kissinger P. T. and W. R. Heineman. *Laboratory Techniques in Electroanalytical Chemistry*. Marcel Dekker, New York, 1984.

[149] Iwahara H., T. Yajima, T. Hibino, and H. Ushida. Performance of Solid Oxide Fuel Cell Using Proton and Oxide Ion Mixed Conductors Based on $BaCe_{1-x}Sm_xO_{3-\alpha}$. *Solid State Ionics*, 57:303–306, 1992.

[150] Mukundan R., E. L. Brosha, S. A. Birdsell, A. L. Costello, F. H. Garzon, and R. S. Willms. Tritium Conductivity and Isotope Effect in Proton-Conducting Perovskites. *J Electrochem Soc*, 146(6):2184–2187, 1999.

[151] Chao Y., S. Yao, W. J. Buttner, and J. R. Stetter. Amperometric sensor for selective and stable hydrogen measurement. *Sensor Actuat B-chem*, 106:784–790, 2005.

[152] DOE Fuel Cell Technologies Office Multi-Year Research, Development, and Demonstration Plan - Section 3.7 Hydrogen Safety, Codes and Standards (updated July 2013). http://energy.gov/sites/prod/files/2014/03/f10/safety-codes.pdf, Accessed December 2014.

[153] Glueckauf E. and G. P Kitt. The hydrogen content of atmospheric air at ground level. *Quarterly Journal of the Royal Meteorological Society*, 83(358):522–528, 1957.

[154] Jak M. J. G., L. N. Raz, L. N. van Rij, J. Schoonman, and I. Reiss. Application of Ultrasonic to a Hydrogen Sensor. *Solid State Ionics*, 143:205–217, 2001.

[155] Iwahara H., Y. Asakura, K. Katahira, and M. Tanaka. Prospect of hydrogen technology using proton-conducting ceramics. *Solid State Ionics*, 168:299–310, 2004.

[156] Kreller C. R., M. S. Wilson, R. Mukundan, R. L. Brosha, and F. H. Garzon. Stability and Conductivity of In^{3+}-Doped SnP_2O_7 with Varying Phosphorous to Metal Ratios. *ECS Electrochemistry Letters*, 2(9):F61–F63, 2013.

[157] Norby T. Solid-state protonic conductors: principles, properties, progress and prospects. *Solid State Ionics*, 125:1–11, 1999.

[158] N. Miura, T. Sato, S. A. Anggraini, H. Ikeda, and S. Zhuiykov. A review of mixed-potential type zirconia-based gas sensors. *Ionics*, 20:901–925, 2014.

[159] Fleming W. J. Physical Principles Governing Nonideal Behavior of the Zirconia Oxygen Sensor. *J. Electrochem Soc.*, 124:21–28, 1977.

[160] Sekhar P. K., J. Zhou, M. B. Post, L. Woo, W. J. Buttner, W. R. Penrose, R. Mukundan, C. R. Kreller, R. S. Glass, F. H. Garzon, and E. L. Brosha. Independent testing and validation of prototype hydrogen sensors. *Int J Hydrogen Energ*, 39:4657–4663, 2014.

[161] Wagner C. The Mechanism of the Decomposition of Nitrous Oxide on Zinc Oxide as Catalyst. *J Chem Phys*, 18, 1950.

[162] Taguchi N. A metal oxide gas sensor. Jpn Patent 45-38200, 1962.

[163] Hübert T., L. Boon-Brett, G. Black, and U. Banach. Hydrogen sensors - A review. *Sensor Actuat B-chem*, 157:329–352, 2011.

[164] Xu Y. and M. A. A. Schoonen. The absolute energy positions of conduction and valence bands of selected semiconducting minerals. *Am Mineral*, 85:543–556, 2000.

[165] Sunu S. S., E. Prabhu, V. Jayaraman, K. I. Gnanasekar, T. K. Seshagiri, and T. Gnanasekaran. Electrical conductivity and gas sensing properties of MoO_3. *Sensor Actuat B-chem*, 101(1-2):161–174, 2004.

[166] Yin W.-J., S. Chen, J.-H. Yang, X.-G. Gong, Y. Yan, and S.-H. Wei. Effective band gap narrowing of anatase TiO_2 by strain along a soft crystal direction. *Appl Phys Lett*, 96, 221901, 2010.

[167] Korotcenkov G. Gas response control through structural and chemical modification of metal oxide films: state of the art and approaches. *Sensor Actuat B-chem*, 107(1):209–232, 2005.

[168] Lu C. and Z. Chen. High-temperature resistive hydrogen sensor based on thin nanoporous rutile TiO_2 film on anodic aluminum oxide. *Sensor Actuat B-chem*, 140(1):109–115, 2009.

[169] Williams D. E. Semiconducting oxides as gas-sensitive resistors. *Sensor Actuat B-chem*, 57(1-3):1–16, 1999.

[170] Shimizu Y. and M. Egashira. Basic Aspects and Challenges of Semiconductor Gas Sensors. *MRS Bulletin*, 24:18–24, 6 1999.

[171] Ihokura K. and J. Watson. *The Stannic Oxide Gas Sensor Principles and Applications*. Taylor & Francis, 1994.

[172] N. Barsan, M. Schweizer-Berberich, and W. Göpel. Fundamental and practical aspects in the design of nanoscaled SnO_2 gas sensors: A status report. *Fresen J Anal Chem*, 365(4):287–304, 1999.

[173] Ahlers S., G. Müller, and T. Doll. A rate equation approach to the gas sensitivity of thin film metal oxide materials. *Sensor Actuat B-chem*, 107(2):587–599, 2005.

[174] G. Tournier and C. Pijolat. Selective filter for SnO_2-based gas sensor: application to hydrogen trace detection. *Sensor Actuat B-chem*, 106(2):553 – 562, 2005.

[175] G. Korotcenkov and B. K. Cho. Engineering approaches for the improvement of conductometric gas sensor parameters: Part 1. improvement of sensor sensitivity and selectivity (short survey). *Sensor Actuat B-chem*, 188:709–728, 2013.

[176] G. Korotcenkov and B. K. Cho. Engineering approaches to improvement of conductometric gas sensor parameters. Part 2: Decrease of dissipated (consumable) power and improvement stability and reliability. *Sensor Actuat B-chem*, 198(0):316–341, 2014.

[177] Batzill M. and U. Diebold. The surface and materials science of tin oxide. *Progress in Surface Science*, 79(2-4):47–154, 2005.

[178] T. Samerjai, N. Tamaekong, C. Liewhiran, A. Wisitsoraat, A. Tuantranont, and S. Phanichphant. Selectivity towards H_2 gas by flame-made Pt-loaded WO_3 sensing films. *Sensor Actuat B-chem*, 157(1):290–297, 2011.

[179] Adamyan A. Z., Z. N. Adamyan, V. M. Aroutiounian, A. H. Arakelyan, K. J. Touryan, and J. A. Turner. Sol-gel derived thin-film semiconductor hydrogen gas sensor. *Int J Hydrogen Energ*, 32(16):4101–4108, 2007.

[180] Korotcenkov G. The role of morphology and crystallographic structure of metal oxides in response of conductometric-type gas sensors. *Mater Sci Eng R-reports*, 61(1-6):1–39, 2008.

[181] Malyshev V. V. and A.V. Pislyakov. Investigation of gas-sensitivity of sensor structures to hydrogen in a wide range of temperature, concentration and humidity of gas medium. *Sensor Actuat B-chem*, 134(2):913–921, 2008.

[182] Simon I., N. Bârsan, M. Bauer, and U. Weimar. Micromachined metal oxide gas sensors: opportunities to improve sensor performance. *Sensor Actuat B-chem*, 73(1):1–26, 2001.

[183] Mozalev A., R. Calavia, R. M. Vazquez, I. Gracia, C. Cane, X. Correig, X. Vilanova, F. Gispert-Guirado, J. Hubalek, and E. Llobet. MEMS-microhotplate-based hydrogen gas sensor utilizing the nanostructured porous-anodic-alumina-supported WO_3 active layer. *Int J Hydrogen Energ*, 38(19):8011–8021, 2013.

[184] Semancik S., R. E. Cavicchi, M. C. Wheeler, J. E. Tiffany, G. E. Poirier, R. M. Walton, J. S. Suehle, B. Panchapakesan, and D. L. DeVoe. Microhotplate platforms for chemical sensor research. *Sens Actuat B-chem*, 77:579–591, 2001.

[185] El Matbouly H., F. Domingue, V. Palmisano, L. Boon-Brett, M. B. Post, C. Rivkin, R. Burgess, and W. J. Buttner. Assessment of commercial micro-machined hydrogen sensors performance metrics for safety sensing applications. *Int J Hydrogen Energ*, 39:4664–4673, 2014.

[186] Sakai G., N. S. Baik, N. Miura, and N. Yamazoe. Gas sensing properties of tin oxide thin films fabricated from hydrothermally treated nanoparticles: Dependence of CO and H_2 response on film thickness. *Sensor Actuat B-chem*, 77(1-2):116–121, 2001.

[187] Galstyan V. E., K. S. Martirosyan, V. M. Aroutiounian, V. M. Arakelyan, A. H. Arakelyan, and P. G. Soukiassian. Investigations of hydrogen sensors made of porous silicon. *Thin Solid Films*, 517(1):239–241, 2008.

[188] Chen K., K. Xie, X. Feng, S. Wang, R. Hu, H. Gu, and Y. Li. An excellent room-temperature hydrogen sensor based on titania nanotube-arrays. *Int J Hydrogen Energ*, 37(18):13602–13609, 2012. ICCE-2011.

[189] Bamsaoud S. F., S. B. Rane, R. N. Karekar, and R. C. Aiyer. Nano particulate SnO_2 based resistive films as a hydrogen and acetone vapour sensor. *Sensor Actuat B-chem*, 153(2):382–391, 2011.

[190] Rashid T.-R., D.-T. Phan, and G.-S. Chung. A flexible hydrogen sensor based on Pd nanoparticles decorated ZnO nanorods grown on polyimide tape. *Sensor Actuat B-chem*, 185(0):777–784, 2013.

[191] Boon-Brett L., J. Bousek, and P. Moretto. Reliability of commercially available hydrogen sensors for detection of hydrogen at critical concentrations: Part II - selected sensor test results. *Int J Hydrogen Energ*, 34(1):562–571, 2009.

[192] Korotcenkov G. Practical aspects in design of one-electrode semiconductor gas sensors: Status report. *Sensor Actuat B-chem*, 121(2):664–678, 2007.

[193] Kilinic N. Resistive hydrogen sensors based on nanostructured metals and metal alloys. *Nanosci Nanotechn Lett*, 5:825–841, 2013.

[194] DiMeo Jr. F., I.-S. Chen, P. Chen, J. Neuner, A. Roerhl, and J. Welch. MEMS-based hydrogen gas sensors. *Sensor Actuat B-chem*, 117(1):10–16, 2006.

[195] Yoshimura K., S. Bao, N. Uchiyama, H. Matsumoto, T. Kanai, S. Nakabayashi, and H. Kanayama. New hydrogen sensor based on sputtered Mg-Ni alloy thin film. *Vacuum*, 83:699–702, 2009.

[196] Jamshidi M. M., K. Alshaltami, F. Akkari, and J. Wright. Development of palladium-based hydrogen thin film sensor using silicon oxide substrate. *Indian Journal of Physics*, 87(5):511–515, 2013.

[197] Hughes R. C. and W. K. Schubert. Thin films of Pd/Ni alloys for detection of high hydrogen concentrations. *J Appl Phys*, 71:542–544, 1992.

[198] Soundarrajan P. Hydrogen-Specific Sensing for Industrial Process Control and Safety. *Sensors Online*, 2008. http://www.sensorsmag.com.

[199] Visel T., P. Soundarrajan, and I. Pavlovsky. Palladium-Nickel Hydrogen Sensor, June 7 2007. US Patent App. 11/551,630.

[200] Lundström I., S. Shivaraman, C. Svensson, and L. Lundkvist. A hydrogen-sensitive MOS field-effect transistor. *App Phys Lett*, 26(2):55–57, 1975.

[201] Lundström I., H. Sundgren, F. Winquist, M. Eriksson, C. Krantz-Rülcker, and A. Lloyd-Spetz. Twenty-five years of field effect gas sensor research in Linköping. *Sensor Actuat B-chem*, 121:247–262, 2007.

[202] Steele M. C., J. W. Hile, and B. A. Maciver. Hydrogen-sensitive palladium gate MOS capacitors. *J Appl Phys*, 47(6):2537–2538, 1976.

[203] Armgarth M., D. Söderberg, and I. Lundström. Palladium and platinum gate metal oxide semiconductor capacitors in hydrogen and oxygen mixtures. *App Phys Lett*, 41:654–655, 1982.

[204] Zhu W., O. K. Tan, and X. Yao. Amorphous $Pb(ZrTi)O_3$ thin film hydrogen gas sensor. *Sensor Actuat B-chem*, 77(1-2):416–420, 2001.

[205] Potje-Kamloth K. Semiconductor junction gas sensors. *Chem Rev.*, 108:367–399, 2001.

[206] Soo M. T., K. Y. Cheong, and A. F. Mohd Noor. Advances of SiC-based MOS capacitor hydrogen sensors for harsh environment applications. *Sensor Actuat B-chem*, 151:39–55, 2010.

[207] Moritz W., V. Fillipov, A. Vasiliev, G. Cherkashinin, and J. Szeponik. A field effect based hydrogen sensor for low and high concentrations. *ECS Transactions*, 3:223–230, 2006.

[208] Lang M., U. Banach, K. Nörthemann, and M. Milstrey A. K. Gerlitzke, R. Kaufer, M. Woratz, T. Hübert, and W. Moritz. Long-term stability of a MEIS low energy hydrogen sensor. *Sensor Actuat B-chem*, 187:395–400, 2013.

[209] H2Scan Application Note "Recycle Gas Hydrogen Measurement," www.h2scan.compdfs/Recycle%20Gas%20in%20Refineries%20and %20Chemical%20Plants.pdf. Accessed 22.02.2015.

[210] Shivaraman M. S., I. Lundström, C. Svensson, and H. Hammarsten. Hydrogen sensitivity of palladium thin-oxide silicon Schottky barriers. *Electronics Letters*, 12:483–484, 1976.

[211] Steele M. and B. MacIver. Palladium/cadmium-sulfide Schottky diodes for hydrogen detection. *Appl Phys Lett*, 28(1-2):687–688, 1976.

[212] Trinchi A., S. Kandasamy, and W. Wlodarski. High temperature field effect hydrogen and hydrocarbon gas sensors based on sic mos devices. *Sensor Actuat B-chem*, 133:705–716, 2008.

[213] Schottky-Dioden Wasserstoffsensor, technical data sheet, neo hydrogen sensor GmbH, www.neohysens.de, 2015, 2010.

[214] Cerchez M., H. Langer, T. Heinzel M. El Achhab, D. Ostermann, H. Lüder, and J. Degenhardt. Dynamics of hydrogen sensing with pt/tio−2 schottky diodes. *Appl Phys Lett*, 103:033522, 1–4, 2013.

[215] Lundström I. Hydrogen Sensitive MOS-Structures Part1: Principles and Applications. *Sensor Actuat B-chem*, 1:403–426, 1981.

[216] Melby J. H., R. F. Davis, and L. M. Porter. Modeling the Electrical Response of Hydrogen Sensors Based on AlGaN/GaN High-Electron-Mobility Transistors. *ECS Journal of Solid State Science and Technology*, 2(11):Q214–Q219, 2013.

[217] Scharnagl K., A. Karthigeyan, M. Burgmair, M. Zimmer, T. Doll, and I. Eisele. Low temperature hydrogen detection at high concentrations: Comparison of platinum and iridium. *Sensor Actuat B-chem*, 80(3):163–168, 2001.

[218] Kosovic M. and N. Edvardsson. Applied sensor fe hydrogen sensor. *Proc. 18th WHEC*, 12:315–321, 2010.

[219] Sauerbrey G. Verwendung von Schwingquarzen zur Wägung dünner Schichten und zur Mikrowägung. *Z Phys*, 155(2):206–222, 1959.

[220] Fanget S., S. Hentz, P. Puget, J. Arcamone, M. Matheron, E. Colinet, P. Andreucci, L. Duraffourg, E. Myers, and M. L. Roukes. Gas sensors based on gravimetric detection - A review. *Sensor Actuat B-chem*, 160(1):804–821, 2011.

[221] Lang H. P. *Solid State Gas Sensing*, chapter Cantilever-Based Gas Sensing, pages 305–328. Springer Science & Business Media, 2009.

[222] Chen D., J. J. Wang, D. H. Li, and Y. Xu. Hydrogen sensor based on pd-functionalized film bulk acoustic resonator. *Sensor Actuat B-chem*, 159(1):234–237, 2011.

[223] Wohltjen H. and R. Dessy. Surface acoustic wave probe for chemical analysis. I. Introduction and instrument description. *Anal Chem*, 51:1458–1475, 1979.

[224] D'Amico A., A. Palma, and E. Verona. Surface acoustic wave hydrogen sensor. *Sensor Actuator*, 3(0):31–39, 1982-1983.

[225] Jakubik W., M. Krzywiecki, E. Maciak, and M. Urbańczyk. Bi-layer nanostructures of CuPc and Pd for resistance-type and SAW-type hydrogen gas sensors. *Sensor Actuat B-chem*, 175(0):255–262, 2012.

[226] Wang C., Y. Wang, S.-Y. Zhang, L. Fan, and X.-J. Shui. Characteristics of SAW hydrogen sensors based on InO_x/128 °YX-$LiNbO_3$ structures at room temperature. *Sensor Actuat B-chem*, 173(0):710–715, 2012.

[227] Viespe C. and C. Grigoriu. SAW sensor based on highly sensitive nanoporous palladium thin film for hydrogen detection. *Microelectron Eng*, 108:218–221, 2013.

[228] Huang F.-C., Y.-Y. Chen, and T.-T. Wu. A room temperature surface acoustic wave hydrogen sensor with Pt coated ZnO nanorods. *Nanotechnology*, 20(6), 2009.

[229] Abe T., N. Iwata, T. Tsuji, T. Mihara, S. Akao, K. Noguchi, N. Nakaso, D. Sim, Y. Ebi, T. Fukiura, H. Tanaka, and K. Yamanaka. Evaluation of response time in ball surface-acoustic-wave hydrogen sensor using digital quadrature detector. *Jpn J Appl Phys 1*, 46(7 B):4726–4728, 2007.

[230] Thundat T., R. J. Warmack, G. Y. Chen, and D. P. Allison. Thermal and ambient-induced deflections of scanning force microscope cantilevers. *Appl Phys Lett*, 64:2894, 1994.

[231] Ma C. and A. Wang. Optical fiber tip acoustic resonator for hydrogen sensing. *Opt. Lett.*, 35(12):2043–2045, Jun 2010.

[232] Pavlovsky I. Hydrogen sensor, January 19, 2010. US Patent 7,647,813.

[233] Suzuki A. *Hydrogen Energy - Challenges and Perspectives, In Tech.* 2012.

[234] Watanabe T., S. Okazaki, H. Nakagawa, K. Murata, and K. Fukuda. A fiber-optic hydrogen gas sensor with low propagation loss. *Sensor Actuat B-chem*, 145(2):781–787, 2010.

[235] Hsu C.-H., C.-C. Chang, C.-M. Tseng, C.-C. Chan, W.-H. Chao, Y.-R. Wu, M.-H. Wen, Y.-T. Hsieh, Yi-C. Wang, Chi-L. Chen, M.-J. Wang, and M.-K. Wu. An ultra-fast response gasochromic device for hydrogen gas detection. *Sensor Actuat B-chem*, 186(0):193–198, 2013.

[236] Hoagland W., and D. K. Benson and R. D. Smith. Novel Wide Area Sensing Technology. Proc. 2nd International Conference on Hydrogen Safety, 2007. In *ICHS*, 2011.

[237] Roberson L. B., J. E. Captain, M. K. Williams, and L. N. C. Tate. Chemochromic detector for sensing gas leakage and process for producing the same, January 3, 2013. US Patent App. 13/615,850.

[238] Huiberts J. N., R. Griessen, J. H. Rector, R. J. Wijngaarden, J. P. Dekker, D. G. De Groot, and N. J. Koeman. Yttrium and lanthanum hydride films with switchable optical properties. *Nature*, 380(6571):231–234, 1996.

[239] Palmisano V., M. Filippi, A. Baldi, M. Slaman, H. Schreuders, and B. Dam. An optical hydrogen sensor based on a Pd-capped Mg thin film wedge. *Int J Hydrogen Energ*, 35(22):12574 – 12578, 2010.

[240] Radeva T., P. Ngene, M. Slaman, R. Westerwaal, H. Schreuders, and B. Dam. Highly sensitive and selective visual hydrogen detectors based on Y_xMg_{1-x} thin films. *Sensor Actuat B-chem*, 203(0):745–751, 2014.

[241] Wang X., Y. Tang, C. Zhou, and B. Liao. Design and optimization of the optical fiber surface plasmon resonance hydrogen sensor based on wavelength modulation. *Opt Commun*, 298-299:88–94, 2013.

[242] Butler M. A. Optical fiber hydrogen sensor. *Appl Phys Lett*, 45(10):1007–1009, 1984.

[243] Dai J., M. Yang, Z. Yang, Z. Li, Y. Wang, G. Wang, Yi Zhang, and Z. Zhuang. Performance of fiber Bragg grating hydrogen sensor coated with Pt-loaded WO_3 coating. *Sensor Actuat B-chem*, 190(0):657–663, 2014.

[244] Butler M. A. Chemical sensing with an optical fiber interferometer. Electrochem Soc Extended Abstracts, 84-2:802, 1984.

[245] Butler M. A. Micromirror optical-fiber hydrogen sensor. *Sensor Actuat B-chem*, 22(2):155–163, 1994.

[246] Bévenot X., A. Trouillet, C. Veillas, H. Gagnaire, and M. Clément. Hydrogen leak detection using an optical fibre sensor for aerospace applications. *Sensor Actuat B-chem*, 67(1-2):57–67, 2000.

[247] Tabib-Azar M., B. Sutapun, R. Petrick, and A. Kazemi. Highly sensitive hydrogen sensors using palladium coated fiber optics with exposed cores and evanescent field interactions. *Sensor Actuat B-chem*, 56(1-2):158–163, 1999.

[248] Perrotton C., R. J. Westerwaal, N. Javahiraly, M. Slaman, H. Schreuders, B. Dam, and P. Meyrueis. A reliable, sensitive and fast optical fiber hydrogen sensor based on surface Plasmon resonance. *Optics Express*, 21(1):382–390, 2013.

[249] Yang M., H. Liu, D. Zhang, and X. Tong. Hydrogen sensing performance comparison of Pd layer and Pd / WO_3 composite thin film coated on side-polished single- and multimode fibers. *Sensor Actuat B-chem*, 149(1):161–164, 2010.

[250] Strohfeldt N., A. Tittl, and H. Giessen. Long-term stability of capped and buffered palladium-nickel thin films and nanostructures for plasmonic hydrogen sensing applications. *Opt Mater Express*, 3(2):194–204, 2013.

[251] Westerwaal R. J., N. Duim, I. Nieuwenhuijse, C. Perrotton, A. Dabirian, J. M. van Leeuwen, V. Palmisano, and B. Dam. Thin film based sensors for a continuous monitoring of hydrogen concentrations. *Sensor Actuat B-chem*, 165(1):88–96, 2012.

[252] Perrotton C., N. Javahiraly, A. Kazemi, and P. Meyrueis. Review of optical fiber sensor technologies for hydrogen leak detection in hydrogen energy storage *Proc. SPIE*, vol. 8026, 2011.

[253] Dessy R. E. and E. W. Richmond. Birefringent single-arm fiber optic enthalpimeter for catalytic reaction monitoring. *Anal Chem*, 64(13):1379–1382, 1992.

[254] Lee B. H., Y. H. Kim, K. S. Park, J. B. Eom, M. J. Kim, B. S. Rho, and H. Y. Choi. Interferometric fiber optic sensors. *Sensors*, 12(3):2467–2486, 2012.

[255] Zhang G., M. Yang, and Y. Wang. Optical fiber-tip Fabry-Perot interferometer for hydrogen sensing. *Opt Comm*, 329(0):34–37, 2014.

[256] Slaman M., B. Dam, M. Pasturel, D. M. Borsa, H. Schreuders, J .H. Rector, and R. Griessen. Fiber optic hydrogen detectors containing Mg-based metal hydrides. *Sensor Actuat B-chem*, 123(1):538–545, 2007.

[257] Villatoro J., D. Luna-Moreno, and D. Monzón-Hernández. Optical fiber hydrogen sensor for concentrations below the lower explosive limit. *Sensor Actuat B-chem*, 110(1):23–27, 2005.

[258] Sekimoto S., H. Nakagawa, S. Okazaki, K. Fukuda, S. Asakura, T. Shigemori, and S. Takahashi. A fiber-optic evanescent-wave hydrogen gas sensor using palladium-supported tungsten oxide. *Sensor Actuat B-chem*, 66(1-3):142–145, 2000.

[259] Luna-Moreno D., D. Monzón-Hernández, J. Villatoro, and Gonćal Badenes. Optical fiber hydrogen sensor based on core diameter mismatch and annealed Pd-Au thin films. *Sensor Actuat B-chem*, 125(1):66–71, 2007.

[260] Trouillet A., E. Marin, and C. Veillas. Fibre gratings for hydrogen sensing. *Meas Sci Technol*, 17(5):1124–1128, 2006.

[261] Kotchourko N., M. Kuznetsov, A. Kotchourko, J. Grune, A. Lelyakin, and T. Jordan. Concentration measurements in a round hydrogen jet using Background Oriented Schlieren (BOS) technique. *Int J Hydrogen Energ*, 39(11):6201–6209, 2014.

[262] Ball A. J. Investigation of Gaseous Hydrogen Leak Detection using Raman Scattering and Laser-induced Breakdown Spectroscopy. Master's thesis, University of Florida, 2005.

[263] Ferraro J. R., K. Nakamoto, and C. W. Brown. Chapter 2 - Instrumentation and Experimental Techniques. In *Introductory Raman Spectroscopy (2nd ed.)*, pages 95–146. Academic Press, San Diego, 2003.

[264] The Use of Spontaneous Raman Scattering for Hydrogen Leak Detection. Technical report, National Aeronautics and Space Administration, 1994.

[265] Adler-Golden S. M., N. Goldstein, F. Bien, M. W. Matthew, M. E. Gersh, W. K. Cheng, and F. W. Adams. Laser Raman sensor for measurement of trace-hydrogen gas. *Appl Optics*, 31(6):831–835, 1992.

[266] Asahi I., S. Sugimoto, H. Ninomiya, T. Fukuchi, and T. Shiina. Remote sensing of hydrogen gas concentration distribution by Raman lidar, Proc. of SPIE, vol. 8526-34, 2012.

[267] Ninomiya H., S. Yaeshima, K. Ichikawa, and T. Fukuchi. Raman lidar system for hydrogen gas detection. *Opt Eng*, 46(9), 2007.

[268] Haber F. Über Schlagwetteranzeige. *Die Naturwissenschaften*, 1:1049–1051, 1913.

[269] www.gassonic.com, Gassonic-Surveyor-PD-2749.pdf. Accessed 18.09.2014.

[270] https://de.trotec.com/en/products/measuring-devices/leak-detection/combination-measuring-devices-for-leak-detection/ld6000-combination-detector-for-leak-detection. Accessed 01.10.2014.

[271] Wan J. K. S., M. S. Ioffe, and M. C. Depew. A novel acoustic sensing system for on-line hydrogen measurements. *Sensor Actuat B-chem*, 32:233–237, 1996.

[272] Fukuoka H., J. Jung, M. Inoue, H. Fujita, and Y. Kato. Absolute Concentration Measurement for Hydrogen. *Energy Procedia*, 29:283–290, 2012.

[273] Sonoyama M., Y. Kato, and H. Fujita. Application of Ultrasonic to a Hydrogen Sensor. *IEEE SENSORS Conference*, pages 2141–2144, 2010.

[274] Mecea V. M. Ultrasonic Hydrogen Detector. *Z Phys Chem*, 183:491–498, 1994.

[275] Jia Y., A. Xue, Z. Zhou, Z. Wu, J. Chen, K. Zhang, J. Zhou, Y. Wang, and H. Chan. Magnetostrictive/piezoelectric drum magnetoelectric transducer for H2 detection. *Int J Hydrogen Energ*, 38:14915–14919, 2013.

[276] IEC 60079-29-2:2015. Explosive atmospheres - Part 29-2: Gas detectors - Selection, installation, use and maintenance of detectors for flammable gases and oxygen.

[277] H2Sense Hydrogen Sensor Data Base, 2015. http://www.h2sense.bam.de/en/data_sheet.

[278] Lakkis S., R. Younes, Y. Alayli, and M. Sawan. Review of recent trends in gas sensing technologies and their miniaturization potential. *Sensor Rev*, 34:24–35, 2014.

[279] Cahn J. W. and R. E. Hanneman. (111) Surface tensions of III-V compounds and their relationship to spontaneous bending of thin crystals. *Surf Sci*, 1(4):387–398, 1964.

[280] Boisen A., S. Dohn, S. S. Keller, S. Schmid, and M. Tenje. Cantilever-like micromechanical sensors. *Reports on Progress in Physics*, 74:036101, 2011.

[281] Shaver P. J. Bimetal strip hydrogen gas detectors. *Rev Sci Instrum*, 40(7):901–905, 1969.

[282] Iannuzzi D., M. Slaman, J. H. Rector, H. Schreuders, S. Deladi, and M.C. Elwenspoek. A fiber-top cantilever for hydrogen detection. *Sensor Actuat B-chem*, 121(2):706–708, 2007.

[283] Hu Z., T. Thundat, and R. J. Warmack. Investigation of adsorption and absorption-induced stresses using microcantilever sensors. *J Appl Phys*, 90(1):427–431, 2001.

[284] Lee E., J. M. Lee, E. Lee, J.-S. Noh, J. H. Joe, B. Jung, and W. Lee. Hydrogen gas sensing performance of Pd–Ni alloy thin films. *Thin Solid Films*, 519(2):880 – 884, 2010.

[285] Patton J. F., S. R. Hunter, M. J. Sepaniak, P. G. Daskos, and D. B. Smith. Rapid response microsensor for hydrogen detection using nanostructured palladium films. *Sensor Actuat A-phys*, 163(2):464–470, 2010.

[286] Baselt D. R., B. Fruhberger, E. Klaassen, S. Cemalovic, C. L. Britton Jr., S. V. Patel, T. E. Mlsna, D. McCorkle, and B. Warmack. Design and performance of a microcantilever-based hydrogen sensor. *Sensor Actuat B-chem*, 88(2):120–131, 2003.

[287] Bogue R. Nanomaterials for gas sensing: A review of recent research. *Sensor Review*, 34:1–8, 2014.

[288] Kochmann S., T. Hirsch, and O. S. Wolfbeis. Graphenes in chemical sensors and biosensors. *Trac-Trends in Analytical Chemisty*, 39:87–113, 2012.

[289] Basu S. and P. Bhattacharyya. Recent developments on graphene and graphene oxide based solid state gas sensors. *Sensor Actuat B-chem*, 173:1–21, 2012.

[290] Gu H., Z. Wang, and Y. Hu. Hydrogen Gas Sensors Based on Semiconductor Oxide Nanostructures. *Sensors*, 12:5517–5550, 2012.

[291] Arafat M. M., B. Dinan, A. Akbar Sheikh, and A. S. M. A. Haseeb. Gas Sensors Based on One Dimensional Nanostructured Metal-Oxides: A Review. *Sensors*, 12:7207–7258, 2012.

[292] Hübert T., L. Boon-Brett, V. Palmisano, and M. A. Bader. Developments in gas sensor technology for hydrogen safety. *Int J Hydrogen Energ*, 39:20474–20483, 2014.

[293] Noh J.-S., J. M. Lee, and W. Lee. Low-Dimensional Palladium Nanostructures for Fast and Reliable Hydrogen Gas Detection. *Sensors*, 11:825–851, 2011.

[294] Fournier C., K. Rajoua, M. L. Doublet, and F. Favier. Palladium-silver mesowires for the extended detection of H_2. *ACS Appl Mater Interfaces.*, 5:310–318, 2013.

[295] Lim Y. T., J.Y. Son, and J.-S. Rhee. Vertical ZnO nanorod array as an effective hydrogen gas sensor. *Ceramics International*, 39:887–890, 2013.

[296] Soriadi N., N. M. Mohamed, and F. A. Hussin. A review of TiO_2 nanotube arrays for hydrogen sensing application. *AIP Conf. Proc.*, 1482:611–616, 2012.

[297] Cho S., J.S. Lee, J. Jun, and J. Jang. High-sensitivity hydrogen gas sensors based on Pd-decorated nanoporous poly(aniline-co-aniline-2-sulfonic acid):poly(4-styrenesulfonic acid). *J. Mater. Chem. A*, 2:1955–1966, 2014.

[298] Arya S. K., S. Krishnan, H. Silva, S. Jean, and S. Bhansali. Advances in materials for room temperature hydrogen sensors. *Analyst*, 137:2743–2756, 2012.

[299] Silva R. A., T. Hashimoto, G.E. Thompson, and C.M. Rangel. Characterization of MEA degradation for an open air cathode PEM fuel cell. *Int J Hydrogen Energ*, 37(8):7299–7308, 2012.

[300] Erickson R.J., J. Jr. Howe, G.W. Kulp, and S.P. Van Keuren. International space station United States orbital segment oxygen generation system. *SAE Int. J. Aerosp.*, 1(April):15–24, 2009.

[301] SAE J2719. Hydrogen Fuel Quality for Fuel Cell Vehicles, version 01.09.2011. Technical Report Part 2, International Organization for Standardization, USA, 2011.

[302] ISO 14687-2:2012. Hydrogen fuel - product specification - part 2: Proton exchange membrane (PEM) fuel cell applications for road vehicles. ISO, Geneva, Switzerland.

[303] CFR 1910.134(b). General industry respiratory protection. Technical Report 29, OSHA.

[304] Ramamoorthy R., P. Dutta, and S. A. Akbar. Oxygen sensors: Materials, methods, designs, and applications. *J Mater Sci*, 38:4271–4282, 2003.

[305] Clark Jr. L. C., R. Wolf, D. Granger, and Z. Taylor. Continuous recording of blood oxygen tensions by polarography. *J Appl Physiol*, 6(3):189–193, 1953.

[306] Warburton P. R., R. S. Sawtelle, A. Watson, and A. Q. Wang. Failure prediction for a galvanic oxygen sensor. *Sensor Actuat B-chem*, 72(3):197–203, 2001.

[307] Buttner W. J., R. Burgess, M. Post, C. Rivkin, L. Boon-Brett, V. Palmisano, and P. Moretto. An Assessment on the Quantification of Hydrogen Releases Through Oxygen Displacement Using Oxygen Sensors. *Int J Hydrogen Energ* 39(35):20484–20490, 2014.

[308] ASTM D7607-11. Standard Test Method for Analysis of Oxygen in Gaseous Fuels (Electrochemical Sensor Method), 2006.

[309] Izu N., W. Shin, I. Matsubara, and N. Murayama. Resistive oxygen gas sensors using ceria-zirconia thick films. *J Ceram Soc Jpn*, 112:s535–s539, 2004.

[310] Rettig F. and R. Moos. Temperature modulated direct thermoelectric gas sensors: Thermal modeling and results for fast hydrocarbon sensors. *Meas Sci Technol*, 20:065205, 2009.

[311] Moos R., N. Izu, F. Rettig, S. Reiss, W. Shin, and I. Matsubara. Resistive oxygen gas sensors for harsh environments. *Sensors*, 11:3439–3465, 2011.

[312] Directive 2008/50/EC of the European Parliament of the council on ambient air quality and cleaner air for Europe. Technical report, Official Journal of the European Union, Brussels, Belgium, 2008.

[313] Verhage A. J. L., J. F. Coolegem, M. J. J. Mulder, M. H. Yildirim, and F. A. de Bruijn. 30,000 h operation of a 70 kW stationary {PEM} fuel cell system using hydrogen from a chlorine factory. *Int J Hydrogen Energ*, 38(11):4714–4724, 2013.

[314] TRGS 900. Technische Regeln für Gefahrstoffe, Arbeitsplatzgrenzwerte. Technical report, International Organization for Standardization, Germany, 2006.

[315] Hodgkinson J. and R. P Tatam. Optical gas sensing: a review. *Meas. Sci. Technol*, 24:012004, 59pp, 2013.

[316] ISO 14687-2:2012. Hydrogen fuel - Product specification - Part 2: Proton exchange membrane (PEM) fuel cell applications for road vehicles. Technical report, Geneva, Switzerland.

[317] Baschuk J. J. and X. Li. Carbon monoxide poisoning of proton exchange membrane fuel cells. *Int. J. Energy Res.*, 25(1):695–713, 2001.

[318] Aroutiounian V. Metal oxide hydrogen, oxygen, and carbon monoxide sensors for hydrogen setups and cells - a review. *Int J Hydrogen Energ*, 32:1145–1158, 2007.

[319] Knight B., T. Clark, W. Buttner, F. DiMeo, and S. Swartz. Development of Sensors for Automotive PEM-based Fuel Cells. DOE Contract No. DE-FC04-01AL67616. Final report, Department of Energy, USA, 2005.

[320] Kirby K. W., A. C. Chu, and K. C. Fuller. Detection of low level carbon monoxide in hydrogen-rich gas streams. *Sensor Actuat B-chem*, 95:224–231, 2003.

[321] Pijolat C., G. Tournier, and J. P. Viricelle. CO detection in H_2 reducing atmosphere with minifuel cell. *Sensor Actuat B-chem*, 156:283–289, 2011.

[322] Mukundan R., E. L. Brosha, and F. H. Garzon. A low temperature sensor for the detection of carbon monoxide in hydrogen. *Solid State Ionics*, 175:497–501, 2004.

[323] Pijolat C., G. Tournier, and Viricelle J. P. Detection of CO in H_2-rich gases with a samarium doped ceria (SDC) sensor for fuel cell applications. *Sensor Actuat B-chem*, 141:7–12, 2009.

[324] Dutta P. K., R. R. Rao, S. L. Swartz, and C. T. Holt. Sensing of carbon monoxide gas in reducing environments. *Sensor Actuat B-chem*, 84:189–193, 2002.

[325] Gopal Reddy C. V., P. K. Dutta, and S. A. Akbar. Detection of CO in a reducing, hydrous environment using CuBr as electrolyte. *Sensor Actuat B-chem*, 92:351–355, 2003.

[326] Liu C., Z. Noda, K. Sasaki, and K. Hayashi. Development of a polyaniline nanofiber-based carbon monoxide sensor for hydrogen fuel cell application. *Int J Hydrogen Energ*, 37(18):13529–13535, 2012.

[327] Simon S., R. Marjunus, T. Stimpel-Lindner, W. Hansch, C. Wilbertz, I. Schmidt, and J. Wartmann. Pt/Au based sensor with a PMMA-Film for detecting CO in hydrogen atmosphere. *Proc. IMCS*, pages 327–330, 2012.

[328] EN 13779:2007. Ventilation for non-residential buildings - Performance requirements for ventilation and room-conditioning systems.

[329] Dong L., J. Wright, B. Peters, B. A. Ferguson, F. K. Tittel, and S. McWhorter. Compact QEPAS sensor for trace methane and ammonia detection in impure hydrogen. *Appl. Phys B*, 107:459–467, 2012.

[330] Xu C. N., N. Miura, Y. Ishida, K. Matsuda, and N. Yamazoe. Selective detection of NH_3 over NO in combustion exhausts by using Au and MoO_3 doubly promoted WO_3 element. *Sensor Actuat B-chem*, 65(1):163–165, 2000.

[331] Ju H. Y., D. G. Fu, L. Zhan, J. Li, X. H. Wang, and F. S. Wang. Influence of humidity on ammonia sensibility of polyaniline. *Acta Polymerica Sinica*, (1):156–163, 2014.

[332] Timmer B., W. Olthuis, and A. v. d. Berg. Ammonia sensors and their applications - a review. *Sensor Actuat B-chem*, 107:666–677, 2005.

[333] http://www.micronas.com/de/produkte/gas-sensoren/gas-86xyb?detailnid=1235. Accessed 16.03.2015.

[334] Menil F., V. Coillard, and C. Lucat. Critical review of nitrogen monoxide sensors for exhaust gases of lean burn engines. *Sensor Actuat B-chem*, 67:1–23, 2000.

[335] Cheng X., Z. Shi, N. Glass, L. Zhang, J. Zhang, D. Song, Z.-S. Liu, H. Wang, and J. Shen. A review of pem hydrogen fuel cell contamination: Impacts, mechanisms, and mitigation. *J Power Sources*, 165(2):739–756, 2007.

[336] Ciccoli R., V. Cigolotti, R. Lo Presti, E. Massi, S. J. McPhail, G. Monteleone, A. Moreno, V. Naticchioni, C. Paoletti, E. Simonetti, and F. Zaza. Molten carbonate fuel cells fed with biogas: Combating H_2S. *Waste Management*, 30(6):1018–1024, 2010.

[337] Haga K., S. Adachi, Y. Shiratori, K. Itoh, and K. Sasaki. Poisoning of SOFC anodes by various fuel impurities. *Solid State Ionics*, 179(27-32):1427–1431, 2008.

[338] Elia A., Lugara P. M., Di Franco C., and Spagnolo V. Photoacoustic techniques for trace gas sensing based on semiconductor laser sources. *Sensors*, 9:9616–9628, 2009.

[339] Zoltan B., Andrea P., and Gabor S. Photoacoustic instruments for practical applications: present, potentials, and future challenges. *Appl Spectrosc Rev*, 46:1–37, 2011.

[340] Tripathi M. M., K. E. Eseller, F-Y. Yueha, and J. P. Singha. An optical sensor for multi-species impurity monitoring in hydrogen fuel. *Sensor Actuat B-chem*, 171-172:416–422, 2012.

[341] NPL. *A Guide to the Measurement of Humidity, The Institute of Measurement and Control.* 1996.

[342] Wiederhold P. R. *Water Vapor Measurement, Methods and Instrumentation.* Marcel Dekker, New York, 1997.

[343] Wernecke R. and J. Wernecke. *Industrial Moisture and Humidity Measurement: A Practical Guide.* Wiley-VCH, 2014.

[344] Kui K. Jiao and X. Li. Water transport in polymer electrolyte membrane fuel cells. *Progress in Energy and Combustion Science*, 37:221–291, 2011.

[345] Pavelko R. G., E. Llobet A. Vasiliev, X. Vilanova, X. Correig, and V. G. Sevastyanov. The Influence of Wide Range Humidity on Hydrogen Detection with Sensors Based on Nano-SnO(2) Materials. In *Olfaction and Electronic nose, Proceedings Book Series: AIP Conference Proceedings, In Pardo, M., Sberveglieri, G. (Eds.)*, pages 29–33, 2009.

[346] Lo C. F., C. Y. Chang, B. H. Chu, S. J. Pearton, A. Dabiran, P. P. Chow, and F. Ren. Effect of humidity on hydrogen sensitivity of Pt-gated AlGaN/GaN high electron mobility transistor based sensors. *Appl Phys Lett*, 96:232106, 2010.

[347] Brokmann G., M. Hintz, B. March, and A. Steinke. Design and Evaluation of Impedance Based Sensors for Micro-condensation Measurement under Field and Climate Chamber Conditions. *Sensors & Transducers*, 14-1:174–184, 2012.

[348] HumiCalc: http://www.thunderscientific.com/software/humicalc1.html. 2014.

[349] Sonntag D. Advancements in the field of hygrometry. *Meteorol. Zeitschrift*, N.F. 3(2):51–66, 1999.

[350] Wagner W. and A. Pruß. The IAPWS Formulation 1995 for the Thermodynamic Properties of Ordinary Water Substance for General and Scientific Use. *J Phys Chem Ref Data*, 31:387–485, 2002.

[351] Okada C. Humidity Sensors: Types, Nanomaterials and Environmental Monitoring. Nova Science; New York, NY, USA, 2011.

[352] Lee C. Y. and G. B. Lee. Humidity sensors: A review. *Sensor Letters*, 3:1–15, 2005.

[353] Goff J. A. Standardization of Thermodynamic Properties of Moist Air. *Heat-piping-air Cond*, 21:118–128, 1949.

[354] Hebestreit A. *Beitrag zur direkten Wassertaupunktbestimmung in Hochdruckgasen.* Thesis Leipzig, 1988.

[355] Hassan N. S. M., W. R. W. Daud, K. Sopian, and J. Sahari. Water management in a single cell proton exchange membrane fuel cells with a serpentine flow field. *Journal of Power Sources*, 193:249–257, 2009.

[356] Bell S., M. de Podesta G. Hinds, M. Stevens, and J. Wilkinson. Humidity, Pressure, and Temperature Measurements in an Interdigitated-Flow PEM Hydrogen Fuel Cell. *Int J Thermophys*, 33:1583–1594, 2012.

[357] Tiebe C., T. Hübert, A. Lorek, and R. Wernecke. HUMItrace-Sensorsystem mit neuartigen planaren Spurenfeuchtesensoren TEMPERATUR 2013 - Verfahren und Geräte in der Temperatur - und Feuchtemesstechnik, 5. / 6. Juni 2013, Berlin, Tagungsband 257-262, ISBN 3-9810021-8-0.

[358] Pohl D. W., E. Irniger, and W. Herrmann. Spectrophonic Humidity-Trace Detection in Hydrogen and Air. *Appl. Phys*, 17:361–365, 1978.

[359] Yeo T. L., T. Sun, and K. T. V. Grattan. Fibre-optic sensor technologies for humidity and moisture measurement. *Sensor Actuat A-phys*, 144:280–295, 2008.

[360] ASTM E104 02(2012). Standard Practice for Maintaining Constant Relative Humidity by Means of Aqueous Solutions. 2012.

[361] Tsushima S. and S. Hirai. In situ diagnostics for water transport in proton exchange membrane fuel cells. *Prog Energ Combust*, 37:204–220, 2011.

[362] Quinn T. J. *Temperature*. Academic Press Ltd, London, 1990.

[363] Fischer J. and B. Fellmuth. Temperature metrology. *Rep. Prog. Phys*, 68:1043–94, 2005.

[364] Bentley R. E. (Ed.). *Handbook of Temperature Measurement*. Springer-Verlag, Singapore, 1998.

[365] Michalski L., K. Eckersdorf, and J. McGhee. *Temperature Measurement*. John Wiley and Sons, Chichester, 1991.

[366] Nicholas J. V. and D. R. White. *Traceable Temperatures*. John Wiley and Sons, Chichester, 2001.

[367] Bernhard F. (ed.). *Technische Temperaturmessung*. Springer-Verlag, Berlin, 1981.

[368] McMillian G. K. *Advanced Temperature Measurement and Control*. Research Triangle Park: International Society of Automation, 2011.

[369] Fellmuth B., C. Gaiser, and J. Fischer. Determination of the Boltzmann constant - status and prospects. *Meas. Sci. Technol.*, 17:R145–R159, 2006.

[370] Fellmuth B. Supplementary Information for the ITS 90: Chapter 1: Introduction, Ed. 2012. http://www.bipm.org/utils/common/pdf/its-90/SInf Chapter 1 Introduction 2013.pdf.

[371] Preston-Thomas H. The International Temperature Scale of 1990 (ITS-90). *Metrologia*, 27(3-10):107, 1990.

[372] Consultative Committee for Thermometry (CCT), 2015, http://www.bipm.org/en/committees/cc/cct/guide-its90.html.

[373] Schooley J. F. *Thermometry*. Boca Raton: CRC Press, Inc., 1988.

[374] Bedford R. E., G. Bonnier, H. Maas, and F. Pavese. *Techniques for Approximating the International Temperature Scale of 1990. Sevres: BIPM.* http://www.bipm.org/en/publications/mep kelvin/its-90 techniques.html.

[375] Rubin L. G. Cryogenic thermometry: a review of progress since 1982. *Cryogenics*, 37:341–356, 1997.

[376] White D. R. *Contribution of Uncertainties in Resistance Measurements to Uncertainty in ITS-90. In Temperature: Its Measurement and Control in Science and Industry,* vol. 7, ed. D. C. Ripple, 321-326. AIP, New York, 2003.

[377] Rudtsch S., G. Ramm, D. Heyer, and R. Vollmert. *Comparison of test and calibration methods for resistance ratio bridges, Proceedings of TEMPMEKO 2004, 9th International Symposium on Temperature and Thermal Measurements in Industry and Science,* 773–80. Zagreb: IMEKO / University of Zagreb, Faculty of Mechanical Engineering and Naval Architecture, 2005.

[378] Strouse G. F. and K. D. Hill. *Performance assessment of resistance ratio bridges used for the calibration of SPRTs. In Temperature: Its Measurement and Control in Science and Industry,* vol. 7, ed. D. C. Ripple, 327–32. AIP, New York, 2003.

[379] White D. R. *A method for calibrating resistance thermometry bridges, Proceedings of TEMPMEKO '96, 6th International Symposium on Temperature and Thermal Measurements in Industry and Science,* ed. P. Marcarino, 129–34. Levrotto and Bella, Torino, 1997.

[380] Nicholas J. V. *A proposal to base the temperature scale from the argon to silver point on thermodynamic functions. In Proceedings of TEMPMEKO '99, 7th International Symposium on Temperature and Thermal Measurements in Industry and Science, ed. J. F. Dubbeldam, and M. J. de Groot, 100-5. Delft: IMEKO / NMi Van Swinden Laboratorium.* 1999.

[381] IEC 60751:2008. Industrial platinum resistance thermometers and platinum temperature sensors.

[382] Curtis D. J. *Thermal hysteresis and stress effects in platinum resistance thermometers. In Temperature: Its Measurement and Control in Science and Industry,* vol. 5, ed. J. F. Schooley, 803–12. AIP, New York, 1982.

[383] Rusby R. L. *The rhodium-iron resistance thermometer: Ten years on. In Temperature: Its Measurement and Control in Science and Industry, vol. 5, ed. J. F. Schooley, 829-833.* AIP, New York, 1982.

[384] Eder F. X. *Arbeitsmethoden der Thermodynamik. Band I. Temperaturmessung.* Springer-Verlag, Berlin, 1981.

[385] Steinhart J. S. and Hart S. R. Calibration Curves for Thermistors. *Deep-sea Res,* 15:497–503, 1968.

[386] Armbruster H., W. P. Kirk, and D. P. Chesire. *Very low temperature thermocouple devices Development and application techniques for temperature measurements. In Temperature: Its Measurement and Control in Science and Industry,* vol. 5, ed. J. F. Schooley, 1025–35. AIP, New York, 1982.

[387] Maeno Y., H. Haucke, and J. Wheathley. Simple differential thermometer for low temperatures using a thermocouple with a SQUID detector. *Rev. Sci. Instrum.,* 54:946–948, 1983.

[388] Blackburn J. A. *Modern instrumentation for scientist and engineers.* Springer-Verlag, New York, 2001.

[389] Krause J. K. and P. R. Swinehart. *Reliable wide range diode thermometry. In Advances in cryogenic engineering,* vol. 31, ed. R. W. Fast, 1247. Plenum Press, New York, 1985.

[390] Wade W. H. and L. J. Slutsky. Quartz Crystal Thermometer. *Rev Sci Instrum,* 33(2):212–213, 1962.

[391] Hammond D. L., C. A. Adams, and P. Schmidt. A linear qurtz crystal temperature sensing element. *Trans. Instrum. Sci. Am.*, 4(4):349–354, 1965.

[392] Benjaminson A. and F. Rowland. *The development of the quartz resonator as a digital temperature sensor with a precision of* 1×10^{-4}. *In Temperature: Its Measurement and Control in Science and Industry,*vol. 4, ed. H. H. Plumb, 701-708. ISA, Pittsburgh, 1972.

[393] Ohsato H., T. Iwataki, and H. Morikoshi. Mechanism of piezoelectricity for langasite based on the framework crystal structure. *Trans. Electrical Electronic Mater*, 13(2):51–59, 2012.

[394] Allison S. W. and G. T. Gillies. *Phosphor thermometry signal analysis and interpretation. In Temperature: Its Measurement and Control in Science and Industry,* vol. 8, ed. C. W. Meyer, 863–6. AIP, New York, 2013.

[395] Consultative Committee for Thermometry (CCT), 2015, http://www.bipm.org/en/committees/cc/cct/guide-its90.html.

[396] Fischer J. Developments in infrared radiation thermometry. In Proceedings of TEMPMEKO '99, 7th International Symposium on Temperature and Thermal Measurements in Industry and Science, ed. J. F. Dubbeldam, and M. J. de Groot, 27-34, Delft: IMEKO / NMi Van Swinden Laboratorium.

[397] Van der Ham E. W. M., Battuello M., P. Bloembergen, R. Bosma, S. Clausen, O. Enouf, and et al. Intercomparison of local temperature scales with transfer radiation thermometers between -50 °C and 300 °C. In *Proceedings of TEMPMEKO 2001, 8th International Symposium on Temperature and Thermal Measurements in Industry and Science, ed. B. Fellmuth, J. Seidel, and G. Scholz* 831-7. VDE Verlag, Berlin, 2002.

[398] Walker R. *Achieving 0.25 mK uncertainty with an integrated-circuit resistance thermometer readout, Proceedings of TEMPMEKO 2001, 8th International Symposium on Temperature and Thermal Measurements in Industry and Science,* ed. B. Fellmuth, J. Seidel, and G. Scholz, 109–114. VDE Verlag, Berlin, 2002.

[399] Candy J. C. and G. C. Temes. *Oversampling delta-sigma data converters: Theory, design, and simulation.* Wiley-IEEE Press, 1991.

[400] Bramley P. and J. Pickering. Better Accuracy in temperature calibration and measurement through a new type of analog-to-digital converter. *Cal Lab Magazine*, (10/11):30–35, 2006.

[401] Riddle J. L., G. T. Furukawa, and H. H. Plumb. *Platinum resistance thermometry.* NBS Monograph 126, U.S. Department of Commerce, 1973.

[402] White D. R. and C. L. Jongenelen. The Immersion Characteristics of Industrial PRTs. *Int J Thermophys*, 31:1685–1695, 2010.

[403] Gaiser C. and B. Fellmuth. *Temperature-measurement errors with capsule-type resistance thermometers. In Temperature: Its Measurement and Control in Science and Industry,* vol. 8, ed. C. W. Meyer, 427–32. Melville. AIP, New York, 2013.

[404] Gupta A. K. and D. G. Lilley. *Advances in Chemical Propulsion.* CRC Press LLC, Boca Raton, 2002.

[405] Gaßmann E. and A. Gries. *Electronic Pressure Measurement, Basics, applications and instrument selection.* Süddeutscher Verlag onpact, 2010.

[406] Gillum D. R. *Industrial Pressure, Level, and Density Measurement.* 2nd ed., 2008.

[407] Barlian A. A., Park W. T., Mallon J. R. Jr., Rastegar A. J., and Pruitt B. L. Review: Semiconductor Piezoresistance for Microsystems, Proc. IEEE 97 (2009), 3, 513-552.

[408] B. Tian, Y. Zhao, Z. Jiang, L. Zhang, N. Liao, Y. Liu, and C. Meng. Fabrication and structural design of micro pressure sensors for tire pressure measurement systems (tpms). *Sensors*, 9:1382–1393, 2009.

[409] Pedersen T., O. Hansen G. Fragiacomo, and E. V. Thomsen. Highly sensitive micromachined capacitive pressure sensor with reduced hysteresis and low parasitic capacitance. *Sensor Actuat A-phys*, A154:35–41, 2011.

[410] Hall effect sensing and application, Sensing and Control, Honeywell Inc. USA, www.sensing.honeywell.com/index.php?ci-id=47847.

[411] Tadigadapa S. and K. Mateti. Piezoelectric MEMS sensors: state-of-the-art and perspectives. *Meas Sci Technol*, 20:092001(30pp), 2009.

[412] Gautschi G. *Piezoelectric Sensorics: Force, Strain, Pressure, Acceleration and Acoustic Emission Sensors, Materials and Amplifiers.* Springer, 2006.

[413] J. Huang, Z. Zhou, X. Wen, and D. Zhang. A diaphragm-type fiber bragg grating pressure sensor with temperature compensation. *Measurement*, 46:1041–1046, 2013.

[414] Albert J., L-Y. Shao, and C. Caucheteur. Tilted fiber Bragg grating sensors. *Laser Photonics Rev.*, 7(1):83–108, 2013.

[415] M. Li, M. Wang, and H. P. Li. Optical mems pressure sensor based on fabry-perot interferometry. *Optics Express*, 14:1497–1504, 2006.

[416] R. Y. Qingxu and X. Zhou. Pressure sensor based on the fiber-optic extrinsic fabry-perot interferometer. *Photonic Sensors*, 1:72–83, 2011.

[417] Urban F., J. Kadlec, R. Vlach, and R. Kuchta. Design of a pressure sensor based on optical fiber Bragg grating lateral deformation. *Sensors*, 10:11212–11225, 2010.

[418] S. Fricke, A. Friedberger, H. Seidel, and U. Schmid. A robust pressure sensor for harsh environmental applications. *Sensors and Actuat A-phys*, 184:16–21, 2012.

[419] Schmidt J. W., K. Jain, A. P. Miiller, W. J. Bowers, and D. A. Olson. Primary pressure standards based on dimensionally characterized piston/cylinder assemblies. *Metrologia*, 43:53–59, 2006.

[420] Chattopadhyay P. *Flowmeters & Flow Measurement.* Books Private Limited, 2006.

[421] Baker R. C. *An Introductory Guide to Flow Measurement.* Wiley, 2002.

[422] Kim H. D., J. H. Lee, K. A. Park, T. Setoguchi, and S. Matsuo. A Study of the Critical Nozzle for Flow Rate Measurement of High-Pressure Hydrogen Gas. *Journal of Thermal Science*, 16(1):28–32, 2007.

[423] Bruun H. H. *Hot-wire Anemometry: Principles and Signal Analysis.* Oxford University Press, 1995.

[424] Hoglen W. and J. Vallentine. Coriolis flow meters improve hydrogen production. *Hydrocarb Process*, pages 72–73, 2007.

[425] Anklin M., W. Drahm, and A. Rieder. Coriolis mass flowmeters: Overview of the current state of the art and latest research. *Flow Meas Instrum*, 17:317–323, 2006.

[426] Warshawsky I., H. F. Hobart, and H. L. Minkin. Small turbine-type flowmeters for liquid hydrogen. NASA Technical memorandum, NASA-TX-52984, Department of Energy, USA, 1971.

[427] Holler G., D. Hrach, A. Fuchs, and G. Brasseur. Application of a combined capacitive flow velocity and density sensor for the measurement of liquid hydrogen mass flow, I2MTC: 2009 IEEE Instrumentation & Measurement Techology Conference. *Cal Lab Magazine*, 1-3(10/11):1408–1412, 2009.

[428] Foreman J. W., E. W. George, and R. D. Lewis. Measurement of Localized Flow Velocities in Gases with a Laser Doppler Flow meter. *Appl Phys Lett*, 7(4):77–78, 1965.

[429] Baker R. C. *Flow Measurement Handbook.* Cambridge University Press, 2000.

[430] Wright J. D., A. N. Johnson, M. R. Moldover, and G. M. Kline. *Gas Flow meter Calibrations with the 34 L and 677 L PVTt Standards.* NIST Special Publication 250-63, USA, 2003.

[431] Wright J. D., J.-P. Kayl, A. N. Johnson, and G. M. Kline. *Gas Flow meter Calibrations with the Working Gas Flow Standard.* NIST Special Publication 250-80, USA, 2009.

[432] Boothroyd T., L. Murnane, and T. Ruane. *Fire Detection and Suppression Systems.* Ifsta, 1996.

[433] Jackson M. A. and I. Robins. Gas sensing for fire detection: Measurements of CO, CO_2, H_2, O_2, and smoke density in european standard fire tests. *Fire Safety Journal*, 22:181–205, 194.

[434] Ming N., J. R. Stetter, and W. J. Buttner. Orthogonal gas sensor arrays with intelligent algorithms for early warning of electrical fires. *Sensor Actuat B-chem*, 130:889–899, 2008.

[435] EN 54-31:2012. Fire detection and fire alarm system - Part 31: Multi-sensor fire detectors - Point detectors using a combination of smoke, carbon monoxide and optionally heat sensors.

[436] Moritz W., V. Fillipov, and A. Vasiliev. Hydrogen Sensor Using Silicon Technology For Fire Detection At Low Ppm Level And For Very High Concentrations, 2nd Intern. Conf on Sensing Technology, New Zeeland, November 26-28, 2007.

[437] Karner D. and D. J. Francfort. *Alternative Fuel -Hydrogen- Pilot Plant Design Report. U.S. Department of Energy Freedom CAR & Vehicle Technologies Program, Appendix F, p.1-3.* Arizona Public Service, 2003.

[438] Anderson C. *Detecting Hydrogen Gas fires: Seeing the unseen, in IHN - Industrial Hygiene News.* www.rimbach.com/scripts/Article, 2013.

[439] *EN 54*. Fire detection and fire alarm systems.

[440] Ordin P. M. *Safety Standard for Hydrogen and Hydrogen Systems, Guidelines for Hydrogen System Design, Materials, Selection, Operations, Storage, and Transportation, NASA report NSS 1740.16 2.1997*. 1997.

[441] NFPA. NFPA glossary of terms, 2014 edition.

[442] EN 45020-2006. Standardization and related activities - General vocabulary (ISO/IEC Guide 2:2004).

[443] Regulation (EU) No 1025/2012 of the European parliament and the council of 25 October 2012 on European standardisation.

[444] NFPA 70. National Electric Code (NEC), ed. 2014.

[445] IEC/EN 60079-20-1:2010. Explosive atmospheres - Part 20-1: Material characteristics for gas and vapour classification - Test methods and data.

[446] IEC/EN 60079-0:2014. Explosive atmospheres - Part 0: Equipment - General requirements.

[447] IEC 60079-10-1:2008. Explosive atmospheres - Part 10-1: Classification of areas - Explosive gas atmospheres.

[448] EN 50194:2009. Electrical apparatus for the detection of combustible gases in domestic premises - Part 1: Test methods and performance requirements.

[449] Keller J. O., M. Gresho, A. Harris, and A. V. Tchouvelev. What is an explosion? *Int J Hydrogen Energ*, 39(35):20426–20433, 2014.

[450] EN 45544:2015-01. Workplace atmospheres - Electrical apparatus used for the direct detection and direct concentration measurement of toxic gases and vapours - Part 1: General requirements and test methods; - Part 2: Performance requirements for apparatus used for exposure measurement; - Part 3: Performance requirements for apparatus used for general gas detection; - Part 4 (2nd ed.) is not finalised the 1999 version is still valid.

[451] EN 50291:2011. Electrical apparatus for the detection of carbon monoxide in domestic premises.

[452] EN 50545:2012. Electrical apparatus for the detection and measurement of toxic and combustible gases in car parks and tunnels.

[453] EN 50104:2011. Electrical apparatus for the detection and measurement of oxygen - Performance requirements and test methods.

[454] IEC/EN 60079-1:2008. Electrical apparatus for explosive gas atmospheres - Part 1: Flameproof enclosures, code “d”.

[455] IEC/EN 60079-7:2006. Explosive atmospheres - Part 7: Equipment protection by increased safety “e”.

[456] IEC/EN 60079-2:2010. Explosive Atmospheres - Part 2: Equipment protection by pressurized enclosure “p”.

[457] IEC/EN 60079-11:2012. Explosive atmospheres - Part 11: Equipment protection by intrinsic safety “i”.

[458] IEC/EN 60079-25:2010. Explosive atmospheres - Part 25: Intrinsically safe electrical systems.

[459] IEC/EN 60079-18:2009. Explosive atmospheres - Part 18: Equipment protection by encapsulation “m”.

[460] IEC/EN 60529:1991. Degrees of protection provided by enclosures (IP code).

[461] IEC/EN 61010-1:2010. Safety requirements for electrical equipment for measurement, control and laboratory use - Part 1: General requirements.

[462] IEC/EN 61326-1:2012. Electrical equipment for measurement, control and laboratory use - EMC requirements - Part 1: General requirements.

[463] EN 50270:2014. Electromagnetic compatibility - Electrical apparatus for the detection and measurement of combustible gases, toxic gases or oxygen.

[464] IEC/EN 60068-2-6:2008. Environmental testing - Part 2-6: Tests - Test Fc: Vibration (sinusoidal).

[465] IEC/EN 61508:2010. Functional safety of electrical/electronic/programmable electronic safety-related systems -
Part 1: General requirements
Part 2: Hardware requirements
Part 3: Software requirements
Part 4: Definitions and abbreviations.

[466] Smith D. J. and K. G. L. Kenneth G. L. Simpson. *Functional Safety. A Straightforward Guide to Applying IEC 61508 and Related Standards, 2nd edition.* Elsevier / Butterworth-Heinemann, Amsterdam, 2004.

[467] IEC/EN 61511-1:2003. Functional safety - Safety instrumented systems for the process industry sector, Part 1: Framework, definitions, system, hardware and software requirements Part 2: Part 2: Guidelines for the application of IEC 61511-1- Informative Part 3: Part 3: Guidance for the determination of the required safety integrity levels.

[468] EN 50271:2011. Electrical apparatus for the detection and measurement of combustible gases, toxic gases or oxygen - Requirements and tests for apparatus using software and/or digital technologies.

[469] EN 50402:2009. Electrical apparatus for the detection and measurement of combustible or toxic gases or vapours or of oxygen - Requirements on the functional safety of fixed gas detection systems.

[470] IEC/EN 60079-29-2:2007. Explosive atmospheres - Part 29-2: Gas detectors - Selection, installation, use and maintenance of detectors for flammable gases and oxygen.

[471] EN 50244:2000. Electrical apparatus for the detection of combustible gases in domestic premises - Guide on the selection, installation, use and maintenance.

[472] IEC/EN 60079-29-3:2013. Explosive atmospheres - Part 29-3: Gas detectors - Guidance on functional safety of fixed gas detection systems.

[473] ISO/TR 15916:2015. Technical Report on basic considerations for the safety of hydrogen systems.

[474] O'Malley K., W. J. Buttner, H. Lopez, J. Cairns, Burgess R., C. Rivkin, and R. Wichert. An Overview of North American Hydrogen Sensor Standards. Technical report, NREL Technical Report NREL/TP-5400-62062, 2015.

[475] 2009 International Fire Code, Country Club Hills, IL: International Code Council, 2009.

[476] NFPA 2011. Hydrogen Technologies Code.

[477] UL 864. Control Units and Accessories for Fire Alarm Systems.

[478] UL 2017. General-Purpose Signaling Devices and Systems.

[479] UL 2075. Gas and Vapor Detectors and Sensors.

[480] Caldes, A., Personal Communication (Interview), November 14, 2013.

[481] SAE J2011. Handbook for Robustness Validation of Automotive Electrical/Electronic Modules 2012, 2011.

[482] RTCA Radio Technical Commission for Aeronautics.

[483] MIL-STD-810. Environmental Engineering Considerations and Laboratory Tests, 2008.

[484] GOST R 52350.29.1-2010. Explosive atmospheres. Part 29-1. Gas detectors. General technical requirements and test methods for the detectors for flammable.

[485] JIS M 7626:1994. Stationary combustible gas alarms.

[486] JIS M 7653:1996. Portable combustible gas detectors.

[487] GB 15322. parts 1 to 6 Combustible gas detectors.

[488] ISO/IEC EN 60079-29-1:2007. Explosive atmospheres - Part 29-1: Gas detectors - Performance requirements of detectors for flammable gases.

[489] JIS M 7626:1994. Stational type combustible gas alarm.

[490] UL 913:2006. Intrinsically Safe Apparatus and Associated Apparatus for Use in Class I, II, and III, Division 1, Hazardous (Classified) Locations.

[491] ISO 9000:2005. Quality management systems - Fundamentals and vocabulary.

[492] EN ISO/IEC 17025:2005. General requirements for the competence of testing and calibration laboratories.

[493] ISO/IEC 17011:2004. Conformity assessment - General requirements for accreditation bodies accrediting conformity assessment bodies.

[494] http://www.iaf.nu/articles/Accreditation-Body-Members-by-Name/52.

[495] EN ISO 9001:2015. Quality management systems - Requirements.

[496] ILAC-P10-01-2013. ILAC Policy on the Traceability of Measurement Results (ILAC-P10-01-2013-Traceability.pdf).

[497] ISO 6142:2001. Gas analysis - Preparation of calibration gas mixtures - Gravimetric method.

[498] ISO 16664:2004-07. Gas analysis - Handling of calibration gases and gas mixtures.

[499] ICH GuidelineQ2A, Q2A, Validation of Analytical Procedures, 1994.

[500] CITAC / EURACHEM Guide to Quality in Analytical Chemistry, 2002.

[501] Namiesnik J. Generation of standards gaseous mixtures. *J. of Chromatography*, 300:79–108, 1984.

[502] ISO 6144:2003. Gas analysis- Preparation of calibration gas mixtures - Static volumetric method.

[503] EN ISO 6145. 1-11 Gas analysis - Preparation of calibration gas mixtures using dynamic volumetric methods.

[504] ISO 6145-10:2008. Gas analysis - Preparation of calibration gas mixtures using dynamic volumetric methods - permeation method.

[505] ISO 6141:2000. Gas analysis - Requirements for certificates for calibration gases and gas mixtures.

[506] ISO 5725-1. Accuracy (trueness and precision) of measurement methods and results - Part 1: General principles and definitions, 1994.

[507] ISO/IEC Guide 98-3, JCGM100:2008. Evaluation of measurement data - Guide to the expression of uncertainty in measurement.

[508] DAkks Deutsche Akkreditierungsstelle, Accreditation with flexible scope of testing laboratories, calibration laboratories and medical laboratories, document 71 SD 0 002, 28.10.2013, 2013.

[509] Grob R. L. and E. F. Barry. *Modern Practice of gas Chromatography, 4th ed.* Wiley Interscience, 2004.

[510] McNair H. and J. Miller. *Basic gas chromatography, 2nd ed.* John Wiley & Sons, Hoboken, New Yersey, 2009.

[511] Hübschmann H. J. *Handbook of GC/MS: Fundamentals and Applications.* Wiley-VCH Verlag GmbH & Co. KGaA, 2008.

[512] Berden G., R. Peeters, and G. Meijer. Cavity ringdown spectroscopy: Experimental schemes and applications. *International Reviews, Physical Chemistry*, 19:565–607, 2000.

[513] Brown A. S., Vargha G. M., M. L. Downey, N. J. Hart, G. G. Ferrier, and K. I. Hall. Methods for the analysis of trace-level impurities in hydrogen for fuel cell applications, NPL report AS 64. Technical report, NPL, 2011.

[514] Kamiski M., R. Kartanowicz, D. Jastrzebski, and M. M. Kuminski. Determination of carbon monoxide, methane and carbon dioxide in refinery hydrogen gases and air by gas chromatography. *J. of Chromatography A*, 989:277–283, 2003.

[515] P+E Power and Energy Inc. Ivyland, PA, USA 0/2013. Hydrogen elimination mass spectrometer. Accessed 7.05.2014, 2013.

[516] Grombik I. and I. Sliwka. Application of pdhid detector in gas chromatographic method of continuous hydrogen concentration measurements in air. *Geophysical Research Abstracts*, 10:2, 2008.

[517] Jones S. E., D. S. Shelton, R. S. Turley, M. J. Lawler, and D. D. Allred. Raman spectroscopic system for quantitative analysis of isotopic hydrogen mixtures for muon catalysis experiments. *Hyperfine Interact*, 101/102:695–698, 1996.

[518] Merkblatt T021 BGI 836. Gaswarneinrichtungen für toxische Gase/Dämpfe und Sauerstoff, Einsatz und Betrieb, Berufsgenossenschaft Rohstoffe Chemische Industrie BG RCI.

[519] Merkblatt Gaswarneinrichtungen für den Explosionsschutz. Technical report, IBG RCI Berufsgenossenschaft Rohstoffe und chemische Industrie, 1-37, 2012.

[520] The selection and use of flammable gas detectors, Published by the Health and Safety Executive 11/04 HSE, UK, 2004.

[521] NFPA 325. Guide to Fire Hazard Properties of Flammable Liquids, Gases, and Volatile Solids.

[522] IEC/EN 60079-20-1:2010. Explosive atmospheres - Part 20-1: Material characteristics for gas and vapour classification - Test methods and data.

[523] NFPA 497. Recommended practice for the Classification of Flammable Liquids, Gases or Vapors and of Hazardous (classified) Locations for Electrical Installations in Chemical Process Areas. .

[524] API RP 500. Recommended Practice for Classification of Locations for electrical Installation at Petroleum Facilities Classified as Class I, Division 1 and Division 2, 2002.

[525] API RP 505. Recommended Practice for Classification of Locations for Electrical Installation at Petroleum Facilities Classified as Class I, Zone 0, Zone 1 and Zone 2, 2002.

[526] Saaty T. L. *Principia Mathematica Decernendi: Mathematical Principles of Decision Making*. Pittsburgh, Pennsylvania: RWS Publications, 2010.

[527] Al-Hawari T., Al-Bo'ol S., and Momani A. Selection of temperature measuring sensors using the analytic hierarchy process. *Jordan Journal of Mechanical and Industrial Engineering*, 5:451–459, 2011.

[528] Deliverable D113 Initial Guidance for Using Hydrogen in ConfinedSpaces - Results from InsHyde, 2009.

[529] Abbas H. F. and W. M. A. Wan Daud. Hydrogen production by methane decomposition: A review. *Int J Hydrogen Energ*, 35:1160–1190, 2010.

[530] Combustion Safety for Steam Reformer Operation, Brussels, www.eiga.eu, Accessed 02.05.2014.

[531] Buttner W., R. Burgess, M. Post, and C. Rivkin. Summary and Findings from the NREL/DOE Hydrogen Sensor Workshop. Technical Report NREL/TP-5600-55645 July 2012.

[532] ISO 16110. Hydrogen generators using fuel processing technologies - Part 1: Safety (2007) - Part 2: Test methods for performance (2010).

[533] Sappey A. D., P. Masterson, E. Huelson, J. Howell, M. Estesa, H. Hofvander, and A. Jobson. Results of closed-loop coal-fired boiler operation using a TDLAS sensor and smart process control software. *Combust Sci Technol*, 183:1282–1295, 2011.

[534] Loloee R., B. Chorpening, S. Beer, and R. N. Ghosh. Hydrogen monitoring for power plant applications using SiC sensors. *Sensor Actuat B-chem*, 129:200–210, 2008.

[535] Carmo M., D. L. Fritz, J. Mergel, and D. Stolten. A comprehensive review on pem water electrolysis. *Int J Hydrogen Energ*, 38:4901–4934, 2013.

[536] ISO 22734: 2012. Hydrogen generators using water electrolysis process - Part 1: Industrial and commercial applications, Part 2: Residential applications.

[537] Gandia L. M., R. Oroz, A. Urua, P. Sanchis, and P. M. Dieguez. Renewable hydrogen production: Performance of an alkaline water electrolyzer working under emulated wind conditions. *Energ Fuel*, 21:1699–1706, 2007.

[538] Grigoriev S. A., P. Millet, S. V. Korobtsev, V. I. Porembskiy, M. Pepic, C. Etievant, C. Puyenchet, and V. N. Fateev. Hydrogen safety aspects related to high-pressure polymer electrolyte membrane water electrolysis. *Int J Hydrogen Energ*, 34:5986–5991, 2009.

[539] Doktor T. Fire and explosion hazards of chlorine-containing systems. *Journal of Hazard Materials*, 10:73–87, 1985.

[540] SAE 2579. Standard for Fuel Systems in Fuel Cell and Other Hydrogen Vehicles, 2013.

[541] Tamura H. and T. Iwase. Present status and future issues of hydrogen storage technologies. *Intern. Sci. Journal for Alternative Energy and Ecology*, 89:111–115, 2010.

[542] Kostival, A., C. Rivkin, W. Buttner, R. Burgess. Pressure Relief Devices for High-Pressure Gaseous Storage Systems: Applicability to Hydrogen Technology, Technical Report, NREL/TP-5400-60175 (November 2013), http://www.nrel.gov/docs/fy14osti/60175.pdf, 2013.

[543] ix35 FCEV Emergency Response Guide, Hyundai Motor Company, 2013.

[544] ISO 13985:2006. Liquid hydrogen - land vehicle fuel tanks.

[545] EMPA Technical offer TT-Ref. 2008-115: Optical hydrogen fuel gauge for metal hydride tanks, 2008.

[546] Marcotte D. and F. Domingue. Embedded sensor for solid-state hydrogen storage devices. *SENSORDEVICES 2011, Proc. 2nd Intern. Conf on Sensor Device Technologies and Applications*, pages 25–28, 2011.

[547] Perrin J., R. Steinberger-Wilckens, and S. C. Trümper. European hydrogen Infrastructure Atlas and Industrial Excess Hydrogen Analysis, PART III: Industrial distribution infrastructure, Document Number: R2H2007PU.1 Date: 3 July 2007.

[548] Melaina M. W., O. Antonia, and M. Penev. Blending Hydrogen into Natural Gas Pipeline Networks: A Review of Key Issues. Technical Report, NREL/TP-5600-51995, March 2013.

[549] FM 6310. Approval Standard for Combustible Gas Detectors, 2001.

[550] FM Approval Standard for Combustible Gas Detectors, Class Number 6320, 2014.

[551] Shirvill L. C., T. A. Roberts, M. Royle, D. B. Willoughby, and T. Gautier. Safety studies on high-pressure hydrogen vehicle refuelling stations: Release into a simulated high-pressure dispensing area. *Int J Hydrogen Energ*, 37:6949–6964., 2012.

[552] Buttner W. J., R. Burgess, C. Rivkin, M. B. Post, L. Boon-Brett, V. Palmisano, and P. Moretto. An assessment on the quantification of hydrogen releases through oxygen displacement using oxygen sensors. *Int J Hydrogen Energ*, 39(35):20484–20490, 2014.

[553] EP1482292B1. Fueling nozzle with integral hydrogen leak sensor.

[554] Nakano S., K. Yokosawa, Y. Goto, and K. Tsukada. Hydrogen gas detection system prototype with wireless sensor networks. *Proc. IEEE Sensors*, pages 159–162, 2005.

[555] Berufsgenossenschaftliche Information für Sicherheit und Gesundheit bei der Arbeit BGI 5108, Wasserstoffsicherheit in Werkstätten, 2009., ed: Berufsgenossenschaft der Strassen-, U-Bahnen und Eisenbahnen, Hamburg, Germany.

[556] U.S. Department of Energy, Office of Energy, Efficiency and Renewable Energy, Fuel Cell Technology Office, 2012 Fuel Cell Technologies Market Report. October 2013, www1.eere.energy.gov/hydrogenandfuelcells/pdfs/2012-market-report.pdf, Accessed 04.06.2014.

[557] IEC EN 62282-3-100:2012. Fuel cell technologies - Part 3-100: Stationary fuel cell power systems - Safety.

[558] NFPA 853:2015. Standard for the Installation of Stationary Fuel Cell Power Systems.

[559] Maeda Y., M. Takahashi, Y. Tamura, J. Suzuki, and S. Watanabe. Test of Vehicle Ignition Due to Hydrogen Gas Leakage, SAE Technical Paper 2006-01-0126, 2006.

[560] Venturi M., R. Fischer, and F. Henkel. Hydrogen Sensors for Automotive Fuel Cell Applications, SAE Technical Paper 2013-01-0497.

[561] GM rescue guide, GM Service technical College 1.04.2009.

[562] Medwin S., The Raymond Corporation, Greene, N.Y., Applying Hydrogen Fuel Cells to Lift Trucks, http://www.raymondcorp.com/images/pdf/FuelCell-WhitePaper.pdf, Accessed 13.02.2015.

[563] www.fuelcelltoday.com/analysis/analyst-views/2013/13-07-03-fuel-cells-and-submarines-sthash.1c7PFIZh.dpuf.

[564] Müller C., S. Fürst, and W. von Klitzing. BMW Group, Germany. Hydrogen Safety: New Challenges Based on BMW Hydrogen 7, 2nd International Conference on Hydrogen Safety, San Sebastian, Spain, September 11-13, 2007.

[565] Safety Standard for Hydrogen and Hydrogen Systems - Guidelines for Hydrogen System Design, Materials Selection, Operations, Storage, and Transportation NASA report NSS 1740.16 2.1997, ch. 6.1.

[566] Ariane 5 Users Manual, Issue5, http://www.arianespace.com/launch-services-ariane5/Ariane5-users-manual-Issue5-July2011.pdf, 2011.

[567] General Monitors, white paper, Hydrogen Detection in Oil Refineries, www.generalmonitors.com. Accessed 24.05.2014.

[568] ATEX Compliant Gas Analyzer System for Hydrogen-cooled Alternators, Data Sheet AK100 Series Issue 7, 2009.

[569] Schwandt C. Solid state electrochemical hydrogen sensor for aluminium and aluminium alloy melts. *Sensor Actuat B-chem*, 187:227–233, 2013.

[570] Soyk, K.-H.. Wasser- and wasserstoffgekühlte Turbogeneratoren. ETG-Fachbericht Nr. 53. VDE-Verlag, 55-64, 1994.

[571] International atomic energy agency, Safety Guide on Design of Reactor Containment Systems for Nuclear Power Plants, IAEA Safety Standards Series No. NS-G-1.14, IAEA, Vienna, 2004.

[572] IAEA-TECDOC-16661. Mitigation of hydrogen hazards in severe accidents in nuclear power plants, International Atomic Energy Agency, Vienna 2011.

[573] Hashemian H. M. *Nuclear Power Plant Instrumentation and Control, 49-66, in Nuclear Power - Control, Reliability 60 and Human Factors, Ed. P. Tsvetkov.* 2010.

[574] Hoffheins B. S., T. E. McKnight, R. J. Lauf, and R. R. Smith. *Evaluation of a hydrogen sensor for nuclear reactor containment monitoring, EPRI TR-106818 October 1997.*

[575] Hach Comp. application note, www.hach.com (Hach Comp. 2011).

[576] Eckardt B. A., M. Blase, and N. Losch. Containment Hydrogen Control and Filtered Venting, Design and Implementation, in Nuclear Safety NEA/CSNI R(2001)20, PSI Report Nr. 01-15, 19 pages, 2001.

[577] Klaucka M., E.-A. Reinecke, S. Kelm, N. Meynet, A. Bentaib, and H.-J. Allele. Passive auto-catalytic recombiners operation in the presence of hydrogen and carbon monoxide: Experimental study and model development. *Nucl Eng Des*, 266:137–147, 2014.

[578] Ganesan R., V. Jayaraman, S. R. Babu, R. Sridharan, and T. Gnanasekaran. Development of Sensors for On-Line Monitoring of Nonmetallic Impurities in Liquid Sodium. *J Nucl Sci Technol*, 48:483–489, 2011.

[579] A. Sree Rama Murthy, A. Ashok Kumar, E. Prabhu, P. C. Clinsha, I. Lakshmigandhan, S. Chandramouli, K. H. Mahendran, K. I. Gnanasekar, V. Jayaraman, B. K. Nashine, K. K. Rajan, and T. Gnanasekaran. Performance of semiconducting oxide based hydrogen sensor for argon cover gas in engineering scale sodium facility. *Nucl Eng Des*, 273:555–559, 2014.

[580] Pavlov D. *Lead-Acid Batteries. Science and Technology.* Elsevier, 2010.

[581] O'Donnell C. and M. Schiemann. *Hydrogen gas management for flooded lead acid batteries.* 2008. www.battcon.com/papersfinal2008/odonnellpaper2008proof_6.pdf.

[582] EN 50272-2:2001. Safety requirements for secondary batteries and battery installations - Part 2: Stationary batteries.

[583] Mak T., R. J. Westerwaal, M. Slaman, H. Schreuders, A. W. van Vugt, M. Victoria, C. Boelsma, and B. Dam. Optical fiber sensor for the continuous monitoring of hydrogen in oil. *Sensor Actuat B-chem*, pages 982–989, 2014.

[584] ASTM D3612-02:2009 Standard Test Method for Analysis of Gases Dissolved in Electrical Insulating Oil by Gas Chromatography.

[585] Hörnsten E. G., I. Lundström, Å. Nordberg, and B. Mathisen. The use of palladium metal oxide semiconductor structures in quantitative studies of H_2 and H_2S in processes related to biogas production. *Bioprocess Engineering*, 6:235–240, 1991.

[586] Zosel J., G. Schiffel, F. Gerlach, K. Ahlborn, U. Sasum, V. Vashook, and U. Guth. Electrode materials for potentiometric hydrogen sensors. *Solide state Ionics*, 177:2301–2304, 2006.

[587] Huck C., A. Poghossian, P. Wagner, and M. J. Schöning. Combined amperometric/field-effect sensor for the detection of dissolved hydrogen. *Sensor Actuat B-chem*, 187:168–173, 2013.

[588] H2Sense, Cost-effective and reliable hydrogen sensors for facilitating the safe use of hydrogen, Project Final Report, 2014.

[589] *Compendium of Chemical Terminology, Gold Book, Version 2.3.3, 1435.* International Union of Pure and Applied Chemistry IUPAC, 2014.

[590] ISO 14912. Gas analysis - concersion of gas mixture composition data. Iso, International Organization for Standardization, Geneva, Switzerland, 2003.

[591] Weast. R. C. (ed.). *Handbook of Chemistry and Physics (53rd edition.).* Cleveland:Chemical Rubber Co., 1972.

[592] Redlich O. and J. N. S. Kwong. On the Thermodynamics of solutions. V:An Equation of State. Fugacities of Gaseous Solutions. *Chem Rev*, 44:233–244, 1949.

[593] Goodwin R. D., D. E. Diller, H. M. Roder, and L. A. Weber. Second and thirs virial coefficients for hydrogen. *Journal of Research of the National Bureau of Standards - A. Physical and Chemistry*, 68A:121–126, 1964.

[594] Lemmon E. W., M. L. Huber, and M. O. McLinden. NIST12 Thermodynamic and Transport Properties of Pure Fluids - NIST Standard Reference Database 23 v. 8.0, Washington DC, 2007.

[595] Nasrafar K. Comparative study of eleven equations of state in predicting the thermodynamic properties of hydrogen. *Int J Hydrogen Energy*, 35:3802–3811, 2010.

[596] Sakoda N., Shindo K., Shinzato K., Kohno M., Takata Y., and Fujii M. Review of the Thermodynamic Properties of Hydrogen Based on Existing Equations of State. *Int J Thermophys*, 31:1276–296, 2010.

[597] Assael M. J. and S. Mixafendi. The viscosity and thermal conductivity of normal hydrogen in the limit of zero. *J Phys Chem Ref Data*, 15(4):1315–1322, 1986.

[598] ISO 5725-1:1994. Accuracy (trueness and precision) of measurement methods and results - Part 1: General principles and definitions.

Index

Printed and bound by CPI Group (UK) Ltd, Croydon, CR0 4YY
22/10/2024
01777611-0008